# Dynamics of Perfect Crystals

# Dynamics of Perfect Crystals

G. Venkataraman

L. A. Feldkamp

V. C. Sahni

The MIT Press
Cambridge, Massachusetts, and London, England

Copyright © 1975 by
The Massachusetts Institute of Technology

This book was set in Monotype Baskerville,
printed on Finch Title 93,
and bound in G.S.B. S/535/83
by The Colonial Press Inc.
in the United States of America

Library of Congress Cataloging in Publication Data
Venkataraman, G.
    Dynamics of Perfect Crystals
    Bibliography: p.
    1. Lattice dynamics. I. Feldkamp, L. A., joint author. II. Sahni, V. C., joint author.
    III. Title.
QC176.8.L3V46   548'.81   74–10959
ISBN 0–262–22019–9

Respectfully dedicated to the memory of Max Born

# Contents

## 5 Experimental Study of Lattice Vibrations

## 6 Survey of Experimental Results

# Preface

The study of atomic motions in crystalline solids is a subject with a long and interesting history and over the years has been explored in all its aspects. The formal basis for the investigation of the vibrations, especially within the framework of the harmonic approximation, was laid principally by Born and his coworkers and has been summarized admirably in the book by Born and Huang. It is unfortunate that this classic was written a few years prior to important advances in the experimental field such as the development of the neutron scattering technique and lasers, and could not witness the consequent revitalization in lattice dynamics. Most of these new developments are still accessible mainly from the original articles and have been reviewed all too briefly, so that the active worker as well as the newcomer to the field is faced with the unenviable task of wading through a jungle of cross references lacking, among other things, in uniformity of notation. A few monographs on lattice dynamics have appeared since Born and Huang's book, but since the spectrum in these is extensive, the dynamics of perfect crystals has not received sufficiently detailed consideration. In our opinion, the need exists for a comprehensive book which starts from fundamentals and explores in a unified fashion the dynamics of perfect crystals with emphasis on developments since the appearance of the Born-Huang classic; the present volume is designed to fill this gap. It is directed principally at the large body of people engaged in obtaining, analyzing and interpreting vibration spectra of solids. It is expected, however, to be of use also to others with peripheral interest in lattice dynamics.

We restrict ourselves to the study of perfect crystals in the harmonic approximation. We are conscious of the fact that, in so doing, we are excluding other equally important topics such as anharmonicity and defects. These, however, are subjects which merit full treatments in their own right. Furthermore, the present volume should be useful even to those interested in the above topics inasmuch as the material covered here provides a good starting point for the discussion of both anharmonicity and defects.

We presuppose a knowledge of basic quantum mechanics and introductory solid state physics. In the context of present-day graduate school curricula, these are quite moderate demands. Indeed, some topics such as the Brillouin zone and the one-dimensional chain, which we would have thought necessary to include a few years ago, are now excluded.

In spite of its appearance, no elaborate grounding in mathematics is necessary for studying the present volume, the only specific requirement being a knowledge of matrices and their manipulation. Familiarity with group theory is necessary for understanding Chapter 3, but this chapter is

not essential for reading the rest of the book (except Sec. 4.4). We regret
that limitations of space prevented our including an introduction to group
theory. However, since courses on group theory are becoming increasingly
common in graduate schools, we hope this omission is not too serious. On
the whole we expect that the present volume will be understandable not
only to the established research worker but also to graduate students.

The plan of the book is as follows:

We begin with a semi-empirical classification of crystals according to
their binding and give a brief discussion of the crystal potential in the
various cases so as to present an idea of the types of forces operating in
crystals.

Chapter 2 reviews the standard Born–von Kármán approach to lattice
dynamics, and the discussion is extended to include internal and external
modes in complex crystals. A detailed example is also offered to illustrate
the calculation of phonon dispersion curves and the classification of singularities in frequency spectra.

Chapter 3 is devoted to the application of group theory to lattice
dynamics. As already mentioned, a study of this chapter requires a previous knowledge of the principles of group theory. Our treatment is based
on the currently popular multiplier-representation method. The principles
involved are illustrated with a specific example.

In Chapter 4 we discuss nonmetallic crystals, in particular the effects of
the electric fields associated with their vibrations, as well as the effects of
electronic polarization. We survey first specific electric-field effects by considering the simple example of zinc blende. A detailed generalized discussion of the influence of electric fields on the long-wavelength modes in an
arbitrary crystal is then given. This is followed by a review of the shell
model together with its ramifications. To set the stage for the last chapter,
these phenomenological models are also examined with the hindsight
provided by recent microscopic theories of lattice dynamics.

We discuss in Chapter 5 the important techniques used to study lattice
vibration spectra, the emphasis being on the principles of the methods.
Typical examples of experimental results are also provided.

The next chapter offers a comprehensive review of the experimental
results obtained thus far, especially in the context of the phenomenological
models discussed in the earlier chapters.

The last chapter, meant mainly for the adventurous reader, reviews
recent developments in the formulation of microscopic theories of lattice
dynamics, as well as theories for quantum crystals. Owing to the complexity of the subject, we avoid derivations and instead restrict ourselves to
explaining the underlying physics. We hope this will offer a sufficient

glimpse of the territory that lies beyond the traditional Born–von Kármán approach.

Two appendixes and an annotated bibliography are also provided; the latter should prove useful to the reader who wishes to pursue the subject in depth.

The present volume is essentially a text with emphasis on pedagogy. It revolves largely around phenomenological models because we expect that in spite of spectacular advances in the development of microscopic theories, phenomenological models will continue to be used for a long time, especially for complicated crystals. We hope, for this reason, that our book will likewise retain its utility.

December 1974
GV
LAF
VCS

# Acknowledgments

The idea of writing this book was conceived in 1968 when two of us (GV and LAF) were at the Department of Nuclear Engineering, University of Michigan. Communication during the work's subsequent translation into reality was entirely through intercontinental correspondence, perhaps a record of sorts! In this effort we have been encouraged and aided by several individuals. Particular mention must be made of Professor S. K. Sinha, who patiently read through several drafts and offered valuable criticisms. Our thanks also go to Professors G. Gilat, S. K. Joshi, A. K. Rajagopal, and R. Srinivasan, all of whom made useful comments and suggestions, especially in areas relating to their fields of specialty. GV and VCS take this opportunity to express gratitude to Dr. R. Ramanna and Dr. P. K. Iyengar for constant encouragement and support and to our colleagues at Trombay for contributing greatly to our knowledge of lattice dynamics over the years. LAF expresses similar thanks to Dr. J. R. Reitz. Thanks are also due to Dr. V. A. Kamath for assistance with the preparation of the manuscript and the figures. Finally, it is a pleasure to record our indebtedness to Mr. M. T. Thomas and Mr. T. Ajaykumar for their patient and skillful typing, but for which this book would hardly have been possible.

# Book Abbreviations

The following abbreviations are used for volumes frequently referred to in the text.

BH
M. Born and K. Huang, *Dynamical Theory of Crystal Lattices* (Clarendon, Oxford, 1954).

Kittel
C. Kittel, *Introduction to Solid State Physics* (Wiley, New York, 1966), Third Edition.

Seitz
F. Seitz, *Modern Theory of Solids* (McGraw-Hill, New York, 1940).

Wallis
*Lattice Dynamics*, edited by R. F. Wallis (Pergamon, New York, 1965).

Nusimovici
*Phonons*, edited by M. A. Nusimovici (Flammarion, Paris, 1971).

Bak
*Phonons and Phonon Interactions*, edited by T. A. Bak (Benjamin, New York, 1964).

Stevenson
*Phonons in Perfect Lattices and Lattices with Imperfections*, edited by R. W. H. Stevenson (Oliver and Boyd, Edinburgh, 1966).

Vienna proceedings
*Inelastic Scattering of Neutrons in Solids and Liquids* (IAEA, Vienna, 1960).

Chalk River proceedings
*Inelastic Scattering of Neutrons in Solids and Liquids* (IAEA, Vienna, 1962).

Bombay proceedings
*Inelastic Scattering of Neutrons* (IAEA, Vienna, 1965).

Copenhagen proceedings
*Neutron Inelastic Scattering* (IAEA, Vienna, 1968).

Grenoble proceedings
*Neutron Inelastic Scattering* (IAEA, Vienna, 1972).

# Dynamics of Perfect Crystals

# 1 Interatomic Forces in Crystals

A crystal is often described as a three-dimensional periodic array of atoms. This statement, however, is true only on the average. If we were to take an instantaneous snapshot of the crystal, we would find that the positions occupied by the atoms as revealed in the photograph are far from constituting a periodic array, especially if the temperature of the crystal is high. Nevertheless, if observations of the atomic positions are made over extended periods of time, it will be noticed that though in constant motion the atoms do move about well-defined positions (we are here ignoring the possibility of diffusion), and it is these *mean* positions that define the periodic lattice. The study of the architecture of this arrangement constitutes the subject of crystallography. In this volume we shall be concerned with the motions that atoms in perfect crystals execute about their mean positions. By a perfect crystal we mean one without any defects, disorders (isotopic or otherwise), etc.

The structure assumed by any given crystal at a specified temperature can be obtained formally by determining the configuration for which the *Helmholtz free energy F* of the system is a minimum. If one restricts attention to 0 K and further ignores zero-point motions (as we shall do frequently), then the mean positions are essentially determined by the minimum of the *crystal potential energy* $\Phi$ *with respect to the atomic positions.* We shall refer to the mean positions for 0 K as the *equilibrium* positions because the forces on the atoms at these positions actually vanish. The potential $\Phi$ depends on the nuclear coordinates* and may therefore be written formally as $\Phi(\{\mathbf{r}(i)\})$, where $\{\mathbf{r}(i)\}$ denotes the set† of nuclear coordinates.‡ As we shall see in the following chapter, small motions of the atoms about their equilibrium positions are determined by the various partial derivatives of $\Phi$ with respect to the atomic positions, the derivatives being evaluated in the equilibrium configuration. Under certain assumptions (to be stated later), these motions can be analyzed in terms of harmonic vibrations, and for this situation one requires only a knowledge of the second derivatives, namely the quantities

$$\left. \frac{\partial^2 \Phi}{\partial r_\alpha(i) \partial r_\beta(j)} \right|_0, \qquad (\alpha, \beta = x, y, z), \qquad (1.1)$$

where $\mathbf{r}(i)$ and $\mathbf{r}(j)$ denote the coordinates of the $i$th and $j$th atoms,

---

*Even though we may occasionally talk of atomic positions or atomic coordinates, we shall always mean the positions occupied by the nuclei. Similarly, atomic and nuclear motions will be used synonymously.
†We shall frequently use the braces $\{\cdots\}$ to denote a set of like quantities.
‡In addition, the crystalline potential energy will in general depend upon externally applied forces such as mechanical stress or electric field. We shall here ignore these external forces.

respectively. The symbol $|_0$ indicates that the derivative is to be evaluated in the equilibrium configuration.

Ideally one would like a uniform prescription for calculating $\Phi$, and thence its derivatives, which is applicable to all types of crystals. This would enable the dynamics of the various types to be discussed in a unified way. Such an approach would have to start at a rather basic level by visualizing the crystal as a system of appropriate nuclei and electrons. This point of view will be discussed further in the last chapter wherein we shall see that while satisfying from a formal perspective, it is not very useful in a practical sense. For purposes of making numerical calculations of crystal dynamics, it is generally more convenient to start with a potential function $\Phi$ which is tailored from empirical considerations to the physical peculiarities of the crystal concerned. Accordingly, we begin our study of crystal dynamics with a brief refresher on the different types of bindings and forces that exist in crystals. It is pertinent to note here that the binding or cohesion of atoms into a crystal is achieved essentially by the outer electrons of the constituent atoms. Effectively, therefore, the $\Phi$ function for the crystal depends upon the nature of the binding that these outer electrons achieve. Here it is customary to recognize (on empirical grounds) the following categories: molecular crystals, metallic crystals, ionic crystals, covalent crystals, and hydrogen-bonded crystals. Often a given crystal may not be assigned uniquely to any one of these categories; nevertheless the classification is useful. In most of the above cases, $\Phi$ can be expressed as a sum of interatomic potentials involving atoms in pairs, triplets, quadruplets, etc. These potentials may be denoted formally as

$$
\begin{aligned}
V^{(2)} &= V(\mathbf{r}(i), \mathbf{r}(j)), \\
V^{(3)} &= V(\mathbf{r}(i), \mathbf{r}(j), \mathbf{r}(k)), \\
V^{(4)} &= V(\mathbf{r}(i), \mathbf{r}(j), \mathbf{r}(k), \mathbf{r}(l)), \text{ etc.}
\end{aligned}
\tag{1.2}
$$

Here the two-body potential $V^{(2)}$ depends only on the coordinates of the atoms $i$ and $j$. Similarly $V^{(3)}$ is a three-body potential which depends on the simultaneous positions of three atoms. Presently we shall comment on the nature and, wherever possible, the form of these potentials in the various types of crystals. Potentials $V^{(3)}$, $V^{(4)}$, etc. will be termed henceforth as many-body potentials and forces derived from them as many-body forces.

## 1.1 Molecular crystals

The essential characteristic of a molecular crystal is that it is built up by assembling "saturated units," units whose internal electronic structure is

not easily altered. Each unit may be either a single atom or a whole molecule. Thus in the category of molecular crystals we include both the rare-gas
solids like solid argon as well as crystals like solid methane and solid
naphthalene. One common feature of these crystals is that the binding of
the various "units" to each other is rather weak compared to that in the
other types of crystals. This is also reflected in their comparatively low
melting points and low cohesive energies.

The rare-gas crystals occupy an especially important place in this group.
As one might expect intuitively in view of the closed-shell configuration of
the rare-gas atoms, the atoms interact via a pair potential $V^{(2)}$ which is
applicable to all the states, solid, liquid, and gas. Semiempirical considerations have led to the following form being proposed[1] for the pair potential
which, we observe, depends only on the *separation* between the atoms.
Explicitly,

$$V^{(2)} \equiv V(r) = 4\varepsilon \left\{ \left( \frac{\sigma}{r} \right)^{12} - \left( \frac{\sigma}{r} \right)^{6} \right\}, \qquad \varepsilon > 0. \tag{1.3}$$

This potential, known as the Lennard-Jones (LJ) potential, is sketched in
Figure 1.1. The first term on the right-hand side describes the repulsion
called into play when one atom tries to penetrate another. The repulsive
part is not as steep as for hard spheres, which is to be expected. Quantum
mechanical considerations reveal that the repulsion arises mainly through
the operation of the Pauli exclusion principle when electron clouds of
different identical saturated units try to penetrate into each other.[2] While
such an analysis suggests the form $Be^{-\alpha r}$ for the repulsive potential ($B$ and $\alpha$
being parameters), an inverse-power potential is also often employed as in
(1.3). The former form is frequently called the Born-Mayer potential.

Of more interest in (1.3) is the second term on the right-hand side which
denotes an attractive potential (by virtue of the negative sign in front of it).
This interaction, known variously as the van der Waals attraction, London
dispersion, etc., arises physically as follows. The electron cloud in an
isolated atom is spherically symmetric. Viewed from outside, therefore, the
atom appears neutral, the nuclear charge being canceled by that of the
surrounding electron cloud. On this basis, we would expect that when two
rare-gas atoms are brought together and do not overlap, there would be no
interaction between them. This, however, would be true only if the electron
clouds of the two atoms continue to be spherically symmetric even when
close to each other. In practice, the electronic charge distributions are not
rigid and can undergo fluctuations which can be visualized as a superposition of appropriate dipoles, quadrupoles, etc. These multipoles in turn can
induce charge distortions in neighboring atoms leading to the creation of

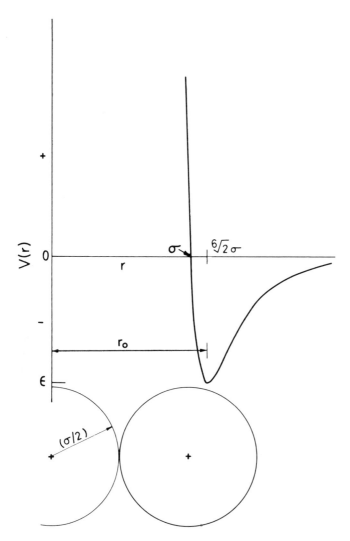

**Figure 1.1**
Sketch of the LJ potential. Observe that $\sigma$ corresponds to the "hard-sphere diameter" of the atom.

fresh multipoles; it is the mutual interaction of these multipoles which produces the attraction.[3] Of the various possibilities, the dipole-dipole interaction is the strongest and varies typically as $-1/r^6$, which explains the structure of the attractive term in (1.3). It is worth noting that given two dipoles of *fixed* magnitudes $p_1$ and $p_2$, say, their interaction goes as $(p_1 p_2/r^3)$. In the present case the value of $p_2$, the induced moment on an atom due to a fluctuation moment $p_1$ on another, itself varies as $(p_1/r^3)$ so that the net interaction goes as $(p_1^2/r^6)$.

The potential function given in (1.3) has been found adequate for describing a wide class of physical properties of rare gases in all states. Table 1.1 summarizes the parameters.[4]

As far as three-body and other higher-body potentials are concerned, it appears that while important from the standpoint of statics (for example, for explaining why the fcc structure is favored as compared to the hcp[5]), they do not play a significant role in determining the dynamics. We shall therefore ignore them.

In crystals like methane, one can broadly divide the forces into two groups, intramolecular and intermolecular. The former are usually very much stronger than the latter, and for purposes of concentrating on the intermolecular forces one could even treat the forces internal to the molecule as being infinitely strong, which is tantamount to assuming the molecular groups to be rigid bodies. The interaction of one molecule with another can be regarded as due to the interaction between the atoms of one with those of the other. Thus we may write the potential $V_{AB}$ between molecules $A$ and $B$ with their centers of mass respectively at $\mathbf{r}(A)$ and $\mathbf{r}(B)$ as

**Table 1.1**
Lennard-Jones Potential parameters for the rare gases (from ref. 4).

| Element | $\varepsilon$(K) | $\sigma$(Å) | $r_0 = 2^{1/6}\sigma$ | $R_0 = 1.09\sigma$[a] | Expt.[b] n.n. dist. (Å) | Structure |
|---------|---------|---------|---------|---------|---------|---------|
| Helium | 10.2 | 2.56 | 2.87 | 2.79 | 3.77 | bcc |
| Neon | 36.3 | 2.82 | 3.16 | 3.07 | 3.16 | fcc |
| Argon | 119.5 | 3.407 | 3.83 | 3.71 | 3.76 | fcc |
| Krypton | 159.0 | 3.61 | 3.04 | 3.94 | 3.99 | fcc |
| Xenon | 228.0 | 3.97 | 4.46 | 4.33 | 4.33 | fcc |

[a]This is the theoretically expected nearest-neighbor distance for a fcc crystal bound solely by a LJ potential. The corresponding lattice constant is given by $a = 1.542\sigma$. Observe that $R_0 < r_0$, a consequence of cohesion.
[b]The distances quoted for all solids except helium are for $p = 0$ and $T = 0$ K. In the case of solid helium the distance quoted is for (bcc) He$^3$ at a molar volume of 24.7 cm$^3$.

$$V_{AB} = \sum_{i \in A} \sum_{j \in B} V(\mathbf{r}(A) + \mathbf{r}(i), \mathbf{r}(B) + \mathbf{r}(j)), \tag{1.4}$$

where $\mathbf{r}(i)$ and $\mathbf{r}(j)$ denote the coordinates of atoms $i$ and $j$ in $A$ and $B$ respectively, relative to their centers of mass. The symbol $i \in A$ attached to the summation implies that $i$ runs over all the atoms in $A$; a similar meaning applies to $j \in B$.*

Based on (1.4), the intermolecular part of the crystal potential energy $\Phi$ can be expressed as

$$\Phi = \tfrac{1}{2} \sum_{A} \sum_{B}{}' V_{AB}, \tag{1.5}$$

where the summations over $A$ and $B$ are carried out over all the molecules in the crystal, omitting, however, the "self terms" of the type $V_{AA}$. This is indicated with a prime.†

Parameterized forms of the two-body potentials $V(r)$ occurring in (1.4) have been proposed by assuming that in molecular crystals, the pair potential between atoms in two *different* molecules must have a steep repulsive term and a weak van der Waals attractive term. In particular, such potential functions have been used by Kitaigorodsky[6] to predict the equilibrium configuration of molecular crystals; as we shall see later, they have also been used for discussing the "rigid-body motions" in molecular crystals.

Kitaigorodsky proposed for $V(r)$ the form

$$V(r) = -\frac{A}{r^6} + Be^{-\alpha r}, \tag{1.6}$$

which is written in the same spirit as the LJ potential and has a similar shape. Sometimes the LJ potential is also used for the same purpose. Note that the use of (1.3) and (1.6) in the context of constructing intermolecular potentials is appropriate only for *nonbonded interactions*. If, for example, a covalent bonding or hydrogen bonding occurs between two molecular groups, then (1.6) cannot be used for describing the interactions between atoms participating in such a bond.

Values for the potential parameters $A$, $B$, and $\alpha$ have been proposed several times based on analyses of various experimental data.[7–9] A set recently compiled is given in Table 1.2, which also includes the LJ parameters used for some of these nonbonded interactions. In this connection we

---

*The symbol $\in$ means "contained in."
†In general a primed summation will be taken to imply the exclusion of that term for which the summed variable matches the fixed index.

**Table 1.2**
Parameters for nonbonded potentials.

$$V(r) = Be^{-\alpha r} - \frac{A}{r^6}$$

or

$$V(r) = \frac{C}{r^{12}} - \frac{A}{r^6}$$

| Atomic pair | $A$ | $B$ | $\alpha$ | $C$ | Ref. |
|---|---|---|---|---|---|
| H–H | 27.3 | 2,654 | 3.74 | — | 7 |
| H–C | 125.0 | 8,766 | 3.67 | — | 7 |
| C–C | 568.0 | 83,630 | 3.60 | — | 7 |
| C–Cl | 1140.0 | 180,000 | 3.62 | — | 8 |
| H–Cl | 250.0 | 18,000 | 3.70 | — | 8 |
| Cl–Cl | 2300.0 | 426,000 | 3.65 | — | 8 |
| H–O | 124.0 | — | — | 25,000 | 9 |
| C–O | 367.0 | — | — | 105,000 | 9 |
| O–O | 367.0 | — | — | 145,000 | 9 |
| H–N | 125.0 | — | — | 27,000 | 9 |
| C–N | 366.0 | — | — | 216,000 | 9 |
| O–N | 365.0 | — | — | 153,000 | 9 |
| N–N | 363.0 | — | — | 161,000 | 9 |

Notes: $V(r)$ calculated using the above parameters are in units of kcal/mole.
This table was kindly compiled for us by Dr. S. K. Sikka.

might note that parameters which deliver a correct structure need not necessarily predict the dynamics satisfactorily.[10] This is understandable because the equilibrium structure or the packing of molecules into a crystal appears to be determined mainly by "atomic size" which is not very sensitive to some of the potential parameters, for example, the well depth (see also Figure 1.1). Cohesive energy and sublimation energy offer a more stringent test of the potential parameters since they involve the well depth and details of the shape of the potential function.

Another point worth drawing attention to is that even if the molecules constituting the crystal are effectively neutral, individual atoms in the molecule can carry nonvanishing charges. As a result one expects Coulomb interactions between such charges, and these are not allowed for in (1.6). The Coulombic contribution to the crystal potential $\Phi$ may be small, but from the standpoint of dynamics it is important since it influences the optical properties of the solid in the infrared region as will be seen in Chapter 4.

The intermolecular potential based on (1.4) and (1.6) can, if necessary, be manipulated into a form that exhibits explicitly both a dependence on the separations of the centers of mass and the relative orientations of the two molecules. Phenomenological forms displaying such features have in fact been proposed. Stockmayer, for example, has suggested the form[11]

$$V(\mathbf{r}) = 4\varepsilon \left[ \left( \frac{\sigma}{r} \right)^{12} - \left( \frac{\sigma}{r} \right)^{6} \right]$$

$$- \frac{\mu_a \mu_b}{r^3} [2 \cos \theta_a \cos \theta_b - \sin \theta_a \sin \theta_b \cos(\phi_a - \phi_b)]. \qquad (1.7)$$

Actually this is a potential function appropriate to the interaction of two polar molecules of dipole moments $\mu_a$ and $\mu_b$; the second term on the right-hand side of (1.7) indicates explicitly the angle-dependent part and arises from the dipolar interactions. (See Figure 1.2 for an explanation of some of the symbols.) For a molecular crystal, a form such as (1.7) may be inappropriate. Extending the Stockmayer idea, one could, however, still have an angle-dependent part associated with higher-order (induced) multipoles. The Kitaigorodsky approach synthesises these effects via atomic interactions. For example, Yasuda[12] has constructed the intermolecular potential for methane by this procedure and has displayed explicitly the radial and angle-dependent parts.

In the study of dynamics, intermolecular potential functions built up from atomic potentials as in (1.6) are more common than potentials like (1.7).

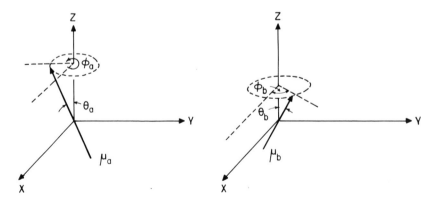

**Figure 1.2**
This figure shows the spatial and angular parameters characterizing the dipoles $\mu_a$ and $\mu_b$ in the Stockmayer potential.

## 1.2 Metallic crystals

The characteristic feature of metallic crystals is the presence of a large number of electrons that are relatively weakly bound. These are made up of the outer electrons of the constituent atoms and to a first approximation may be regarded as nearly free, at least in many nontransition metals. Thus Na, for example, may be viewed as a collection of $Na^+$ ions immersed in a sea of free electrons made up of the $3s$ electrons of the constituent sodium atoms. It is these outer electrons that lead to the high conductivity of metals. Not surprisingly, they also help to bind the ions together as if by a uniformly pervading glue. This fact, which has been known for a long time,[13] also makes it difficult at first sight to visualize $\Phi$ in terms of pair atomic potentials $V(r)$ and other many-body potentials. Indeed until recently when the many-body effects associated with conduction electrons became clarified, the concept of $\Phi$ as a superposition of atomic pair potentials $V(r)$ was largely speculative, although it was frequently invoked in phenomenological treatments of the dynamics of metals. Now, however, the picture can be more clearly sketched. One can view $\Phi$ as being primarily made up of a superposition of two-body potentials between the ion cores, their mutual interactions being modified by the dielectric properties of the surrounding medium, that is, the conduction electron sea. In other words, just as in electrostatics one writes the Coulomb interaction between charges $Z_1 e$ and $Z_2 e$ placed in a medium of dielectric constant $\varepsilon$ as

$$\frac{Z_1 Z_2 e^2}{\varepsilon r} = \frac{Z_1 Z_2 e^2}{r} - \frac{Z_1 Z_2 e^2}{r}\left(1 - \frac{1}{\varepsilon}\right),$$

one can write for (the nearly free electron) metals,

$$V(r) = \frac{Z^2 e^2}{r} - V^{el}(r), \tag{1.8}$$

where $V^{el}$ represents the modification due to the conduction electrons and can be expressed in terms of the dielectric properties of the electronic medium. As we shall see later, the dielectric properties can be expressed in terms of a *spatially dispersive* dielectric function $\varepsilon(r)$ or, equivalently, in terms of its Fourier transform $\varepsilon(Q)$. Analytical forms for $V^{el}(r)$ have been proposed using the pseudopotential concept.[14] Irrespective of the form, one striking feature is that asymptotically

$$V(r) \sim \text{const}\frac{\cos(2k_F r)}{(k_F r)^3}, \qquad (k_F = \text{Fermi wave vector}), \tag{1.9}$$

which is oscillatory, unlike the usual two-body potentials. Thus it is legitimate to view $\Phi$ as being made up primarily of superposition of two-body potentials (with oscillatory tails), the potentials representing *screened Coulomb interaction* between the ions.

An equivalent but slightly different interpretation has been offered for the effective potential $V(r)$ and may be understood as follows.[15] Suppose for a moment we have a gas of electrons,* and we introduce in it an impurity of charge $Ze$. This causes a disturbance to the electronic charge distribution in the neighborhood which we shall denote by $eZ^{el}(r)$. Many-body theory in fact shows that $Z^{el}(r)$ is oscillatory,[16,17] the oscillations typically extending to a few Angstroms if the density of the electron gas is of the order of conduction electron density in metals. The form of $Z^{el}(r)$ is governed by the polarization properties of the electron gas or, equivalently, by its dielectric properties. Furthermore,

$$\int Z^{el}(r) \, d\mathbf{r} = -Z, \qquad (1.10)$$

where the integral is taken over a limited space around the impurity. The above equation implies that the impurity $Ze$ together with the induced ripple $eZ^{el}(r)$, or the *screening charge* as it is sometimes called, appears more or less like a neutral object from some distance away. An external test charge (say in the shape of another ion) thus samples not merely the potential due to the impurity but that of the surrounding screening charge as well. In other words, the test charge samples the potential from the net *neutral pseudoatom*.[18]

Thus $V(r)$ can be regarded either as a screened ion-ion interaction or as the interaction between a neutral pseudoatom and an ion, where the pseudoatom concept has built into it the features related to screening. We shall make some use of this concept in the last chapter.

While two-body potentials account for the bulk of the crystal potential energy $\Phi$, a small but important residue remains which is dictated essentially by the conduction electron density. This contribution, which we shall denote as $E(v)$, cannot be absorbed into the pair potential $V(r)$. While depending parametrically on the nuclear coordinates, $E(v)$ is primarily a function of the volume per atom, and for this reason the forces contributed by it are often referred to as *volume-dependent forces*.[19] In the sense that $E(v)$ cannot be written as the sum of pair potentials and yet is dependent on the nuclear coordinates, one could also look upon it as a many-body potential.

---

*Strictly speaking, it must be a gas of electrons existing against a *uniform* background of positive charge. Such a system is often referred to as an *electron gas*.

Although the idea of a pair potential $V(r)$ for metals has now become better established, one still does not work with it in terms of a specific analytic form. Part of the reason, as we shall see later, is that most calculations are carried out using the Fourier transform $V(Q)$ rather than $V(r)$ itself. Parameterized forms are therefore generally proposed for $V(Q)$ rather than for the potential itself (see Chapter 7).

It is pertinent here to make a brief reference to alloys. An alloy may be defined as a combination of two or more monatomic metals that has metallic properties. Frequently alloys can be formed over a range of composition and involve in such cases at least partial randomness in atomic arrangement. From the standpoint of binding, the key fact is their metallic behavior, that is, the fact that there are weakly bound electrons as in monatomic metals. One could therefore represent the crystal potential once again as a superposition of appropriate screened ion-pair potentials plus a volume-dependent term.

## 1.3 Ionic crystals

An ionic crystal, very simply, is a neutrally charged lattice made up of positive and negative ions. The familiar example is NaCl, wherein the ions are Na$^+$ (electronic configuration $1s^2 2s^2 2p^6$) and Cl$^-$ ($1s^2 2s^2 2p^6 3s^2 3p^6$). At first sight it would appear that the physical reason for NaCl to be ionic is that, by being so, both ions can assume the rare-gas electronic configuration which is well known to be stable. Closer examination of the ionization energy of the alkali atom and the electron affinity of the halogen atom shows that a simple electron transfer leading to rare-gas configurations is energetically unfavorable since it involves an expenditure of energy.[20] In practice Nature compensates for this by ensuring that in the solid the nearest neighbors of every ion are those of opposite charge. The energy of Coulomb attraction thus available offsets the difficulties of electron transfer; in other words, the structure assumed provides the necessary energy compensation. It is clear therefore that the dominant forces are electrostatic.

The question of expressing the crystal potential energy in terms of suitable interatomic and other potentials has received considerable attention, especially from the standpoint of calculating the cohesive energy. The model usually considered in this context is that originally due to Born in which the interatomic pair potential $V(r)$ is regarded as made up of two parts, one of which represents short-range repulsive interaction while the other represents the Coulomb interaction between the ions; the latter is clearly of long range by virtue of its $(1/r)$ dependence. The short-range potential is usually represented in the Born-Mayer form $Be^{-\alpha r}$ discussed previously. This potential is regarded as negligible beyond the first neighbor so that

$$V(|\mathbf{r}(i) - \mathbf{r}(j)|) = Be^{-\alpha|\mathbf{r}(i)-\mathbf{r}(j)|} + \frac{Z_i Z_j e^2}{|\mathbf{r}(i) - \mathbf{r}(j)|}$$

(if $i, j$ are nearest neighbors)

$$= \frac{Z_i Z_j e^2}{|\mathbf{r}(i) - \mathbf{r}(j)|}$$

(for $i, j$ = second neighbors and beyond).     (1.11)

The representation

$$\Phi = \tfrac{1}{2} \sum_i \sum_j{}' V(|\mathbf{r}(i) - \mathbf{r}(j)|), \tag{1.12}$$

where the prime denotes the exclusion of terms $(i = j)$, is capable of giving a fairly satisfactory account of the cohesive energy of at least the simple ionic crystals like the alkali halides.[21] One might wonder whether there is a London-type contribution to $V(r)$, associated with the interaction of closed-shell configurations. Indeed there is, but this is so small relative to the contributions appearing in (1.11) that it is usually ignored.

Returning to (1.12), a careful examination reveals that even in the case of classic ionic crystals, $\Phi$ as given has some shortcomings. In particular it does not allow for the possibility of three-body forces which are now known to exist in ionic crystals, including even the alkali halides.[22] An expression for the three-body potential (appropriate to NaCl-type crystals) based on Heitler-London description of the electronic wave functions has been derived by Lundqvist.[23] This will be discussed later in Sec. 4.6 in connection with the effects of three-body forces on the dynamics of alkali halides.

Besides three-body forces, $\Phi$ has a contribution associated with *distortion* of the electronic charge cloud, that is, electronic polarization. Though the latter may not be significant for cohesive energy calculations, it is nevertheless important from the standpoint of dynamics, as we shall see in Chapter 4. The contribution to $\Phi$ from polarization cannot be expressed in a simple analytic form and requires rather elaborate models which will be discussed when we are ready to consider the dynamics.

## 1.4 Covalently bonded crystals

The concept of the covalent bond really belongs to the realm of chemistry and molecular physics. The simplest example known is the familiar bond in molecular hydrogen. From an electronic standpoint, the covalent bond is essentially built up from electrons of *opposite* spins occupying the *same* orbital state. The Pauli principle does not prevent the electrons being at the same space point, and this leads to a certain amount of charge pile-up along the bond. This is in fact a characteristic feature which bestows a highly directional character to the bond.

Many crystals exhibit covalent bonds. In molecular crystals, for instance, the intramolecular binding is usually predominantly covalent in character even though intermolecular interaction is due to nonbonded interactions built up from potentials of the type (1.6). A few crystals can be regarded as built up completely by the propagation of covalent bonds. Such crystals are sometimes referred to as valency crystals. The classic example is diamond, often regarded as the perfect covalent solid. In addition, Si and to a lesser extent Ge, Te, and Se fall in the same category.

Owing to the propagation of highly directed bonds, the potential energy function $\Phi$ of a covalent crystal cannot be expressed in terms of two-body potentials alone. To appreciate the reason for this, consider a nonlinear triatomic molecule with two strong covalent bonds as illustrated in Figure 1.3. The potential energy of this molecule will evidently change when the length of either the bond 12 or the bond 23 is changed. Clearly, therefore, the molecular potential energy must include contributions from the two-body potentials $V(r_{12})$ and $V(r_{23})$ describing the energies of the two covalent bonds. By comparison, the interaction energy of the atomic pair (13) must be relatively small and, if required, could be represented in terms of a Kitaigorodsky-type potential function as in (1.6), characteristic of weakly interacting atoms.

Two-body potentials considered above do not by themselves exhaust the total potential energy of the molecule in Figure 1.3. The potential energy

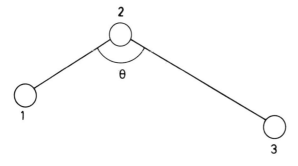

**Figure 1.3**
Sketch of a nonlinear triatomic molecule in which atom pairs (1 2) and (2 3) are covalently bonded.

must also depend on the *relative disposition* of the two bonds and must change when the bond angle $\theta$ is changed. The energy associated with *bond bending*, that is, change of the angle $\theta$, naturally depends on the simultaneous positions of all the three atoms, which implies a three-body contribution to the molecular potential energy. Extending this argument, it is easy to see that $\Phi$ for a covalent crystal must necessarily be composed of two-body and other many-body potentials.

A familiar analytic representation of $V^{(2)}$ appropriate to a covalent bond is the Morse function.[24] It is given by

$$V(r) = D\{e^{-2a(r-r_0)} - 2e^{-a(r-r_0)}\},\tag{1.13}$$

where $D$, $a$, and $r_0$ are parameters.

Analytic forms for three- and many-body potentials appropriate to valence bonds are not available. In the study of dynamics, one could overcome this lacuna by representing the *changes* in the many-body potential with respect to its value for the equilibrium configuration, in terms of appropriate parameters and coordinates. For example,

$$V(\mathbf{r}(i), \mathbf{r}(j), \mathbf{r}(k)) - V(\mathbf{x}(i), \mathbf{x}(j), \mathbf{x}(k)),$$

where the $\mathbf{x}$'s denote the equilibrium positions, is expressed in terms of a bond-bending coordinate $\Delta\theta_{ijk}$ which denotes the distortion of the bond angle $\theta_{ijk}$. Thus one writes

$$V(\mathbf{r}(i), \mathbf{r}(j), \mathbf{r}(k)) - V(\mathbf{x}(i), \mathbf{x}(j), \mathbf{x}(k)) = k_1\Delta\theta_{ijk} + k_2(\Delta\theta_{ijk})^2 + \cdots,\tag{1.14}$$

where $k_1, k_2, \ldots$ are parameters. Similarly, changes in the four-body potential can be expressed as a power series in the torsional angles in the group of the four relevant atoms. The same procedure can be extended to other many-body potentials. Further discussion of this topic will be given in the next chapter. It may be noted in passing that while this approach of describing $\Phi$ is suitable for discussing dynamics, it is evidently not useful for discussing the cohesive energy of the equilibrium crystal, since for the latter the *absolute* values rather than *changes* are important.

## 1.5 Hydrogen-bonded crystals

In the formation of molecules and crystals containing hydrogen, the H atom is usually attracted to only one atom, forming in the process a covalent bond. Frequently, however, the hydrogen is strongly attracted to two atoms, forming as it were a bond with both of them. Such a linkage is termed a *hydrogen bond*. Formally, a hydrogen bond may be said to exist whenever a

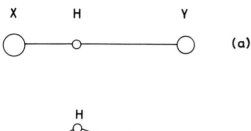

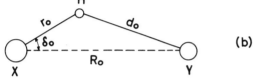

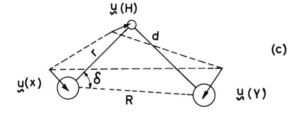

**Figure 1.4**
(a) Linear H bond; (b) equilibrium configuration of a bent H bond; (c) H bond in a deformed configuration obtained via the displacements $\mathbf{u}(X)$, $\mathbf{u}(H)$, and $\mathbf{u}(Y)$.

hydrogen atom is shared between two electronegative atoms. The hydrogen bond configuration is usually denoted as $X$—H - - - $Y$. The *donor* atom $X$ is O, N (or $N^+$), F, and occasionally S, Cl, P, or C, while the *acceptor* atom $Y$ is usually O (or $O^-$), N, $F^-$, $Cl^-$, $Br^-$, $I^-$, or S. Normally the distance H - - - $Y$ is larger than the distance $X$—H, but examples exist in which the two are equal.

The hydrogen bond is generally presumed to be linear, that is, the H atom is assumed to lie on the line joining $X$ and $Y$. (See Figure 1.4a.) Experimental investigations, especially during the last few years have, however, revealed that the bond is usually bent as illustrated in Figure 1.4b. Quite a wide range of values of $R_0$ and $\delta_0$ are found to occur in crystals. In O—H - - - O bonds, for which the maximum data are available, $R_0$ varies between 2.4 Å and 3.1 Å, while $\delta_0$ varies between $0°$ and $30°$.

In view of the wide prevalence of the hydrogen bond, and especially because of its importance in biological systems, considerable work has been

done on the potential function appropriate to isolated bonds such as
O—H - - - O and N—H - - - O. To the extent such bonds occur in crystals,
one can employ these functions to describe the contribution to $\Phi$ from these
bonds. (One crystal made up entirely of H bonds is ice.)

For linear hydrogen bonds $X$—H - - - $Y$, Lippincott and Schroeder[25]
proposed that the potential energy could be expressed as the sum of four
contributions:

$$V = V_1 + V_2 + V_3 + V_4; \tag{1.15}$$

here $V_1$ represents the $X$—H interaction, $V_2$ the H - - - $Y$ interaction and
$V_3$, $V_4$ the repulsive and attractive interactions between the donor and the
acceptor groups. The expressions for $V_1, \ldots, V_4$ are

$$V_1 = D_1\{1 - \exp[-n_1(r - r_{01})^2/2r]\}; \tag{1.16a}$$

$$V_2 = -CD_2 \exp[-n_2(R - r - r_{02})^2/2C(R - r)]; \tag{1.16b}$$

$$V_3 = A \exp(-bR); \tag{1.16c}$$

$$V_4 = -B/R^6. \tag{1.16d}$$

In these, $D_1$ and $r_{01}$ are the dissociation energy and equilibrium distance of
the unmodified $X$—H bond, that is, a bond where H is attached solely to
$X$; $D_2$ and $r_{02}$ are similar parameters for the unmodified $Y$—H bond; $C$ is
a factor, usually less than 1, which takes into account the weakness of the
H - - - $Y$ bond relative to a $Y$—H bond. The parameter $n_1$ is defined by
$n_1 = (k_1 r_{01}/D_1)$, where $k_1$ is the force constant for the stretching vibration
of the unmodified $X$—H group; $n_2$ is defined similarly. Using suitable
values for the parameters, Lippincott and Schroeder found that they could
obtain a fairly good fit with several observed structural, spectroscopic and
other data.

For bent bonds, Chidambaram and Sikka[26] modified the potential
function of Lippincott and Schroeder as

$$V = V_1 + V_2' + V_3 + V_4, \tag{1.17}$$

where $V_1$, $V_3$, $V_4$ are defined as before and $V_2'$ is defined by

$$V_2' = -CD_2 \exp[-n_2(d - r_{02})^2/2Cd]. \tag{1.18}$$

Supplementing the above equations was the requirement that the equilibrium position of the hydrogen atom for a given $R_0$ and $\delta_0$ is given by

$$\left(\frac{\partial V}{\partial r}\right)_{R_0,\delta_0} = 0.$$

Using the values $C = 0.715$, $A = 7.87 \times 10^6$ kcal/mole, $b = 4.8$ Å$^{-1}$,

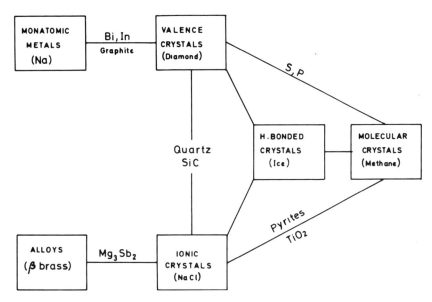

**Figure 1.5**
Schematic classification of crystals according to binding. Each box specifies a particular type and includes a canonical example. Between boxes are examples of crystals which show an intermediate behavior. The present figure is an adaptation of one due to Seitz (ref. 28).

$r_{01} = r_{02} = 0.957$ Å, $n_1 = n_2 = 9.06$ Å$^{-1}$, $D_1 = D_2 = 110.8$ kcal/mole, $B = 2652$ kcal Å$^6$/mole, Chidambaram and Sikka were able to explain the experimentally observed range of bending angles as a function of the O—O distance for the O—H --- O bonds.* While these authors were interested in the equilibrium values $r_0$ and $\delta_0$ for the bonds, the function (1.17–1.18) can also be used for discussing dynamics. Essentially one must compute the difference $V(r, \delta, R) - V(r_0, \delta_0, R_0)$ as a function of the displacements $\mathbf{u}(X)$, $\mathbf{u}(Y)$, and $\mathbf{u}(H)$ of $X$, $Y$, and H, respectively. This gives the contribution to $\Phi$ arising from a dynamic distortion of the bond. By summing up such terms associated with all the bonds in the crystal, the total hydrogen-bond contribution to $\Phi$ may be evaluated.

The potential functions (1.15–1.16) and (1.17–1.18) do not provide for the possibility that net charges could reside on the members of a H bond. However, the resulting electrostatic effects pertaining to a given bond could perhaps be implicitly absorbed into the potential parameters. On the

*Recently a similar potential function has been proposed for the N–H --- O bond which is of importance in many biological structures.[27]

other hand, the (possible) Coulomb interaction between charges belonging to different bonds cannot be so disposed of and would have to be treated separately as in the case of intermolecular potentials built up from the nonbonded potentials.

In this chapter we have tried to identify the different types of bindings that exist in crystals and to show how $\Phi$ may be visualized and constructed for the various cases. It must of course be realized that frequently a given crystal may not fall uniquely into a specific category. To emphasize this, we present in Figure 1.5 an adaptation of a figure due to Seitz.[28] The figure displays the classification we have been discussing, each box containing a canonical example. Listed between boxes are examples of solids that exhibit intermediate character. In spite of such possible ambiguities, the classification scheme suggested can be useful. Once the type of binding has been identified, one can tailor a suitable $\Phi$ function and then set the stage for calculating the dynamics. How exactly the dynamics may be computed given a knowledge of $\Phi$ will be the topic of the following chapter.

## References

1. J. E. Lennard-Jones, Proc. Roy. Soc. (London) **A106**, 463 (1924).

2. BH, p. 5.

3. BH, p. 8 ff.; an elegent semiclassical discussion is given in Appendix 39 of M. Born, *Atomic Physics* (Hafner, New York, 1962).

4. G. L. Pollack, Rev. Mod. Phys. **36**, 748 (1964); R. A. Guyer, Solid State Phys. **23**, 413 (1969).

5. L. Jansen, Advan. Quantum Chem. **2**, 119 (1965).

6. A. I. Kitaigorodsky, Tetrahedron **14**, 230 (1961); J. Chim. Phys. **63**, 9 (1966).

7. D. E. Williams, J. Chem. Phys. **47**, 4680 (1967).

8. P. A. Reynolds, J. K. Kjems, and J. W. White, in Grenoble proceedings, p. 195.

9. H. Scheraga, Advan. Phys. Org. Chem. **6**, 103 (1968).

10. G. S. Pawley, in Grenoble proceedings, p. 175.

11. See J. O. Hirschfelder, C. F. Curtiss, and R. B. Bird, *Molecular Theory of Gases and Liquids* (Wiley, New York, 1954), p. 211.

12. H. Yasuda, Progr. Theoret. Phys. (Kyoto) **45**, 1361 (1971).

13. E. P. Wigner and F. Seitz, Phys. Rev. **43**, 804 (1933); see also Solid State Phys. **1**, 97 (1955).

14. W. A. Harrison, *Pseudopotentials in the Theory of Metals* (Benjamin, New York, 1966), p. 44.

15. W. Cochran, in Bombay proceedings, Vol. I, p. 3.

16. J. Ziman, *Principles of the Theory of Solids* (Cambridge University Press, London, 1964), p. 138.

17. J. S. Langer and S. H. Vosko, J. Phys. Chem. Solids **12**, 196 (1959).

18. J. Ziman, Advan. Phys. **13**, 89 (1964).

19. K. Fuchs, Proc. Roy. Soc. (London) **A153**, 622 (1936).

20. BH, p. 3.

21. Kittel, Chapter 3.

22. P. O. Löwdin, Advan. Phys. **5**, 1 (1956).

23. S. O. Lundqvist, Arkiv. Fysik **12**, 263 (1957).

24. P. M. Morse, Phys. Rev. **34**, 57 (1929).

25. E. R. Lippincott and R. Schroeder, J. Chem. Phys. **23**, 1099 (1955).

26. R. Chidambaram and S. K. Sikka, Chem. Phys. Letters **2**, 162 (1968).

27. R. Chidambaram, R. Balasubramanian, and G. N. Ramachandran, Biochim. Biophys. Acta **221**, 182 (1970).

28. Seitz, Chapter 1.

# 2 Born–von Kármán Formalism

## 2.1 Adiabatic approximation

We address ourselves in this chapter to the task of setting up a formalism for analyzing small amplitude motions of atoms in crystals, given the concept of a crystal potential function $\Phi$ which describes the binding of atoms. Prior to this it is necessary to establish that the concept of a $\Phi$ function is indeed valid and relevant for discussing nuclear motions. The problem is not trivial, since on the face of it dynamics and binding do not appear to be directly related, as the former involves *nuclear* motions while the latter involves *electronic* motions. The required justification was first provided by Born and Oppenheimer[1] in the context of molecular vibrations. They showed that describing dynamics via a potential function is possible provided a simplifying assumption, called the *adiabatic approximation*, is made. Their arguments can be extended readily to crystals. Since excellent discussions of the Born-Oppenheimer approximation are available in the literature,[2-5] we shall here restrict ourselves to considering its implications.

Let

$$H = \sum_i (\mathbf{p}_i^2)/2m_i + \sum_j (\mathbf{p}_j^2)/2m + V_1(R) + V_2(r) + V_3(r; R) \tag{2.1}$$

be the Hamiltonian of the crystal, viewed as a collection of nuclei and electrons. The first two terms denote respectively the kinetic energies of the nuclei and of the electrons; $V_1(R)$ represents the potential energy of the nuclei, $V_2(r)$ that of the electrons and $V_3(r; R)$ the energy of electron-nuclear interactions; $R$ and $r$ are collective symbols, that is, they represent the sets $\{\mathbf{R}_1, \mathbf{R}_2, \ldots\}$ and $\{\mathbf{r}_1, \mathbf{r}_2, \ldots\}$. The adiabatic approximation consists in asserting that the eigenfunction $\Psi(r, R)$ of $H$ can be approximated by

$$\Psi_{n\lambda}(r, R) \approx \phi_{n\lambda}(R)\psi_n(r, R). \tag{2.2}$$

Here $\psi_n(r, R)$ is the solution of the Schrödinger equation for the electronic system:

$$[\sum_j (\mathbf{p}_j^2)/2m + V_2(r) + V_3(r; R)]\psi_n(r, R) = E_n(R)\psi_n(r, R); \tag{2.3}$$

$\phi_{n\lambda}(R)$ is an eigenfunction of the nuclear Hamiltonian and is obtained by solving the equation

$$[\sum_i (\mathbf{p}_i^2)/2m_i + E_n(R) + V_1(R)]\phi_{n\lambda}(R)$$
$$= [\sum_i (\mathbf{p}_i^2)/2m_i + \Phi_n(R)]\phi_{n\lambda}(R) = E_{n\lambda}\phi_{n\lambda}(R). \tag{2.4}$$

Stated differently, the essence of the adiabatic approximation is that electron and nuclear motions may be considered separately via (2.3) and (2.4). Furthermore there exists an effective potential

$$\Phi_n(R) = E_n(R) + V_1(R) \tag{2.5}$$

which governs nuclear motions when the electronic system is in its $n$th state. The potential function $\Phi$ whose empirical construction was discussed in the preceding chapter refers to the ground state, that is, it is to be identified with $\Phi_0(R)$ of the above equation.

We may ask under what conditions the adiabatic assumption holds. The answer is implicit in Eq. (2.4), the spirit of which is that as the nuclei move the electronic system continually readjusts itself according to Eq. (2.3) *without changing its quantum number.* In the process, the electronic system contributes an energy $E_n(R)$ to the crystal potential $\Phi_n(R)$. This is possible only when the nuclei move very slowly compared to the electrons so that the latter can make continuous readjustments so as to remain in the same (electronic) state. In other words, for the approximation to hold, the frequencies of nuclear motions must be much smaller than the charac-

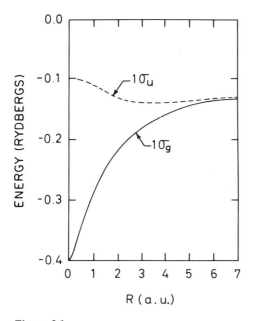

**Figure 2.1**
Energy of the two lowest electronic levels of $H_2^+$ as a function of the internuclear separation. Distances are given in atomic units (1 a.u. $= 0.528$Å) and the energy in Rydbergs (1 Ry $\approx 13.6$ eV). (After Slater, ref. 3.)

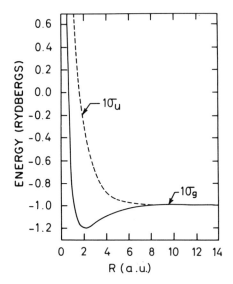

**Figure 2.2**
Lowest energy levels of $H_2^+$ as a function of internuclear separation. The curves are obtained from those of Figure 1 by adding the nuclear repulsion $(e^2/R)$. The curve marked $1\sigma_u$ corresponds to $\Phi_0$ and that labeled $1\sigma_u$ corresponds to $\Phi_1$. (After Slater, ref. 3.)

teristic electronic transition frequencies. This is invariably the case in insulators, where, owing to the energy gap between the filled and the unfilled electron states, the electronic (transition) frequencies are high. Hence the adiabatic approximation is expected to be quite good for insulators. On the other hand, one would expect it to be very poor for metals since there is no corresponding energy gap. Surprisingly, however, it works even for metals,[6,7] but for a very different reason. Because of the Pauli principle, only a few electrons near the Fermi level can undergo real transitions; the adiabatic assumption is thus good for most of the electrons, making the concept of a $\Phi_0$ function acceptable for metals also.

To illustrate the implications of the adiabatic assumption, we consider the example of the $H_2^+$ molecule.[3] Figure 2.1 shows the electronic energies for $n = 0$ and $n = 1$ as a function of the internuclear separation $R$, obtained by solving Eq. (2.3) for this ion. A plot of $\Phi_0$ and $\Phi_1$ corresponding to these states is shown in Figure 2.2. From this figure the following conclusions may be drawn: (1) The equilibrium configuration of the ion corresponds to the minimum in $\Phi_0(R)$. (2) Small oscillations performed by the ion about this configuration may be described by approximating $\Phi_0(R)$ to a parabola. (3) The energy of such oscillations, $\sim 10^{-2} Ry$, is not sufficient to excite the electron to the first excited state. In other words, nuclear motions do not change the quantum state of the electron and therefore the adiabatic assumption (implied in 2) is quite satisfactory.

The essential virtue of the approximation is that once $\Phi_0$ has been constructed, the restoring forces contributed by it to nuclear displacements can be visualized in mechanistic terms. In the $H_2{}^+$ problem, for instance, the molecular vibrational frequency may be discussed as if the molecule were held together by a spring, with a spring constant equal to $(d^2\Phi_0/dR^2)_0$. The reader might recall that elementary solid state texts often introduce crystal dynamics via a one-dimensional chain held together by harmonic springs. The justification for such a mechanical model is in fact the Born-Oppenheimer approximation. In Chapter 7 we shall see how the derivatives of $\Phi_0$ (which are the quantities required in the study of dynamics) may be explicitly calculated from basic considerations using (2.3) without the need for phenomenological description as outlined in the previous chapter.

## 2.2 Lattice dynamical equations

Given the adiabatic assumption and the existence of the potential function $\Phi_0$, the next problem is the formulation and solution of the equations of nuclear motions. Here we shall follow Born and von Kármán[8] who approached the problem essentially as mechanical. We shall suppose that the crystal is perfect, infinite, and free from stresses. Further it will be assumed to be at 0 K, and zero point effects will be ignored. Later we will comment on the need for some of these assumptions and how they may be relaxed.

### 2.2.1 Crystallographic preliminaries

As a preliminary, we review some aspects of the equilibrium configuration, that is, where all the atoms are in their respective equilibrium positions, forming a periodic array. This array may be generated by the repetition of a smaller unit called the cell. The smallest cell that generates the crystal is referred to as the *primitive cell*. While this cell can be chosen in infinitely many ways, conventions exist specifying the choice appropriate to each of the 14 Bravais lattices.[9] The rhombohedral cell associated with the fcc lattice[10] is a familiar example. Since the lattice structure underlying all crystals must belong to one of the possible Bravais lattices, this convention scheme covers all crystal structures.

It is relevant to note here the distinction between the primitive cell defined above and the *unit cell* employed by crystallographers. The latter is chosen so as to reflect the point-group symmetry of the crystal class. Although in many instances the two cells are the same, frequently they are not. The chief advantage of the primitive cell (in spite of its possibly awkward shape) is that it possesses the fewest degrees of freedom, a property that proves advantageous in dynamical problems. A later example will clarify this point.

Let $n$ be the number of atoms in the primitive cell. If $\mathbf{x}\binom{l}{k}$ is the equilibrium position of the $k$th atom in the $l$th cell then we can write

$$\mathbf{x}\begin{pmatrix} l \\ k \end{pmatrix} = \mathbf{x}(l) + \mathbf{x}(k), \tag{2.6}$$

where

$$\mathbf{x}(l) = l_1 \mathbf{a}_1 + l_2 \mathbf{a}_2 + l_3 \mathbf{a}_3, \tag{2.7}$$

$l_1, l_2, l_3$ being integers, and $\mathbf{a}_1, \mathbf{a}_2, \mathbf{a}_3$ are the basis vectors of the crystal lattice. By convention these vectors are chosen to coincide with three edges of the primitive cell sharing a corner. The lattice generated by (2.7) is a Bravais lattice; it will often be referred to as the *direct lattice*. The triplet $(l_1, l_2, l_3)$ will be denoted collectively by $l$ in formal treatment. It may be noted that for each of the $n$ possible $\mathbf{x}(k)$'s we can generate a Bravais lattice identical with that obtained from (2.7) but displaced from it by $\mathbf{x}(k)$. Each of these will be referred to as a *sublattice*. The crystal thus consists of $n$ interpenetrating sublattices, each of which is a Bravais lattice. While the atoms in any one sublattice must belong to the same chemical species, those on different sublattices need not, although they may in some cases. Thus, whereas in Mg the two interpenetrating hexagonal lattices are occupied by atoms of the same species, in NaCl the two fcc lattices are occupied by dissimilar atoms.

Given the lattice (2.7), we may define a *reciprocal lattice* by

$$\mathbf{G}(h) = h_1 \mathbf{b}_1 + h_2 \mathbf{b}_2 + h_3 \mathbf{b}_3, \tag{2.8}$$

where $h_1, h_2, h_3$ are integers and $\mathbf{b}_1, \mathbf{b}_2, \mathbf{b}_3$ are the basis vectors of the reciprocal lattice, related to $\mathbf{a}_1, \mathbf{a}_2, \mathbf{a}_3$ through

$$\mathbf{a}_i \cdot \mathbf{b}_j = 2\pi \delta_{ij}, \qquad (i, j = 1, 2, 3); \tag{2.9}$$

$\delta_{ij}$ is the Kronecker delta. It follows that

$$\mathbf{G}(h) \cdot \mathbf{x}(l) = 2\pi \times \text{integer}. \tag{2.10}$$

We might also note that the volume of the primitive cell is given by

$$v = |\mathbf{a}_1 \cdot (\mathbf{a}_2 \times \mathbf{a}_3)| \tag{2.11}$$

while that of the corresponding cell of the reciprocal lattice by

$$|\mathbf{b}_1 \cdot (\mathbf{b}_2 \times \mathbf{b}_3)| = (2\pi)^3/v. \tag{2.12}$$

A crystal may possess symmetries besides the translational symmetry considered in (2.7). A comprehensive account of the total symmetry of the crystal can be given in terms of the *space group*. Formally, the space group (symbol $G$) is the set of *elements* or *operations* which when applied to the crystal bring it into coincidence with itself. A typical space group element $\mathscr{S}_m$ is compounded of a proper or improper rotation $\mathbf{S}$ (that is, a pure rotation or a combination of a pure rotation and an inversion or reflection), a lattice translation $\mathbf{x}(m)$, and perhaps a fractional translation $\mathbf{v}(S)$. Such fractional translations are associated with screw and glide operations. Space groups that do not have any essential fractional translations are referred to as *symmorphic* while those with such translations are termed *nonsymmorphic*.

The combined operation $\mathscr{S}_m$ is usually represented by the Seitz notation $\{\mathbf{S} \mid \mathbf{x}(m) + \mathbf{v}(S)\}$ and its effect on the crystal is defined through

$$\mathscr{S}_m \mathbf{x} \binom{l}{k} = \mathbf{x}' \binom{l}{k} = \mathbf{S}\mathbf{x} \binom{l}{k} + \mathbf{x}(m) + \mathbf{v}(S). \tag{2.13}$$

Here $\mathbf{S}\mathbf{x}\binom{l}{k}$ is the vector resulting from the application of $\mathbf{S}$ to $\mathbf{x}\binom{l}{k}$, and its components are obtainable from the matrix product

$$\begin{pmatrix} S_{xx} & S_{xy} & S_{xz} \\ S_{yx} & S_{yy} & S_{yz} \\ S_{zx} & S_{zy} & S_{zz} \end{pmatrix} \begin{pmatrix} x_x\binom{l}{k} \\ x_y\binom{l}{k} \\ x_z\binom{l}{k} \end{pmatrix},$$

where $S_{xx}, S_{xy}, \dots$ denote the elements of the orthogonal matrix corresponding to $\mathbf{S}$.* The interpretation of (2.13) is in the *active sense*, that is, $\mathscr{S}_m$ involves a physical movement of the point $\mathbf{x}\binom{l}{k}$ to a new point $\mathbf{x}'\binom{l}{k}$. If $\mathscr{S}_m$ is to be a symmetry operation, then clearly $\mathbf{x}'\binom{l}{k}$ must coincide with one of the atomic positions, $\mathbf{x}\binom{L}{K}$ say, occupied by the same chemical species as that at the original site; this must be true for every untransformed and corresponding transformed site.

The effect of repeated application of space group operations is easily determined using (2.13) which gives

$$\{\mathbf{S}_i \mid \mathbf{v}(S_i) + \mathbf{x}(m_i)\}\{\mathbf{S}_j \mid \mathbf{v}(S_j) + \mathbf{x}(m_j)\}$$
$$= \{\mathbf{S}_i\mathbf{S}_j \mid \mathbf{S}_i\mathbf{v}(S_j) + \mathbf{S}_i\mathbf{x}(m_j) + \mathbf{v}(S_i) + \mathbf{x}(m_i)\}. \tag{2.14}$$

---

*While the Cartesian axes to which the elements of $\mathbf{S}$ are referred can be chosen arbitrarily, it is convenient and useful to make the choice in conformity with some accepted convention such as recommended by the Standards Committee of IRE.[11] The latter has prescribed the orientation of the Cartesian axes with respect to the crystallographic axes as given, for example, in the International Tables of X-ray Crystallography.[12]

The totality of elements $\{\mathscr{S}_m\}$ constitute the space group. Since the crystal is regarded as infinite, the order of the space group is also infinite. Recalling the group postulates, the equation defining the inverse of $\mathscr{S}_m$ is

$$\mathscr{S}_m\mathscr{S}_m^{-1} = \mathscr{S}_m^{-1}\mathscr{S}_m = \{\mathbf{E} \mid 0\}, \tag{2.15}$$

where $\mathbf{E}$ denotes the identity and is represented by $\mathbb{1}_3$, the unit matrix of order 3. Using (2.14) and (2.15), we obtain

$$\mathscr{S}_m^{-1} = \{\mathbf{S}^{-1} \mid -\mathbf{S}^{-1}\mathbf{v}(S) - \mathbf{S}^{-1}\mathbf{x}(m)\}. \tag{2.16}$$

Among the elements $\mathscr{S}_m$, some are of the type $\mathscr{E}_m = \{\mathbf{E} \mid \mathbf{x}(m)\}$ whose effect is to produce merely a lattice translation:

$$\mathscr{E}_m\mathbf{x}\binom{l}{k} = \mathbf{x}(l) + \mathbf{x}(m) + \mathbf{x}(k).$$

The set $\{\mathscr{E}_m\}$ of translational symmetry elements constitutes a group $\mathscr{T}$ called the *translational group*. $\mathscr{T}$ is a subgroup of $G$. Further it is *abelian*, that is, $\mathscr{E}_m\mathscr{E}_n = \mathscr{E}_n\mathscr{E}_m$, as is easily verified.

Besides the groups $G$ and $\mathscr{T}$, we must also take note of the point group of the crystal, $G_0$, which consists of the set of elements $\{\mathbf{S}\}$. In contrast to $G$ and $\mathscr{T}$, the order of $G_0$ is finite. It is shown in texts on crystallography that there exist only 32 crystallographic point groups, the largest of which has only 48 elements. In group-theoretical parlance, $G_0$ is a subgroup of $G$ for symmorphic space groups while for nonsymmorphic groups it is not. It is worth noting that the point group of the Bravais lattice (2.7) is in general of higher symmetry than $G_0$. Further, the reciprocal and the direct lattices share the same point-group symmetry though not the same translational symmetry.*

### 2.2.2 Potential energy expansion
We next consider a distorted configuration in which the atoms have positions

$$\mathbf{r}\binom{l}{k} = \mathbf{x}\binom{l}{k} + \mathbf{u}\binom{l}{k}, \tag{2.17}$$

where $\mathbf{u}$ denotes a small displacement. The kinetic energy is then given by

$$T = \tfrac{1}{2}\sum_l\sum_k\sum_\alpha m_k\dot{u}_\alpha^2\binom{l}{k}, \tag{2.18}$$

where $m_k$ is the mass of the atom in the $k$th sublattice.† The $l$ summation is

*Recall in this context that the fcc lattice and its reciprocal, namely, the bcc lattice, belong to the same point group $O_h$ but have different translational symmetry.
†We ignore the possibility that different isotopes may be present in a sublattice. Isotopic fluctuations usually can be accommodated by using a suitable average value for $m_k$. This is reasonable for all but the light nuclei.

over all the primitive cells; $k$ runs from 1 to $n$; $\alpha$ is the Cartesian component index. The dot over $\mathbf{u}$ denotes differentiation with respect to time.

In specifying the potential energy we shall assume that the electronic system is in its lowest state and shall drop the (electronic) subscript on $\Phi_0$. Since the displacements are assumed to be small, $\Phi$ may be expressed in a Taylor series as follows:

$$\Phi = \Phi^{(0)} + \Phi^{(1)} + \Phi^{(2)} + \Phi^{(3)} + \cdots, \tag{2.19}$$

where

$$\Phi^{(0)} = \Phi\left(\left\{\mathbf{x}\binom{l}{k}\right\}\right), \tag{2.20a}$$

$$\begin{aligned}\Phi^{(1)} &= \sum_{lk\alpha} \frac{\partial \Phi}{\partial u_\alpha\binom{l}{k}}\bigg|_0 u_\alpha\binom{l}{k} \\ &= \sum_{lk\alpha} \phi_\alpha\binom{l}{k} u_\alpha\binom{l}{k},\end{aligned} \tag{2.20b}$$

$$\begin{aligned}\Phi^{(2)} &= \tfrac{1}{2}\sum_{lk\alpha}\sum_{l'k'\beta} \frac{\partial^2 \Phi}{\partial u_\alpha\binom{l}{k}\,\partial u_\beta\binom{l'}{k'}}\bigg|_0 u_\alpha\binom{l}{k} u_\beta\binom{l'}{k'} \\ &= \tfrac{1}{2}\sum_{lk\alpha}\sum_{l'k'\beta} \phi_{\alpha\beta}\binom{l\ \ l'}{k\ \ k'} u_\alpha\binom{l}{k} u_\beta\binom{l'}{k'},\end{aligned} \tag{2.20c}$$

$$\begin{aligned}\Phi^{(3)} &= \tfrac{1}{6}\sum_{lk\alpha}\sum_{l'k'\beta}\sum_{l''k''\gamma} \frac{\partial^3 \Phi}{\partial u_\alpha\binom{l}{k}\,\partial u_\beta\binom{l'}{k'}\,\partial u_\gamma\binom{l''}{k''}}\bigg|_0 u_\alpha\binom{l}{k} u_\beta\binom{l'}{k'} u_\gamma\binom{l''}{k''} \\ &= \tfrac{1}{6}\sum_{lk\alpha}\sum_{l'k'\beta}\sum_{l''k''\gamma} \phi_{\alpha\beta\gamma}\binom{l\ \ l'\ \ l''}{k\ \ k'\ \ k''} u_\alpha\binom{l}{k} u_\beta\binom{l'}{k'} u_\gamma\binom{l''}{k''},\end{aligned} \tag{2.20d}$$

$\cdots$

Our objective in making the power series expansion (2.19) is to restrict it to the second-order term. As will be seen shortly, this leads to harmonic vibrations of the atoms and is therefore referred to as the *harmonic approximation*. From a practical standpoint, the expansion (2.19) is meaningful only for substances wherein the rms displacements due to zero-point motion are small compared to interatomic distances. While this is probably true for most solids, it certainly does not hold for solid helium where the rms displacement is $\sim 30\%$ of the lattice constant.* Helium therefore requires a totally different approach and will be considered in the last chapter.

### 2.2.3 Properties of expansion coefficients
Let us now consider some of the properties of the expansion coefficients in $\Phi^{(1)}$ and $\Phi^{(2)}$. In the equilibrium configuration, the force on every atom

---

*For comparison, the rms displacement due to zero-point motion in Ge is only $\sim 2\%$.

must vanish. This immediately leads to the result

$$\phi_\alpha\begin{pmatrix} l \\ k \end{pmatrix} = 0 \text{ for every } \alpha, k, l, \tag{2.21}$$

whence $\Phi^{(1)} = 0$. Thus in the harmonic approximation,

$$\Phi \approx \Phi^{(0)} + \Phi^{(2)}$$

$$= \Phi^{(0)} + \tfrac{1}{2} \sum_{l k \alpha} \sum_{l' k' \beta} \phi_{\alpha\beta}\begin{pmatrix} l\ l' \\ k\ k' \end{pmatrix} u_\alpha\begin{pmatrix} l \\ k \end{pmatrix} u_\beta\begin{pmatrix} l' \\ k' \end{pmatrix}. \tag{2.22}$$

We shall shortly see that the coefficients $\phi_{\alpha\beta}\begin{pmatrix} l l' \\ k k' \end{pmatrix}$ are to be interpreted as force constants. Some of their properties are listed below.

**Permutation symmetry.** From the reasonable assumption that $\Phi$ is a well-behaved function, it follows that the order of differentiation in (2.20c) is immaterial. This leads to the result

$$\phi_{\alpha\beta}\begin{pmatrix} l\ l' \\ k\ k' \end{pmatrix} = \phi_{\beta\alpha}\begin{pmatrix} l'\ l \\ k'\ k \end{pmatrix}. \tag{2.23}$$

**Crystal symmetry.** Crystal symmetry bestows some important properties on the force constants. Let $\mathscr{S}_m$ acting on the undistorted crystal produce the transformation

$$\mathbf{x}\begin{pmatrix} l \\ k \end{pmatrix} \rightarrow \mathbf{x}\begin{pmatrix} L \\ K \end{pmatrix} = \mathscr{S}_m \mathbf{x}\begin{pmatrix} l \\ k \end{pmatrix}.$$

Applied to the distorted crystal, $\mathscr{S}_m$ carries the displacement $\mathbf{u}\begin{pmatrix} l \\ k \end{pmatrix}$ to the site $\begin{pmatrix} L \\ K \end{pmatrix}$ as $\mathbf{S}\mathbf{u}\begin{pmatrix} l \\ k \end{pmatrix}$, because

$$\mathscr{S}_m \mathbf{r}\begin{pmatrix} l \\ k \end{pmatrix} = \mathbf{S}\mathbf{x}\begin{pmatrix} l \\ k \end{pmatrix} + \mathbf{S}\mathbf{u}\begin{pmatrix} l \\ k \end{pmatrix} + \mathbf{v}(S) + \mathbf{x}(m)$$

$$= \mathscr{S}_m \mathbf{x}\begin{pmatrix} l \\ k \end{pmatrix} + \mathbf{S}\mathbf{u}\begin{pmatrix} l \\ k \end{pmatrix}$$

$$= \mathbf{x}\begin{pmatrix} L \\ K \end{pmatrix} + \mathbf{S}\mathbf{u}\begin{pmatrix} l \\ k \end{pmatrix}.$$

Thus the displacement $\mathbf{u}'\begin{pmatrix} L \\ K \end{pmatrix}$ of $\begin{pmatrix} L \\ K \end{pmatrix}$ after the operation $\mathscr{S}_m$ is $\mathbf{S}\mathbf{u}\begin{pmatrix} l \\ k \end{pmatrix}$. Similarly, if the site $\begin{pmatrix} l' \\ k' \end{pmatrix}$ is carried by $\mathscr{S}_m$ to $\begin{pmatrix} L' \\ K' \end{pmatrix}$, then

$$\mathbf{u}\begin{pmatrix} l' \\ k' \end{pmatrix} \xrightarrow{\mathscr{S}_m} \mathbf{u}'\begin{pmatrix} L' \\ K' \end{pmatrix} = \mathbf{S}\mathbf{u}\begin{pmatrix} l' \\ k' \end{pmatrix}. \tag{2.24}$$

Now the potential energy of the pair $\begin{pmatrix} l\ l' \\ k\ k' \end{pmatrix}$ before the transformation must equal that of the equivalent pair $(^L_K \, ^{L'}_{K'})$ after the transformation. Therefore

$$\sum_{\alpha\beta} u_\alpha \binom{l}{k} \phi_{\alpha\beta} \binom{l \; l'}{k \; k'} u_\beta \binom{l'}{k'} = \sum_{\mu\nu} u'_\mu \binom{L}{K} \phi_{\mu\nu} \binom{L \; L'}{K \; K'} u'_\nu \binom{L'}{K'}. \qquad (2.25)$$

Remembering $u'_\mu(^L_K) = \sum_\gamma S_{\mu\gamma} u_\gamma(^l_k)$ and $u'_\nu(^{L'}_{K'}) = \sum_\delta S_{\nu\delta} u_\delta(^{l'}_{k'})$, we obtain from (2.25)

$$\phi_{\alpha\beta} \binom{l \; l'}{k \; k'} = \sum_{\mu\nu} S_{\mu\alpha} S_{\nu\beta} \, \phi_{\mu\nu} \binom{L \; L'}{K \; K'}, \qquad (2.26a)$$

or, in matrix notation,

$$\boldsymbol{\phi} \binom{l \; l'}{k \; k'} = \tilde{\mathbf{S}} \boldsymbol{\phi} \binom{L \; L'}{K \; K'} \mathbf{S}, \qquad (2.26b)$$

where $\boldsymbol{\phi}$ is a $3 \times 3$ matrix with elements $\phi_{\alpha\beta}$ and $\sim$ implies transposition. Since $\mathbf{S}$ is *orthogonal*, (2.26b) may be inverted to yield

$$\boldsymbol{\phi} \binom{L \; L'}{K \; K'} = \mathbf{S} \boldsymbol{\phi} \binom{l \; l'}{k \; k'} \tilde{\mathbf{S}}. \qquad (2.26c)$$

Certain operations $\mathscr{S}_m$ may leave a given pair of sites $(^l_k)$ and $(^{l'}_{k'})$ unchanged. Equation (2.26) then becomes

$$\boldsymbol{\phi} \binom{l \; l'}{k \; k'} = \mathbf{S} \boldsymbol{\phi} \binom{l \; l'}{k \; k'} \tilde{\mathbf{S}}. \qquad (2.27)$$

This relation is extremely useful in determining the structure of the force-constant matrix, that is, in determining which if any elements vanish and in establishing possible interrelations among the nonvanishing elements. A practical example of this will be given in Sec. 2.9.

A few other special cases of (2.26) are also worth recording.

1. Let $\mathscr{S}_m = \mathscr{E}_m$. Then

$$\phi_{\alpha\beta} \binom{l \; l'}{k \; k'} = \phi_{\alpha\beta} \binom{l + m \;\; l' + m}{k \qquad k'}. \qquad (2.28a)$$

In particular,

$$\phi_{\alpha\beta} \binom{l \; l'}{k \; k'} = \phi_{\alpha\beta} \binom{l - l' \;\; 0}{k \qquad k'} \qquad (2.28b)$$

$$= \phi_{\alpha\beta} \binom{0 \;\; l' - l}{k \qquad k'}. \qquad (2.28c)$$

2. In some crystals (for example, NaCl), every atom is situated at a center of inversion. By this we mean that the crystal reproduces itself if inverted about

any atom. Such crystals have the element $\{\mathbf{I} \mid 0\}$ in the space group, where $\mathbf{I}$ stands for inversion and is represented by the matrix $-\mathbb{1}_3$. The centrosymmetric property of the crystal is expressed through the equation

$$\{\mathbf{I} \mid 0\}\mathbf{x}\binom{l}{k} = -\mathbf{x}\binom{l}{k} \text{ for all } l \text{ and } k.$$

In particular, if

$$\mathbf{x}\binom{l}{k} \xrightarrow{I} \mathbf{x}\binom{L}{k}$$

and

$$\mathbf{x}\binom{l'}{k'} \xrightarrow{I} \mathbf{x}\binom{L'}{k'},$$

then

$$\mathbf{x}(l') - \mathbf{x}(l) + \mathbf{x}(k') - \mathbf{x}(k)$$
$$= -[\mathbf{x}(L') - \mathbf{x}(L)] - \mathbf{x}(k') + \mathbf{x}(k). \qquad (2.29)$$

Use now the notation $\bar{l} = l' - l$ and $\bar{L} = L' - L$. It then follows from (2.26) that

$$\mathbf{I}\boldsymbol{\phi}\begin{pmatrix} 0 & \bar{l} \\ k & k' \end{pmatrix}\tilde{\mathbf{I}} = \boldsymbol{\phi}\begin{pmatrix} 0 & \bar{l} \\ k & k' \end{pmatrix} \qquad \text{(on multiplying)},$$

$$= \boldsymbol{\phi}\begin{pmatrix} 0 & \bar{L} \\ k & k' \end{pmatrix} \qquad \text{(from (2.26))}. \qquad (2.30)$$

3. In some crystals (for example, diamond, $CaF_2$) inversion interchanges like atoms in different sublattices. Let $k$ and $k'$ denote the sublattices so interchanged, that is,

$$\mathbf{x}\binom{l}{k} \xrightarrow{I} \mathbf{x}\binom{L}{k'}$$
$$\mathbf{x}\binom{l'}{k'} \xrightarrow{I} \mathbf{x}\binom{L'}{k}.$$

It follows that $(l' - l) = -(L' - L)$. Using (2.26) we then have

$$\boldsymbol{\phi}\begin{pmatrix} 0 & l' - l \\ k & k' \end{pmatrix} = \boldsymbol{\phi}\begin{pmatrix} 0 & L' - L \\ k' & k \end{pmatrix}$$

$$= \boldsymbol{\phi}\begin{pmatrix} l' - l & 0 \\ k' & k \end{pmatrix} \qquad \text{(using (2.28))}.$$

With (2.23) this then leads to

$$\Phi\begin{pmatrix} 0 & l' - l \\ k & k' \end{pmatrix} = \tilde{\Phi}\begin{pmatrix} 0 & l' - l \\ k & k' \end{pmatrix}. \tag{2.31}$$

In other words, the force-constant matrix connecting the two sublattices is symmetric.

4. Let $\mathscr{P}_m = \{\mathbf{P} \mid \mathbf{x}(m)\}$ be a space group operation that carries a *given* pair of atoms at $\binom{l}{k}$ and $\binom{l'}{k}$ to new sites $\binom{L}{k}$ and $\binom{L'}{k}$, respectively, with the additional constraint

$$(l' - l) = -(L' - L) = \Lambda.$$

From (2.26),

$$
\begin{aligned}
\mathbf{P}\Phi\begin{pmatrix} 0 & l' - l \\ k & k \end{pmatrix}\tilde{\mathbf{P}} &= \Phi\begin{pmatrix} 0 & -\Lambda \\ k & k \end{pmatrix} \\
&= \Phi\begin{pmatrix} \Lambda & 0 \\ k & k \end{pmatrix} \quad \text{(using (2.28))} \\
&= \tilde{\Phi}\begin{pmatrix} 0 & \Lambda \\ k & k \end{pmatrix} \quad \text{(using (2.23)).}
\end{aligned} \tag{2.32}
$$

This relation is helpful, for example, in the determination of the structure of $\Phi$ for second neighbors in the diamond lattice.

In addition to the properties discussed above, there are other constraints on the force constants due to the invariance of the potential energy under infinitesimal translation and rotation of the crystal as a whole. These are best discussed after we have set up the equations of motion.

### 2.2.4 Equations of motion
Our next task is to formulate the dynamical problem starting from the harmonic Hamiltonian

$$H = T + \Phi$$

$$= \sum_\alpha p_\alpha^2\binom{l}{k}\bigg/ 2m_k + \Phi^{(0)} + \tfrac{1}{2}\sum_{lk\alpha}\sum_{l'k'\beta}\phi_{\alpha\beta}\begin{pmatrix} l & l' \\ k & k' \end{pmatrix} u_\alpha\binom{l}{k} u_\beta\binom{l'}{k'}, \tag{2.33}$$

where $p_\alpha\binom{l}{k}$ is the momentum conjugate to $u_\alpha\binom{l}{k}$. This formulation will first be done classically, the transition to quantum mechanics being made later.

Using Hamilton's equations, we obtain directly from (2.33) the following equations of motion:

$$-\dot{p}_\alpha\begin{pmatrix} l \\ k \end{pmatrix} = -m_k \ddot{u}_\alpha\begin{pmatrix} l \\ k \end{pmatrix}$$

$$= \sum_{l'k'\beta} \phi_{\alpha\beta}\begin{pmatrix} l & l' \\ k & k' \end{pmatrix} u_\beta\begin{pmatrix} l' \\ k' \end{pmatrix}, \quad \text{for } \alpha = x, y, z;$$

$$k = 1, \ldots, n;$$

$$l \rightarrow \text{over the entire crystal.} \quad (2.34)$$

In obtaining (2.34), a slight manipulation of the dummy indices is required, followed by a use of (2.23). From (2.34) it is clear that $\phi_{\alpha\beta}\begin{pmatrix} l & l' \\ k & k' \end{pmatrix}$ is the negative of the force exerted on atom $\begin{pmatrix} l \\ k \end{pmatrix}$ in the $\alpha$ direction due to unit displacement of the atom $\begin{pmatrix} l' \\ k' \end{pmatrix}$ in the $\beta$ direction. The quantity $(-\phi_{\alpha\beta})$ is often referred to as the *force constant* and is usually expressed in the units dynes/cm. However, we shall employ the term interchangeably both for $\phi_{\alpha\beta}$ as well as $(-\phi_{\alpha\beta})$. It is these constants which are traditionally symbolized in introductory texts by springs obeying Hooke's law.

### 2.2.5 Translational and rotational sum rules

The constraints imposed on the force constants by infinitesimal translational and rotational invariance will now be derived from (2.34). Suppose that the (undistorted)crystal is given an infinitesimal translation $\varepsilon_\beta$. All the atoms have now the same displacement $\varepsilon_\beta$. Since this does not call into play any restoring forces, we have from (2.34)

$$0 = -\left[\sum_{l'k'} \phi_{\alpha\beta}\begin{pmatrix} l & l' \\ k & k' \end{pmatrix}\right]\varepsilon_\beta,$$

whence

$$\sum_{l'k'} \phi_{\alpha\beta}\begin{pmatrix} l & l' \\ k & k' \end{pmatrix} = 0, \qquad (2.35a)$$

since $\varepsilon_\beta$ is arbitrary. Equation (2.35a) is true for all $l$, $\alpha$, $\beta$, and $k$. In view of (2.28), it may also be written as

$$\sum_{l'k'} \phi_{\alpha\beta}\begin{pmatrix} 0 & l' \\ k & k' \end{pmatrix} = 0. \qquad (2.35b)$$

This sum rule is of great importance in practical calculations as it enables the determination of the "self force constant" $\phi_{\alpha\beta}\begin{pmatrix} 0 & 0 \\ k & k \end{pmatrix}$. Physically, this quantity represents the negative of the force on the atom $\begin{pmatrix} 0 \\ k \end{pmatrix}$ in the $\alpha$ direction when it is given a unit displacement in the $\beta$ direction, all the other atoms being held in their respective equilibrium positions. This pull may be visualized as exerted by all the *other* atoms; the force thus exerted (per unit displacement) is given by

$$\phi_{\alpha\beta}\begin{pmatrix} 0 & 0 \\ k & k \end{pmatrix} = -\sum_{l'k'}' \phi_{\alpha\beta}\begin{pmatrix} 0 & l' \\ k & k' \end{pmatrix}, \tag{2.35c}$$

where the prime implies the exclusion of the ($l' = 0$, $k' = k$) term. Note that in view of (2.23) we also have

$$\sum_{l'k'} \phi_{\beta\alpha}\begin{pmatrix} l' & 0 \\ k' & k \end{pmatrix} = 0. \tag{2.35d}$$

Next we consider the constraint arising out of rotational invariance. To obtain this, we imagine an infinitesimal rotation $\boldsymbol{\Omega}$ to be performed on the undistorted crystal. In matrix representation,[13]

$$\boldsymbol{\Omega} = \begin{pmatrix} 1 & 0 & 0 \\ 0 & 1 & 0 \\ 0 & 0 & 1 \end{pmatrix} + \begin{pmatrix} 0 & -\theta_z & \theta_y \\ \theta_z & 0 & -\theta_x \\ -\theta_y & \theta_x & 0 \end{pmatrix}, \tag{2.36}$$

where $\theta_\alpha$ denotes an infinitesimal rotation about the $\alpha$ axis. The displacement components due to the rotation are therefore given by

$$u_\alpha\begin{pmatrix} l \\ k \end{pmatrix} = \left[ \boldsymbol{\theta} \times \mathbf{x}\begin{pmatrix} l \\ k \end{pmatrix} \right]_\alpha. \tag{2.37}$$

For purposes of illustration, let us suppose that the rotation is about the $x$ axis. Using (2.37) in (2.34) and remembering further that no restoring forces are present and that $\theta_x$ is arbitrary, we obtain[14]

$$0 = \sum_{l'k'} \left\{ \phi_{\alpha z}\begin{pmatrix} 0 & l' \\ k & k' \end{pmatrix} x_y\begin{pmatrix} l' \\ k' \end{pmatrix} - \phi_{\alpha y}\begin{pmatrix} 0 & l' \\ k & k' \end{pmatrix} x_z\begin{pmatrix} l' \\ k' \end{pmatrix} \right\}. \tag{2.38}$$

Similar constraints may be obtained by considering the rotations $\theta_y$ and $\theta_z$. All these may be expressed compactly via the Levi-Civita symbol $\varepsilon_{\alpha\beta\gamma}$, defined as follows:

$\varepsilon_{\alpha\beta\gamma} = 0$ if any two of the Cartesian indices $\alpha$, $\beta$, $\gamma$ are equal;

$\quad = 1$ if $(\alpha, \beta, \gamma)$ corresponds to a cyclic order of $(x, y, z)$;

$\quad = -1$ if $(\alpha, \beta, \gamma)$ corresponds to a noncyclic order of $(x, y, z)$. $\quad$ (2.39)

Using this symbol, (2.38) may be generalized to

$$\sum_{l'k'} \sum_{\mu\nu} \phi_{\alpha\nu}\begin{pmatrix} 0 & l' \\ k & k' \end{pmatrix} x_\mu\begin{pmatrix} l' \\ k' \end{pmatrix} \varepsilon_{\beta\mu\nu} = 0, \qquad \text{for all } \alpha, \beta, \text{ and } k. \tag{2.40}$$

### 2.2.6 Solution of the equations of motion

Returning to the equations of motion, crystal periodicity suggests that the solutions must be such that the displacements of corresponding atoms in different cells be equivalent, apart from a possible phase factor. Accordingly we seek wavelike solutions of the type

$$u_\alpha\binom{l}{k} = \frac{1}{\sqrt{m_k}} U_\alpha(k|\mathbf{q}) \exp\{i[\mathbf{q}\cdot\mathbf{x}(l) - \omega(\mathbf{q})t]\}. \tag{2.41}$$

Here $\mathbf{q}$ is the wave vector and $\omega(\mathbf{q})$ the angular frequency* associated with the wave. The factor $(m_k)^{-1/2}$ is introduced for later convenience. It is to be noted that the (complex) wave amplitude is independent of $l$ but depends on $\mathbf{q}$. In trying a complex solution of the type (2.41), we understand that when all the independent solutions are superposed the displacements must be real.

Substitution of (2.41) into the equations of motion leads to the following $3n$ simultaneous equations in the wave amplitudes $\{U_\alpha(k \mid \mathbf{q})\}$:

$$\omega^2(\mathbf{q})U_\alpha(k \mid \mathbf{q}) = \sum_{k'\beta} D_{\alpha\beta}\binom{\mathbf{q}}{kk'} U_\beta(k' \mid \mathbf{q}), \qquad \alpha = x, y, z; k = 1, \ldots, n, \tag{2.42a}$$

where

$$D_{\alpha\beta}\binom{\mathbf{q}}{kk'} = \frac{1}{\sqrt{m_k m_{k'}}} \sum_{l'} \phi_{\alpha\beta}\binom{l\ l'}{k\ k'} \exp\{i\mathbf{q}\cdot[\mathbf{x}(l') - \mathbf{x}(l)]\}$$
$$= \frac{1}{\sqrt{m_k m_{k'}}} \sum_{\bar{l}} \phi_{\alpha\beta}\binom{0\ \bar{l}}{k\ k'} \exp\{i\mathbf{q}\cdot\mathbf{x}(\bar{l})\} \quad \text{(using (2.28))}. \tag{2.43}$$

Equation (2.42a) may be expressed compactly in matrix notation as

$$\omega^2(\mathbf{q})\mathbf{U}(\mathbf{q}) = \mathbf{D}(\mathbf{q})\mathbf{U}(\mathbf{q}), \tag{2.42b}$$

where $\mathbf{D}(\mathbf{q})$ is a $3n$-dimensional square matrix and $\mathbf{U}(\mathbf{q})$ is a $3n$-component column matrix:

$$\mathbf{D}(\mathbf{q}) = \begin{pmatrix} \mathbf{D}\binom{\mathbf{q}}{11} & \cdots & \mathbf{D}\binom{\mathbf{q}}{1n} \\ \cdots & \cdots & \cdots \\ \cdots & \cdots & \cdots \\ \mathbf{D}\binom{\mathbf{q}}{n1} & \cdots & \mathbf{D}\binom{\mathbf{q}}{nn} \end{pmatrix}, \mathbf{D}\binom{\mathbf{q}}{kk'} = \begin{pmatrix} D_{xx}\binom{\mathbf{q}}{kk'} & D_{xy}\binom{\mathbf{q}}{kk'} & D_{xz}\binom{\mathbf{q}}{kk'} \\ D_{yx}\binom{\mathbf{q}}{kk'} & D_{yy}\binom{\mathbf{q}}{kk'} & D_{yz}\binom{\mathbf{q}}{kk'} \\ D_{zx}\binom{\mathbf{q}}{kk'} & D_{zy}\binom{\mathbf{q}}{kk'} & D_{zz}\binom{\mathbf{q}}{kk'} \end{pmatrix}$$
$$\tag{2.44}$$

$$\mathbf{U}(\mathbf{q}) = \begin{pmatrix} U_x(k = 1 \mid \mathbf{q}) \\ U_y(k = 1 \mid \mathbf{q}) \\ \cdots \cdots \\ U_z(k = n \mid \mathbf{q}) \end{pmatrix} \tag{2.45}$$

*The frequency is often expressed by the quantity $\nu = \omega/2\pi$ (usual units: THz = $10^{12}$ Hz) or by $\nu/c$ (expressed in cm$^{-1}$ ("wave numbers")). The conversions are 1 THz = $2\pi \times 10^{12}$ rad/sec $\approx 33.36$ cm$^{-1}$.
Further, $\nu = 1$ THz corresponds to $h\nu = \hbar\omega \approx 4.136$ meV.

$\mathbf{D}(\mathbf{q})$ is often termed the *dynamical matrix*.

From (2.42) it is clear that the problem of determining the frequencies appropriate to waves of wave vector $\mathbf{q}$ is essentially one of solving the eigenvalue problem (2.42). As is usual, the eigenvalues are obtained by solving the *characteristic* or *secular* equation

$$\| \mathbf{D}(\mathbf{q}) - \omega^2(\mathbf{q}) \mathbb{1}_{3n} \| = 0 \tag{2.46}$$

which results when we insist that (2.42) must yield nontrivial solutions for the wave amplitudes. Solving (2.46) yields $3n$ eigenvalues which we label $\omega_j^2(\mathbf{q})$, $(j = 1, 2, \ldots, 3n)$. Not all of these need be distinct; depending on the symmetry of the crystal and the value of $\mathbf{q}$, some may be degenerate.*

It is easily proved using (2.23) and (2.28) that

$$D_{\alpha\beta}\begin{pmatrix} \mathbf{q} \\ kk' \end{pmatrix} = D_{\beta\alpha}^*\begin{pmatrix} \mathbf{q} \\ k'k \end{pmatrix}. \tag{2.47}$$

In other words, $\mathbf{D}(\mathbf{q})$ is Hermitian. It follows then from a well-known theorem of matrix algebra that its eigenvalues must be real.

Once the eigenvalues $\omega_j^2(\mathbf{q})$ are known, the components of the corresponding eigenvectors $U_\alpha(k \mid {}_j^\mathbf{q})$ may be extracted from (2.42a) after substituting the eigenvalue into that equation. In this connection we may draw upon the results of matrix algebra[15] to make the following remarks:

1. Since $\mathbf{D}(\mathbf{q})$ is Hermitian, its eigenvectors may be chosen to be orthonormal. This point will be elaborated upon shortly.

2. If a given eigenvalue $\omega_j^2(\mathbf{q})$ is nondegenerate, then since the $3n$ equations (2.42) are homogeneous, $\mathbf{U}({}_j^\mathbf{q})$ can be determined only to within an arbitrary constant. In other words, after choosing one component of $\mathbf{U}({}_j^\mathbf{q})$ arbitrarily, we may solve for the other $(3n - 1)$.

3. In case there is degeneracy, the situation is slightly complicated. For example, if there is a two-fold degeneracy then it is always possible to construct two eigenvectors $\mathbf{U}({}_{j_1}^\mathbf{q})$ and $\mathbf{U}({}_{j_2}^\mathbf{q})$ which are orthonormal. However, these two eigenvectors are not unique; in fact by suitably combining $\mathbf{U}({}_{j_1}^\mathbf{q})$ and $\mathbf{U}({}_{j_2}^\mathbf{q})$, it is possible to construct an infinite number of pairs of vectors which are equally acceptable as eigenvectors. (The presence of a double degeneracy permits only $3n - 2$ components of each eigenvector to be solved for.) In general, if there is a $\nu$-fold degeneracy, then it is always possible to construct $\nu$ orthonormal eigenvectors belonging to that eigenvalue. However, there are a $(\nu - 1)$-fold infinity of such eigenvector sets.

While solving for the eigenvectors, one would like to impose on them a suitable normalizing condition. Here we shall apply such a constraint not on $\mathbf{U}({}_j^\mathbf{q})$ but on a dimensionless quantity $\mathbf{e}({}_j^\mathbf{q})$ with components $e_\alpha(k|{}_j^\mathbf{q})$ and

*This topic will be pursued further in the following chapter.

related to $\mathbf{U}(^{\mathbf{q}}_j)$ by

$$U_\alpha\left(k\left|^{\mathbf{q}}_j\right.\right) = A\left(^{\mathbf{q}}_j\right)e_\alpha\left(k\left|^{\mathbf{q}}_j\right.\right), \tag{2.48}$$

where $A$ is independent of $\alpha$, $k$, and time, and has the same dimensions as $\mathbf{U}$, that is, $M^{1/2}L$. It is clear that $\mathbf{e}(^{\mathbf{q}}_j)$ is as acceptable an eigenvector of $\mathbf{D}(\mathbf{q})$ as is $\mathbf{U}(^{\mathbf{q}}_j)$.

Remembering that the eigenvectors of $\mathbf{D}(\mathbf{q})$ can be chosen orthonormal and that the normalization constraint is to be imposed on $\mathbf{e}$ rather than $\mathbf{U}$, we can express the orthonormality condition as

$$\mathbf{e}^\dagger\left(^{\mathbf{q}}_j\right)\mathbf{e}\left(^{\mathbf{q}}_{j'}\right) = \delta_{jj'}, \tag{2.49a}$$

where the symbol † implies transposition followed by complex conjugation. In component form (2.49a) reads

$$\sum_{\alpha k} e_\alpha^*\left(k\left|^{\mathbf{q}}_j\right.\right)e_\alpha\left(k\left|^{\mathbf{q}}_{j'}\right.\right) = \delta_{jj'}. \tag{2.49b}$$

In passing we note that the solution of the secular equation (2.46) is nowadays invariably carried out on a computer. Routines commonly employed for this purpose deliver also the normalized eigenvectors as a byproduct.

### 2.2.7 Vector-space interpretation of eigenvectors and their transformations

It is helpful to cast some of the foregoing results in the language of multi-dimensional vector spaces.* Now the eigenvectors $\mathbf{e}(^{\mathbf{q}}_j)$ can be regarded as vectors in an abstract $3n$-dimensional space. In this scheme, the components $e_\alpha(k|^{\mathbf{q}}_j)$ denote the components of $\mathbf{e}(^{\mathbf{q}}_j)$ with respect to the following $3n$ unit vectors:

$$\begin{pmatrix}1\\0\\0\\ \vdots \\0\\0\end{pmatrix}, \begin{pmatrix}0\\1\\0\\ \vdots \\0\\0\end{pmatrix}, \ldots, \begin{pmatrix}0\\0\\0\\ \vdots \\0\\1\end{pmatrix} \qquad 3n \text{ elements.}$$

The orthonormality condition (2.49) may now be viewed as the statement that in this abstract space the vectors $\mathbf{e}(^{\mathbf{q}}_j)$ are of unit length and are mutually "perpendicular."

*A good introductory account of vector spaces with an orientation of relevance to us is available in *Symmetry* by R. W. McWeeny (MacMillan, New York, 1963).

Let us now construct a $3n$-dimensional matrix $\mathbf{e}(\mathbf{q})$ by arranging the $3n$ eigenvectors as below (remember that $\mathbf{e}\binom{\mathbf{q}}{j}$ is a column matrix with $3n$ rows):

$$\mathbf{e}(\mathbf{q}) = \left[ \mathbf{e}\binom{\mathbf{q}}{1}, \cdots, \mathbf{e}\binom{\mathbf{q}}{3n} \right]. \tag{2.50}$$

Using this matrix, we can write the $3n$ equations

$$\mathbf{D}(\mathbf{q})\mathbf{e}\binom{\mathbf{q}}{j} = \omega_j^2(\mathbf{q})\mathbf{e}\binom{\mathbf{q}}{j}, \qquad (j = 1, \ldots, 3n), \tag{2.51}$$

compactly as

$$\mathbf{D}(\mathbf{q})\mathbf{e}(\mathbf{q}) = \mathbf{e}(\mathbf{q})\mathbf{\Omega}(\mathbf{q}), \tag{2.52}$$

where $\mathbf{\Omega}(\mathbf{q})$ is the diagonal matrix

$$\Omega_{jj'}(\mathbf{q}) = \omega_j^2(\mathbf{q})\delta_{jj'}. \tag{2.53}$$

Now in view of (2.49) we have

$$\mathbf{e}^\dagger(\mathbf{q})\mathbf{e}(\mathbf{q}) = \mathbb{1}_{3n}. \tag{2.54}$$

This enables us to rewrite (2.52) as

$$\mathbf{e}^\dagger(\mathbf{q})\mathbf{D}(\mathbf{q})\mathbf{e}(\mathbf{q}) = \mathbf{\Omega}(\mathbf{q}). \tag{2.55}$$

Furthermore, from (2.54),*

$$\mathbf{e}^\dagger(\mathbf{q})\mathbf{e}(\mathbf{q})\mathbf{e}^{-1}(\mathbf{q}) = \mathbf{e}^{-1}(\mathbf{q})$$

or

$$\mathbf{e}(\mathbf{q})\mathbf{e}^\dagger(\mathbf{q}) = \mathbb{1}_{3n}. \tag{2.56}$$

From (2.54) and (2.56) it is clear that $\mathbf{e}$ is unitary. Thus the "geometrical" significance of (2.55) is that $\mathbf{e}(\mathbf{q})$ is a unitary transformation in the $3n$-dimensional space which brings the dynamical matrix into a diagonal form, that is, it is the principal-axis transformation[16] of the present problem.

Let us now express (2.56) in component form. This gives

$$\sum_j e_\alpha\left(k\middle|\begin{matrix}\mathbf{q}\\j\end{matrix}\right)e_\beta^*\left(k'\middle|\begin{matrix}\mathbf{q}\\j\end{matrix}\right) = \delta_{\alpha\beta}\delta_{kk'} \tag{2.57}$$

which is called the *completeness relation* for eigenvectors and is analogous to the relation $\sum_i \psi_i(x)\psi_i^*(x') = \delta(x - x')$ satisfied by quantum-mechanical wave functions. The "geometrical" significance of (2.57) is that the $3n$

---

*Since $\mathbf{e}$ is assembled from orthonormal vectors, it is nonsingular.

vectors $\{e(^q_j)\}$ form a complete set and are therefore suitable as basis vectors in the $3n$-dimensional space.

### 2.2.8 Reality of the eigenfrequencies—constraint on the force constants

We remarked earlier that the eigenvalues of $\mathbf{D}(\mathbf{q})$ are real since it is a Hermitian matrix. This implies that $\omega_j(\mathbf{q})$ is either real or purely imaginary. A purely imaginary solution for the frequency would imply that the displacements increase indefinitely with time (see Eq. (2.41)), leading to destruction of the crystal. For stable crystals, therefore, we can admit only the possibility that $\omega_j^2(\mathbf{q})$ is positive.*

The mathematical requirement for $\omega_j^2(\mathbf{q})$ to be positive is that the principal minors of $\mathbf{D}(\mathbf{q})$ must be positive.[17] From a physical point of view this imposes some restrictions on the force constants. Since our concern is with stable crystals, we shall assume that this requirement is automatically met.

### 2.2.9 Cyclic boundary conditions

Until now we have analyzed the problem of determining the vibrational frequencies and amplitudes associated with waves corresponding to a given wave vector $\mathbf{q}$. Next comes the task of specifying the allowed values of $\mathbf{q}$ or, equivalently, the wavelengths that can be sustained by the lattice. This will clearly depend upon the boundary conditions, and for an infinite crystal, it is customary to apply constraints due to Born and von Kármán,[8] usually termed the *periodic* or *cyclic boundary conditions*. These are similar to the box normalization of momentum eigenfunctions in quantum mechanics, and should be familiar to the reader in the context of the one-dimensional chain and of band theory as discussed in elementary solid state texts.

Briefly, the periodic boundary conditions are applied as follows: We imagine the infinite crystal to be subdivided into large *macrocells*, each containing a very large number $N$ of primitive cells, where $N$ is typically of the order of the number of cells in the finite specimens used in the laboratory. For convenience (although this is not necessary), we choose the macrocell to have the same shape as the primitive cell and with edges $L\mathbf{a}_1$, $L\mathbf{a}_2$, $L\mathbf{a}_3$. It follows that $N = L^3$. The cyclic conditions are now applied to the macrocell and for this reason it is frequently referred to as a *cyclic* crystal.

The cyclic conditions demand that the displacements of equivalent atoms in different macrocells be identical, that is,

$$\mathbf{u}\begin{pmatrix} l_1 l_2 l_3 \\ k \end{pmatrix} = \mathbf{u}\begin{pmatrix} l_1 + L, l_2 + L, l_3 + L \\ k \end{pmatrix}.$$

---

*In fact this is a condition for stability.[17]

In conjunction with (2.41) this implies that $\mathbf{q}$ may be represented as[18]

$$\mathbf{q} = \zeta_1 \mathbf{b}_1 + \zeta_2 \mathbf{b}_2 + \zeta_3 \mathbf{b}_3, \tag{2.58a}$$

where

$$\zeta_i = n_i/L, \qquad (i = 1, 2, 3), \tag{2.58b}$$

and $n_i$ is an integer. The values of $\mathbf{q}$ defined by (2.58) form in reciprocal space a fine mesh of points with a density $(Nv/8\pi^3)$. Not all the $\mathbf{q}$ values defined above are physically distinct since, as we shall prove shortly, $\omega_j^2(\mathbf{q}) = \omega_j^2(\mathbf{q} + \mathbf{G})$. All the distinct values of $\mathbf{q}$ are therefore obtained by restricting attention to just one primitive cell of the reciprocal lattice or, equivalently, to the $\mathbf{q}$ values distributed in the central *Brillouin zone* (BZ), which also has the same volume, $(2\pi)^3/v$. The latter cell is to be preferred by virtue of being more symmetrical. It follows that the total number of distinct or *allowed values* of $\mathbf{q}$ is exactly equal to $N$. Observe that the size chosen for the cyclic crystal is arbitrary. The latter is merely a convenient artifact for sampling the vibrational frequencies of an infinite crystal.*

It is appropriate here to take note of some of the geometrical symmetries of the BZ.[9,19–21] Firstly, the BZ has inversion symmetry. Furthermore, the BZ as a geometrical object is left invariant by all the operations of the point group of the direct lattice. This implies that the BZ is also invariant under the point group $G_0$ of the crystal since $G_0$ is always a subgroup of the point group of the lattice.† It is the latter property that is of importance in our discussions.

Questions may be raised (and indeed have been[22]) regarding the applicability to finite crystals of the results deduced for infinite crystals on the basis of cyclic conditions. It turns out, however, that if $\delta$ denotes the range of interatomic forces and $\mathcal{L}$ is a linear dimension of the finite crystal, then, provided $(\delta/\mathcal{L})$ is small, surface effects are negligible, and for all practical purposes the finite specimen may be replaced by a cyclic crystal of equivalent volume. Effectively one ascribes to the finite crystal the vibration frequencies of an infinite crystal.‡

From now on we shall consider only cyclic crystals, and the index $l$ will label only the $N$ cells in such crystals. Of these, $l = 0$ (that is, $l_1 = l_2 = l_3 = 0$) will refer to the cell at the origin.

Some useful theorems pertaining to cyclic crystals are collected in Appendix 1.

---

*It also permits the normalization to finite volume of the extensive properties of the crystal, which otherwise would have infinite value.
†Recall earlier remarks about the distinction between these two point groups.
‡Ionic crystals exhibit electrostatic forces that make the range $\delta$ very large. The usual arguments[23] assume $\delta$ to be small and hence do not apply here. Nevertheless it has been shown[24] that cyclic conditions may be used, provided $N$ is large.

## 2.2.10 Superposition of solutions

Knowing how the frequencies for a given $\mathbf{q}$ may be calculated and having enumerated the allowed values of $\mathbf{q}$, we can now write the most general solution as a superposition of the independent solutions as follows:

$$u_\alpha\binom{l}{k} = \frac{1}{\sqrt{Nm_k}} \sum_{\mathbf{q}} \sum_j e_\alpha\left(k \Big| \begin{matrix} \mathbf{q} \\ j \end{matrix}\right) A\binom{\mathbf{q}}{j} \exp\{i[\mathbf{q}\cdot\mathbf{x}(l) - \omega_j(\mathbf{q})t]\}. \qquad (2.59)$$

The factor $(1/\sqrt{N})$ has been added for convenience. It must be pointed out that corresponding to each eigenvalue $\omega_j^2(\mathbf{q})$ are two solutions, namely $\pm\omega_j(\mathbf{q})$, for the frequency; and therefore two possible lattice waves

$$\exp\{i[\mathbf{q}\cdot\mathbf{x}(l) - \omega_j(\mathbf{q})t]\}, \qquad\qquad\qquad\qquad (a)$$

$$\exp\{i[\mathbf{q}\cdot\mathbf{x}(l) + \omega_j(\mathbf{q})t]\}. \qquad\qquad\qquad\qquad (b)$$

Now (a) can be interpreted either as a wave traveling along $\mathbf{q}$ with frequency $+\omega_j(\mathbf{q})$ or as a wave along $-\mathbf{q}$ with frequency $-\omega_j(\mathbf{q})$. Likewise, (b) represents the waves $(\mathbf{q}, -\omega_j(\mathbf{q}))$ and $(-\mathbf{q}, +\omega_j(\mathbf{q}))$. On the other hand, an independent consideration of $-\mathbf{q}$ yields the solutions

$$\exp\{i[-\mathbf{q}\cdot\mathbf{x}(l) - \omega_j(-\mathbf{q})t]\}, \qquad\qquad\qquad (c)$$

$$\exp\{i[-\mathbf{q}\cdot\mathbf{x}(l) + \omega_j(-\mathbf{q})t]\}, \qquad\qquad\qquad (d)$$

representing the waves $(-\mathbf{q}, +\omega_j(-\mathbf{q}))$ and $(-\mathbf{q}, -\omega_j(-\mathbf{q}))$, respectively. The property $\omega_j(\mathbf{q}) = \omega_j(-\mathbf{q})$ (to be discussed later) implies that solutions (a) and (d) are equivalent as are (b) and (c). Hence it is sufficient to restrict attention to positive frequencies as has been done in (2.59). By combining the time factor with $A$, (2.59) is often written

$$u_\alpha\binom{l}{k} = \frac{1}{\sqrt{Nm_k}} \sum_{\mathbf{q}j} e_\alpha\left(k \Big| \begin{matrix} \mathbf{q} \\ j \end{matrix}\right) Q\binom{\mathbf{q}}{j} \exp[i\mathbf{q}\cdot\mathbf{x}(l)]. \qquad (2.60)$$

This equation amounts to a Fourier expansion for the displacements. Seen in this light, Eq. (2.42) represents the dynamical problem corresponding to a particular Fourier component, with $D_{\alpha\beta}(^{\mathbf{q}}_{kk'})$ representing the appropriate Fourier component of the force constants. Lattice sums of the type entering $\mathbf{D}(\mathbf{q})$ are therefore sometimes referred to as Fourier sums.

Summarizing the discussion given thus far, we have seen that if we restrict the potential energy to second-order terms in the displacements, then the atomic motions can be analyzed in terms of vibrational waves associated with the harmonic vibration of atoms about their respective equilibrium positions. There are $3n$ wavelike solutions for each wave vector $\mathbf{q}$, and the most general solution is a superposition of the totality of the $3nN$ independent solutions associated with all the wave vectors in the BZ.

An important feature of this analysis is that the periodicity of the crystal enables us to focus attention on the Fourier components one at a time. In practical terms, the problem of solving an infinite set of coupled equations as in (2.34) has been changed to that of repeated solving of the set of $3n$ equations in (2.42). If the complete solution for the cyclic crystal is desired, (2.42) must be solved for each of the $N$ values of $\mathbf{q}$, though such a fine mesh is seldom required.

### 2.2.11 Continuity of solutions—dispersion curves
Since the allowed values of $\mathbf{q}$ form a fine mesh, it is possible to represent the $3nN$ frequencies graphically by continuous curves, as a function of $\mathbf{q}$. Corresponding to every direction in $\mathbf{q}$-space are $3n$ curves $\omega = \omega_j(\mathbf{q})$, $(j = 1, 2, \ldots, 3n)$. Such curves are called *dispersion curves* or *dispersion relations*, by analogy with optics where a similar plot describes the dispersion of light in the medium. The curves we are considering reflect likewise the dispersion of vibrational waves by the lattice. Each member of the manifold of $3n$ curves (for a given direction $\mathbf{q}$) is referred to as a *branch*. There are thus $3n$ branches to the dispersion curves. In general the branches are distinct. Along special directions, however, some of these may be degenerate because of symmetry. This topic will be pursued further in the next chapter.

### 2.2.12 Physical significance of eigenvectors
We wish to comment briefly on the physical significance of the eigenvectors. From (2.59) it is clear that $\mathbf{e}\left(\begin{smallmatrix}\mathbf{q}\\j\end{smallmatrix}\right)$ determines the direction and phase and $A\left(\begin{smallmatrix}\mathbf{q}\\j\end{smallmatrix}\right)$ the amplitude of vibration of the atoms in the wave $(\mathbf{q}, \omega_j(\mathbf{q}))$. More explicitly, if we write

$$e_\alpha\left(k\left|\begin{matrix}\mathbf{q}\\j\end{matrix}\right.\right) = \left|e_\alpha\left(k\left|\begin{matrix}\mathbf{q}\\j\end{matrix}\right.\right)\right| \exp\left\{i\delta_\alpha\left(k\left|\begin{matrix}\mathbf{q}\\j\end{matrix}\right.\right)\right\},$$

then the quantities $|e_\alpha(k|_j^{\mathbf{q}})|$ are direction cosines of the displacement, while $\delta_\alpha(k|_j^{\mathbf{q}})$ determines the phase. In other words, the components of the eigenvectors essentially determine the *pattern* of displacement of the atoms in a particular mode of vibration. For this reason $\mathbf{e}(k|_j^{\mathbf{q}})$ is sometimes referred to as the *polarization vector*. In special situations, the polarization vectors for all the atoms are either parallel (or perpendicular) to $\mathbf{q}$. The mode is then said to be longitudinal (or transverse).

To illustrate the significance of the eigenvectors, we consider one of those for cubic ZnS pertaining to a special point in the BZ.* There are two atoms

---

*Actually the point $W$ in Figure 2.19 which represents also the BZ for the zinc blende structure.

per cell and the eigenvector under consideration has the form

$$
e\begin{pmatrix} \mathbf{q} \\ j \end{pmatrix} = \begin{pmatrix} e_x(\mathrm{Zn}|_j^{\mathbf{q}}) \\ e_y(\mathrm{Zn}|_j^{\mathbf{q}}) \\ e_z(\mathrm{Zn}|_j^{\mathbf{q}}) \\ e_x(\mathrm{S}|_j^{\mathbf{q}}) \\ e_y(\mathrm{S}|_j^{\mathbf{q}}) \\ e_z(\mathrm{S}|_j^{\mathbf{q}}) \end{pmatrix} = \frac{1}{\sqrt{2}} \begin{pmatrix} 0 \\ 0 \\ 0 \\ 1 \\ 0 \\ e^{-i\pi/2} \end{pmatrix}.
\tag{2.61}
$$

In this mode, the Zn atoms are stationary while the S atoms move circularly in the $XZ$ plane.

### 2.2.13 Summary of important results
Let us at this stage collect several important properties of $\mathbf{D}(\mathbf{q})$, its eigen-values and its eigenvectors. Most of these are easily proved.

**Properties of $\mathbf{D}(\mathbf{q})$.**

1.  $\mathbf{D}(\mathbf{q}) = \mathbf{D}^\dagger(\mathbf{q}).$  (2.62)

2.  $\mathbf{D}(\mathbf{q}) = \mathbf{D}^*(-\mathbf{q}).$  (2.63)

3.  $\mathbf{D}(\mathbf{q}) = \mathbf{D}(\mathbf{q} + \mathbf{G}).$  (2.64)

4.  $\mathbf{D}(0), \mathbf{D}(\mathbf{G}/2)$ are real.

5. If atoms in the sublattices $k$ and $k'$ are interchanged by inversion, then

$$
\mathbf{D}\begin{pmatrix} \mathbf{q} \\ k'k \end{pmatrix} = \mathbf{D}^*\begin{pmatrix} \mathbf{q} \\ kk' \end{pmatrix}
\tag{2.65}
$$

for that pair $(kk')$. (Hint: Use (2.31)).
6. If $\mathbf{S}$ is an element of the point group $G_0$ of the crystal, then

$$
\mathbf{D}(\mathbf{Sq}) = \mathbf{\Gamma}\mathbf{D}(\mathbf{q})\mathbf{\Gamma}^\dagger,
\tag{2.66}
$$

where $\mathbf{\Gamma}$ is an appropriately defined unitary transformation. This property will be established in the following chapter.

**Properties of $\omega_j^2(\mathbf{q})$.**

1.  $\omega_j^2(\mathbf{q})$ is real.

2.  $\omega_j^2(\mathbf{q}) = \omega_j^2(\mathbf{q} + \mathbf{G}).$  (2.67)

Given (2.64), it is obvious that the sets of eigenvalues $\{\omega_j^2(\mathbf{q})\}$ and $\{\omega_j^2(\mathbf{q} + \mathbf{G})\}$ must be identical. By suitably labeling the members of the two sets, (2.67) may be ensured.

3.  $\omega_j^2(\mathbf{q}) = \omega_j^2(-\mathbf{q}).$  (2.68)

Replace $\mathbf{q}$ by $-\mathbf{q}$ in (2.51) and take the complex conjugate, remembering that the eigenvalues are real. Using (2.63) we then find that the sets $\{\omega_j^2(\mathbf{q})\}$ and $\{\omega_j^2(-\mathbf{q})\}$ are identical. Once again by suitable labeling, (2.68) may be fulfilled.

4. $\quad \omega_j^2(\mathbf{q}) = \omega_j^2(\mathbf{Sq}).$ \hfill (2.69)

This is a consequence of (2.66). Remembering that the eigenvalues are unchanged by a unitary transformation, suitable labels may be given to the eigenvalues pertaining to $\mathbf{q}$ and $\mathbf{Sq}$ to obtain (2.69).

**Properties of $\mathbf{e}\binom{\mathbf{q}}{j}$.**

1. $\quad \displaystyle\sum_{k\alpha} e_\alpha^*\left(k\left|\begin{matrix}\mathbf{q}\\j\end{matrix}\right.\right) e_\alpha\left(k\left|\begin{matrix}\mathbf{q}\\j'\end{matrix}\right.\right) = \delta_{jj'}.$ \hfill (2.70)

2. $\quad \displaystyle\sum_{j} e_\alpha\left(k\left|\begin{matrix}\mathbf{q}\\j\end{matrix}\right.\right) e_\beta^*\left(k'\left|\begin{matrix}\mathbf{q}\\j\end{matrix}\right.\right) = \delta_{\alpha\beta}\delta_{kk'}.$ \hfill (2.71)

3. $\quad \mathbf{e}\binom{\mathbf{q}}{j} = \mathbf{e}\binom{\mathbf{q} + \mathbf{G}}{j}.$ \hfill (2.72)

This is a consequence of (2.64) and (2.67). Assuming that $\omega_j^2(\mathbf{q})$ is non-degenerate, the eigenvectors corresponding to $\mathbf{q}$ and $\mathbf{q} + \mathbf{G}$ can differ only to the extent of an arbitrary phase factor of modulus unity. Without loss of generality, this phase factor may be chosen equal to unity, whence (2.72) follows. If $\omega_j(\mathbf{q})$ is degenerate, then $\mathbf{e}\binom{\mathbf{q}+\mathbf{G}}{j}$ must be a normalized linear combination of the set of $\mathbf{e}\binom{\mathbf{q}}{j}$'s belonging to $\omega_j(\mathbf{q})$. This linear combination can always be chosen to satisfy (2.72).

4. $\quad \mathbf{e}\binom{\mathbf{q}}{j} = \mathbf{e}^*\binom{-\mathbf{q}}{j}.$ \hfill (2.73)

Starting from (2.63) and by using arguments similar to those just employed, it can be proved that, irrespective of whether $\omega_j(\mathbf{q})$ is degenerate or not,

$$\mathbf{e}^*\binom{-\mathbf{q}}{j} = e^{i\phi}\mathbf{e}\binom{\mathbf{q}}{j}.$$

Once again the phase factor may be set equal to unity without loss of generality, whence (2.73) follows.

5. $\quad \mathbf{e}\binom{\mathbf{Sq}}{j} = e^{i\phi}\mathbf{\Gamma}\mathbf{e}\binom{\mathbf{q}}{j}.$ \hfill (2.74)

This property* is closely related to (2.66) and (2.69) and is considered further in the next chapter.

---

*The simultaneous fulfillment of (2.73) and (2.74) poses a minor complication, mentioned briefly in Chapter 3.

6. The eigenvectors for a crystal whose structure is a Bravais lattice can be chosen real.

## 2.2.14 Variants of the eigenvalue equation

A few minor variations of Eq. (2.42) are possible and in fact often found in the literature.[25–27] In one of these, the trial solution used in place of (2.41) has the form

$$u_\alpha\binom{l}{k} = \mathcal{U}_\alpha(k|\mathbf{q}) \exp\{i[\mathbf{q}\cdot\mathbf{x}(l) - \omega(\mathbf{q})t]\} \tag{2.75}$$

which leads to the eigenvalue equation

$$m\omega^2(\mathbf{q})\mathcal{U}(\mathbf{q}) = \mathbf{B}(\mathbf{q})\mathcal{U}(\mathbf{q}). \tag{2.76}$$

Here the elements of $\mathbf{B}(\mathbf{q})$ and $\mathbf{m}$ are given by

$$B_{\alpha\beta}\binom{\mathbf{q}}{kk'} = \sum_{l'} \phi_{\alpha\beta}\binom{0 \; l'}{k \; k'} \exp\{i\mathbf{q}\cdot\mathbf{x}(l')\} \tag{2.77}$$

$$m_{\alpha\beta}(kk') = m_k\delta_{\alpha\beta}\delta_{kk'}. \tag{2.78}$$

$\mathcal{U}(\mathbf{q})$, as usual, is a column matrix with elements $\mathcal{U}_\alpha(k \mid \mathbf{q})$.

Sometimes the trial solution is written in the form

$$u_\alpha\binom{l}{k} = \mathcal{U}_\alpha(k \mid \mathbf{q}) \exp\left\{i\left[\mathbf{q}\cdot\mathbf{x}\binom{l}{k} - \omega(\mathbf{q})t\right]\right\} \tag{2.79}$$

leading to the eigenvalue equation

$$m\omega^2(\mathbf{q})\mathcal{U}(\mathbf{q}) = \mathbf{M}(\mathbf{q})\mathcal{U}(\mathbf{q}), \tag{2.80}$$

where

$$M_{\alpha\beta}\binom{\mathbf{q}}{kk'} = \sum_{l'} \phi_{\alpha\beta}\binom{0 \; l'}{k \; k'} \exp\left\{i\mathbf{q}\cdot\left[\mathbf{x}\binom{l'}{k'} - \mathbf{x}\binom{0}{k}\right]\right\}. \tag{2.81}$$

Both (2.76) and (2.80) are examples of the generalized eigenvalue problem[28]

$$\mathbf{AX} = \lambda\mathbf{BX}, \tag{2.82}$$

where $\mathbf{A}$ and $\mathbf{B}$ are Hermitian matrices of the same dimension, $\lambda$ is an eigenvalue of $\mathbf{A}$ relative to $\mathbf{B}$, and $\mathbf{X}$ is the corresponding eigenvector. The familiar form $\mathbf{AX} = \lambda\mathbf{X}$ results from (2.82) when $\mathbf{B}$ is the unit matrix.

Let us further suppose that $\mathbf{B}$ is positive definite. Now the eigenvectors $\mathbf{X}$ of (2.82) cannot be chosen orthonormal in the usual sense. However, they may be chosen according to the modified orthonormality condition

$$\mathbf{X}^\dagger(i)\mathbf{BX}(j) = \delta_{ij}. \tag{2.83}$$

This result may be applied to (2.76) and (2.80). For example, following (2.48) let us write

$$\mathscr{U}\binom{\mathbf{q}}{j} = A\binom{\mathbf{q}}{j}\mathbf{E}\binom{\mathbf{q}}{j}, \tag{2.84}$$

where the components of $\mathbf{E}$ are dimensionless. The "orthogonality" condition on the $\mathbf{E}$'s may now be stated as

$$\mathbf{E}^{\dagger}\binom{\mathbf{q}}{j}\mathbf{m}\mathbf{E}\binom{\mathbf{q}}{j'} = \text{const} \times \delta_{jj'}.$$

The constant may be seen to have the dimensions of mass. A convenient and popular choice is to take the constant as $M$, the mass of the primitive cell. Thus the "orthonormality" condition on the eigenvectors $\mathbf{E}\binom{\mathbf{q}}{j}$ is given by

$$\sum_{\alpha k} E_{\alpha}^{*}\left(k\left|\begin{matrix}\mathbf{q}\\j\end{matrix}\right.\right)m_k E_{\alpha}\left(k\left|\begin{matrix}\mathbf{q}\\j'\end{matrix}\right.\right) = M\delta_{jj'}. \tag{2.85}$$

In the interests of notational simplicity, we shall from now on always employ the vector $\mathbf{U}$ for describing the amplitude components of the nuclear displacements, irrespective of the form of the dynamical matrix. It is to be understood, of course, that in every case $\mathbf{U}$ is to be appropriately defined with regard to scale and phase factors as in Eqs. (2.75) and (2.79).

It is helpful also to record here representations of the dynamical matrix in which the contribution from the self term is exhibited explicitly. For example, from (2.81) we have

$$M_{\alpha\beta}(\substack{\mathbf{q}\\kk'}) = \sum_{\substack{l'\\(l'\neq 0 \text{ if } k'=k)}} \phi_{\alpha\beta}(\substack{0\ l'\\k\ k'}) \exp\{i\mathbf{q}\cdot[\mathbf{x}(\substack{l'\\k'}) - \mathbf{x}(\substack{0\\k})]\} - \delta_{kk'}\sum_{l''k''}{}' \phi_{\alpha\beta}(\substack{0\ l''\\k\ k''})$$

$$= \hat{M}_{\alpha\beta}(\substack{\mathbf{q}\\kk'}) - \delta_{kk'}\sum_{k''}\hat{M}_{\alpha\beta}(\substack{0\\kk'}). \tag{2.86}$$

Observe that the above result would not be affected by the addition of the same arbitrary constant to $\hat{M}_{\alpha\beta}(\substack{\mathbf{q}\\kk})$ and $\hat{M}_{\alpha\beta}(\substack{0\\kk})$. Analogously, we also have

$$D_{\alpha\beta}\binom{\mathbf{q}}{kk'} = \hat{D}_{\alpha\beta}\binom{\mathbf{q}}{kk'} - \delta_{kk'}\sum_{k''}\hat{D}_{\alpha\beta}\binom{0}{kk''} \tag{2.87}$$

and

$$B_{\alpha\beta}\binom{\mathbf{q}}{kk'} = \hat{B}_{\alpha\beta}\binom{\mathbf{q}}{kk'} - \delta_{kk'}\sum_{k''}{}_{\alpha}\hat{B}_{\beta}\binom{0}{kk''}. \tag{2.88}$$

### 2.2.15 Modes at the Brillouin zone center

The form of the dispersion curves $\omega = \omega_j(\mathbf{q})$ depends on the crystal structure and on the nature of the interatomic forces. Irrespective of these, however, a cyclic crystal always has three zero-frequency modes at $\mathbf{q} = 0$ corresponding to the lateral translation of the crystal along three mutually perpendicular directions. As a result, three branches of the dispersion curves always approach zero frequency as $\mathbf{q} \to 0$; this feature is repeated at every reciprocal lattice point owing to (2.67). These three branches are referred to as *acoustic branches* for reasons to be explained later. The remaining $(3n - 3)$ branches generally tend to finite frequencies as $\mathbf{q} \to 0$.* Historically, such modes in crystals of the NaCl structure were found to interact strongly with light.† Based on this, the $(3n - 3)$ branches have come collectively to be labeled *optic branches*, whether or not these modes interact with light in the $\mathbf{q} \to 0$ limit.

Consider now Eq. (2.80) for the case $\mathbf{q} = 0$. If we assume the solution

$$U_\beta \left( k \middle| \begin{matrix} 0 \\ j \end{matrix} \right) = \delta_{\alpha\beta} \varepsilon_\alpha \qquad \text{for all } k, \tag{2.89}$$

then (2.80) leads to

$$\sum_\beta \left[ \sum_{l'k'} \phi_{\alpha\beta} \begin{pmatrix} 0 & l' \\ k & k' \end{pmatrix} \right] \delta_{\alpha\beta} \varepsilon_\beta = m_k \omega_j^2 (0) \varepsilon_\alpha.$$

The quantity in the square bracket vanishes owing to the sum rule (2.35). We see, therefore, that a zero-frequency mode for $\mathbf{q} = 0$ with displacement components as in (2.89) is a solution of (2.80) and corresponds to a uniform translation of the crystal in the $\alpha$ direction. Two other zero-frequency modes with displacements orthogonal to that of (2.89) may be derived similarly.

An interesting corollary to the above result is that the $\mathbf{q} = 0$ optic modes correspond to motions which leave the center of mass of the primitive cell fixed. This may be seen by invoking the orthogonality relation (2.85), which in the present case yields

$$
\begin{aligned}
0 &= \sum_{\beta k} E_\beta \left( k \middle| \begin{matrix} 0 \\ \text{ac} \end{matrix} \right) m_k E_\beta \left( k \middle| \begin{matrix} 0 \\ \text{optic} \end{matrix} \right) \\
&= \text{const} \times \sum_{\beta k} U_\beta \left( k \middle| \begin{matrix} 0 \\ \text{ac} \end{matrix} \right) m_k U_\beta \left( k \middle| \begin{matrix} 0 \\ \text{optic} \end{matrix} \right) \\
&= \text{const} \times \sum_{\beta k} \varepsilon_\alpha \delta_{\alpha\beta} m_k U_\beta \left( k \middle| \begin{matrix} 0 \\ \text{optic} \end{matrix} \right) \\
&= \text{const} \times \varepsilon_\alpha \left[ \sum_k m_k U_\alpha \left( k \middle| \begin{matrix} 0 \\ \text{optic} \end{matrix} \right) \right].
\end{aligned}
$$

*One can, however, invent models where the limiting frequency for some of these branches is zero. See, for example, M. A. Nusimovici and J. L. Birman, Phys. Rev. **156**, 925 (1961), Fig. 4.
†This topic will be pursued further in later chapters.

Since $\varepsilon_\alpha$ is arbitrary, the quantity in the square brackets must vanish; this proves our assertion since that quantity is a measure of the center of mass displacement. Thus the $\mathbf{q} = 0$ optic modes essentially involve the vibrations of the various sublattices against each other and in such a manner that the center of mass of each primitive cell is undisturbed. The reader will naturally recall in this context the familiar example of the $q = 0$ modes in linear diatomic chains.

It is interesting to note that whereas in the molecular case there are six zero-frequency modes (three corresponding to translations and three to rotations), in the case of cyclic crystals there are only three zero-frequency modes. The reason for this lies in the periodic form (2.41) of the solution which must be fulfilled for any mode. While a uniform displacement of the crystal is consistent with this form, a rotation is not. This can be seen from (2.37) which shows that as $l$ increases so does $u_\alpha\binom{l}{k}$; this is clearly inconsistent with the periodicity requirement. In molecules, however, rotations are permitted since no periodicity requirement must be met.

The foregoing remarks concerning the absence of rotational modes apply not only to three-dimensional crystals but also to those of two dimensions. For one-dimensional chains, however, one zero-frequency mode corresponding to rotations about the chain axis is possible, provided the chain has finite extension perpendicular to its axis. Molecular polyethylene $[CH_2]_n$, a portion of whose dispersion curves[29] is displayed in Figure 2.3,

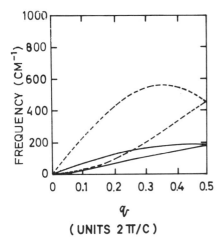

**Figure 2.3**
Low-frequency region of the dispersion curves for a polyethylene chain. (After Lynch, ref. 29.)

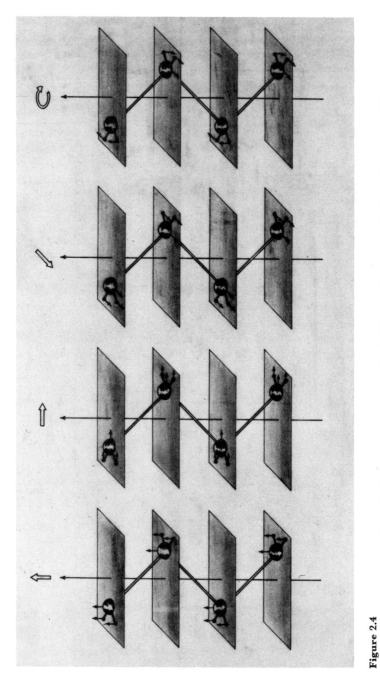

**Figure 2.4**
Displacement of the atoms of the polyethylene chain in the $q = 0$ limit. Three modes correspond to lateral translation of the chain, while the fourth corresponds to a rotation about the chain axis. All have zero frequency.

provides an example. The interesting feature here is that there are four branches for which $\omega \to 0$ as $q \to 0$ instead of the usual three. The atomic displacements in the $q \to 0$ limit for these four branches are shown in Figure 2.4 from which it may be seen that, in addition to the usual translational modes, there is also a rotational mode. The latter is permitted since the displacements do not vary with the position of the cell and are therefore consistent with the periodicity requirements. It must be emphasised that our observations concerning the absence of rotational modes in three-dimensional crystals apply specifically to cyclic crystals. Being like a giant molecule, a finite crystal will necessarily have three zero-frequency rotational modes.

## 2.3 Frequency distribution function

### 2.3.1 Definition

For many purposes, it is necessary to consider the spectrum of the $3nN$ vibrational frequencies. This spectrum, also known as the *frequency distribution function*, is defined as $(1/\Delta\omega)$ times the fraction of vibrational frequencies between $\omega$ and $(\omega + \Delta\omega)$ in the limit $\Delta\omega \to 0$. More formally,

$$g(\omega) = \lim_{\Delta\omega \to 0} \frac{1}{3nN(\Delta\omega)} \sum_{j}^{\omega \leqslant \omega_j \leqslant (\mathbf{q}) \leqslant \omega + \Delta\omega} \sum_{\mathbf{q}} \quad (2.90)$$

The understanding is that there is a contribution of unity to the sum if $\omega_j(\mathbf{q})$ lies in the range indicated, zero otherwise. The normalization condition is

$$\int_0^\infty g(\omega)d\omega = 1. \quad (2.91)$$

Although $\omega$ is allowed formally to range up to $\infty$ in (2.91), it extends in practice only up to the highest vibrational frequency $\omega_{max}$.

An equivalent definition of $g(\omega)$ is

$$g(\omega) = \frac{1}{3nN} \sum_{j} \sum_{\mathbf{q}} \delta\{\omega - \omega_j(\mathbf{q})\}, \quad (2.92)$$

where $\delta(x)$ is the Dirac $\delta$-function. By integrating this expression over a small interval $\Delta\omega$, it is readily verified to be equivalent to (2.90).

Noting that the allowed values of $\mathbf{q}$ are closely spaced, the summation over $\mathbf{q}$ in (2.90) may be replaced by an integration via the prescription (see Appendix 1, Eq. (A1.8))

$$\sum_{\mathbf{q}} \to \frac{Nv}{(2\pi)^3} \int_{BZ} d\mathbf{q},$$

which then gives

$$g(\omega) = \lim_{\Delta\omega \to 0} \frac{1}{3nN(\Delta\omega)} \frac{Nv}{(2\pi)^3} \sum_j \int_{\omega \leqslant \omega_j(\mathbf{q}) \leqslant \omega + \Delta\omega} d\mathbf{q}. \tag{2.93}$$

The volume integral in $\mathbf{q}$ may be converted to a surface integral as follows: Let $dS$ be an elementary area on a constant-frequency surface $S$ corresponding to frequency $\omega$ (for the $j$th branch). The volume of a right prism with base $dS$ and bounded by constant-frequency surfaces $\omega$ and $\omega + \Delta\omega$ is $dS\Delta q$, where $\Delta q = \Delta\omega/|\mathrm{grad}_\mathbf{q}\, \omega_j(\mathbf{q})|$ denotes the height of the prism. Thus

$$\int_{\omega \leqslant \omega_j(\mathbf{q}) \leqslant \omega + \Delta\omega} d\mathbf{q} = \int_S dS\, \Delta q = \Delta\omega \int_S \frac{dS}{|\mathrm{grad}\, \omega_j(\mathbf{q})|}.$$

With this result, (2.93) becomes

$$g(\omega) = \sum_j g_j(\omega) = \frac{v}{3n(2\pi)^3} \sum_j \int \frac{dS}{|\mathrm{grad}\, \omega_j(\mathbf{q})|}. \tag{2.94}$$

This form is useful for later discussions.

### 2.3.2 Calculation of $g(\omega)$

The frequency distribution is an important quantity, especially in relation to several thermodynamic properties; much effort has therefore been directed towards devising methods for its computation. The earliest effort is that due to Debye[30] who proposed the approximation

$$g(\omega) = \sum_{i=1}^{3} A_i \omega^2,$$

where the $A_i$'s are appropriate constants. Later, various analytical methods[31] were devised, but these have now yielded to computer calculations.

The basic idea employed in the computer calculations is *root sampling*. One first fills the BZ with a suitably chosen mesh $M_c$ of points at each of which the secular determinant (2.46) is solved. The roots or frequencies are then sorted into channels of suitable width $\Delta\omega$, giving $g(\omega)$ in the form of a histogram. Computing time sets a limit on accuracy which increases as the mesh $M_c$ is made finer.

To overcome this limitation, Gilat and Dolling[32] proposed an *extrapolation technique*. As in the root sampling method, one first divides the BZ into cells with their centers defining a coarse mesh $M_c$. Each cell is next subdivided with a fine mesh $M_f$. Then $\mathbf{D}(\mathbf{q})$ is diagonalized exactly at the points of $M_c$, and the frequencies appropriate to the points of $M_f$ are obtained via the extrapolation

$$\omega_j(\mathbf{q} + \delta\mathbf{q}) = \omega_j(\mathbf{q}) + \delta\mathbf{q}\cdot\text{grad } \omega_j(\mathbf{q}), \tag{2.95}$$

where the gradient is computed by first-order perturbation theory. In effect, the sampling size is increased from $M_c$ to $(M_c \times M_f)$ points without a corresponding increase in computer time.

The method is not free from limitations. First, it is inapplicable wherever there is degeneracy. Fortunately, such degeneracies are restricted to symmetry points and directions in the BZ, and to a large extent can be avoided by a suitable choice of $M_c$. Where they cannot the gradient must be evaluated by degenerate perturbation theory.

Another difficulty is associated with points where grad $\omega_j(\mathbf{q})$ vanishes. Once again this is not a serious problem since the number of such points in the BZ is small.

More recently, Gilat and coworkers[33] have developed a technique in which instead of sampling over a fine mesh $M_f$ in each coarse cell, they extract "all" frequencies by analytical integration. This is done by filling each coarse cell with constant-frequency surfaces which are then replaced by parallel planes normal to $[\text{grad } \omega_j(\mathbf{q})]_{\mathbf{q}=\mathbf{q}_c}$, where $\mathbf{q}_c$ denotes the center of the cell. This technique enhances considerably the accuracy of calculation.

### 2.3.3 Singularities in frequency spectra

A number of frequency spectra have been calculated using these methods, and a feature common to all of them is that they exhibit singularities. These appear as discontinuities in $g'(\omega) = [dg(\omega)/d\omega]$ or in the higher derivatives of the frequency spectrum. In general they arise whenever *either* grad $\omega_j(\mathbf{q})$ vanishes completely (see Eq. (2.94)) *or* one or more components of the gradient change sign discontinuously, while the remaining components vanish. A point in $\mathbf{q}$ space where such behavior is exhibited is referred to as a *critical point* (CP), and the corresponding frequency $\omega_j(\mathbf{q})$ as a *critical frequency*.

A systematic discussion of critical points and the singularities in $g(\omega)$ which they produce is best given by considering the shape of the constant-frequency surfaces in the neighborhood of the critical points, as was first done by van Hove[34] and Phillips.[35] This analysis involves sophisticated topological considerations outside our scope. We shall therefore merely summarize their important results.

In examining the critical points of a given crystal, one first carries out an *ordered labeling* by indexing the branches 1, 2, . . . , $3n$, strictly in order of increasing frequency. Some examples are given in Figure 2.5.

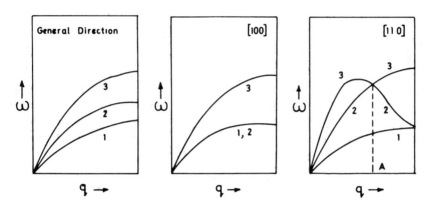

**Figure 2.5**
Ordered labeling of the branches for a fcc crystal like argon. Sketched are the dispersion curves along three representative directions. Along [100] the two lower branches are degenerate. Observe the crossover at $A$.

The singularities are next discussed with reference to the partial spectrum $g_j(\omega)$ for the $j$th branch by examining the nature of the constant-frequency surfaces in the neighborhood of the critical points occurring in that branch. Based on such considerations, CP's are classified into (1) analytical critical points (ACP), (2) fluted critical points (FCP), and (3) singular critical points (SCP).

**ACP.** Analytical critical points occur on nondegenerate branches and are characterized by the vanishing of all components of grad $\omega_j(\mathbf{q})$ at that point. Furthermore, it is possible to expand $\omega_j(\mathbf{q})$ in the neighborhood of that point in a Taylor series in $\boldsymbol{\phi} = \mathbf{q} - \mathbf{q}_0$, where $\mathbf{q}_0$ is the CP. Thus we may write, with an appropriate choice of local axes,

$$\omega_j(\mathbf{q}) - \omega_j(\mathbf{q}_0) = \varepsilon = b_1\phi_1^2 + b_2\phi_2^2 + b_3\phi_3^2. \tag{2.96}$$

The number $i$ of negative $b$'s in this form is known as the *index* of the CP; the point itself is labeled $P_i$. It is obvious that $i$ can take only the values 0, 1, 2, or 3, so that the only possible ACP's are $P_0$, $P_1$, $P_2$, or $P_3$.

The various possible shapes appropriate to the quadratic (2.96) are well known from analytical geometry (see Figure 2.6). Based on these shapes, the contributions to the frequency spectra are:[34]

*Minimum*: $P_0$ $(i = 0)$

$$g_j(\omega) = \begin{cases} C + O(\varepsilon) & \varepsilon < 0 \\ C + D\varepsilon^{1/2} + O(\varepsilon) & \varepsilon > 0 \end{cases}$$

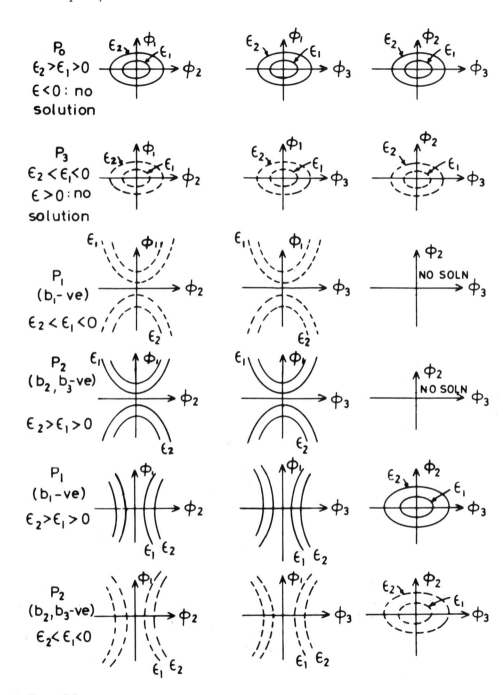

**Figure 2.6**
Constant-frequency contours encountered in the neighborhood of various types of ACP's.

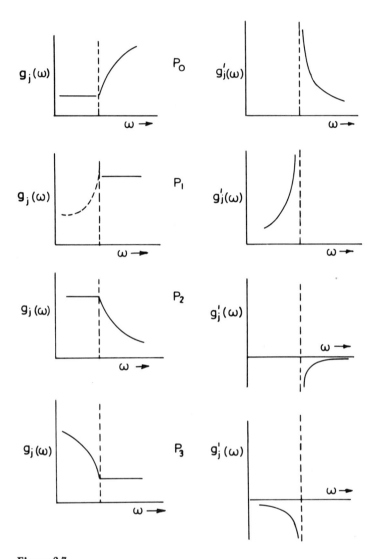

**Figure 2.7**
Schematic plots of the contributions to $g_j(\omega)$ and $g'_j(\omega)$ from the various types of ACP.
The frequency corresponding to the dotted line is the critical frequency. The derivatives
better reveal the singularity.

*Saddle point*: $P_1$ $(i = 1)$

$$g_j(\omega) = \begin{cases} C - D\varepsilon^{1/2} + O(\varepsilon) & \varepsilon < 0 \\ C + O(\varepsilon) & \varepsilon > 0 \end{cases}$$

*Saddle point*: $P_2$ $(i = 2)$

$$g_j(\omega) = \begin{cases} C + O(\varepsilon) & \varepsilon < 0 \\ C - D\varepsilon^{1/2} + O(\varepsilon) & \varepsilon > 0 \end{cases}$$

*Maximum*: $P_3$ $(i = 3)$

$$g_j(\omega) = \begin{cases} C + D\varepsilon^{1/2} + O(\varepsilon) & \varepsilon < 0 \\ C + O(\varepsilon) & \varepsilon > 0. \end{cases} \tag{2.97}$$

Here $C$ and $D$ are constants while $O(\varepsilon)$ denotes a term of order $\varepsilon$ for $\varepsilon \to 0$. The singularities are illustrated in Figure 2.7. We see from above that the index number has the utility of providing a ready categorization of the singular structure produced in the frequency spectrum.

**FCP.** A fluted critical point is a point where two or more branches are degenerate and grad $\omega_j(\mathbf{q})$ vanishes for all the members of the degenerate set at that point. Such CP's occur at symmetry points in the BZ, and their existence may be deduced from symmetry considerations. For such points it is not possible to write a Taylor expansion as in (2.96). However, Phillips[35] has shown that it is still possible to assign an index $i$ by employing geometrical considerations. The general idea is to determine the number of positive and negative sectors in the neighborhood of the CP's, a sector being defined as the solid angle taken with its apex at the CP $\mathbf{q}_0$ and within which $\varepsilon$ is of the same sign. The numbers $(P, N)$ of positive and negative sectors are called *sector numbers*. Once these are known, the index $i$ of the CP can be specified using rules given by Phillips. Figure 2.8 offers an example of division into sectors. It may be noted that even though the constant-frequency contours in this figure are considerably fluted or warped, the general features are similar to those for saddle points and maxima as discussed earlier.

Given the sector numbers, the index $i$ is assigned as follows: (We quote rules which cover three- as well as lower-dimensional cases).
1. If the sector numbers are $(1, 0)$ or $(0, 1)$, then $i = 0$ or $l$, respectively, where $l$ is the dimensionality of the crystal.
2. In two dimensions, a $(n, n)$ point* has $i = 1$.
3. In three dimensions, usually either $P$ or $N$ will be greater than 1. If $P > 1$, then $i = 2$ while if $N > 1$, $i = 1$. In general, however, both $P$ and

*Lower-case letters are used for $l < 3$.

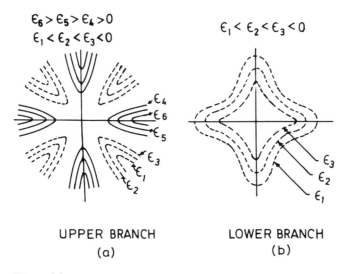

$$\epsilon_6 > \epsilon_5 > \epsilon_4 > 0$$
$$\epsilon_1 < \epsilon_2 < \epsilon_3 < 0$$

$$\epsilon_1 < \epsilon_2 < \epsilon_3 < 0$$

UPPER BRANCH      LOWER BRANCH
(a)            (b)

**Figure 2.8**
Frequency contours in the neighborhood of a FCP in a two-dimensional crystal. The degeneracy at the FCP is twofold. The patterns represent a possible situation. The sector numbers for the upper and lower branches are respectively (4, 4) and (0, 1).

$N$ could be greater than 1, in which case the point must be reckoned as both an $i = 1$ and as an $i = 2$ point.

As in the case of ACP, the index $i$ of a FCP can only be 0, 1, 2, or 3. Of these, $i = 0$ and $i = 3$ have closed surfaces like those for ACP's of type $P_0$ and $P_3$, though the surfaces are warped. On account of this topological equivalence, the FCP's of indices 0 and 3 are also labeled as $P_0$(FCP) and $P_3$(FCP). Regarding FCP's of indices 1 and 2, several sector number combinations are possible, but usually no distinction is made between them and they too are designated simply as $P_1$(FCP) or $P_2$(FCP) as appropriate.

The FCP's of various indices make the same contribution to $g_j(\omega)$ as do the ACP's of corresponding indices. It is worth remarking that the geometrical method of indexing is equally applicable to the ACP's.

**SCP.** Singular critical points are those for which one or more components of grad $\omega_j(\mathbf{q})$ change sign discontinuously while the remaining components vanish. Such CP's occur where the dispersion curves for two branches cross, as with the point $A$ in Figure 2.5. Like the other types, SCP's can also be indexed with sector numbers and represented as $P_0$, $P_1$, $P_2$, and $P_3$. To distinguish them from the other types, the identification SCP and the number of discontinuous components of grad $\omega_j(\mathbf{q})$ changing sign are included in parentheses.

Crossover of branches leading to SCP can occur not only at points of

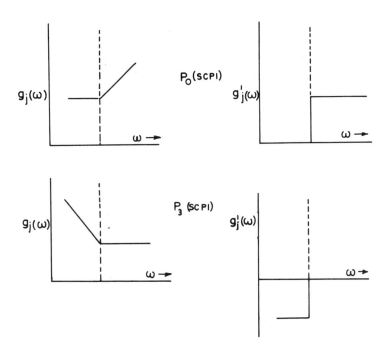

**Figure 2.9**
Contribution to $g_j(\omega)$ and $g'_j(\omega)$ from $P_0(\text{SCP } 1)$ and $P_3(\text{SCP } 1)$, corresponding to singular minima and maxima respectively.

symmetry in the BZ, but elsewhere accidentally. Such accidental degeneracies* can occur at isolated points in the zone or along closed curves.

A very special case of SCP is the zone center. Here the acoustic branches meet, and all three components of grad $\omega_j(\mathbf{q})$ change sign discontinuously as one passes through the origin in a given direction. The sector numbers appropriate to the acoustic branches are evidently $(1, 0)$, so that the SCP at $\mathbf{q} = 0$ for the acoustic branches would be classified as $P_0$ (SCP 3).

Unlike ACP and FCP, SCP produce mild singularities in $g(\omega)$. The most important ones are the singular minimum $P_0$ (SCP 1) and the singular maximum $P_3$ (SCP 1), for which $g_j(\omega)$ and $g'_j(\omega)$ have the forms sketched in Figure 2.9. The other types, namely, $P_0$ (SCP $n$), $P_3$ (SCP $n$), $n \geqslant 2$, and $P_1$ (SCP $n$), $P_2$ (SCP $n$), $n > 0$, may be ignored.

### 2.3.4 Morse relations
One important consequence that emerges from the topological analysis of CP's is that the existence of some CP in a given branch requires the existence

---

*This terminology is defined more formally in the next chapter (see Sec. 3.4).

of others and that the total numbers of different kinds of CP's are related. The relations that must hold are referred to as *Morse relations* and are given below:

two dimensions                three dimensions

$n_0 \geqslant 1,$                    $N_0 \geqslant 1,$

$n_1 - n_0 \geqslant 1,$               $N_1 - N_0 \geqslant 1,$

$n_2 - n_1 + n_0 = 0,$            $N_2 - N_1 + N_0 \geqslant 1,$

$$N_3 - N_2 + N_1 - N_0 = 0. \qquad (2.98)$$

Here $n_i$ and $N_i$ stand for the total number of CP's of index $i$ in two and three dimensions, respectively. In computing these, care must be exercised in assigning the proper *topological weights w*. The latter is defined as the number of times a CP of index $i$ is to be counted in computing $n_i$ or $N_i$. The number $n_i$ ($N_i$) itself is obtained as the product of $w$ with the multiplicity, the latter denoting the number of points in the BZ related to the one under consideration by symmetry.*

The rules[35] for assigning $w$ are summarized below:

1. If the sector number is $(1, 0)$ or $(0, 1)$, then $w = 1$.
2. In two dimensions, a $(n, n)$ point has $w = (n - 1)$.
3. In three dimensions, if $P > 1$ (that is, $i = 2$), then $w = (P - 1)$, while if $N > 1$ (that is, $i = 1$), then $w = (N - 1)$. If both $P$ and $N$ are greater than unity, the point must be counted *both* as an $i = 1$ point with $w = (N - 1)$ and as an $i = 2$ point with $w = (P - 1)$. (See earlier remarks regarding assignment of $i$). Note that since ACP's of types $P_0, P_1, P_2, P_3$ always have sector numbers $(1, 0)$, $(1, 2)$, $(2, 1)$ and $(0, 1)$, respectively, the weight for any ACP is 1.

While locating the CP's of a given crystal, it must be ensured that the Morse relations are satisfied. This does not of course imply that *all* the CP's have been found. However, it at least guarantees that the set of CP's obtained satisfies the necessary requirement embodied in the Morse relations.

In the above discussion we considered the singularities occurring in $g_j(\omega)$, the frequency spectrum associated with a particular branch $j$. The singularities appearing in $g(\omega) = \sum_j g_j(\omega)$ will clearly consist of singularities arising from all the branches. It is possible, of course, that in the process of superposition of contributions from the various branches, cancellation or compensation may occur if there are contacts between branches.

---

*In the language of Sec. 3.6, the multiplicity is the same as the order of the star of $\mathbf{q}$.

In a later section we shall illustrate some of these ideas by considering the frequency spectrum of a specific crystal.

We might point out here that much of the original interest in singularities of frequency spectra arose from the belief that from a knowledge of them it might be possible to synthesize a reasonably accurate $g(\omega)$. However, with the development of computer techniques for calculating $g(\omega)$, there was a natural decline of interest in CP's and singularities. We have nevertheless discussed them here because the methods of classification considered presently are applicable also to the so-called combined frequency spectra which determine second-order optical spectra (see Chapter 5).

## 2.4 Acoustic waves and elasticity

We noted earlier that, of the $3n$ branches of the dispersion curves, the three acoustic branches approach zero frequency as $\mathbf{q} \to 0$ and that for $\mathbf{q} = 0$ the motions of atoms correspond simply to lateral translations of the crystal. For $\mathbf{q}$ slightly greater than zero the atoms in neighboring cells move in unison, but over distances large compared to cell dimensions there is a phase difference. Vibrational waves of this kind are well known in acoustics and in the theory of elasticity, in which one ignores the microscopic structure of matter and treats it as an elastic homogeneous solid. In fact it is because the long-wavelength vibrational waves of the acoustic branches have the same character as sound waves in elastic media that the branches are termed acoustic. By comparing the solutions of the microscopic vibrational problem with the corresponding results of the macroscopic theory, it is possible to express the parameters of the macroscopic theory, namely, the elastic constants, in terms of microscopic force constants. Our aim here is to indicate this connection.

### 2.4.1 Some definitions
Let us first remind ourselves of the definitions of stress and strain.[36] A body in which one part exerts a force on neighboring parts is said to be in a state of stress. If the forces acting on a surface element $dS$ in the body are the same irrespective of the position of $dS$ as long as the element is always oriented in the same manner with respect to a fixed set of axes, then the body is said to be under a *homogeneous stress*.

Consider now a homogeneously stressed body in which all parts are in static equilibrium, and imagine within this a cube of unit edge. Following the convention in this field, we shall temporarily label the axes $OX_1$, $OX_2$, and $OX_3$. This representative cube will be acted on by various forces due to

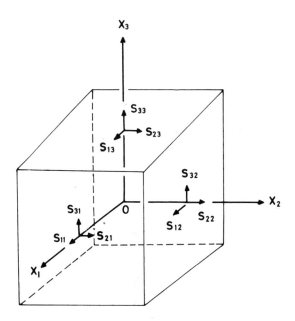

**Figure 2.10**
Forces on the faces of a unit cube in a homogeneously stressed body.

stress, and we denote by $S_{\alpha\beta}$,* ($\alpha$, $\beta$ = 1, 2, 3), the force along the $+\alpha$ direction (that is, along $OX_\alpha$) transmitted across the face normal to $OX_\beta$ by the medium on the $+$ side of the face (see Figure 2.10). The quantities $S_{\alpha\beta}$ are called stress components, being the elements of the second-rank stress tensor **S**.

Since the cube is by assumption in static equilibrium, the moments about $OX_1$, $OX_2$, and $OX_3$ vanish. This gives $S_{\alpha\beta} = S_{\beta\alpha}$. In other words, the stress tensor is symmetric.

In the case of inhomogeneous stress, the stress varies within the body and at a given point $P$ may be specified by considering a small area $dS$ around $P$ and then examining the force transmitted across it using the convention above.

In passing we observe that pressure $p$ is defined by

$$p = -\tfrac{1}{3}(S_{11} + S_{22} + S_{33}) \tag{2.99}$$

(note negative sign).

A body subjected to stress in general undergoes a deformation. The com-

---

*We shall follow BH in our notation. Nye[36] uses $\sigma$ for stress.

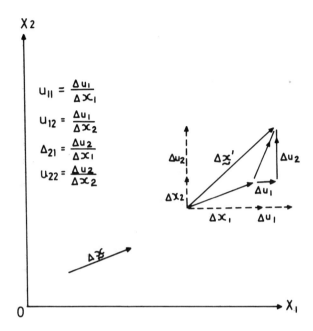

**Figure 2.11**
A two-dimensional example illustrating the meaning of the strain components.

ponents $u_{\alpha\beta}$ of the *infinitesimal strain tensor* $\mathbf{u}$ are defined by the derivative

$$u_{\alpha\beta} = \frac{\partial u_\alpha}{\partial x_\beta}, \tag{2.100}$$

where $\mathbf{u}$ denotes the displacement of the point around which the strain is being examined. Observe that the strain parameters are dimensionless. Figure 2.11 illustrates strain components for a two-dimensional system.

If the strain $u_{\alpha\beta}$ is independent of the position within the medium, then the latter is said to have undergone a *homogeneous deformation*. The particle displacements produced by homogeneous deformation in a crystal are given by

$$x_\alpha\binom{l}{k} \rightarrow x'_\alpha\binom{l}{k} = x_\alpha\binom{l}{k} + u_\alpha(k) + \sum_{\beta=1}^{3} u_{\alpha\beta}x_\beta\binom{l}{k}, \tag{2.101}$$

$$k = 1, 2, \ldots, n.$$

The quantities $u_\alpha(k)$ denote the relative shifts of the sublattices. It is relevant to observe that after a homogeneous deformation the crystal is still a lattice (though perhaps with different symmetry). The internal shifts $u_\alpha(k)$ are not arbitrary. Rather they are determined by the condition that the strain energy density $\Psi$ be stationary with respect to these shifts. In other words, when a crystal is homogeneously deformed by external forces, the various sublattices shift relatively such that[37]

$$\frac{\partial \Psi}{\partial u_\alpha(k)} = 0 \qquad \text{for all } \alpha \text{ and } k. \tag{2.102}$$

The quantities $u_\alpha(k)$ are sometimes referred to as *internal strains*.

Returning to external strains, it is more convenient to deal with a symmetric tensor $s$ with elements related to those of $u$ by

$$s_{\alpha\beta} = \tfrac{1}{2}(u_{\alpha\beta} + u_{\beta\alpha}). \tag{2.103}$$

### 2.4.2 Elastic constants
For small stresses, the strain is proportional to stress (Hooke's law). The reverse statement is also true and can be written formally as

$$S_{\alpha\gamma} = \sum_{\beta=1}^{3} \sum_{\lambda=1}^{3} C_{\alpha\gamma,\beta\lambda} s_{\beta\lambda}, \qquad (\alpha, \gamma = 1, 2, 3). \tag{2.104}$$

The quantities $C_{\alpha\gamma,\beta\lambda}$ are called *elastic constants*; they are typically $\sim 10^{11}$–$10^{12}$ dynes/cm$^2$.

The 81 elastic constants $\{C_{\alpha\gamma,\beta\lambda}\}$ form a fourth-rank tensor connecting the second-rank stress and strain tensors. They have the properties[36]

$$C_{\alpha\gamma,\beta\lambda} = C_{\gamma\alpha,\beta\lambda}, \qquad C_{\alpha\gamma,\beta\lambda} = C_{\alpha\gamma,\lambda\beta} \tag{2.105}$$

which reduce the number of independent constants from 81 to 36.

Given Hooke's law, it can be shown that $\Psi$ is given by

$$\Psi = \Psi_0 + \tfrac{1}{2} \sum_{\alpha\beta\gamma\lambda} C_{\alpha\gamma,\beta\lambda} s_{\alpha\gamma} s_{\beta\lambda},^* \tag{2.106}$$

where

$$C_{\alpha\gamma,\beta\lambda} = \left. \frac{\partial^2 \Psi}{\partial s_{\alpha\gamma} \partial s_{\beta\lambda}} \right|_0, \tag{2.107}$$

and $\Psi_0$ is the energy density in the absence of strain.

Observe the similarity of (2.106) to the harmonic potential energy (2.22). The difference is that here we are dealing with a macroscopic system characterized by macroscopic constants. As in the case of force constants,

---

*This result is true only to first order. If the strain energy is to be in a form that is invariant with respect to rigid rotations, then the parameters that must be used are[38]

$$\bar{u}_{\alpha\beta} = \frac{1}{2}\left\{ u_{\alpha\beta} + u_{\beta\alpha} + \sum_\gamma u_{\gamma\alpha} u_{\gamma\beta} \right\} = \bar{u}_{\beta\alpha}.$$

we have the permutational symmetry $(\alpha\gamma) \leftrightarrow (\beta\lambda)$, that is,

$$C_{\alpha\gamma,\beta\lambda} = C_{\beta\lambda,\alpha\gamma} \tag{2.108}$$

which reduces the number of independent constants from 36 to 21. Crystal symmetry reduces the number even further, often causing many of the constants to vanish. For example, cubic crystals have only 12 nonvanishing elastic constants, of which only 3 are independent.

It is customary to simplify the notation by contracting pairs of indices into single indices according to the following pattern:

| $\alpha\gamma$ | 11 | 22 | 33 | (23, 32) | (31, 13) | (12, 21) |
|---|---|---|---|---|---|---|
| $\downarrow$ | $\downarrow$ | $\downarrow$ | $\downarrow$ | $\downarrow$ | $\downarrow$ | $\downarrow$ |
| $i$ | 1 | 2 | 3 | 4 | 5 | 6 |

Based on the above scheme, called the Voigt notation, we define

$$\begin{aligned}
\mathsf{s}_i &= (u_{\alpha\gamma} + u_{\gamma\alpha}) = 2\mathsf{s}_{\alpha\gamma} = 2\mathsf{s}_{\gamma\alpha}, && (\alpha \neq \gamma), \\
\mathsf{s}_i &= u_{\alpha\gamma} = \mathsf{s}_{\alpha\gamma}, && (\alpha = \gamma).
\end{aligned} \tag{2.109}$$

Equation (2.104) then becomes

$$S_i = \sum_{j=1}^{6} C_{ij}\mathsf{s}_j, \qquad (i = 1, 2, \dots, 6), \tag{2.110}$$

where, owing to (2.105) and (2.108),

$$C_{ij} = C_{ji}. \tag{2.111}$$

### 2.4.3 Elastic equations
Suppose now that a crystal is inhomogeneously stressed and the stress is then released. Elastic restoring forces set up vibrations, the equations governing these being

$$\begin{aligned}
\rho \ddot{u}_\alpha &= \sum_{\gamma\beta\lambda} C_{\alpha\gamma,\beta\lambda} \frac{\partial \mathsf{s}_{\beta\lambda}}{\partial x_\gamma} \\
&= \sum_{\gamma\beta\lambda} C_{\alpha\gamma,\beta\lambda} \frac{\partial^2 u_\beta}{\partial x_\lambda \partial x_\gamma}, \qquad (\alpha = 1, 2, 3), 
\end{aligned} \tag{2.112}$$

where $\rho$ is the density. We try to solve these by wavelike solutions

$$\mathbf{u}(\mathbf{x}) = \mathbf{u}(\mathbf{q}) \exp\{i[\mathbf{q}\cdot\mathbf{x} - \omega(\mathbf{q})t]\}. \tag{2.113}$$

This gives

$$\rho\omega^2(\mathbf{q})\mathbf{u}(\mathbf{q}) = \mathbf{A}(\mathbf{q})\mathbf{u}(\mathbf{q}), \tag{2.114}$$

where

$$A_{\alpha\beta}(\mathbf{q}) = \sum_{\gamma\lambda} C_{\alpha\gamma,\beta\lambda} q_\gamma q_\lambda. \qquad (2.115)$$

The (real) matrix $\mathbf{A}(\mathbf{q})$ is the analogue of the dynamical matrix $\mathbf{M}(\mathbf{q})$ considered earlier. Upon demanding nontrivial solutions for $\mathbf{u}(\mathbf{q})$, we get the secular determinant

$$\| \rho^{-1} A_{\alpha\beta}(\mathbf{q}) - \omega^2(\mathbf{q}) \delta_{\alpha\beta} \| = 0 \qquad (2.116)$$

which, when solved, yields the squared frequencies of the three elastic waves of wave vector $\mathbf{q}$. Once the frequencies are known, the amplitude vectors $\mathbf{u}\binom{\mathbf{q}}{j}$ ($j = 1, 2, 3$) can be solved for.

The elements of the symmetric matrix $\mathbf{A}(\mathbf{q})$ are given below:

$$A_{11}(\mathbf{q}) = C_{11} q_1^2 + C_{66} q_2^2 + C_{55} q_3^2 + 2C_{16} q_1 q_2 + 2C_{56} q_2 q_3 + 2C_{15} q_1 q_3,$$

$$A_{12}(\mathbf{q}) = C_{16} q_1^2 + C_{26} q_2^2 + C_{45} q_3^2 + (C_{12} + C_{66}) q_1 q_2$$
$$\qquad\quad + (C_{46} + C_{25}) q_2 q_3 + (C_{14} + C_{56}) q_1 q_3,$$

$$A_{13}(\mathbf{q}) = C_{15} q_1^2 + C_{46} q_2^2 + C_{35} q_3^2 + (C_{14} + C_{56}) q_1 q_2$$
$$\qquad\quad + (C_{45} + C_{36}) q_2 q_3 + (C_{13} + C_{55}) q_1 q_3,$$

$$A_{22}(\mathbf{q}) = C_{66} q_1^2 + C_{22} q_2^2 + C_{44} q_3^2 + 2C_{26} q_1 q_2 + 2C_{24} q_2 q_3 + 2C_{46} q_1 q_3,$$

$$A_{23}(\mathbf{q}) = C_{56} q_1^2 + C_{24} q_2^2 + C_{34} q_3^2 + (C_{25} + C_{46}) q_1 q_2$$
$$\qquad\quad + (C_{23} + C_{44}) q_2 q_3 + (C_{45} + C_{36}) q_1 q_3,$$

$$A_{33}(\mathbf{q}) = C_{55} q_1^2 + C_{44} q_2^2 + C_{33} q_3^2 + 2C_{45} q_1 q_2 + 2C_{34} q_2 q_3$$
$$\qquad\quad + 2C_{35} q_1 q_3. \qquad (2.117)$$

Observe that since $\mathbf{A}(\mathbf{q})$ is quadratic in $\mathbf{q}$, $\omega_j^2(\mathbf{q})$ is likewise, which means that $\omega_j(\mathbf{q})$ is a linear function of $|\mathbf{q}|$. This linear dependence is frequently written

$$\omega_j(\mathbf{q}) = c_j(\hat{\mathbf{q}})|\mathbf{q}|, \qquad j = 1, 2, 3, \qquad (2.118)$$

$c_j(\hat{\mathbf{q}})$ denoting the velocity of acoustic waves corresponding to the $j$th branch in the direction $\hat{\mathbf{q}} = (\mathbf{q}/|\mathbf{q}|)$.

### 2.4.4 Connection between elastic constants and force constants— long-wavelength method

We wish to consider now the microscopic counterpart of the equations (2.114). This will enable us to connect elastic constants with force constants. We shall employ a perturbation method due to Born;[39] since full details are available in the literature, only an outline will be presented here.

We start from the set of equations (Sec. 2.2.14)

$$\omega_j^2(\mathbf{q})\mathbf{m}\mathbf{U}\!\begin{pmatrix}\mathbf{q}\\j\end{pmatrix} = \mathbf{M}(\mathbf{q})\mathbf{U}\!\begin{pmatrix}\mathbf{q}\\j\end{pmatrix}. \tag{2.119}$$

We next replace $\mathbf{q}$ by $\eta\mathbf{q}$ and expand each $\mathbf{q}$-dependent quantity in a power series in the expansion parameter $\eta$, the latter being allowed eventually to take the value unity.* Thus we write

$$\mathbf{M}(\eta\mathbf{q}) = \mathbf{M}(\mathbf{q}=0) + \eta\sum_\gamma\left[\frac{\partial\mathbf{M}(\eta\mathbf{q})}{\partial q_\gamma}\right]_{\mathbf{q}=0}q_\gamma$$

$$+ \frac{\eta^2}{2}\sum_{\gamma\lambda}\left[\frac{\partial^2\mathbf{M}(\eta\mathbf{q})}{\partial q_\gamma\partial q_\lambda}\right]_{\mathbf{q}=0}q_\gamma q_\lambda + \cdots \tag{2.120a}$$

$$= \mathbf{M}^{(0)} + i\eta\sum_\gamma\mathbf{M}_\gamma^{(1)}q_\gamma + \frac{\eta^2}{2}\sum_{\gamma\lambda}\mathbf{M}_{\gamma\lambda}^{(2)}q_\gamma q_\lambda + \cdots. \tag{2.120b}$$

The various matrices are all of dimension $3n$ and are labeled as usual by the Cartesian indices $(\alpha,\beta)$ and the sublattice indices $(k,k')$. The elements of the matrices are given by

$$M_{\alpha\beta}^{(0)}(kk') = \sum_{l'}\phi_{\alpha\beta}\begin{pmatrix}0 & l'\\k & k'\end{pmatrix}, \tag{2.121}$$

$$M_{\alpha\beta,\gamma}^{(1)}(kk') = -\sum_{l'}\phi_{\alpha\beta}\begin{pmatrix}0 & l'\\k & k'\end{pmatrix}x_\gamma\begin{pmatrix}0 & l'\\k & k'\end{pmatrix}, \tag{2.122}$$

$$M_{\alpha\beta,\gamma\lambda}^{(2)}(kk') = -\sum_{l'}\phi_{\alpha\beta}\begin{pmatrix}0 & l'\\k & k'\end{pmatrix}x_\gamma\begin{pmatrix}0 & l'\\k & k'\end{pmatrix}x_\lambda\begin{pmatrix}0 & l'\\k & k'\end{pmatrix}, \tag{2.123}$$

where

$$x_\gamma\begin{pmatrix}0 & l'\\k & k'\end{pmatrix} = x_\gamma\begin{pmatrix}0\\k\end{pmatrix} - x_\gamma\begin{pmatrix}l'\\k'\end{pmatrix}. \tag{2.124}$$

Note that all the matrices are real. Some properties required for further discussion are:[39]

$$\mathbf{M}^{(0)} = \bar{\mathbf{M}}^{(0)}, \tag{2.125}$$

$$\mathbf{M}_\gamma^{(1)} = -\bar{\mathbf{M}}_\gamma^{(1)}, \tag{2.126}$$

$$\mathbf{M}_{\gamma\lambda}^{(2)} = \bar{\mathbf{M}}_{\gamma\lambda}^{(2)}, \tag{2.127}$$

$$\sum_{k'}M_{\alpha\beta}^{(0)}(kk') = 0 \qquad \text{(for all }\alpha,\beta\text{ and }k\text{)}, \tag{2.128}$$

$$\sum_{k}M_{\alpha\beta}^{(0)}(kk') = 0 \qquad \text{(for all }\alpha,\beta\text{ and }k'\text{)}. \tag{2.129}$$

*Note we are considering a perturbation series expansion in which the direction of $\mathbf{q}$ is fixed and its magnitude is changing.

$$\sum_{k'} M^{(1)}_{\alpha\beta,\gamma}(kk') \;=\; \sum_{k'} M^{(1)}_{\alpha\gamma,\beta}(kk'), \tag{2.130}$$

$$\sum_{kk'} M^{(1)}_{\alpha\beta,\gamma}(kk') \;=\; 0. \tag{2.131}$$

Let us now also express the amplitude and the vibration frequency in power series:

$$\mathbf{U}\!\left(\begin{matrix}\eta\mathbf{q}\\j\end{matrix}\right) = \mathbf{U}^{(0)}\!\left(\begin{matrix}\mathbf{q}\\j\end{matrix}\right) + i\eta\mathbf{U}^{(1)}\!\left(\begin{matrix}\mathbf{q}\\j\end{matrix}\right) + \frac{\eta^2}{2}\mathbf{U}^{(2)}\!\left(\begin{matrix}\mathbf{q}\\j\end{matrix}\right) + \cdots, \tag{2.132a}$$

$$\omega_j(\eta\mathbf{q}) = \eta\omega_j^{(1)}(\mathbf{q}) + \frac{\eta^2}{2}\,\omega_j^{(2)}(\mathbf{q}) + \cdots. \tag{2.132b}$$

It is worth noting that (2.132b) contains no term independent of $\eta$. Omission of such a term ensures that the *results to be obtained pertain specifically to the acoustic branches*.

On substituting (2.132) and (2.120) in (2.119) and equating various powers of $\eta$, we get to second order

$$\mathbf{M}^{(0)}\mathbf{U}^{(0)} = 0, \tag{2.133a}$$

$$\mathbf{M}^{(0)}\mathbf{U}^{(1)} = -\sum_{\gamma} \mathbf{M}^{(1)}_{\gamma}q_{\gamma}\mathbf{U}^{(0)}, \tag{2.133b}$$

$$\mathbf{M}^{(0)}\mathbf{U}^{(2)} = 2[\omega_j^{(1)}(\mathbf{q})]^2 m\mathbf{U}^{(0)} - \sum_{\gamma\lambda} \mathbf{M}^{(2)}_{\gamma\lambda}q_{\gamma}q_{\lambda}\mathbf{U}^{(0)}$$
$$+\; 2\sum_{\gamma} \mathbf{M}^{(1)}_{\gamma}q_{\gamma}\mathbf{U}^{(1)}. \tag{2.133c}$$

For notational convenience, we sometimes suppress the indices $(\mathbf{q}, j)$ associated with the amplitude factors. We shall now solve (2.133a) to (2.133c) successively. Observe that the second-order equation is the first in which frequency appears.

Consider first the zero-order equations. These have nontrivial solutions of the form

$$\mathbf{U}^{(0)}\!\left(\begin{matrix}\mathbf{q}\\j\end{matrix}\right) = \left.\begin{matrix}\begin{pmatrix} u_x(\begin{smallmatrix}\mathbf{q}\\j\end{smallmatrix}) \\ u_y(\begin{smallmatrix}\mathbf{q}\\j\end{smallmatrix}) \\ u_z(\begin{smallmatrix}\mathbf{q}\\j\end{smallmatrix}) \\ \vdots \\ u_x(\begin{smallmatrix}\mathbf{q}\\j\end{smallmatrix}) \\ u_y(\begin{smallmatrix}\mathbf{q}\\j\end{smallmatrix}) \\ u_z(\begin{smallmatrix}\mathbf{q}\\j\end{smallmatrix}) \end{pmatrix}\end{matrix}\right\}\begin{matrix}3n\\\text{elements}\end{matrix} = \mathbf{u}\!\left(\begin{matrix}\mathbf{q}\\j\end{matrix}\right)\left.\begin{pmatrix}1\\1\\\vdots\\1\end{pmatrix}\right\}n\,\text{elements},\; (j=1,2,3);$$
$$\tag{2.134}$$

here $\mathbf{u}(\begin{smallmatrix}\mathbf{q}\\j\end{smallmatrix})$ is a constant vector, at present arbitrary. It will turn out that we can remove this arbitrariness with the help of the second-order equation.

That (2.134) is a solution of the zero-order equation follows immediately upon using (2.128), which itself follows directly from the sum rule (2.35b). The argument should be familiar to the reader from Sec. 2.2.15, and as then we note that there are in fact 3 independent solutions to the zero-order equation. The constant vectors associated with these will be denoted $\mathbf{u}\binom{\mathbf{q}}{1}$, $\mathbf{u}\binom{\mathbf{q}}{2}$, $\mathbf{u}\binom{\mathbf{q}}{3}$.*

Next we consider the first-order equation. For this to be soluble, a necessary condition is that†

$$\tilde{\mathbf{U}}^{(0)H}\binom{\mathbf{q}}{j'}\left\{-\sum_{\gamma}\mathbf{M}_{\gamma}^{(1)}q_{\gamma}\mathbf{U}^{(0)}\binom{\mathbf{q}}{j}\right\} = 0, \ (j,j' = 1, 2, 3), \tag{2.135}$$

where $\tilde{\mathbf{U}}^{(0)H}$ is the solution of the homogeneous equation (2.133a) without additional restrictions. This can be proved using (2.131). The first-order solution may now be expressed as (for details see BH)

$$\mathbf{U}^{(1)} = -\mathbf{\Gamma}\left(\sum_{\gamma}\mathbf{M}_{\gamma}^{(1)}q_{\gamma}\mathbf{U}^{(0)}\right) \tag{2.136}$$

with

$$\mathbf{\Gamma} = \begin{pmatrix} 0 & 0 & 0 & \cdots & 0 \\ 0 & 0 & 0 & \cdots & 0 \\ 0 & 0 & 0 & \cdots & 0 \\ \vdots & \vdots & \vdots & & \\ 0 & 0 & 0 & & \mathbf{\Gamma}' \end{pmatrix} ; \quad \mathbf{\Gamma}' = \begin{bmatrix} \mathbf{M}^{(0)}(22) & \cdots & \mathbf{M}^{(0)}(2n) \\ \vdots & \cdots & \vdots \\ \mathbf{M}^{(0)}(n2) & \cdots & \mathbf{M}^{(0)}(nn) \end{bmatrix}^{-1} ;$$

$$\mathbf{M}^{(0)}(kk') = \begin{bmatrix} M_{11}^{(0)}(kk') & M_{12}^{(0)}(kk') & M_{13}^{(0)}(kk') \\ M_{21}^{(0)}(kk') & M_{22}^{(0)}(kk') & M_{23}^{(0)}(kk') \\ M_{31}^{(0)}(kk') & M_{32}^{(0)}(kk') & M_{33}^{(0)}(kk') \end{bmatrix} . \tag{2.137}$$

In the above, the Cartesian indices $\alpha\beta$ have been written as 11, 12, ... instead of $xx$, $xy$, ... , since we are currently exploring the elastic region and wish to make contact with the expressions of Sec. 2.4.3.

Equation (2.136) has an interesting implication which was pointed out by Born and Huang.[39] Earlier we noted that, under the action of a

---

*From now on, we shall whenever necessary assign the values 1, 2, and 3 for the branch index $j$ to characterize acoustic branches.

†Given a system of equations

$$\mathbf{AX} = \mathbf{W}, \tag{a}$$

where $\mathbf{X}$ and $\mathbf{W}$ are column matrices while $\mathbf{A}$ is a real square matrix, and further that nontrivial solutions exist for the homogeneous equations

$$\mathbf{AX} = 0, \tag{b}$$

a necessary and sufficient condition for the solution of (a) to exist is that each solution $\mathbf{X}^H$ of the transposed homogeneous equation be orthogonal to $\mathbf{W}$, that is, $\tilde{\mathbf{X}}^H\mathbf{W} = 0$, or $\sum_i X_i^H W_i = 0$.[40]

homogeneous deformation, an internal strain is set up which adjusts itself suitably to provide the necessary balance. Equation (2.136) in fact represents such a balance under conditions appropriate to the propagation of an acoustic wave. Here the zero-order part of the wave, that is,

$$u_\alpha^{(0)}(\mathbf{x}) = u_\alpha\left(\begin{matrix}\mathbf{q}\\j\end{matrix}\right)\exp\left\{i\left[\eta\mathbf{q}\cdot\mathbf{x} - \eta\omega^{(1)}\left(\begin{matrix}\mathbf{q}\\j\end{matrix}\right)t\right]\right\},$$

can be thought of as causing homogeneous deformation over regions small compared with the wavelength, the strain components being given by

$$u_{\alpha\beta} = \frac{\partial u_\alpha}{\partial x_\beta} = i\eta q_\beta u_\alpha\left(\begin{matrix}\mathbf{q}\\j\end{matrix}\right)\exp\left\{i\left[\eta\mathbf{q}\cdot\mathbf{x} - \eta\omega^{(1)}\left(\begin{matrix}\mathbf{q}\\j\end{matrix}\right)t\right]\right\}. \tag{2.138a}$$

On the other hand, particle displacements associated with the first-order part are given by

$$u_\alpha^{(1)}(k) = i\eta\, U_\alpha^{(1)}\left(k\bigg|\begin{matrix}\mathbf{q}\\j\end{matrix}\right)\exp\left\{i\left[\eta\mathbf{q}\cdot\mathbf{x} - \eta\omega^{(1)}\left(\begin{matrix}\mathbf{q}\\j\end{matrix}\right)t\right]\right\} \tag{2.138b}$$

(see (2.132a)). Treating the exponentials in (2.138) as constants, we have in $u_{\alpha\beta}$ and $u_\alpha^{(1)}(k)$, respectively, the external and internal strains appropriate to the region under consideration arising out of the elastic wave; Eq. (2.136) merely expresses the balance between these two quantities.

Returning to the algebra, we substitute for $\mathbf{U}^{(1)}$ from (2.136) in the second-order equation; we then get

$$\mathbf{M}^{(0)}\mathbf{U}^{(2)} = 2[\omega_j^{(1)}(\mathbf{q})]^2\mathbf{m}\mathbf{U}^{(0)} - \sum_{\gamma\lambda}(\mathbf{M}_{\gamma\lambda}^{(2)}\mathbf{U}^{(0)})q_\gamma q_\lambda$$
$$-2\sum_{\gamma\lambda}(\mathbf{M}_\gamma^{(1)}\mathbf{\Gamma}\mathbf{M}_\lambda^{(1)}\mathbf{U}^{(0)})q_\gamma q_\lambda. \tag{2.139}$$

For this to be soluble we require that

$$\tilde{\mathbf{U}}^{(0)H}\left(\begin{matrix}\mathbf{q}\\j'\end{matrix}\right)\left\{2[\omega_j^{(1)}(\mathbf{q})]^2\mathbf{m}\mathbf{U}^{(0)}\left(\begin{matrix}\mathbf{q}\\j\end{matrix}\right) - \sum_{\gamma\lambda}\left(\mathbf{M}_{\gamma\lambda}^{(2)}\mathbf{U}^{(0)}\left(\begin{matrix}\mathbf{q}\\j\end{matrix}\right)\right)q_\gamma q_\lambda\right.$$
$$\left. - 2\sum_{\gamma\lambda}\left((\mathbf{M}_\gamma^{(1)}\mathbf{\Gamma}\mathbf{M}_\lambda^{(1)}\mathbf{U}^{(0)}\left(\begin{matrix}\mathbf{q}\\j\end{matrix}\right)\right)q_\gamma q_\lambda\right\} = 0, \quad (j, j' = 1, 2, 3). \tag{2.140a}$$

Remembering the form of the zero-order solution we write the above equation in component form and obtain

$$\sum_\alpha u_\alpha^H\left(\begin{matrix}\mathbf{q}\\j'\end{matrix}\right)\left[2[\omega_j^{(1)}(\mathbf{q})]^2\sum_k m_k u_\alpha\left(\begin{matrix}\mathbf{q}\\j\end{matrix}\right) - \sum_\beta\left(\sum_{\gamma\lambda}\left\{\sum_{kk'}M_{\alpha\beta,\gamma\lambda}^{(2)}(kk')\right\}q_\gamma q_\lambda\right)u_\beta\left(\begin{matrix}\mathbf{q}\\j\end{matrix}\right)\right.$$
$$- 2\sum_\beta\left(\sum_{\gamma\lambda}\left\{\sum_{\mu\nu}\sum_{kk'}\sum_{k''k'''}M_{\alpha\mu,\gamma}^{(1)}(k''k)\Gamma_{\mu\nu}(kk')\right.\right.$$
$$\left.\left.\times M_{\nu\beta,\lambda}^{(1)}(k'k''')\right\}q_\gamma q_\lambda\right)u_\beta\left(\begin{matrix}\mathbf{q}\\j\end{matrix}\right)\right] = 0, \quad (j, j' = 1, 2, 3). \tag{2.140b}$$

(Remember $\Gamma_{\mu\nu}(kk') = 0$ for all $\mu$ and $\gamma$ if either $k$ or $k' = 1$; see (2.137)). Since the $u_\alpha^H\binom{\mathbf{q}}{j'}$'s are arbitrary, the term in the large square brackets above must vanish giving, with (2.126),

$$\left(\sum_k m_k\right)[\omega_j^{(1)}(\mathbf{q})]^2 u_\alpha\binom{\mathbf{q}}{j} = \sum_\beta \left[ \tfrac{1}{2} \sum_{\gamma\lambda} \left\{ \sum_{kk'} M_{\alpha\beta,\gamma\lambda}^{(2)}(kk') \right\} q_\gamma q_\lambda \right.$$
$$- \sum_{\gamma\lambda} \left\{ \sum_{kk'} \sum_{\mu\nu} \left( \sum_{k''} M_{\mu\alpha,\gamma}^{(1)}(kk'') \right) \Gamma_{\mu\nu}(kk') \right.$$
$$\left. \left. \times \left( \sum_{k'''} M_{\nu\beta,\lambda}^{(1)}(k'k''') \right) \right\} q_\gamma q_\lambda \right] u_\beta\binom{\mathbf{q}}{j}.$$

Dividing throughout by $v$, the volume of the primitive cell, we obtain

$$\rho[\omega_j^{(1)}(\mathbf{q})]^2 u_\alpha\binom{\mathbf{q}}{j} = \sum_\beta \left\{ \sum_{\gamma\lambda} [\alpha\beta, \gamma\lambda] q_\gamma q_\lambda + \sum_{\gamma\lambda} (\alpha\gamma, \beta\lambda) q_\gamma q_\lambda \right\} u_\beta\binom{\mathbf{q}}{j},$$

(2.141)

where

$$[\alpha\beta, \gamma\lambda] = \frac{1}{2v}\left(\sum_{kk'} M_{\alpha\beta,\gamma\lambda}^{(2)}(kk')\right)$$

(2.142)

and

$$(\alpha\gamma, \beta\lambda) = \frac{1}{v} \sum_{k''k'''} [\tilde{\mathbf{M}}_\gamma^{(1)} \mathbf{\Gamma} \mathbf{M}_\lambda^{(1)}]_{\alpha\beta}(k''k''')$$
$$= -\frac{1}{v}\sum_{kk'}\sum_{\mu\nu}\left(\sum_{k''} M_{\mu\alpha,\gamma}^{(1)}(kk'')\right)\Gamma_{\mu\nu}(kk')\left(\sum_{k'''} M_{\nu\beta,\lambda}^{(1)}(k'k''')\right).$$

(2.143)

Observe in passing that in the case of a crystal with one atom per cell there exists for every atom at $\mathbf{x}(l)$ another at $-\mathbf{x}(l)$, from which the property $\mathbf{M}_\gamma^{(1)} = 0$ follows. The *brackets* $(\alpha\gamma, \beta\lambda)$ therefore vanish and (2.141) assumes a correspondingly simpler form.

The square and the round brackets have the properties[39]

$$[\alpha\beta, \gamma\lambda] = [\beta\alpha, \gamma\lambda] = [\alpha\beta, \lambda\gamma] = [\beta\alpha, \lambda\gamma],$$

(2.144)

$$(\alpha\gamma, \beta\lambda) = (\gamma\alpha, \beta\lambda) = (\beta\lambda, \alpha\gamma).$$

(2.145)

By solving equation (2.141), the microscopic equation for elastic waves, we can obtain the acoustic frequencies $\omega_j(\mathbf{q})$ and amplitudes $\mathbf{u}\binom{\mathbf{q}}{j}$.* Of course, if frequency and amplitude had been our only objectives, then these could have been directly obtained from (2.119). We performed a perturbation expansion mainly to obtain equations directly comparable to

---

*The reader will now appreciate the remark made earlier that $\mathbf{u}\binom{\mathbf{q}}{j}$ is determined by the second-order equation.

the macroscopic equations (2.114). From this comparison Born and Huang[39] demonstrated the following relation between the elastic and force constants.†

$$C_{\alpha\gamma,\beta\lambda} = [\alpha\beta, \gamma\lambda] + [\beta\gamma, \alpha\lambda] - [\beta\lambda, \alpha\gamma] + (\alpha\gamma, \beta\lambda), \qquad (2.146)$$

provided the square brackets have the property

$$[\alpha\beta, \gamma\lambda] = [\gamma\lambda, \alpha\beta]. \qquad (2.147)$$

Remembering (2.144), it can be seen that (2.147) implies 15 new relations among the square brackets. The significance of these *Born-Huang relations* will be discussed shortly.

The relations (2.146) offer a means of determining the force constants if the total number of independent force constants does not exceed the number of independent elastic constants. Some years ago, before detailed information about vibrational spectra became available, one often assumed that interatomic forces were of short range. This resulted in a relatively small number of independent force constants, which were then determined from the experimental values of $C_{ij}$. Nowadays one seldom relies exclusively on elastic constants.

### 2.4.5 Significance of the Born-Huang relations

It is necessary to underscore the preceding discussion with several comments.

The first of these concerns the equilibrium conditions of the crystal. In Sec. 2.2 we assumed that our crystal was at 0 K and free from stresses and then discussed the oscillations of the atoms about their equilibrium positions. The motivation for these assumptions was that we wished to consider the oscillations in a crystal in equilibrium. Here we wish to discuss the rider concerning the stress, without which restriction our specification of equilibrium in an infinite crystal would not be complete.

Consider first the problem of specifying crystal equilibrium in the case of a one-dimensional chain. At first sight it might appear adequate to state that crystal equilibrium merely implies that every atom is in equilibrium, that is, the force on every atom vanishes. However, if we now consider two chains, one with a lattice spacing $a_0$ appropriate to a "natural" crystal and another with a spacing $a > a_0$ corresponding to a strained crystal, it becomes clear that the specification of *crystal equilibrium* considered above is not adequate since the *atoms* in both chains are in equilibrium. The requirement that the force on every atom vanish must therefore be supplemented by the condition that the crystal must also be free from stresses.

To visualize an infinitely extended crystal that is free from stresses we

†Remember that the square and the round brackets involve mainly the force constants.

proceed as follows: Consider first a finite crystal which is *not* subjected to any external stresses. This means that stresses vanish throughout the crystal. We now imagine that we take a portion from the interior where surface effects are negligible and from it generate an infinite crystal by repeating indefinitely. The latter will have the same atomic arrangements as that in the interior of the finite crystal and it is this configuration that is termed the equilibrium configuration. Any other atomic arrangement would, in the light of a finite crystal, correspond to a stressed condition and one in which the configuration would not be determined entirely by forces intrinsic to the crystal. In other words, the configuration could not be obtained by minimizing $\Phi$ in the usual fashion. Rather $\Phi$ would have to be minimized subject to some constraints representing the presence of arbitrary stresses. It is to avoid this arbitrariness that we specified that the configuration correspond to vanishing stresses.

We noted earlier that given a knowledge of the crystal potential, one could determine (in principle) the equilibrium configuration, that is, the full crystal structure including all the $\mathbf{x}(k)$'s, by minimizing $\Phi$. Once the structural parameters are so obtained, it becomes superfluous to mention that the crystal is free from stresses. However, when dealing with a formal potential function $\Phi$, the additional specification is necessary.

Even in a formal treatment it is natural to expect some constraints on the force constants as a manifestation of the condition corresponding to vanishing stresses. It turns out that indeed this is the case. It has been demonstrated[41,42] that starting with a strained crystal and invoking the invariance of the strain energy to rigid rotations, the following conditions on the square brackets are obtained

$$[\alpha\beta, \gamma\lambda] - S_{\gamma\lambda}\delta_{\alpha\beta} = [\gamma\lambda, \alpha\beta] - S_{\alpha\beta}\delta_{\gamma\lambda}, \tag{2.148}$$

with the symmetries (2.144) continuing to hold. The $S_{\alpha\beta}$ are the initial homogeneous stresses acting on the medium. The relations given in (2.148) are extensions of the Born-Huang relations (2.147), and will be referred to as the *generalized Huang conditions*. Let us write some of these explicitly. We have

$$[22, 11] - [11, 22] = S_{11} - S_{22}, \tag{2.149a}$$

$$[33, 22] - [22, 33] = S_{22} - S_{33}, \tag{2.149b}$$

$$[22, 12] - [12, 22] = S_{12},^{*} \tag{2.149c}$$

*The stress $S_{12}$ can also be written as

$$S_{12} = [11, 12] - [12, 11]. \tag{2.149f}$$

If we adopt (2.149c) as primary then (2.149f) simply determines $[11, 12]$ in terms of the quantities of (2.149c), and so on.

$$[22, 23] - [23, 22] = S_{23}, \qquad\qquad\qquad (2.149d)$$

$$[22, 31] - [31, 22] = S_{31}. \qquad\qquad\qquad (2.149e)$$

These five expressions exhibit the connection between the anisotropic stresses, namely, $S_{11} - S_{22}$, $S_{22} - S_{33}$, $S_{12}$, $S_{23}$, $S_{31}$, and the force constants. The corresponding relations for the stress-free case are obtained simply by setting the various stress components in (2.149) equal to zero. In other words, five of the Born-Huang conditions (2.147) represent the restrictions on the force constants corresponding to the vanishing of anisotropic stresses in the lattice. (It is helpful to remember that the Born-Huang conditions arose when we made a comparison between the microscopic expression for sound propagation and the macroscopic expression, the latter corresponding to a crystal free from initial stress. In this way we ensured that the force constants pertained to the equilibrium configuration in the sense explained earlier. The above discussion shows more explicitly the significance of the Born-Huang relations vis-à-vis the vanishing of stresses.) Notice that isotropic pressure eludes the formalism since no expression for $p = -\frac{1}{3}(S_{11} + S_{22} + S_{33})$ can be written solely in terms of the square brackets.

Summarizing, in the case of an infinite crystal the equilibrium configuration corresponds to one in which (1) the forces on all atoms vanish and (2) there are no stresses. If in formulating the problem we leave the force constants as parameters to be determined, then they must be so determined as to be consistent with the Born-Huang relations. In this way we ensure, among other things, that the parameters characterize a crystal free from anisotropic stresses. Isotropic stress, that is, pressure, eludes the formalism and must continue to be wished away! From a practical standpoint, what this implies is that the information (for example, lattice parameters and elastic constants) used for deriving the force constants must pertain to a crystal not subjected to pressure.

We have remarked that five of the Born-Huang conditions can be interpreted as restrictions corresponding to the vanishing of anisotropic stresses. The question naturally arises as to what the rest connote. This cannot be answered in a simple fashion. However, it has been pointed out[42–45] that since the generalized Huang conditions are basically determined by assuming rotational invariance, these conditions as well as their special form (2.147) represent constraints imposed by rotational invariance. In this context, Lax[44] has remarked that the sum rule for force constants deduced in Sec. 2.2.5 by considering infinitesimal rotational invariance constitutes a necessary but not sufficient condition. Lax also speculates that the Born-Huang conditions *may* be complete. There is, however, no proof of this. Hence there exists at present no general method of assuring that force constants being used in a practical calculation satisfy completely all requirements of rotational invariance. A conservative alternative is to formulate the dynamics in terms of bond-stretching, bond-bending and torsional force constants[46] (see Sec. 2.8.2) in the same spirit as using internal coordinates in molecular vibrations calculations, automatically eliminating the need for considering the constraints due to rotational invariance. Alternately, one could construct explicitly the crystal potential energy $\Phi$ in terms of two-body potentials, three-body potentials, etc., and start from a configuration that corresponds to the equilibrium configuration prescribed by the potential itself. The force constants are then evaluated by taking the appropriate derivatives of $\Phi$ in this configuration. In this case, all invariances and equilibrium constraints are contained in the determination of crystal parameters so that the force constants automatically satisfy all invariance requirements, and there is no need (except perhaps as a check) to ensure their fulfillment.

### 2.4.6 Cauchy relations

In crystals whose potential energy can be *completely* expressed in terms of two-body central potentials of the form $V_{kk'}(|\mathbf{r}|)$ and in which every atom is at a center of inversion, the elastic constants acquire symmetries in addition to those discussed earlier. These are referred to as *Cauchy relations* and are given by[47]

$$C_{23} = C_{44}, \; C_{31} = C_{55}, \; C_{12} = C_{66},$$

$$C_{14} = C_{56}, \; C_{25} = C_{64}, \; C_{36} = C_{45}. \tag{2.150}$$

These reduce the number of independent elastic constants from 21 to 15.

The Cauchy relations for cubic crystals have received frequent attention. The independent elastic constants for cubic crystals are $C_{11}$, $C_{12}$, and $C_{44}$ ($=C_{66}$). If in addition Cauchy relations hold, then

$$C_{12} = C_{44}, \tag{2.151}$$

whence only two elastic constants are required to describe the elastic properties.

**NaCl-type crystals.** The elastic constants of a number of crystals having the NaCl structure very nearly fulfill the Cauchy relations. This is expected because every atom is at a center of symmetry and because the forces are believed to be central (see Eq. (1.12)). There are, however, several exceptions as may be seen from Table 2.1. The deviations imply the existence of many-body forces, which could arise either from anisotropy in the electron distributions or from covalent bonding.

**Table 2.1**
Elastic constants for some crystals with the sodium chloride structure. All data pertain to 300 K and are expressed in $10^{12}$ dynes/cm².

| Crystal | $C_{11}$ | $C_{12}$ | $C_{44}$ |
|---|---|---|---|
| NaCl | 0.487 | 0.124 | 0.126 |
| RbBr | 0.317 | 0.042 | 0.039 |
| RbI | 0.256 | 0.031 | 0.029 |
| CsBr | 0.300 | 0.078 | 0.076 |
| CsI | 0.246 | 0.067 | 0.062 |
| MgO | 2.86 | 0.87 | 1.48 |
| AgCl | 0.592 | 0.364 | 0.0616 |
| AgBr | 0.563 | 0.330 | 0.072 |
| CaO | 2.03 | 0.61 | 0.76 |

**Table 2.2**
Elastic constants for some cubic metals at 300 K in units of $10^{12}$ dynes/cm².

| Metal | $C_{11}$ | $C_{12}$ | $C_{44}$ |
|---|---|---|---|
| K | 0.046 | 0.037 | 0.026 |
| Fe | 2.37 | 1.41 | 1.16 |
| W | 5.01 | 1.98 | 1.15 |
| Cu | 1.684 | 1.214 | 0.754 |
| Al | 1.08 | 0.62 | 0.28 |
| Pb | 0.48 | 0.41 | 0.14 |

**Cubic metals.** A number of metals exhibit the fcc and the bcc structures, and in these every atom is at a center of inversion. Hence if the crystal is held together entirely by central forces, then the Cauchy relations must hold. In fact this was expected but the expectations are belied by experiment as Table 2.2 shows. The reason for the failure of the Cauchy relation in metals was first pointed out by Fuchs[48] who noted that the crystal potential $\Phi$ has a contribution $E(v)$ which depends on the volume per atom. The latter arises from the fact that the electronic contribution $E_0(R)$ to the crystal potential includes the kinetic energy of the electrons (see (2.3) and (2.5)), which in a free-electron model is given by[49]

$$U = \frac{3}{5}\rho^{(0)}E_F$$
$$= \frac{3}{5}\frac{\rho^{(0)}\hbar^2}{2m}(3\pi^2\rho^{(0)})^{2/3},$$

where $E_F$ is the Fermi energy and $\rho^{(0)} = nZ/v$ is the number of electrons per unit volume ($Z$ = valency). The forces are thus not entirely central and the Cauchy relation is therefore violated.

One interesting consequence of Fuchs' observation is the following. Suppose we consider the following sound waves:

1. a transverse (shear) wave propagating along [110] and with particle displacements along [1$\bar{1}$0];

2. a transverse wave propagating along [001] with particle displacements along [100].

Using (2.114) and (2.117), the sound velocities for these cases are easily shown to be given by

$$[(C_{11} - C_{12})/2\rho]^{1/2} \text{ and } [C_{44}/\rho]^{1/2}, \qquad (2.152a)$$

respectively. Now, noting that $\Phi$ has a part built up from pair potentials and a volume-dependent part, we may write

$$C_{ij} = C_{ij}^{p} + C_{ij}^{v}. \qquad (2.153)$$

However, the shear waves considered above do not involve any change in the atomic volume. The strain energy is thus contributed only by the forces generated from pair potentials. Therefore from a detailed consideration of the forces, one has for the velocities

$$\left(\frac{C_{11}^{p} - C_{12}^{p}}{2\rho}\right)^{1/2} \text{ and } \left(\frac{C_{44}^{p}}{\rho}\right)^{1/2}. \qquad (2.152b)$$

Comparing with (2.152a), we find

$$C_{44} = C_{44}^{p}, \text{ that is, } C_{44}^{v} = 0$$

$$C_{11} - C_{12} = C_{11}^{p} - C_{12}^{p}. \qquad (2.154)$$

These are referred to sometimes as *Fuchs' relations*.

### 2.4.7 Comments on piezoelectric crystals

In piezoelectric crystals, an acoustic strain sets up a macroscopic electric field. The mechanical and electrical effects are not separable and must be treated together. The formulation of Sec. 2.4.4 does not allow for the simultaneous presence of these two effects and for this reason is not applicable to piezoelectric crystals. The necessary extensions to cover such cases will be considered in Chapter 4.

### 2.5 Lattice dynamics at finite temperatures—quasi-harmonic approximation

We wish to comment now on what might seem to be rather severe assumptions made at the beginning of Sec. 2.2, namely that the crystal is free from stresses, is held at 0 K and that zero-point effects are negligible. Our main purpose in making these assumptions was to guarantee that the equilibrium configuration of the crystal corresponds to the minimum of the *potential* energy. The significance of insisting on the absence of stresses in this context has already been explained in the previous section. Since the force on every atom vanishes, $\Phi^{(1)} = 0$ and the potential energy expansion has the form

$$\Phi_0 = \Phi_0^{(0)} + \Phi_0^{(2)} + \Phi_0^{(3)} + \Phi_0^{(4)} + \cdots, \tag{2.155}$$

where we have indicated the temperature with a subscript to emphasize the fact that the expansion is made about the positions occupied by the atoms at 0 K. In constructing a harmonic theory of the vibrations, we retained just the first two terms on the right-hand side. (In fact $\Phi^{(0)}$ does not appear in the dynamics; it merely provides an energy base line). The terms $\Phi^{(3)}$ and beyond contribute to anharmonicity.

It is interesting that if the crystal were truly harmonic, then there would be no thermal expansion and, further, the thermal conductivity would be infinite.[50] Since these properties are clearly contrary to experiment, we conclude that anharmonic effects are always present, indeed even at 0 K. We observe in passing that if the crystal were truly harmonic, then due to the absence of thermal expansion the force constants would not change with temperature and consequently the vibration frequencies would be temperature independent.

The question next arises as to how we set up a theory to discuss vibrations of a crystal at a finite temperature $T$ so as to compare with experiment. In attempting to answer this question, we note that with increasing temperature the lattice expands, and the mean positions of the atoms change from the set $\{\mathbf{x}^0\binom{l}{k}\}$ at 0 K to the set $\{\mathbf{x}^T\binom{l}{k}\}$, the latter being determined by the minimum of the *free energy*.[51]

The vibrations of atoms at finite temperatures take place about the new mean positions. To describe these, Leibfried and Ludwig[52] propose that $\Phi_T$ be expanded as in (2.155), but about the mean positions appropriate to $T$. We consider the expansion

$$\Phi_T = \Phi_T^{(0)} + \Phi_T^{(1)} + \Phi_T^{(2)} + \Phi_T^{(3)} + \cdots, \tag{2.156}$$

where

$$\Phi_T^{(0)} = \Phi_T\left(\left\{\mathbf{x}^T\binom{l}{k}\right\}\right), \tag{2.157a}$$

$$\Phi_T^{(1)} = \sum_{lk\alpha} \frac{\partial \Phi_T}{\partial r_\alpha\binom{l}{k}}\bigg|_{\substack{\text{mean}\\\text{posn. at } T}} u_\alpha^T\binom{l}{k} = \sum_{lk\alpha} \phi_\alpha^T\binom{l}{k} u_\alpha^T\binom{l}{k}, \tag{2.157b}$$

$$
\begin{aligned}
\Phi_T^{(2)} &= \tfrac{1}{2} \sum_{l k \alpha} \sum_{l' k' \beta} \left. \frac{\partial^2 \Phi_T}{\partial r_\alpha\binom{l}{k} \partial r_\beta\binom{l'}{k'}} \right|_{\substack{\text{mean}\\\text{posn. at } T}} u_\alpha^T \binom{l}{k} u_\beta^T \binom{l'}{k'} \\
&= \tfrac{1}{2} \sum_{l k \alpha} \sum_{l' k' \beta} \phi_{\alpha\beta}^T \binom{l\ \ l'}{k\ \ k'} u_\alpha^T \binom{l}{k} u_\beta^T \binom{l'}{k'},
\end{aligned}
\tag{2.157c}
$$

. . . .

As at 0 K, a harmonic approximation is obtained by retaining only $\Phi_T^{(0)}$ and $\Phi_T^{(2)}$.* The dynamical problem then goes through exactly as before with, however, the force constants being taken as $\phi_{\alpha\beta}^T\binom{l\ l'}{k\ k'}$. This scheme of expanding the potential energy about the mean positions appropriate to a finite temperature and then employing the usual truncation is referred to as the *quasi-harmonic approximation*. The use of force-constant models to analyze experimental lattice-dynamical data obtained at finite temperatures is made in this spirit. Unlike those in the strictly harmonic theory, the frequencies in the quasi-harmonic approximation are temperature dependent. This temperature dependence is intuitively understandable since the force constants change by virtue of the derivatives being evaluated about the mean positions, which change with temperature. In fact the temperature dependence of the force constants may be estimated as follows. Express $\Phi_T$ as a power series in the changes $\delta \mathbf{r}^T\binom{l}{k} = [\mathbf{x}^T\binom{l}{k} - \mathbf{x}^0\binom{l}{k}] + \mathbf{u}^T\binom{l}{k} = \delta \mathbf{x}\binom{l}{k} + \mathbf{u}^T\binom{l}{k}$, measured with respect to the equilibrium positions at 0 K. This gives†

$$
\begin{aligned}
\Phi_T &= \Phi_0^{(0)} + \sum_{l k \alpha} \phi_\alpha^0 \binom{l}{k} \delta r_\alpha^T \binom{l}{k} \\
&\quad + \tfrac{1}{2} \sum_{l k \alpha} \sum_{l' k' \beta} \phi_{\alpha\beta}^0 \binom{l\ \ l'}{k\ \ k'} \delta r_\alpha^T \binom{l}{k} \delta r_\beta^T \binom{l'}{k'} \\
&\quad + \tfrac{1}{6} \sum_{l k \alpha} \sum_{l' k' \beta} \sum_{l'' k'' \gamma} \phi_{\alpha\beta\gamma}^0 \binom{l\ \ l'\ \ l''}{k\ \ k'\ \ k''} \delta r_\alpha^T \binom{l}{k} \delta r_\beta^T \binom{l'}{k'} \delta r_\gamma^T \binom{l''}{k''} + \cdots
\end{aligned}
$$

Comparing the above with (2.156) and isolating in both the terms of second-order in $\mathbf{u}^T$, we get

$$
\phi_{\alpha\beta}^T \binom{l\ \ l'}{k\ \ k'} = \phi_{\alpha\beta}^0 \binom{l\ \ l'}{k\ \ k'} + \sum_{l'' k'' \gamma} \phi_{\alpha\beta\gamma}^0 \binom{l\ \ l'\ \ l''}{k\ \ k'\ \ k''} \delta x_\gamma \binom{l''}{k''} + \cdots
\tag{2.158}
$$

---

*Unlike $\Phi_0^{(1)}$, $\Phi_T^{(1)}$ does not always vanish. It does so for crystals with one atom per cell. For crystals containing more than one sublattice, however, $\Phi_T^{(1)}$ is generally nonzero though small compared to $\Phi_T^{(2)}$. It is therefore treated as a perturbation along with $\Phi_T^{(3)}$ and the other higher-order terms.[52]

As earlier, expansion (2.156) is meaningful only if the rms displacements from the mean positions $\mathbf{X}^T$ are small compared to interatomic separations. Such an assumption may be questionable for "hot" solids, that is, for solids approaching melting temperatures.

†The superscript on the coefficients $\Phi_\alpha$ indicates the temperature.

in which the force constants $\phi_{\alpha\beta}^T$ depend on temperature through the thermal expansion $\delta\mathbf{x}$. In other words, the quasi-harmonic approximation takes implicit (partial) account of anharmonic effects. (Note the presence of $\phi_{\alpha\beta\gamma}^0$ and higher-order terms in (2.158)).

In brief, the use of the force-constant formalism as outlined in Sec. 2.2 for describing the vibrations in a crystal at finite temperatures implies making the quasi-harmonic approximation. This approximation is at best a prescription of limited use. For one thing it is not easy via this approach to establish a correspondence between the quasi-harmonic force constants and the elastic constants as in Sec. 2.4.4. When finite-temperature effects are taken into account, the elastic constants are defined as the derivatives of the free energy[51] rather than of the strain energy as in (2.107). Furthermore, one must distinguish between isothermal and adiabatic elastic constants. In practice, experimentalists often do not worry about

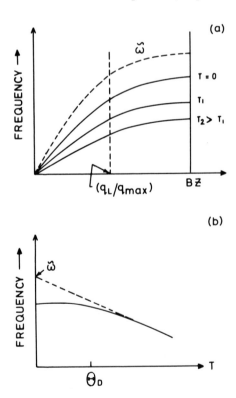

**Figure 2.12**
(a) Temperature-dependent dispersion curves. (b) Linear extrapolation to obtain the "harmonic" frequency $\tilde{\omega}$ for a given $(q/q_{max})$. (After Leibfried, ref. 54.)

such subtleties and use the formula (2.146) to relate the quasi-harmonic force constants to the measured elastic constants appropriate to that temperature.* The procedure on the whole is not very satisfactory, and Leibfried[54] therefore advocates the following (in our notation):

"If one were able to measure the quasi-harmonic dispersion curves one would expect a behaviour as sketched in Fig. 2.12. The curves change relatively little through thermal expansion for $T < \Theta_D$ (Debye temperature). Then the change is linear with temperature because the thermal expansion depends linearly on $T$. If one plots the frequency for a given $q_L$ reduced by thermal expansion one obtains the harmonic value by linear extrapolation to $T = 0$. An interpretation of the dispersion curves by temperature dependent force constants would be quite deceiving. The same would apply to explaining the temperature dependent elastic moduli by temperature dependent force constants. . . . The only correct information about the potential is given by the harmonic values." This advice to extrapolate to 0 K before data analysis is rarely heeded!

## 2.6 Quantum aspects of lattice vibrations

Our approach to the dynamical problem thus far has been entirely classical, and we now wish to make the transition to quantum mechanics. By analogy with the one-dimensional harmonic oscillator, we expect that the frequencies would continue to be given by (2.51) but that the discussion of the amplitudes and of the energy states would have to be modified. To see this explicitly, it is necessary to transform the Hamiltonian (2.33) to a more convenient form. In doing so, we do not yet specify whether the Hamiltonian is classical or quantum mechanical. The transformation is thus common to both.

The transformation will be carried out in two stages; first we exploit the periodicity of the crystal and go from $u_\alpha(^l_k)$ to new coordinates $\zeta_\alpha(k \mid \mathbf{q})$ which are in the Bloch form.[55] Thus we write

$$\zeta_\alpha(k \mid \mathbf{q}) = \sqrt{m_k/N} \sum_l u_\alpha\binom{l}{k} e^{-i\mathbf{q}\cdot\mathbf{x}(l)}, \qquad (2.159)$$

where $\mathbf{q}$, as before, is determined via cyclic boundary conditions to be a

---

*If one is seeking such a relation in the context of analyzing dispersion curve data obtained by neutron spectroscopy (Sec. 5.1), then to be "consistent" one must use the isothermal elastic constant since neutron inelastic scattering seems to measure isothermal constants.[53] On the other hand, ultrasonic experiments deliver only adiabatic constants and the desired isothermal constants would have to be computed through the use of an appropriate interrelation. See, for example, ref. 36, p. 186.

vector in the BZ. The factor $(m_k/N)^{1/2}$ is introduced for convenience. The transformation (2.159) is unitary. To appreciate this, we write the above as

$$\zeta_\alpha(k \mid \mathbf{q}) = \sqrt{m_k} \sum_l R(\mathbf{q} \mid l) u_\beta\binom{l}{k}.$$

It is then easily seen using the result (A1.5) of Appendix 1 that the matrix **R** with elements

$$R(\mathbf{q} \mid l) = \frac{1}{\sqrt{N}} e^{-i\mathbf{q}\cdot\mathbf{x}(l)}$$

has the property $\mathbf{RR}^\dagger = \mathbb{1}_N$.

The relation inverse to (2.159) is

$$u_\alpha\binom{l}{k} = \frac{1}{\sqrt{Nm_k}} \sum_{\mathbf{q}} \zeta_\alpha(k \mid \mathbf{q}) e^{i\mathbf{q}\cdot\mathbf{x}(l)}, \tag{2.160}$$

where the summation over $\mathbf{q}$ ranges over the BZ.

Since the displacements are real,[55]

$$\zeta_\alpha(k \mid \mathbf{q}) = \zeta_\alpha^*(k \mid -\mathbf{q}). \tag{2.161}$$

From (2.160) we also have the momentum:

$$p_\alpha\binom{l}{k} = \sqrt{\frac{m_k}{N}} \sum_{\mathbf{q}} \dot{\zeta}_\alpha(k \mid \mathbf{q}) e^{i\mathbf{q}\cdot\mathbf{x}(l)}. \tag{2.162}$$

Use of (2.160) and (2.162) in (2.33), together with the result (A1.5) and Eq. (2.161) then leads to the following result:†

$$H = \tfrac{1}{2} \sum_{\mathbf{q}k\alpha} \dot{\zeta}_\alpha^*(k \mid \mathbf{q})\dot{\zeta}_\alpha(k \mid \mathbf{q})$$

$$+ \tfrac{1}{2} \sum_{\mathbf{q}kk'\alpha\beta} \zeta_\alpha^*(k \mid \mathbf{q}) D_{\alpha\beta}\binom{\mathbf{q}}{kk'} \zeta_\beta(k' \mid \mathbf{q}). \tag{2.163}$$

The Hamiltonian (2.163) is the sum of two Hermitian forms.‡ From matrix algebra[56,57] we have this result: Given two Hermitian forms $\sum_{\alpha\beta} X_\alpha'^* A_{\alpha\beta} X_\beta'$ and $\sum_{\alpha\beta} X_\alpha^* B_{\alpha\beta} X_\beta$, the latter being positive definite, it is possible to find a transformation **E** which simultaneously reduces them to sums of squares. Moreover, if **A** and **B** commute then **E** can be chosen to be unitary. These requirements are fulfilled in our case.

---

†We suppress $\Phi^{(0)}$ in $H$.

‡An expression $\mathscr{H}(\mathbf{X}^*, \mathbf{X}) = \sum_{\alpha\beta} X_\alpha^* H_{\alpha\beta} X_\beta$, where $H_{\alpha\beta}$ are elements of a Hermitian matrix, is said to be a *Hermitian* form. In the special case of a real symmetric matrix **A**, the quantity $\mathscr{A}(\mathbf{X}, \mathbf{X}) = \sum_{\alpha\beta} X_\alpha A_{\alpha\beta} X_\beta$ is called a *quadratic* form.

We consider a further unitary transformation $\zeta \to Q$ defined by

$$\zeta_\alpha(k \mid \mathbf{q}) = \sum_j e_\alpha\left(k \left| {\mathbf{q} \atop j}\right.\right) Q\left({\mathbf{q} \atop j}\right), \tag{2.164}$$

where at present we do not identify the coefficients $e_\alpha(k \mid {\mathbf{q} \atop j})$ on the right with eigenvector components introduced earlier. The inverse of (2.164) is

$$Q\left({\mathbf{q} \atop j}\right) = \sum_{k\alpha} \zeta_\alpha(k \mid \mathbf{q}) e_\alpha^*\left(k \left| {\mathbf{q} \atop j}\right.\right). \tag{2.165}$$

Correspondingly,

$$\dot{Q}\left({\mathbf{q} \atop j}\right) = \sum_{k\alpha} \dot{\zeta}_\alpha(k \mid \mathbf{q}) e_\alpha^*\left(k \left| {\mathbf{q} \atop j}\right.\right). \tag{2.166}$$

Using the unitary transformation (2.164) in (2.163) we get

$$H = \tfrac{1}{2} \sum_{\mathbf{q}j} \dot{Q}^*\left({\mathbf{q} \atop j}\right) \dot{Q}\left({\mathbf{q} \atop j}\right)$$
$$+ \tfrac{1}{2} \sum_{\mathbf{q}jj'} Q^*\left({\mathbf{q} \atop j}\right) \left\{ \sum_{\substack{\alpha\beta \\ kk'}} e_\alpha^*\left(k \left| {\mathbf{q} \atop j}\right.\right) D_{\alpha\beta}\left({\mathbf{q} \atop kk'}\right) e_\beta\left(k' \left| {\mathbf{q} \atop j'}\right.\right) \right\} Q\left({\mathbf{q} \atop j'}\right). \tag{2.167}$$

Now, in view of the discussion above, we may insist the transformation be such that

$$H = \tfrac{1}{2} \sum_{\mathbf{q}j} \left\{ \dot{Q}^*\left({\mathbf{q} \atop j}\right) \dot{Q}\left({\mathbf{q} \atop j}\right) + \omega_j^2(\mathbf{q}) Q^*\left({\mathbf{q} \atop j}\right) Q\left({\mathbf{q} \atop j}\right) \right\}. \tag{2.168}$$

Comparison of the last two equations shows that

$$\sum_{kk'\alpha\beta} e_\alpha^*\left(k \left| {\mathbf{q} \atop j}\right.\right) D_{\alpha\beta}\left({\mathbf{q} \atop kk'}\right) e_\beta\left(k' \left| {\mathbf{q} \atop j'}\right.\right) = \omega_j^2(\mathbf{q})\delta_{jj'}, \tag{2.169}$$

which is equivalent to (2.51). In other words, the eigenvalue equation (2.51) arises in the context of a transformation problem irrespective of whether we regard $H$ as classical or quantum, confirming our intuition that frequencies are the same in both the classical and quantum pictures.

The quantity $Q({\mathbf{q} \atop j})$ is called a *normal coordinate*, the transformation $u \to Q$ a normal-coordinate transformation, and the frequencies $\omega_j(\mathbf{q})$ normal-mode frequencies.

Based on (2.160) and (2.164), we have

$$u_\alpha\left({l \atop k}\right) = \frac{1}{\sqrt{Nm_k}} \sum_{\mathbf{q}j} e_\alpha\left(k \left| {\mathbf{q} \atop j}\right.\right) Q\left({\mathbf{q} \atop j}\right) e^{i\mathbf{q}\cdot\mathbf{x}(l)},$$

which is the same as (2.60) which was obtained via a consideration of wavelike solutions. The inverse of this result is

$$Q\binom{\mathbf{q}}{j} = \frac{1}{\sqrt{N}} \sum_{lk\alpha} \sqrt{m_k} e_\alpha^*\left(k \middle| \begin{matrix}\mathbf{q}\\j\end{matrix}\right) u_\alpha\binom{l}{k} e^{-i\mathbf{q}\cdot\mathbf{x}(l)}. \tag{2.170}$$

A useful corollary which can be deduced from (2.170) and (2.73) using the reality of the $u$'s is

$$Q\binom{\mathbf{q}}{j} = Q^*\binom{-\mathbf{q}}{j}. \tag{2.171}$$

It is also useful to note that the dynamical equation (2.169) corresponding to the Fourier component $\mathbf{q}$ can be deduced from the partial Hamiltonian

$$H(\mathbf{q}) = \dot{\zeta}^\dagger(\mathbf{q})\dot{\zeta}(\mathbf{q}) + \zeta^\dagger(\mathbf{q})\mathbf{D}(\mathbf{q})\zeta(\mathbf{q}), \tag{2.172}$$

where $\zeta(\mathbf{q})$ is a $3n$-component column matrix with elements $\zeta_\alpha(k \mid \mathbf{q})$.

We have already remarked that the Hamiltonian (2.33) and its transformed version (2.168) are common to both the classical and quantum cases. In the former the displacements and their conjugate momenta are classical quantities; in the latter they are operators satisfying the following commutation rules:

$$\left[u_\alpha\binom{l}{k}, u_\beta\binom{l'}{k'}\right] = \left[p_\alpha\binom{l}{k}, p_\beta\binom{l'}{k'}\right] = 0,$$
$$\left[u_\alpha\binom{l}{k}, p_\beta\binom{l'}{k'}\right] = i\hbar\delta_{\alpha\beta}\delta_{ll'}\delta_{kk'}. \tag{2.173}$$

Correspondingly,

$$\left[Q^*\binom{\mathbf{q}}{j}, Q\binom{\mathbf{q'}}{j'}\right] = \left[\dot{Q}^*\binom{\mathbf{q}}{j}, \dot{Q}\binom{\mathbf{q'}}{j'}\right] = 0,$$
$$\left[\dot{Q}^*\binom{\mathbf{q}}{j}, Q\binom{\mathbf{q'}}{j'}\right] = i\hbar\delta_{\mathbf{qq'}}\delta_{jj'},$$
$$\left[Q\binom{\mathbf{q}}{j}, \dot{Q}^*\binom{\mathbf{q'}}{j'}\right] = i\hbar\delta_{\mathbf{qq'}}\delta_{jj'}. \tag{2.174}$$

Note that the star in (2.174) really implies Hermitian conjugation. Also $\dot{Q}^*$ is the momentum conjugate to the coordinate $Q$.

The form of the Hamiltonian (2.168) is very suggestive. We recognize it as the sum of the Hamiltonians of $3nN$ one-dimensional oscillators of squared frequencies $\omega_j^2(\mathbf{q})$ ($\mathbf{q}$ ranges over the $N$ allowed values; $j = 1, \ldots, 3n$). To proceed further we introduce new operators $a_{\mathbf{q}j}$ and $a_{\mathbf{q}j}^+$ by[31,58,59]

$$Q\binom{\mathbf{q}}{j} = (\hbar/2\omega_j(\mathbf{q}))^{1/2}[a_{\mathbf{q}j} + a_{-\mathbf{q}j}^\dagger], \qquad (2.175)$$

$$\dot{Q}\binom{\mathbf{q}}{j} = -i(\hbar\omega_j(\mathbf{q})/2)^{1/2}[a_{\mathbf{q}j} - a_{-\mathbf{q}j}^\dagger], \qquad (2.176)$$

with $a_{\mathbf{q}j}$ and $a_{\mathbf{q}j}^\dagger$ satisfying the commutation relations

$$[a_{\mathbf{q}j}, a_{\mathbf{q}'j'}] = [a_{\mathbf{q}j}^\dagger, a_{\mathbf{q}'j'}^\dagger] = 0 \qquad (2.177a)$$

$$[a_{\mathbf{q}j}, a_{\mathbf{q}'j'}^\dagger] = \delta_{jj'}\delta_{\mathbf{q}\mathbf{q}'}. \qquad (2.177b)$$

These operators are analogous to those introduced in texts on quantum mechanics in connection with the one-dimensional oscillator. In terms of these operators,

$$H = \sum_{\mathbf{q}j} \hbar\omega_j(\mathbf{q})[\tfrac{1}{2} + a_{\mathbf{q}j}^\dagger a_{\mathbf{q}j}], \qquad (2.178)$$

a form that is very useful, especially as the properties of the operators $a$ and $a^\dagger$ are well known.[31,58,59] The latter include the following:

$$a_{\mathbf{q}j}\,|\,n\rangle = [n\binom{\mathbf{q}}{j}]^{1/2}|n'\rangle \qquad (2.179a)$$

$$a_{\mathbf{q}j}^\dagger\,|\,n\rangle = [1 + n\binom{\mathbf{q}}{j}]^{1/2}|n''\rangle. \qquad (2.179b)$$

Here $|n\rangle$ denotes the crystal state in which the oscillator with frequency $\omega_{j_1}(\mathbf{q}_1)$ is excited to the state $n\binom{\mathbf{q}_1}{j_1}$, oscillator $\omega_{j_2}(\mathbf{q}_2)$ to the state $n\binom{\mathbf{q}_2}{j_2}$, and so on; $|n'\rangle$ denotes a state in which all oscillator quantum numbers are the same as in $|n\rangle$ except that pertaining to $\omega_j(\mathbf{q})$ which has a value $n\binom{\mathbf{q}}{j} - 1$; $|n''\rangle$ is also similar to $|n\rangle$ but with $n\binom{\mathbf{q}}{j}$ replaced by $n\binom{\mathbf{q}}{j} + 1$. In view of (2.179), the operators $a$ and $a^\dagger$ are referred to as *annihilation* and *creation* operators respectively.

Some useful consequences of (2.179) are

$$a_{\mathbf{q}j}^\dagger a_{\mathbf{q}j}\,|\,n\rangle = n\binom{\mathbf{q}}{j}\Big|n\Big\rangle; \qquad (2.180)$$

$$\langle a_{\mathbf{q}j}^\dagger a_{\mathbf{q}j}\rangle_T = \Big\langle n\binom{\mathbf{q}}{j}\Big\rangle_T = \frac{1}{\exp[\beta\hbar\omega_j(\mathbf{q})] - 1}; \qquad (2.181)$$

$$\langle a_{\mathbf{q}j} a_{\mathbf{q}j}^\dagger\rangle_T = 1 + \Big\langle n\binom{\mathbf{q}}{j}\Big\rangle_T, \qquad (2.182)$$

where $\beta = 1/k_B T$ and

$$\langle O \rangle_T = \frac{\sum_i \langle i|e^{-\beta H} O|i\rangle}{\sum_i \langle i|e^{-\beta H}|i\rangle} \tag{2.183}$$

denotes the thermodynamic average of the operator $O$. The above results are very useful in the evaluation of many thermodynamic quantities. The annihilation and creation operator formalism is also useful in describing the deexcitation and excitation of the vibrational oscillators by external probes (for example, slow neutrons).

The foregoing quantum treatment admits of two equivalent interpretations.

1. The crystal may be regarded as a collection of $3nN$ *independent distinguishable oscillators* with squared frequencies $\omega_j^2(\mathbf{q})$. Each oscillator is capable of being in various stationary states labeled by the quantum numbers $n\binom{\mathbf{q}}{j} = 0, 1, 2, \ldots$ and having energies $E\binom{\mathbf{q}}{j} = [n\binom{\mathbf{q}}{j} + \frac{1}{2}]\hbar\omega_j(\mathbf{q})$. The action of $a^\dagger$ and $a$ is to excite or deexcite the oscillators.

2. The system may be considered as a *gas of indistinguishable particles called phonons*. This interpretation is in analogy to the quantum aspects of the electromagnetic field. A phonon can be in various stationary states labeled by the quantum numbers $(\mathbf{q}, j)$ (note!). The number of phonons in the state $(\mathbf{q}, j)$ is given by $n\binom{\mathbf{q}}{j}$ and the operators $a^\dagger$ and $a$ respectively create and annihilate phonons in a particular state.* Like photons, phonons obey Bose statistics. In the harmonic theory the phonons do not interact since the underlying oscillators are independent. Thus the phonon gas consists of noninteracting Bose particles.

We can therefore visualize the cyclic crystal as $3nN$ oscillators each capable of being excited to an infinity of levels, or as a system with a family of $3nN$ states, each of which can be occupied by an infinite number of particles.

Although we are here primarily concerned with a harmonic theory of crystal vibrations, it is pertinent to take note of some of the consequences of anharmonicity which enter the picture when terms beyond $\Phi^{(2)}$ in the expansion (2.19) are retained. The most important consequence is that the harmonic frequencies $\omega_j(\mathbf{q})$ are modified to

$$\omega_j'(\mathbf{q}) = \omega_j(\mathbf{q}) + \Delta\omega_j(\mathbf{q}) + i\Gamma(\mathbf{q}, j), \tag{2.184}$$

$\Delta\omega$ and $\Gamma$ both real.[61] The factor $i\Gamma$ implies damping of the vibrational

---

*The particle aspects of phonons are discussed by Jensen.[60] The labeling of phonon states is analogous to the labeling of electron states in band theory. The labels used in the latter are $\mathbf{k}$ for wave vector, $s$ for band index, and $\alpha$ for spin index. However, electrons being Fermions, the occupation number $n(\mathbf{k}, s, \alpha)$ must be either 0 or 1.

waves with time; in other words, phonons do not have an infinite lifetime as in harmonic theory but acquire a finite lifetime. This may be interpreted as being due to phonon-phonon collisions. Thus when anharmonic effects are included, the phonon gas becomes an *interacting* Bose system. The quantity $(\omega_j(\mathbf{q}) + \Delta\omega_j(\mathbf{q}))$ is termed the *renormalized* phonon frequency and is temperature dependent.

## 2.7 External and internal vibrations in complex crystals

In the preceding discussion we formulated the dynamics by viewing the composition of the primitive cell entirely in terms of individual atoms. A number of crystals contain well-bound atomic clusters such as neutral molecules and ionic clusters. In dealing with the dynamics of such crystals it is advantageous to recognize a priori the existence of the molecular groups.* As a first approximation, such groups are regarded as rigid bodies capable both of translations and rotations. The dynamics are then analyzed in terms of the possible translations of both the molecular groups and the atoms with individual identity and in terms of the rotations of the molecular groups. This leads to the so-called *external modes*; the corresponding branches of the dispersion curves are referred to as *external branches*.

This approach is admittedly approximate since the molecular groups are not rigid and can execute vibrations as in the free state. However, the effects of nonrigidity can be treated separately by considering the coupling of the vibrations of the different groups, leading to a description of the *internal modes* and the *internal branches*. Thus the dynamical problem can be simplified by separately considering the external and the internal degrees of freedom of the molecular groups.

As an illustration we shall consider a specific crystal, namely $NH_4Cl$. Figure 2.13 shows the primitive cell; it contains one $Cl^-$ ion and one $NH_4^+$ ion, resulting in 18 degrees of freedom. Of these, 3 correspond to $Cl^-$ translations, 6 correspond to the external degrees of freedom of the $NH_4^+$ (three translations and three rotations) and the remaining 9 to the internal degrees of freedom of $NH_4^+$. In the approach of Sec. 2.2, all the atoms would be treated on an individual basis and the dispersion relations would therefore consist of 18 branches. These are illustrated in Figure 2.14 for a typical direction.[62] On the other hand, the method suggested above focuses attention separately on the external and internal branches so that one need deal with only 9 branches at a time. An additional ad-

---

*We shall use the term *molecular group* to cover both molecules like methane and benzene and ionic radicals like $NH_4^+$ and $SO_4^{--}$.

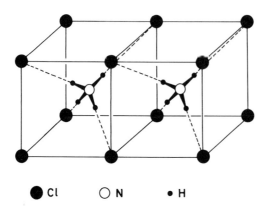

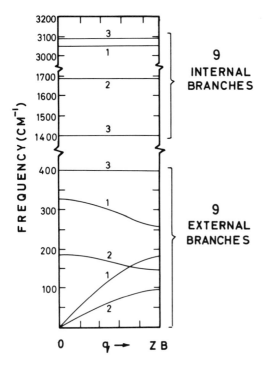

**Figure 2.13**
Primitive cell of $NH_4Cl$ in CsCl phase.

**Figure 2.14**
Dispersion curves for $NH_4Cl$ along [100]. The external branches are as calculated by Parlinski (ref. 62). The internal branches are speculative, based on the measured $\mathbf{q} = 0$ frequencies (Table 2.3). These branches are expected to be flat owing to weak inter-molecular coupling. (Electric-field effects associated with internal branches are not shown.) The degeneracies of the branches are indicated.

vantage is that the motions of the atoms in the ammonium ion can be described more physically in terms of rotations and internal vibrations.

Such a clear-cut separation is meaningful only when the forces inside molecular groups are much stronger than those binding atoms in different groups. It is, of course, possible to elaborate such an analysis by allowing for coupling between the external and the internal modes via some perturbation scheme. At this point one can no longer talk of pure external and internal modes, and one recovers more or less complete identification with the standard Born–von Kármán approach where all the degrees of freedom are taken together and not analyzed piecemeal.

### 2.7.1 External modes

We now wish to formulate the external-mode problem along the lines of Sec. 2.2. Let us suppose that the crystal has (I) $\mu$ independent atoms labeled $\kappa = 1, 2, \ldots, \mu$ and (II) $\nu$ nonlinear molecular groups labeled $\kappa = (\mu + 1), \ldots, (\mu + \nu)$. The crystal thus has $(\mu + \nu)$ sublattices* and has $(3\mu + 6\nu)$ external degrees of freedom per cell. As a result there will be $(3\mu + 6\nu)$ external branches, of which 3 are acoustic and the remaining optic (see also Figure 2.14). If $\lambda$ of the $\nu$ molecular groups are linear, then there are only $(3\mu + 6\nu - \lambda)$ external branches, since linear groups have only two rotational degrees of freedom.

The displacement coordinates of the present problem are (1) $\mathbf{u}^{t}\binom{l}{\kappa}$, which describes the center of mass translation, and (2) $\mathbf{u}^{r}\binom{l}{\kappa}$, which describes the (infinitesimal) angular displacement. Clearly $\mathbf{u}^{r} = 0$ for $\kappa = 1, 2, \ldots, \mu$. In terms of the above coordinates, the displacement of the $k$th atom in the group $\binom{l}{\kappa}$ is given by

$$\mathbf{u}\begin{pmatrix} l \\ \kappa \\ k \end{pmatrix} = \mathbf{u}^{t}\binom{l}{\kappa} + \left[ \mathbf{u}^{r}\binom{l}{\kappa} \times \mathbf{x}(k) \right], \qquad (2.185)$$

where $\mathbf{x}(k)$ denotes the equilibrium position of the atom concerned with respect to the center of mass of its parent molecular group.

Though both $\mathbf{u}^{t}$ and $\mathbf{u}^{r}$ are vectors, there is an important difference in their transformation properties. Whereas $\mathbf{u}^{t}$ is an ordinary or *polar* vector, $\mathbf{u}^{r}$ is an *axial* or *pseudo* vector. In other words, $\mathbf{u}^{t}$ changes sign under inversion while $\mathbf{u}^{r}$ does not. As a result, under an orthogonal transformation $\mathbf{S}$, $\mathbf{u}^{r}$ behaves as follows:[63]

$$\mathbf{u}^{r} \rightarrow (\mathbf{u}^{r})' = \mathbf{S}\mathbf{u}^{r} \qquad \text{if } \mathbf{S} \text{ is a proper rotation,}$$
$$= -\mathbf{S}\mathbf{u}^{r} \qquad \text{if } \mathbf{S} \text{ is an improper rotation.} \qquad (2.186a)$$

---

*Usually each sublattice is thought of as being occupied by only one atom. We generalize the concept here by supposing that the sites are occupied by vibrating units, that is, atoms and molecular groups.

Since $\|\mathbf{S}\| = \pm 1$ according as the rotation is proper or improper, (2.186a) may be written

$$(\mathbf{u}^r)' = \|\mathbf{S}\|\mathbf{S}\mathbf{u}^r. \tag{2.186b}$$

Two conventions are used in specifying the displacement components. In the first, they are specified with reference to a set of Cartesian axes fixed to the crystal. This frame will henceforth be called the fixed frame. In the second, called the principal-axes frame, the displacements of each vibrating unit are referred to a set of local axes oriented so as to coincide with the principal axes of the moment-of-inertia tensor. For $\kappa = 1, 2, \ldots, \mu$, these local axes can clearly be chosen as we wish, and for convenience they are chosen to be parallel to the crystal-fixed axes.

In the fixed frame, the harmonic Hamiltonian* may be written as

$$
\begin{aligned}
H &= T + V \\
&= \tfrac{1}{2}\sum_l \sum_{\kappa=1}^{(\mu+\nu)} \sum_\alpha m_\kappa \left[\dot{u}_\alpha^t\!\left(\begin{smallmatrix} l \\ \kappa \end{smallmatrix}\right)\right]^2 + \tfrac{1}{2}\sum_l \sum_{\kappa=(\mu+1)}^{(\mu+\nu)} \sum_{\alpha\beta} \mathscr{I}_{\alpha\beta}(\kappa)\dot{u}_\alpha^r\!\left(\begin{smallmatrix} l \\ \kappa \end{smallmatrix}\right)\dot{u}_\beta^r\!\left(\begin{smallmatrix} l \\ \kappa \end{smallmatrix}\right) \\
&\quad + \tfrac{1}{2}\sum_{ll'}\sum_{\kappa\kappa'}\sum_{\alpha\beta}\sum_{i,i'=t,r} \phi_{\alpha\beta}^{ii'}\!\left(\begin{smallmatrix} l & l' \\ \kappa & \kappa' \end{smallmatrix}\right) u_\alpha^i\!\left(\begin{smallmatrix} l \\ \kappa \end{smallmatrix}\right) u_\beta^{i'}\!\left(\begin{smallmatrix} l' \\ \kappa' \end{smallmatrix}\right).
\end{aligned} \tag{2.187}
$$

Here $m_\kappa$ is the mass of the $\kappa$th unit and $\mathscr{I}_{\alpha\beta}(\kappa)$ is the $\alpha\beta$th element of its inertia tensor. The quantity $\phi_{\alpha\beta}^{ii'}$ is the force constant of the present problem.

The external force constants $\phi_{\alpha\beta}^{ii'}$ have many properties similar to the Born–von Kármán constants. These include

1. $\phi_{\alpha\beta}^{ii'}\!\left(\begin{smallmatrix} l & l' \\ \kappa & \kappa' \end{smallmatrix}\right) = \phi_{\beta\alpha}^{i'i}\!\left(\begin{smallmatrix} l' & l \\ \kappa' & \kappa \end{smallmatrix}\right);$ $\qquad\qquad\qquad$ (2.188)

2. $\phi_{\alpha\beta}^{ii'}\!\left(\begin{smallmatrix} l & l' \\ \kappa & \kappa' \end{smallmatrix}\right) = \phi_{\alpha\beta}^{ii'}\!\left(\begin{smallmatrix} l+L & l'+L \\ \kappa & \kappa' \end{smallmatrix}\right).$ $\qquad\qquad$ (2.189)

The transformation of the force constants under crystal point-group rotations exhibits a slight complication. Under $\mathscr{S}_m$, the displacements $\mathbf{u}^i\!\left(\begin{smallmatrix} l \\ \kappa \end{smallmatrix}\right)$ and $\mathbf{u}^{i'}\!\left(\begin{smallmatrix} l' \\ \kappa' \end{smallmatrix}\right)$ are carried over to $\left(\begin{smallmatrix} L \\ \kappa \end{smallmatrix}\right)$ and $\left(\begin{smallmatrix} L' \\ \kappa' \end{smallmatrix}\right)$ (say) as

$$
\begin{aligned}
\mathbf{u}^i\!\left(\begin{smallmatrix} l \\ \kappa \end{smallmatrix}\right) &\xrightarrow{\mathscr{S}_m} C^i(S)\,\mathbf{S}\,\mathbf{u}^i\!\left(\begin{smallmatrix} l \\ \kappa \end{smallmatrix}\right), \\
\mathbf{u}^{i'}\!\left(\begin{smallmatrix} l' \\ \kappa' \end{smallmatrix}\right) &\xrightarrow{\mathscr{S}_m} C^{i'}(S)\,\mathbf{S}\,\mathbf{u}^{i'}\!\left(\begin{smallmatrix} l' \\ \kappa' \end{smallmatrix}\right),
\end{aligned} \tag{2.190}
$$

---

*As in standard lattice dynamics, the harmonic approximation is not meaningful in crystals where molecular groups exhibit large-amplitude rotational motions. A notable example is solid hydrogen, which exhibits a free rotation spectrum. This requires quite different methods.[64] The same applies also to plastic crystals.

where

$$C^i(S) = 1 \qquad \text{for } i = t,$$
$$\phantom{C^i(S)} = \|S\| \qquad \text{for } i = r. \tag{2.191}$$

Result (2.190) incorporates the transformation law (2.186), and the resulting force-constant transformation is expressed by

$$\phi^{ii'}\begin{pmatrix} l & l' \\ \kappa & \kappa' \end{pmatrix} = C^i(S)\tilde{S}\phi^{ii'}\begin{pmatrix} L & L' \\ K & K' \end{pmatrix}SC^{i'}(S). \tag{2.192}$$

The equations of motion are easily deduced from (2.187). They are:[65]

$$m_\kappa \ddot{u}_\alpha^t\begin{pmatrix} l \\ \kappa \end{pmatrix} = -\left[ \sum_{l'}\sum_{\kappa'}\sum_\beta \phi_{\alpha\beta}^{tt}\begin{pmatrix} l & l' \\ \kappa & \kappa' \end{pmatrix}u_\beta^t\begin{pmatrix} l' \\ \kappa' \end{pmatrix} \right.$$
$$\left. + \sum_{l'}\sum_{\kappa'\in \text{II}}\sum_\beta \phi_{\alpha\beta}^{tr}\begin{pmatrix} l & l' \\ \kappa & \kappa' \end{pmatrix}u_\beta^r\begin{pmatrix} l' \\ \kappa' \end{pmatrix} \right] \qquad \text{for all } l, \kappa, \alpha; \tag{2.193a}$$

$$\mathscr{I}_{\alpha\alpha}(\kappa)\ddot{u}_\alpha^r\begin{pmatrix} l \\ \kappa \end{pmatrix} = -\left[ \sum_{l'}\sum_{\kappa'}\sum_\beta \phi_{\alpha\beta}^{rt}\begin{pmatrix} l & l' \\ \kappa & \kappa' \end{pmatrix}u_\beta^t\begin{pmatrix} l' \\ \kappa' \end{pmatrix} \right.$$
$$\left. + \sum_{l'}\sum_{\kappa'\in \text{II}}\sum_\beta \phi_{\alpha\beta}^{rr}\begin{pmatrix} l & l' \\ \kappa & \kappa' \end{pmatrix}u_\beta^r\begin{pmatrix} l' \\ \kappa' \end{pmatrix} + \sum_{\beta(\neq\alpha)} \mathscr{I}_{\alpha\beta}(\kappa)\ddot{u}_\beta^r\begin{pmatrix} l \\ \kappa \end{pmatrix} \right]$$
$$\text{for all } l, \alpha \text{ and } \kappa \in \text{II}. \tag{2.193b}$$

In the above equations, as well as in those to follow, the $\kappa'$ summation ranges over all sublattices unless specified as $\kappa' \in \text{II}$ in which case it is restricted to the range $(\mu + 1), \ldots, (\mu + v)$.

From (2.193) the translational and rotational sum rules may be deduced as before, and these are[65]

1. $\sum_{l'\kappa'} \phi_{\alpha\beta}^{tt}\begin{pmatrix} 0 & l' \\ \kappa & \kappa' \end{pmatrix} = 0 \qquad \text{for all } \alpha, \beta, \kappa,$

2. $\sum_{l'\kappa'} \phi_{\alpha\beta}^{rt}\begin{pmatrix} 0 & l' \\ \kappa & \kappa' \end{pmatrix} = 0 \qquad \text{for all } \alpha, \beta, \text{ and } \kappa \in \text{II},$

3. $\sum_{l'\kappa'}\sum_{\mu\nu} \phi_{\alpha\nu}^{tt}\begin{pmatrix} 0 & l' \\ \kappa & \kappa' \end{pmatrix}x_\mu\begin{pmatrix} l' \\ \kappa' \end{pmatrix}\varepsilon_{\beta\mu\nu} + \sum_{l'}\sum_{\kappa'\in\text{II}} \phi_{\alpha\beta}^{tr}\begin{pmatrix} 0 & l' \\ \kappa & \kappa' \end{pmatrix} = 0$

$$\text{for all } \alpha, \beta, \text{ and } \kappa,$$

4. $\sum_{l'\kappa'}\sum_{\mu\nu} \phi_{\alpha\nu}^{rt}\begin{pmatrix} 0 & l' \\ \kappa & \kappa' \end{pmatrix}x_\mu\begin{pmatrix} l' \\ \kappa' \end{pmatrix}\varepsilon_{\beta\mu\nu} + \sum_{l'}\sum_{\kappa'\in\text{II}} \phi_{\alpha\beta}^{rr}\begin{pmatrix} 0 & l' \\ \kappa & \kappa' \end{pmatrix} = 0$

$$\text{for all } \alpha, \beta, \text{ and } \kappa \in \text{II}. \tag{2.194}$$

To solve (2.193) we try the solution*

$$u_\alpha^i\begin{pmatrix} l \\ \kappa \end{pmatrix} = U_\alpha^i(\kappa|\mathbf{q})\exp\{i[\mathbf{q}\cdot\mathbf{x}(l) - \omega(\mathbf{q})t]\}, \tag{2.195}$$

---

*Observe that as per our convention we use $\mathbf{U}$ for the amplitude vector. See remarks in Sec. 2.2.14.

which then gives

$$\omega^2(\mathbf{q})m_\kappa U_\alpha^t(\kappa|\mathbf{q}) = \sum_{\kappa'}\sum_\beta B_{\alpha\beta}^{tt}\left(\begin{matrix}\mathbf{q}\\ \kappa\kappa'\end{matrix}\right)U_\beta^t(\kappa'|\mathbf{q})$$

$$+ \sum_{\kappa'\in\mathrm{II}}\sum_\beta B_{\alpha\beta}^{tr}\left(\begin{matrix}\mathbf{q}\\ \kappa\kappa'\end{matrix}\right)U_\beta^r(\kappa'|\mathbf{q}) \qquad \text{for all } \alpha \text{ and } \kappa;$$

$$(2.196a)$$

$$\omega^2(\mathbf{q})\mathscr{I}_{\alpha\alpha}(\kappa)U_\alpha^r(\kappa|\mathbf{q}) = \sum_{\kappa'}\sum_\beta B_{\alpha\beta}^{rt}\left(\begin{matrix}\mathbf{q}\\ \kappa\kappa'\end{matrix}\right)U_\beta^t(\kappa'|\mathbf{q})$$

$$+ \sum_{\kappa'\in\mathrm{II}}\sum_\beta B_{\alpha\beta}^{rr}\left(\begin{matrix}\mathbf{q}\\ \kappa\kappa'\end{matrix}\right)U_\beta^r(\kappa'|\mathbf{q})$$

$$- \omega^2(\mathbf{q})\sum_{\beta\neq\alpha}\mathscr{I}_{\alpha\beta}(\kappa)U_\beta^r(\kappa|\mathbf{q})$$

$$\text{for } \kappa\in\mathrm{II} \text{ and all } \alpha. \qquad (2.196b)$$

The dynamical matrix element $B_{\alpha\beta}^{ii'}\left(\begin{smallmatrix}\mathbf{q}\\ \kappa\kappa'\end{smallmatrix}\right)$ above is defined by

$$B_{\alpha\beta}^{ii'}\left(\begin{matrix}\mathbf{q}\\ \kappa\kappa'\end{matrix}\right) = \sum_{l'}\phi_{\alpha\beta}^{ii'}\left(\begin{matrix}0 & l'\\ \kappa & \kappa'\end{matrix}\right)\exp\left[i\mathbf{q}\cdot\mathbf{x}(l')\right]. \qquad (2.197)$$

In matrix notation (2.196) reads

$$\omega^2(\mathbf{q})\mathbf{m}\mathbf{U}(\mathbf{q}) = \mathbf{B}(\mathbf{q})\mathbf{U}(\mathbf{q}), \qquad (2.196c)$$

where $\mathbf{m}$ and $\mathbf{B}(\mathbf{q})$ are $(3\mu + 6\nu)$-dimensional matrices and $\mathbf{U}(\mathbf{q})$ is a $(3\mu + 6\nu)$-component column matrix. The elements of $\mathbf{B}(\mathbf{q})$ and $\mathbf{U}(\mathbf{q})$ have already been specified while those of $\mathbf{m}$ are given by

$$m_{\alpha\beta}^{ii'}(\kappa\kappa') = m_\kappa\delta_{\alpha\beta}\delta_{\kappa\kappa'}\delta_{ii'} \qquad \text{for } i = t$$
$$= \mathscr{I}_{\alpha\beta}(\kappa)\delta_{\kappa\kappa'}\delta_{ii'} \qquad \text{for } i = r. \qquad (2.198)$$

The eigenvalues $\omega_j^2(\mathbf{q})$, $j = 1, \ldots, (3\mu + 6\nu)$, are obtained by solving the secular equation

$$\|\mathbf{B}(\mathbf{q}) - \omega^2(\mathbf{q})\mathbf{m}\| = 0. \qquad (2.199)$$

The normalization and the orthogonality relations for the eigenvectors follow the pattern of the generalized eigenvalue problem discussed in Sec. 2.2.14.

We introduce $e_\alpha^i(\kappa|_j^{\mathbf{q}})$ via the relation

$$U_\alpha^i\left(\kappa\left|\begin{matrix}\mathbf{q}\\ j\end{matrix}\right.\right) = A\left(\begin{matrix}\mathbf{q}\\ j\end{matrix}\right)e_\alpha^i\left(\kappa\left|\begin{matrix}\mathbf{q}\\ j\end{matrix}\right.\right) \qquad (2.200)$$

such that $A$ has the dimensions of length. Note that $\mathbf{e}^t$ is dimensionless while $\mathbf{e}^r$ has dimensions $L^{-1}$. The orthonormality requirement is then specified by

$$\mathbf{e}^\dagger\begin{pmatrix}\mathbf{q}\\j\end{pmatrix}\mathbf{me}\begin{pmatrix}\mathbf{q}\\j'\end{pmatrix} = M\delta_{jj'}, \tag{2.201}$$

where $M$ is the mass of the primitive cell.

Having solved the eigenvalue problem for one $\mathbf{q}$, the determination of the allowed $\mathbf{q}$ values, the superposition of their respective solutions and the transition to quantum mechanics follows as before.

One important feature of the present treatment is that the resulting normal modes in general are not entirely translational or rotational. In other words, the mixing produced by the rotational-translational coefficients $\phi_{\alpha\beta}^{tr}$ and $\phi_{\alpha\beta}^{rt}$ permits both $e_\alpha^t(\kappa|\substack{\mathbf{q}\\j})$ and $e_\alpha^r(\kappa|\substack{\mathbf{q}\\j})$ to be nonvanishing for a given $(\mathbf{q}, j)$. The mode thus involves small amplitude linear oscillations as well as small amplitude angular oscillations. The latter are often referred to as *torsional oscillations* or *librations*.

The frequency distribution is defined as before with appropriate modifications. We have

$$g(\omega) = \frac{1}{N(3\mu + 6\nu)}\sum_{\mathbf{q}j}\delta\{\omega - \omega_j(\mathbf{q})\}, \tag{2.202}$$

with the usual normalization (2.91). For many purposes, it is convenient to think of $g(\omega)$ as made up of translational and librational parts. Such a separation can be achieved by the following definitions:[65]

$$g^t(\omega) = \frac{1}{MN(3\mu + 6\nu)}\sum_{\mathbf{q}j}\left\{\sum_{\alpha\kappa}e_\alpha^{t*}\left(\kappa\Big|\substack{\mathbf{q}\\j}\right)m_\kappa e_\alpha^t\left(\kappa\Big|\substack{\mathbf{q}\\j}\right)\right\}\delta\{\omega - \omega_j(\mathbf{q})\}, \tag{2.203a}$$

$$g^r(\omega) = \frac{1}{MN(3\mu + 6\nu)}\sum_{\mathbf{q}j}\left\{\sum_{\alpha,\beta,\kappa\in\mathrm{II}}e_\alpha^{r*}\left(\kappa\Big|\substack{\mathbf{q}\\j}\right)\mathcal{I}_{\alpha\beta}(\kappa)e_\beta^r\left(\kappa\Big|\substack{\mathbf{q}\\j}\right)\right\}$$
$$\times \delta\{\omega - \omega_j(\mathbf{q})\}. \tag{2.203b}$$

The "orthonormality" condition (2.201) ensures

$$g(\omega) = g^t(\omega) + g^r(\omega). \tag{2.204}$$

The dynamics of the external modes can also be formulated in the principal-axes frame. One advantage of doing so is that the inertia tensor $\mathcal{I}^P(\kappa)$ (the superscript makes the frame explicit) now has a diagonal form, that is, $\mathcal{I}_{\alpha\beta}^P(\kappa) = \mathcal{I}_{\alpha\alpha}^P(\kappa)\delta_{\alpha\beta}$.

Let $\mathbf{A}(\kappa)$ be the orthogonal transformation which carries a set of axes associated with $\kappa$ from an orientation parallel to the fixed frame to the principal axes of the unit concerned. For $\kappa \in I$, we make the choice $\mathbf{A}(\kappa) = \mathbb{1}_3$ as remarked earlier. For $\kappa \in II$, the transformation matrix is specified in terms of Eulerian angles.[66] In terms of $\mathbf{A}(\kappa)$, the new displacement coordinates are

$$\mathbf{u}^{i,P}\begin{pmatrix} l \\ \kappa \end{pmatrix} = \mathbf{A}(\kappa)\mathbf{u}^i\begin{pmatrix} l \\ \kappa \end{pmatrix}, \tag{2.205}$$

and the energies and the equations of motion must be expressed in terms of these. The eigenvalue problem then becomes

$$\mathbf{B}^P(\mathbf{q})\mathbf{U}^P(\mathbf{q}) = \omega^2(\mathbf{q})\mathbf{m}^P\mathbf{U}^P(\mathbf{q}), \tag{2.206}$$

where

$$\begin{aligned}
\mathbf{B}^{ii',P}\begin{pmatrix} \mathbf{q} \\ \kappa\kappa' \end{pmatrix} &= \mathbf{A}(\kappa)\mathbf{B}^{ii'}\begin{pmatrix} \mathbf{q} \\ \kappa\kappa' \end{pmatrix}\tilde{\mathbf{A}}(\kappa') \\
&= \sum_{l'}\left[\mathbf{A}(\kappa)\boldsymbol{\phi}^{ii'}\begin{pmatrix} 0 & l' \\ \kappa & \kappa' \end{pmatrix}\tilde{\mathbf{A}}(\kappa')\right]\exp\{i\mathbf{q}\cdot\mathbf{x}(l')\},
\end{aligned} \tag{2.207}$$

and

$$\begin{aligned}
m_{\alpha\beta}^{ii',P}(\kappa\,\kappa') &= m_\kappa\delta_{\alpha\beta}\delta_{\kappa\kappa'}\delta_{ii'} && \text{for } i = t, \\
&= \mathscr{I}_{\alpha\alpha}^P(\kappa)\delta_{\alpha\beta}\delta_{\kappa\kappa'}\delta_{ii'} && \text{for } i = r.
\end{aligned} \tag{2.208}$$

The quantity in the square brackets in (2.207) is the modified force-constant tensor $\boldsymbol{\phi}^{ii',P}(\begin{smallmatrix} 0 & l' \\ \kappa & \kappa' \end{smallmatrix})$. Equation (2.206) delivers the same eigenvalues as (2.196), since the matrices in the two equations are related by a similarity transformation.

## 2.7.2 Internal vibrations

In discussing the internal branches, it is necessary to assume that (1) the isolated atoms in the crystal are stationary, and (2) the molecular groups do not rotate or translate as a whole. Vibrational energy is thus assumed to be transmitted through the crystal entirely by waves built up from the internal vibrations of the individual molecules through intermolecular interactions. To illustrate the point, consider Table 2.3 which displays the vibrational frequencies of an isolated $NH_4^+$ ion[66] and the $\mathbf{q} = 0$ frequencies of the internal branches measured by optical techniques.[67,68] It is clear from the close correspondence of the two sets of frequencies that the internal branches are primarily built up from the coupling of the corresponding vibrations of the various $NH_4^+$ ions. For example, the non-degenerate branch with limiting frequency 3048 cm$^{-1}$ (see Figure 2.14) can be visualized as arising from the coupling of the Einstein oscillators of frequency 3033 cm$^{-1}$. This is analogous to the formation of the $3s$ band in

**Table 2.3**
Frequencies of the $NH_4^+$ ion.

| Vibration | Isolated ion[a] cm$^{-1}$ | In crystal cm$^{-1}$ |
|---|---|---|
| $\nu_1$ (nondegenerate) | 3033 | 3048[b] |
| $\nu_2$ (doubly degenerate) | 1684 | 1683[c] |
| $\nu_3$ (triply degenerate) | 3134 | 3086[c] |
| $\nu_4$ (triply degenerate) | 1397 | 1402[c] |

[a]From ref. 66.
[b]From ref. 67.
[c]From ref. 68.

crystalline sodium due to the coupling of the $3s$ levels of the isolated sodium atoms.[69] Of course, the assumption that each internal branch has a unique parentage in terms of the normal modes of the isolated molecule is an oversimplification. In situations where there are closely lying frequencies in the single molecule, mode mixing may occur in the formation of the internal branches. In the most general situation, every internal branch would be a hybrid of all the unperturbed vibrations.

In general, if $n_\kappa$ denotes the number of atoms in the $\kappa$th molecular group (assumed to be nonlinear), then there will in all be $\sum_{\kappa \text{ell}} (3n_\kappa - 6)$ internal branches, all of which are optical branches. The total number of branches must equal $3n$ so that

$$(3\mu + 6\nu) + \sum_{\kappa \text{ell}} (3n_\kappa - 6) = 3n, \qquad (2.209)$$

where $n$ as before denotes the total number of atoms in the cell.

To bring out some of the physical features discussed above, it is helpful to consider the problem of a one-dimensional chain of identical (nonlinear) molecules, each molecule containing $p$ atoms. The equations of motion as usual are

$$m_k \ddot{u}_\alpha \binom{l}{k} = - \sum_{l',k',\beta} \phi_{\alpha\beta}\binom{l \ \ l'}{k \ \ k'} u_\beta \binom{l'}{k'} \qquad (2.210)$$

which, upon assuming wavelike solutions, take the form

$$\omega^2(q) m_k U_\alpha(k|q) = \sum_{k',\beta} B_{\alpha\beta}\binom{q}{kk'} U_\beta(k'|q), \qquad (\alpha = x, y, z; \ k = 1, \ldots, p),$$
$$(2.211)$$

where

$$B_{\alpha\beta}\binom{q}{kk'} = \sum_{l'} \phi_{\alpha\beta}\binom{0 \ \ l'}{k \ \ k'} e^{iqx_z(l')}. \qquad (2.212)$$

In the above, $k$ labels the atoms within a molecule and $q$ is a wave vector directed along the chain axis (taken here as the $z$ axis).

As it stands, Eq. (2.210) is quite general. We now wish to solve it under the assumption that the atomic displacements are linear combinations of those associated with the internal motions of the molecule. In the present case, each molecule has $(3p - 6)$ internal vibration modes. We may therefore expand the displacement amplitudes $U_\alpha(k|q)$ in the following form:

$$U_\alpha(k|q) = \sum_{\xi=1}^{(3p-6)} a_\xi(q)e_\alpha(k|\xi), \qquad (2.213)$$

where the $e_\alpha(k|\xi)$'s denote the eigenvector components associated with the $\xi$th normal mode of the isolated molecule and corresponding to one of the internal vibrations. These eigenvector components are found from the eigenvalue equation for a free molecule:

$$\omega_\xi^2 m_k e_\alpha(k|\xi) = \sum_{k'\beta} \phi_{\alpha\beta}\begin{pmatrix} 0 & 0 \\ k & k' \end{pmatrix} e_\beta(k'|\xi). \qquad (2.214)$$

Equation (2.213) thus allows for mixing between all the internal vibrations, brought about by intermolecular forces.

Next we substitute (2.213) in (2.211), multiply by $e_\alpha^*(k|\xi)$, and sum over $(\alpha, k)$. Remembering

$$\sum_{\alpha k} e_\alpha^*(k|\xi)m_k e_\alpha(k|\xi') = M\delta_{\xi\xi'},$$

where $M$ is the mass of the molecule (see (2.85)), we find

$$\sum_{\xi'=1}^{(3p-6)} [\mathscr{B}_{\xi\xi'}(q) - \delta_{\xi\xi'}\omega_j^2(q)M]a_{j\xi'}(q) = 0, \qquad (2.215)$$

where

$$\mathscr{B}_{\xi\xi'}(q) = \sum_{\alpha\beta kk'} e_\alpha^*(k|\xi)B_{\alpha\beta}\begin{pmatrix} q \\ kk' \end{pmatrix} e_\beta(k'|\xi'). \qquad (2.216)$$

Requirement of nontrivial solutions for the coupling coefficients $a_{\xi'}(q)$ then leads to the secular determinant

$$\|\mathscr{B}_{\xi\xi'}(q) - \delta_{\xi\xi'}\omega^2(q)M\| = 0 \qquad (2.217)$$

which delivers the dispersion curves for the internal branches.

If a given internal branch $j$ is entirely built up from a particular vibration $\xi$, then assuming there is no degeneracy we may write

$$U_\alpha\left(k\Big|{q\atop j}\right) = a_{j\xi}(q)e_\alpha(k|\xi).$$  (2.218)

Equation (2.217) then reduces to

$$\begin{aligned}
M\omega^2(q) = \mathcal{B}_{\xi\xi}(q) &= \sum_{\alpha\beta kk'} e_\alpha^*(k|\xi)B_{\alpha\beta}\left({q\atop kk'}\right)e_\beta(k'|\xi) \\
&= \sum_{\alpha\beta kk'} e_\alpha^*(k|\xi)\phi_{\alpha\beta}\left({0\ \ 0\atop k\ \ k'}\right)e_\beta(k'|\xi) \\
&\quad + \sum_{\alpha\beta kk'} e_\alpha^*(k|\xi)\left[\sum_{l'}{}' \phi_{\alpha\beta}\left({0\ \ l'\atop k\ \ k'}\right)e^{iqx_z(l')}\right]e_\beta(k'|\xi) \\
&= M\omega_\xi^2 + \sum_{\alpha\beta kk'} e_\alpha^*(k|\xi)\left[\sum_{l'}{}' \phi_{\alpha\beta}\left({0\ \ l'\atop k\ \ k'}\right)e^{iqx_z(l')}\right]e_\beta(k'|\xi).
\end{aligned}$$  (2.219)

This equation shows explicitly how the intermolecular coupling modifies the frequency $\omega_\xi$ of the isolated molecule. Depending on the situation, one could go a little beyond (2.218) and allow for mixing among a select subset of normal modes having frequencies close to each other. Equation (2.213) represents the most general situation where mixing occurs among the entire set of vibrations of the isolated molecule. The reader familiar with band theory may have noticed that Eq. (2.213) is written in the same spirit as one writes the wave function in the tight-binding approximation, that is, as a linear combination of the atomic orbitals of the constituent atoms of the crystal.[70]

An interesting feature occurs when there are several identical molecular groups in the primitive cell. Suppose, for example, we have two identical molecules per cell in our chain. Consider again the branch built up from the nondegenerate frequency $\omega_\xi$. In this case, Eq. (2.219) is modified to

$$\left\|\begin{matrix} \mathcal{B}_{\xi_1\xi_1} - \omega^2 2M & \mathcal{B}_{\xi_1\xi_2} \\ \mathcal{B}_{\xi_2\xi_1} & \mathcal{B}_{\xi_2\xi_2} - \omega^2 2M \end{matrix}\right\| = 0,$$  (2.220)

where $\xi_1$ and $\xi_2$ denote explicitly the normal modes of the two molecules. By assumption $\omega_{\xi_1}^2 = \omega_{\xi_2}^2 = \omega_\xi^2$. From (2.214) we see that *two* branches are built up from a single frequency $\omega_\xi$. These are usually close to each other, and the phenomenon is sometimes referred to as *Davydov splitting*.[71] Extending the argument, if there are $\sigma$ identical molecules in the cell, then each nondegenerate vibrational frequency of the isolated molecule gives rise to $\sigma$ corresponding branches in the crystal.

The generalization of the above treatment to complex crystals of the type considered earlier is straightforward. We assume that only the atoms in the molecular groups move, and write their equations of motion as

$$m_k \ddot{u}_\alpha \begin{pmatrix} l \\ \kappa \\ k \end{pmatrix} = - \sum_{\substack{k' \in \kappa' \\ \kappa' \in \text{II} \\ \beta, l'}} \phi_{\alpha\beta} \begin{pmatrix} l & l' \\ \kappa & \kappa' \\ k & k' \end{pmatrix} u_\beta \begin{pmatrix} l' \\ \kappa' \\ k' \end{pmatrix}, \qquad (\alpha = x, y, z; \ k \in \kappa; \ \kappa \in \text{II}).$$

$$(2.221)$$

The $\mathbf{q}$-space version is

$$m_k \omega^2(\mathbf{q}) U_\alpha(k\kappa|\mathbf{q}) = \sum_{\substack{k' \in \kappa' \\ \kappa' \in \text{II} \\ \beta}} B_{\alpha\beta} \begin{pmatrix} \mathbf{q} \\ \kappa\kappa' \\ kk' \end{pmatrix} U_\beta(k'\kappa'|\mathbf{q}). \qquad (2.222)$$

After performing an expansion analogous to (2.213) we eventually get

$$\| \mathscr{B}_{\xi\xi'}(\mathbf{q}) - \omega^2(\mathbf{q}) M \delta_{\xi\xi'} \| = 0, \qquad (2.223)$$

where $M$ now denotes the mass of the primitive cell, and

$$\mathscr{B}_{\xi\xi'}(\mathbf{q}) = \sum_{\kappa\kappa'} \sum_{\substack{k \in \kappa \\ k' \in \kappa'}} \sum_{\alpha\beta} e_\alpha^*(k\kappa|\xi) B_{\alpha\beta} \begin{pmatrix} \mathbf{q} \\ \kappa\kappa' \\ kk' \end{pmatrix} e_\beta(k'\kappa'|\xi'). \qquad (2.224)$$

The present treatment differs from conventional lattice dynamics firstly in that (2.221) permits interactions only between *molecular groups* and secondly that $\mathbf{U}$ is expanded in a restricted manner in terms of the eigenvectors of the isolated molecules. These constraints enable the focusing of explicit attention on the internal branches.

## 2.8 Force constants in some special situations

Thus far we have treated the force constants as formal quantities subject only to the requirements of crystal symmetry and the translational and rotational sum rules. For numerical calculations of the dispersion curves, it is necessary to know their values. Several approaches are possible in obtaining these. One is to treat the Born–von Kármán constants as parameters and determine their values from experimental data, for example, elastic constants (see Sec. 2.4). Force constants so determined subject only to the constraints mentioned above are referred to as *tensor* force constants.

In some cases, however, a knowledge of the binding may be employed to construct $\Phi$, leading to special forms for the force constants. We shall consider a few examples.

### 2.8.1 Axially symmetric and central force constants

Frequently one assumes that the forces between a given pair of atoms $\left(\begin{smallmatrix} l & l' \\ k & k' \end{smallmatrix}\right)$ are derivable from a two-body potential $V_{kk'}(r)$ which depends only on the magnitude of the separation between the atoms. In this case,[72]

$$\phi_{\alpha\beta}\begin{pmatrix} l & l' \\ k & k' \end{pmatrix} = -\frac{\partial^2 V_{kk'}(r)}{\partial r_\alpha \partial r_\beta}\bigg|_0 \qquad (2.225a)$$

$$= \left[-\frac{1}{r}V'_{kk'}(r)\left\{\delta_{\alpha\beta} - \frac{r_\alpha r_\beta}{r^2}\right\} - \frac{r_\alpha r_\beta}{r^2}V''_{kk'}(r)\right]_{r=|\mathbf{x}(\begin{smallmatrix} l & l' \\ k & k' \end{smallmatrix})|}. \qquad (2.225b)$$

We observe that the force constants are characterized by two parameters, namely, the first and second derivatives of the pair potential, even though symmetry considerations based on (2.27) may allow more parameters. If the potential is completely known then so are the force constants. Otherwise the two derivatives may be treated as parameters to be determined from experiment. In passing we observe that the two terms in (2.225b) represent respectively the contributions associated with the tangential and radial forces between the concerned pair of atoms.

Suppose now that *all* atomic pairs in the crystal interact via two-body scalar potentials. Then

$$\Phi = \frac{1}{2}\sum_{kk'}\sum_{ll'} V_{kk'}\left(\left|\mathbf{r}\begin{pmatrix} l' \\ k' \end{pmatrix} - \mathbf{r}\begin{pmatrix} l \\ k \end{pmatrix}\right|\right), \qquad (lk \neq l'k')$$

$$= \frac{N}{2}\sum_{kk'}\sum_{l'} V_{kk'}(y) \qquad \text{(for a cyclic crystal; } y \neq 0), \qquad (2.226)$$

where $y = |\mathbf{r}(\begin{smallmatrix} l' \\ k' \end{smallmatrix}) - \mathbf{r}(\begin{smallmatrix} 0 \\ k \end{smallmatrix})|$. In this case all force constants can be obtained via (2.225). Furthermore, important constraints can be derived by examining the equilibrium condition of the crystal. This is specified by the minimum of $\Phi$ with respect to the structural constants and may be deduced as follows: Let $\mathbf{a}_1, \mathbf{a}_2, \mathbf{a}_3$ denote the basis vectors and $\alpha, \beta, \gamma$ the angles between $(\mathbf{a}_2, \mathbf{a}_3)$, $(\mathbf{a}_3, \mathbf{a}_1)$ and $(\mathbf{a}_1, \mathbf{a}_2)$, respectively. Then $y$ can be expressed as

$$y = |[l'_1 + \lambda(k') - \lambda(k)]\mathbf{a}_1 + [l'_2 + \mu(k') - \mu(k)]\mathbf{a}_2$$
$$+ [l'_3 + \nu(k') - \nu(k)]\mathbf{a}_3|, \qquad (2.227)$$

where $l'_1, l'_2, l'_3$ are integers and $\lambda(k), \ldots,$ are fractions. The $(3n + 6)$ quantities, namely, $a_1, a_2, a_3, \alpha, \beta, \gamma$ and $\lambda(k), \mu(k), \nu(k)$ $(k = 1, \ldots . n)$, are now regarded as variational parameters,* and the equilibrium con-

*Actually there are only $(3n + 3)$ independent variational parameters since $\lambda(1), \mu(1), \nu(1)$ depend on the choice of the origin, which is arbitrary.

figuration is specified by the $(3n + 6)$ equations

$$\left(\frac{\partial \Phi}{\partial p_i}\right)_{\text{eqbr.}} = 0, \qquad (i = 1, \ldots, 3n + 6),$$

where $p_i$ denotes the $i$th variational parameter. Remembering (2.226), these equations can be written as

$$\sum_{kk'} \sum_{l'} \left[\frac{dV_{kk'}(y)}{dy} \cdot \frac{\partial y}{\partial p_i}\right]_{\text{eqbr.}} = 0. \tag{2.228}$$

There are thus $(3n + 3)$ constraints in the form of sum rules involving the first derivatives of $V$. A simple application of (2.228) is to the fcc lattice leading to the result[73]

$$6\sqrt{2}\left(\frac{dV}{dr}\right)_1 + 6\left(\frac{dV}{dr}\right)_2 + 12\sqrt{6}\left(\frac{dV}{dr}\right)_3 + \cdots = 0, \tag{2.229}$$

where the subscripts indicate the neighbors at which the derivatives must be evaluated.

One clarifying remark must be made concerning Eq. (2.228). Suppose that the two-body potentials are known for all values of $r$ and for all pairs $(k\,k')$ (for example, they could be LJ-type potentials). Equations (2.228) can then be used to determine the equilibrium configuration. The force constants can subsequently be obtained by evaluating the appropriate second derivatives of $\Phi$ in this configuration. In this case, no parameters are involved. On the other hand, one could have the situation in which the structure is known and $\Phi$ is expressible as in (2.226), but the two-body potentials are not completely known. In such a context parameterization becomes necessary, and (2.228) leads to useful constraints on the parameters.

A comment is also necessary concerning the nomenclature. According to currently accepted usage, if one employs (2.225) without insisting that the crystal is held together entirely by potentials of the form $V_{kk'}(r)$, then the potentials $V$ are said to be *axially symmetric*,[74] and the corresponding constants, axially symmetric force constants. If, however, the crystal is held together entirely by two-body potentials, then the force constants are said to be *central force constants*.

In the case of both central and axially symmetric forces, the dynamical matrix (2.86) can be expressed entirely as a reciprocal-space sum as follows. Define first $V_{kk'}(Q)$, the Fourier transform of $V_{kk'}(r)$, through

$$V_{kk'}(Q) = \int V_{kk'}(r)e^{-i\mathbf{Q}\cdot\mathbf{r}}\, d\mathbf{r}. \tag{2.230}$$

The inverse relation is (see Eq. (A1.12))

$$V_{kk'}(r) = \frac{1}{(2\pi)^3} \int V_{kk'}(Q) e^{iQ \cdot r} \, dQ.$$                    (2.231)

Remembering (2.225a) and (A1.10), we can write Eq. (2.86) as

$$M_{\alpha\beta}\binom{q}{kk'} = \frac{1}{v}\left\{ \sum_{G} (q + G)_\alpha (q + G)_\beta V_{kk'}(|q + G|) e^{iG \cdot [x(k) - x(k')]} \right.$$
$$\left. - \delta_{kk'} \sum_{k''} \sum_{G} G_\alpha G_\beta V_{kk''}(G) e^{iG \cdot [x(k) - x(k'')]} \right\},$$                    (2.232)

a form that is sometimes found useful.

### 2.8.2 Valence force constants

We have previously noted that in covalently bonded crystals it is convenient to describe distortions in terms of molecular-type deformation coordinates such as bending and stretching. Such coordinates are referred to as *valence coordinates*, and among those commonly employed are the following:

1. Bond stretching—a change in the distance between a pair of atoms along the line joining the pair.
2. Bond bending—a change in the angle between two bonds with a common atom.
3. Torsion—a change in the dihedral angle involved in a sequence of four atoms as illustrated in Figure 2.15.

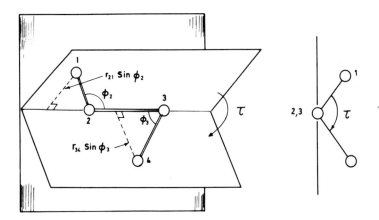

**Figure 2.15**
1–4 is a sequence of four atoms bonded as shown. *Torsion* is the name applied to the change in dihedral angle between planes (123) and (234).

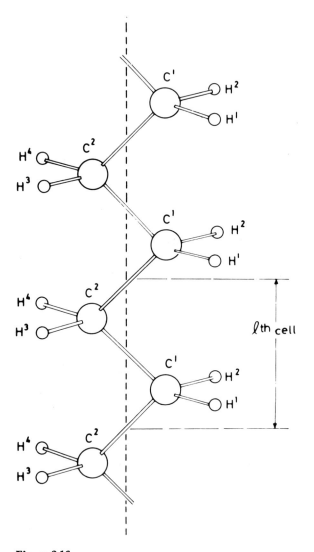

**Figure 2.16**
Portion of a polyethylene chain. Observe that bending of $C^1$-$C^2$-$C^1$ angle involves the motion of atoms in adjacent cells.

Other valence coordinates sometimes employed are out-of-plane bending, scissoring, rocking, wagging, and twisting.[75,76]

When the harmonic part of the deformation energy is expressed in terms of valence coordinates, the force constants that appear are the *valence force constants*. We now wish to discuss the connection between these and the more familiar Born–von Kármán constants.

Let $S_i(l)$ denote the $i$th valence coordinate associated with the cell $l$, and let a total of $\mu$ coordinates be associated with the cell. The number $\mu$ must be at least equal to $3n$ so as to bestow the right number of degrees of freedom, but it can also be larger, in which case it is possible to discover redundancies reducing the number of independent coordinates to $3n$. It merits pointing out that even though a valence coordinate is assigned formally to a particular cell, it can involve the motion of atoms in several cells. An example is the bending of the $C^1$–$C^2$–$C^1$ angle in polyethylene (see Figure 2.16) which clearly involves the motion of carbon atoms in adjacent cells.

In terms of the valence coordinates (which, note, are real), the harmonic part of the potential energy can be written

$$\Phi^{(2)} = \tfrac{1}{2} \sum_{ll'} \sum_{ij} S_i(l) F_{ij}(l, l') S_j(l'). \tag{2.233}$$

The quantities $F_{ij}(l, l')$ denote the valence force constants and are analogous to $\phi_{\alpha\beta}\binom{l\ l'}{k\ k'}$.

The relationship between the valence and the Cartesian displacement coordinates can be expressed as

$$S_i(l) = \sum_{l'k'\beta} B_{i,\beta k'}(l, l') u_\beta \binom{l'}{k'}. \tag{2.234}$$

The coefficients $B_{i,\beta k'}(l, l')$ may be evaluated by noting that to first order in Cartesian displacements

$$S_i(l) = \sum_{l'k'\beta} \left[ \partial S_i(l) / \partial u_\beta \binom{l'}{k'} \right]_0 u_\beta \binom{l'}{k'}. \tag{2.235}$$

Comparing the last two equations we get

$$B_{i,\beta k'}(l, l') = \left[ \partial S_i(l) / \partial u_\beta \binom{l'}{k'} \right]_0. \tag{2.236}$$

In other words, the elements $B_{i,\beta k'}(l, l')$ ($\beta = x, y, z$; all other indices fixed) are merely the components of the gradient $S_i(l)$ with respect to $\mathbf{u}\binom{l'}{k'}$. Recalling the definition of the gradient, it is seen that the direction of grad $S_i(l)$ is that in which a given displacement of $\binom{l'}{k'}$ produces the greatest increase in $S_i(l)$. Further, grad $S_i(l)$ is equal to the increase in $S_i(l)$, produced by unit displacement of $\binom{l'}{k'}$ in the most effective direction. These considerations enable the systematic determination of the coefficients in (2.234).[75]

Returning to the problem of relating the two types of force constants,

$$\Phi^{(2)} = \tfrac{1}{2} \sum_{ll'} \sum_{ij} S_i(l) F_{ij}(l, l') S_j(l')$$

$$= \tfrac{1}{2} \sum_{ll'} \sum_{ij} \sum_{\alpha\beta} \sum_{kk'} F_{ij}(l, l') \left[ \sum_{\lambda} B_{i,\alpha k}(l, \lambda) u_\alpha \binom{\lambda}{k} \right]\left[ \sum_{\lambda'} B_{j,\beta k}(l', \lambda') u_\beta \binom{\lambda'}{k'} \right]$$

$$= \tfrac{1}{2} \sum_{\lambda\lambda'} \sum_{\alpha\beta} \sum_{kk'} u_\alpha \binom{\lambda}{k} \left[ \sum_{ll'} \sum_{ij} B_{i,\alpha k}(l, \lambda) F_{ij}(l, l') B_{j,\beta k'}(l', \lambda') \right] u_\beta \binom{\lambda'}{k'}.$$

Comparing this with the more familiar expression

$$\Phi^{(2)} = \tfrac{1}{2} \sum_{\lambda\lambda'} \sum_{kk'} \sum_{\alpha\beta} u_\alpha \binom{\lambda}{k} \phi_{\alpha\beta}\binom{\lambda\;\lambda'}{k\;k'} u_\beta \binom{\lambda'}{k'},$$

we obtain

$$\phi_{\alpha\beta}\binom{\lambda\;\lambda'}{k\;k'} = \sum_{ll'} \sum_{ij} B_{i,\alpha k}(l, \lambda) F_{ij}(l, l') B_{j,\beta k'}(l', \lambda'). \tag{2.237}$$

By virtue of the translational symmetry of the crystal, the valence constants exhibit the obvious property

$$F_{ij}(l, l') = F_{ij}(l - l', 0) = F_{ij}(0, l' - l). \tag{2.238}$$

Further, under space group operations, every bond in the (equilibrium) crystal is uniquely transformed into some other equivalent bond. Correspondingly, every $S_i(l)$ is uniquely mapped to some $S_I(L)$ and every $S_j(l')$ to some $S_J(L')$. This leads to the symmetry

$$F_{ij}(l, l') \xrightarrow{\mathscr{S}_m} F_{IJ}(L, L'). \tag{2.239}$$

It is interesting to observe that the invariance of the infinitesimal translation or rotation does not place any restriction on the valence force constants. This follows from the fact that $S_i(l) = 0$ for both uniform translation and for rotation, whence $\Phi^{(2)} = 0$ for such operations, quite independently of the values of the valence constants. In spite of this apparent arbitrariness of the valence constants, the correct sum rules can be recovered on transforming to the Born–von Kármán constants, since the necessary constraints are built into the transformation (2.234). To illustrate, consider

$$\sum_{\lambda'k'} \phi_{\alpha\beta}\binom{\lambda\;\lambda'}{k\;k'} = \sum_{ll'} \sum_{ij} B_{i,\alpha k}(l, \lambda) F_{ij}(l, l') \left[ \sum_{\lambda'k'} B_{j,\beta k'}(l', \lambda') \right].$$

Assuming the crystal to be given a uniform translation so that $u_\alpha\binom{l}{k} = \varepsilon_\beta \delta_{\alpha\beta}$ for all $\binom{l}{k}$, we get from (2.234)

$$\sum_{l'k'} B_{i,\beta k'}(l, l') = 0,$$

remembering that $S_i(l) = 0$ in a uniform translation. Combining these two equations then leads to the familiar translational sum rule.

The explicit determination of the coefficients in (2.234) for various types of valence coordinates is well discussed in the molecular literature.[75] We quote below some of the important results.

**Bond stretching.** Let $\Delta r(12)$ denote the stretching coordinate associated with the bond $(1, 2)$ illustrated in Figure 2.17a. In the spirit of (2.234) we can write

$$\Delta r(12) = \mathbf{B}_{\Delta r,1} \cdot \mathbf{u}(1) + \mathbf{B}_{\Delta r,2} \cdot \mathbf{u}(2). \tag{2.240}$$

Application of the rules mentioned earlier shows that

$$\mathbf{B}_{\Delta r,1} = \hat{\mathbf{e}}_{21}, \qquad \mathbf{B}_{\Delta r,2} = \hat{\mathbf{e}}_{12}, \tag{2.241}$$

$\hat{\mathbf{e}}_{ab}$ being a unit vector directed from $a$ to $b$.

**Bond bending.** Let $\Delta\phi$ denote the change in the bond angle $\phi$ shown in Figure 2.17b. As earlier,

$$\Delta\phi = \sum_{i=1,2,3} \mathbf{B}_{\Delta\phi,i} \cdot \mathbf{u}(i), \tag{2.242}$$

(a)

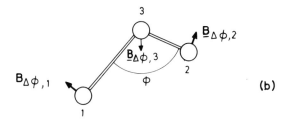

(b)

**Figure 2.17**
(a) illustrates the stretching of the bond (12). (b) depicts the bending of the bond angle between two bonds sharing an atom.

$$\mathbf{B}_{\Delta\phi,1} = \frac{\cos\phi\,\hat{\mathbf{e}}_{31} - \hat{\mathbf{e}}_{32}}{r_{31}\sin\phi}, \tag{2.243a}$$

$$\mathbf{B}_{\Delta\phi,2} = \frac{\cos\phi\,\hat{\mathbf{e}}_{32} - \hat{\mathbf{e}}_{31}}{r_{32}\sin\phi}, \tag{2.243b}$$

$$\mathbf{B}_{\Delta\phi,3} = \frac{(r_{31} - r_{32}\cos\phi)\hat{\mathbf{e}}_{31} + (r_{32} - r_{31}\cos\phi)\hat{\mathbf{e}}_{32}}{r_{31}r_{32}\sin\phi}. \tag{2.243c}$$

**Torsion.** As usual, express $\Delta\tau$, the torsional coordinate, as (see Figure 2.15)

$$\Delta\tau = \sum_{i=1}^{4}\mathbf{B}_{\Delta\tau,i}\cdot\mathbf{u}(i). \tag{2.244}$$

The coefficients are found to be

$$\mathbf{B}_{\Delta\tau,1} = \frac{\hat{\mathbf{e}}_{21} \times \hat{\mathbf{e}}_{23}}{r_{12}\sin^2\phi_2}, \tag{2.245a}$$

$$\mathbf{B}_{\Delta\tau,2} = \frac{(r_{23} - r_{12}\cos\phi_2)(\hat{\mathbf{e}}_{23} \times \hat{\mathbf{e}}_{21})}{r_{12}r_{23}\sin^2\phi_2} + \frac{\cos\phi_3(\hat{\mathbf{e}}_{32} \times \hat{\mathbf{e}}_{34})}{r_{23}\sin^2\phi_3}, \tag{2.245b}$$

$$\mathbf{B}_{\Delta\tau,3} = \frac{(r_{32} - r_{43}\cos\phi_3)(\hat{\mathbf{e}}_{32} \times \hat{\mathbf{e}}_{34})}{r_{32}r_{43}\sin^2\phi_3} + \frac{\cos\phi_2(\hat{\mathbf{e}}_{23} \times \hat{\mathbf{e}}_{21})}{r_{32}\sin^2\phi_2}, \tag{2.245c}$$

$$\mathbf{B}_{\Delta\tau,4} = \frac{\hat{\mathbf{e}}_{34} \times \hat{\mathbf{e}}_{32}}{r_{43}\sin^2\phi_3}. \tag{2.245d}$$

All the above results can be expressed in lattice dynamical notation by suitably indexing the atoms. For example, if $1 \to \binom{l}{k}$, $2 \to \binom{l'}{k'}$, and $S_i(l) \to$ bond 12, then from (2.236) and (2.241)

$$B_{i,\alpha k}(l, l) = \hat{x}_\alpha\binom{l\ \ l'}{k\ \ k'}, \qquad (\alpha = x, y, z),$$

$$B_{i,\beta k'}(l, l') = \hat{x}_\beta\binom{l'\ \ l}{k'\ \ k}, \qquad (\beta = x, y, z),$$

where $\hat{\mathbf{x}}\binom{l\ \ l'}{k\ \ k'}$ is a unit vector in the direction $\mathbf{x}\binom{l}{k} - \mathbf{x}\binom{l'}{k'}$. The coefficients given above are the only nonzero ones that occur in the representation of the stretching of the bond $\binom{l\ \ l'}{k\ \ k'}$ with respect to Cartesian displacement components. Similar expressions may be written down from the other formulas.

### 2.8.3 External force constants in terms of Born–von Kármán constants

As above, simplifying assumptions can also be made regarding the force constants governing external modes. In particular, based on (1.5) it is possible to express intermolecular force constants in terms of interatomic force constants. To do this we first note that

$$
u_\alpha \begin{pmatrix} l \\ \kappa \\ k \end{pmatrix} = u_\alpha^t \begin{pmatrix} l \\ \kappa \end{pmatrix} + \left[ \mathbf{u}^r \begin{pmatrix} l \\ \kappa \end{pmatrix} \times \mathbf{x}(k) \right]_\alpha
$$

$$
= u_\alpha^t \begin{pmatrix} l \\ \kappa \end{pmatrix} + \sum_{\mu\nu} \varepsilon_{\alpha\mu\nu} u_\mu^r \begin{pmatrix} l \\ \kappa \end{pmatrix} x_\nu(k),
\tag{2.246}
$$

using the Levi-Civita symbol. Now the second-order term in the potential energy is given by

$$
\Phi^{(2)} = \tfrac{1}{2} \sum_{ll'} \sum_{\kappa\kappa'} \sum_{kk'} \sum_{\alpha\beta} \phi_{\alpha\beta} \begin{pmatrix} l & l' \\ \kappa & \kappa' \\ k & k' \end{pmatrix} u_\alpha \begin{pmatrix} l \\ \kappa \\ k \end{pmatrix} u_\beta \begin{pmatrix} l' \\ \kappa' \\ k' \end{pmatrix}
$$

$$
= \tfrac{1}{2} \sum_{ll'} \sum_{\kappa\kappa'} \sum_{kk'} \sum_{\alpha\beta} \phi_{\alpha\beta} \begin{pmatrix} l & l' \\ \kappa & \kappa' \\ k & k' \end{pmatrix} \left[ u_\alpha^t \begin{pmatrix} l \\ \kappa \end{pmatrix} + \sum_{\mu\nu} \varepsilon_{\alpha\mu\nu} u_\mu^r \begin{pmatrix} l \\ \kappa \end{pmatrix} x_\nu(k) \right]
$$

$$
\times \left[ u_\beta^t \begin{pmatrix} l' \\ \kappa' \end{pmatrix} + \sum_{\gamma\delta} \varepsilon_{\beta\gamma\delta} u_\gamma^r \begin{pmatrix} l' \\ \kappa' \end{pmatrix} x_\delta(k') \right].
\tag{2.247}
$$

Comparing this with the expression (2.187) given earlier, we find[65]

$$
\phi_{\alpha\beta}^{tt} \begin{pmatrix} l & l' \\ \kappa & \kappa' \end{pmatrix} = \sum_{kk'} \phi_{\alpha\beta} \begin{pmatrix} l & l' \\ \kappa & \kappa' \\ k & k' \end{pmatrix},
\tag{2.248a}
$$

$$
\phi_{\alpha\beta}^{tr} \begin{pmatrix} l & l' \\ \kappa & \kappa' \end{pmatrix} = \sum_{kk'} \sum_{\gamma\delta} \phi_{\alpha\gamma} \begin{pmatrix} l & l' \\ \kappa & \kappa' \\ k & k' \end{pmatrix} \varepsilon_{\gamma\beta\delta} x_\delta(k'),
\tag{2.248b}
$$

$$
\phi_{\alpha\beta}^{rt} \begin{pmatrix} l & l' \\ \kappa & \kappa' \end{pmatrix} = \sum_{kk'} \sum_{\mu\nu} \phi_{\mu\beta} \begin{pmatrix} l & l' \\ \kappa & \kappa' \\ k & k' \end{pmatrix} \varepsilon_{\mu\alpha\nu} x_\nu(k),
\tag{2.248c}
$$

$$
\phi_{\alpha\beta}^{rr} \begin{pmatrix} l & l' \\ \kappa & \kappa' \end{pmatrix} = \sum_{kk'} \sum_{\gamma\delta} \sum_{\mu\nu} \phi_{\mu\gamma} \begin{pmatrix} l & l' \\ \kappa & \kappa' \\ k & k' \end{pmatrix} \varepsilon_{\mu\alpha\nu} \varepsilon_{\gamma\beta\delta} x_\nu(k) x_\delta(k').
\tag{2.248d}
$$

One application of these formulas is to molecular crystals where the interatomic force constants can be related to interatomic pair potentials, assumed to be of either the LJ or Kitaigorodsky types. Another is to the evaluation of the Coulomb interaction between two ionic molecular groups. This problem will be mentioned again in Chapter 4.

## 2.9 An example—solid argon

As an illustration of a dispersion curve calculation, we shall consider in this section a simple example, that of solid argon. This crystal has the fcc structure and belongs to the space group $O_h^5(Fm3m)$, with the underlying point group $O_h(m3m)$. The cubic unit cell and the rhombohedral primitive cell are shown in Figure 2.18. The BZ is illustrated in Figure 2.19 and can be generated out of the *irreducible prism* shown with bold lines by subjecting the prism to the operations of $O_h$.

### 2.9.1 Dispersion curves

In setting up the dynamical matrix we shall make the reasonable approximation that the interatomic forces are negligible beyond the second neighbors. Table 2.4 lists the coordinates of the first and the second neighbors of the atom at the origin. Consider the force-constant matrix $\boldsymbol{\phi}(l, l')$ with $l \to (0, 0, 0)$ and $l' \to (1, 0, 0,)$. Let us write this as

$$\boldsymbol{\phi}(000, 100) = \begin{pmatrix} xx & xy & xz \\ yx & yy & yz \\ zx & zy & zz \end{pmatrix}. \tag{2.249}$$

Noting that the mirror reflection $\boldsymbol{\sigma}(1\bar{1}0)$ in the $(1\bar{1}0)$ plane leaves the bond (0—1) invariant and using (2.27), we find,

$$\boldsymbol{\phi}(000, 100) = \boldsymbol{\sigma}(1\bar{1}0)\ \boldsymbol{\phi}(000, 100)\ \tilde{\boldsymbol{\sigma}}(1\bar{1}0). \tag{2.250}$$

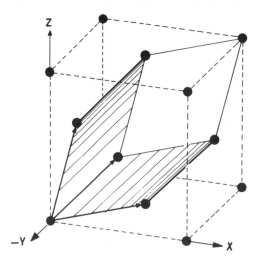

**Figure 2.18**
Cubic and primitive cells of solid argon. The basis vectors are shown by heavy lines.

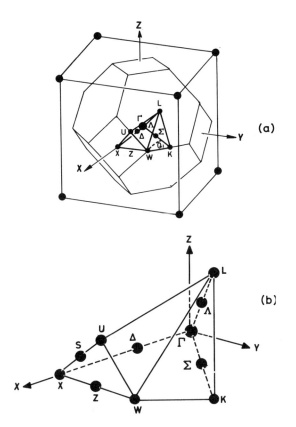

**Figure 2.19**
Brillouin zone for argon. The symmetry points and directions are labeled as is customary
in solid state physics (see ref. 9). In (b) is illustrated the irreducible prism from which the
entire BZ can be generated through application of the point group operations. The
dotted line from $W$ to $Q_1$ indicates the line of contact along which there is an accidental
degeneracy of the middle and lower branches.

**Table 2.4**
Coordinates of the first and second neighbors of the atom at the origin.

| Atom No. | Cartesian Coordinates | $(l_1, l_2, l_3)$ | Atom No. | Cartesian Coordinates | $(l_1, l_2, l_3)$ |
|---|---|---|---|---|---|
| 1 | $a(\ \tfrac{1}{2},\ \tfrac{1}{2},\ 0)$ | $(\ 1,\ 0,\ 0)$ | 7 | $a(\ 0,\ -\tfrac{1}{2},\ -\tfrac{1}{2})$ | $(\ 0,\ -1,\ 0)$ |
| 2 | $a(-\tfrac{1}{2},\ \tfrac{1}{2},\ 0)$ | $(\ 0,\ 1,\ -1)$ | 8 | $a(\ 0,\ \tfrac{1}{2},\ -\tfrac{1}{2})$ | $(\ 1,\ 0,\ -1)$ |
| 3 | $a(-\tfrac{1}{2},\ -\tfrac{1}{2},\ 0)$ | $(-1,\ 0,\ 0)$ | 9 | $a(\ \tfrac{1}{2},\ 0,\ \tfrac{1}{2})$ | $(\ 0,\ 0,\ 1)$ |
| 4 | $a(\ \tfrac{1}{2},\ -\tfrac{1}{2},\ 0)$ | $(\ 0,\ -1,\ 1)$ | 10 | $a(-\tfrac{1}{2},\ 0,\ \tfrac{1}{2})$ | $(-1,\ 1,\ 0)$ |
| 5 | $a(\ 0,\ \tfrac{1}{2},\ \tfrac{1}{2})$ | $(\ 0,\ 1,\ 0)$ | 11 | $a(-\tfrac{1}{2},\ 0,\ -\tfrac{1}{2})$ | $(\ 0,\ 0,\ -1)$ |
| 6 | $a(\ 0,\ -\tfrac{1}{2},\ \tfrac{1}{2})$ | $(-1,\ 0,\ 1)$ | 12 | $a(\ \tfrac{1}{2},\ 0,\ -\tfrac{1}{2})$ | $(\ 1,\ -1,\ 0)$ |
| 13 | $a(\ 1,\ 0,\ 0)$ | $(\ 1,\ -1,\ 1)$ | 16 | $a(\ 0,\ -1,\ 0)$ | $(-1,\ -1,\ 1)$ |
| 14 | $a(\ 0,\ 1,\ 0)$ | $(\ 1,\ 1,\ -1)$ | 17 | $a(\ 0,\ 0,\ 1)$ | $(-1,\ 1,\ 1)$ |
| 15 | $a(-1,\ 0,\ 0)$ | $(-1,\ 1,\ -1)$ | 18 | $a(\ 0,\ 0,\ -1)$ | $(\ 1,\ -1,\ -1)$ |

Note: Atoms above the dotted line correspond to the first neighbors, those below to the second neighbors.

With the matrix representation

$$\sigma(1\bar{1}0) = \begin{pmatrix} 0 & 1 & 0 \\ 1 & 0 & 0 \\ 0 & 0 & 1 \end{pmatrix},$$

(2.250) leads to the interrelations

$xx = yy$, $xy = yx$, $yz = xz$, and $zx = zy$.

Similarly, consideration of the reflection in the plane (001), which also leaves the bond (0—1) invariant, yields

$xz = -xz = 0$, $zx = -zx = 0$.

Further examination along these lines shows that no more simplification is possible. The tensor $\phi(000, 100)$ thus has three independent components and may be expressed as

$$\phi(000, 100) = -\begin{pmatrix} \alpha_1 & \gamma_1 & 0 \\ \gamma_1 & \alpha_1 & 0 \\ 0 & 0 & \beta_1 \end{pmatrix}, \tag{2.251}$$

where the negative sign has been introduced for convenience. The force-constant matrices for the remaining 11 first neighbors may be obtained from (2.26) by using operators which transform the bond (0—1). For example, bond (0—8) is obtainable from (0—1) by $\mathbf{C}_4[010]$, an anticlockwise rotation by $90°$ about $[010]$. Noting that

$$\mathbf{C}_4[010] = \begin{pmatrix} 0 & 0 & 1 \\ 0 & 1 & 0 \\ -1 & 0 & 0 \end{pmatrix},$$

(2.26) leads us to

$$\phi(000,\, 10\bar{1}) = \mathbf{C}_4[010]\phi(000,\, 100)\tilde{\mathbf{C}}_4[010] = -\begin{pmatrix} \beta_1 & 0 & 0 \\ 0 & \alpha_1 & -\gamma_1 \\ 0 & -\gamma_1 & \alpha_1 \end{pmatrix}.$$

Table 2.5 summarizes the matrices appropriate to the first neighbors. The second neighbors are dealt with in a similar fashion, and the results are given in Table 2.6. The self term $\phi(000,\, 000)$ is determined by the sum rule (2.35c) to be

$$\phi(000,\, 000) = (8\alpha_1 + 4\beta_1 + 2\alpha_2 + 4\beta_2)\mathbb{1}_3.$$

It is useful to emphasize that the self force constant is *not* given by

$$\left(\frac{\partial^2 V}{\partial r_\alpha \partial r_\beta}\right)$$

evaluated at $r = 0$. It is easily verified that the force-constant tensors as determined above are consistent with the rotational sum rule (2.40)*

**Table 2.5**
Force-constant matrices $\phi(0,\, l)$ corresponding to the first neighbors of the atom at the origin.

| Atom pair | $\phi(0, l)$ | Atom pair | $\phi(0, l)$ | Atom pair | $\phi(0, l)$ |
|---|---|---|---|---|---|
| 0–1 | $-\begin{pmatrix} \alpha_1 & \gamma_1 & 0 \\ \gamma_1 & \alpha_1 & 0 \\ 0 & 0 & \beta_1 \end{pmatrix}$ | 0–5 | $-\begin{pmatrix} \beta_1 & 0 & 0 \\ 0 & \alpha_1 & \gamma_1 \\ 0 & \gamma_1 & \alpha_1 \end{pmatrix}$ | 0–9 | $-\begin{pmatrix} \alpha_1 & 0 & \gamma_1 \\ 0 & \beta_1 & 0 \\ \gamma_1 & 0 & \alpha_1 \end{pmatrix}$ |
| 0–2 | $-\begin{pmatrix} \alpha_1 & -\gamma_1 & 0 \\ -\gamma_1 & \alpha_1 & 0 \\ 0 & 0 & \beta_1 \end{pmatrix}$ | 0–6 | $-\begin{pmatrix} \beta_1 & 0 & 0 \\ 0 & \alpha_1 & -\gamma_1 \\ 0 & -\gamma_1 & \alpha_1 \end{pmatrix}$ | 0–10 | $-\begin{pmatrix} \alpha_1 & 0 & -\gamma_1 \\ 0 & \beta_1 & 0 \\ -\gamma_1 & 0 & \alpha_1 \end{pmatrix}$ |
| 0–3 | $-\begin{pmatrix} \alpha_1 & \gamma_1 & 0 \\ \gamma_1 & \alpha_1 & 0 \\ 0 & 0 & \beta_1 \end{pmatrix}$ | 0–7 | $-\begin{pmatrix} \beta_1 & 0 & 0 \\ 0 & \alpha_1 & \gamma_1 \\ 0 & \gamma_1 & \alpha_1 \end{pmatrix}$ | 0–11 | $-\begin{pmatrix} \alpha_1 & 0 & \gamma_1 \\ 0 & \beta_1 & 0 \\ \gamma_1 & 0 & \alpha_1 \end{pmatrix}$ |
| 0–4 | $-\begin{pmatrix} \alpha_1 & -\gamma_1 & 0 \\ -\gamma_1 & \alpha_1 & 0 \\ 0 & 0 & \beta_1 \end{pmatrix}$ | 0–8 | $-\begin{pmatrix} \beta_1 & 0 & 0 \\ 0 & \alpha_1 & -\gamma_1 \\ 0 & -\gamma_1 & \alpha_1 \end{pmatrix}$ | 0–12 | $-\begin{pmatrix} \alpha_1 & 0 & -\gamma_1 \\ 0 & \beta_1 & 0 \\ -\gamma_1 & 0 & \alpha_1 \end{pmatrix}$ |

*In general it is useful to impose (2.40) explicity because sometimes this may reveal redundancies among the force constants. See, for example, A. H. A. Penny, Proc. Camb. Phil. Soc. **44,** 423 (1948).

**Table 2.6**
Force-constant matrices $\boldsymbol{\phi}(0, l)$ corresponding to the second neighbors of the atom at the origin.

| Atom pair | $\boldsymbol{\phi}(0, l)$ | Atom pair | $\boldsymbol{\phi}(0, l)$ | Atom pair | $\boldsymbol{\phi}(0, l)$ |
|---|---|---|---|---|---|
| 0–13 | $-\begin{pmatrix} \alpha_2 & & \\ & \beta_2 & \\ & & \beta_2 \end{pmatrix}$ | 0–15 | $-\begin{pmatrix} \alpha_2 & & \\ & \beta_2 & \\ & & \beta_2 \end{pmatrix}$ | 0–17 | $-\begin{pmatrix} \beta_2 & & \\ & \beta_2 & \\ & & \alpha_2 \end{pmatrix}$ |
| 0–14 | $-\begin{pmatrix} \beta_2 & & \\ & \alpha_2 & \\ & & \beta_2 \end{pmatrix}$ | 0–16 | $-\begin{pmatrix} \beta_2 & & \\ & \alpha_2 & \\ & & \beta_2 \end{pmatrix}$ | 0–18 | $-\begin{pmatrix} \beta_2 & & \\ & \beta_2 & \\ & & \alpha_2 \end{pmatrix}$ |

It is pertinent to digress briefly and note that by virtue of (2.23) the self force-constant tensor $\boldsymbol{\phi}\binom{l\ \ l}{k\ \ k}$ must always be symmetric. On the other hand, $\boldsymbol{\phi}\binom{l\ \ l}{k\ \ k}$ deduced during computations by an application of the sum rule (2.35) (as in the above example) need not in general display this behavior. However, as Czachor[77] has pointed out, such a situation may arise only if the site group* of the atom concerned belongs to one of the following point groups: $C_1$, $C_i$, $C_2$, $C_s$, $C_{2h}$. In these cases the symmetry (2.23) for the self force-constants must be externally imposed which then leads to some constraints on the force constants.

Once the force-constant matrices have been determined, the elements of the dynamical matrix may be evaluated by performing the sum over $l$ in (2.43). This gives

$$D_{\alpha\alpha}(\mathbf{q}) = \frac{1}{m_\mathrm{A}}\left[ A - 4\alpha_1 \cos\left(\frac{aq_\alpha}{2}\right)\left\{\cos\left(\frac{aq_\beta}{2}\right) + \cos\left(\frac{aq_\gamma}{2}\right)\right\}\right.$$

$$- 4\beta_1 \cos\left(\frac{aq_\beta}{2}\right)\cos\left(\frac{aq_\gamma}{2}\right) - 2[\alpha_2 \cos\,(aq_\alpha) + \beta_2\{\cos\,(aq_\beta)$$

$$\left. + \cos(aq_\gamma)\}\right], \ (\alpha = x, y, z; \alpha \neq \beta \neq \gamma);$$

$$D_{\alpha\beta}(\mathbf{q}) = \frac{4}{m_\mathrm{A}}\gamma_1 \sin\left(\frac{aq_\alpha}{2}\right)\sin\left(\frac{aq_\beta}{2}\right), \ (\alpha \neq \beta)$$

$$A = (8\alpha_1 + 4\beta_1 + 2\alpha_2 + 4\beta_2),$$

where $m_\mathrm{A}$ denotes the mass of the argon atom.

To obtain numerical values for the force constants we assume the interatomic potential to be of the LJ form with parameters as given in Table 1.1. Taking the lattice spacing as 5.311 Å (measured value at 4.2 K), the

---

*The site group associated with an atom is the set of those elements of the crystallographic point group which when applied about the atom in question leave the crystal invariant.

force constants are found from (2.225) to be

$\alpha_1 = 577.382, \quad \beta_1 = -18.001, \quad \gamma_1 = 595.383.$
$\alpha_2 = -50.732, \quad \beta_2 = 8.416$

(all in units of dynes/cm).

It is worth drawing attention to the fact that the lattice spacing used above does not correspond to the equilibrium spacing appropriate to the LJ potential employed, in the sense $a \neq 1.542\sigma$ (see Table 1.1). For this reason, the force constants introduced here must be viewed as axially symmetric rather than central.*

Figure 2.20 shows the results obtained by diagonalizing $\mathbf{D}(\mathbf{q})$ for $\mathbf{q}$ corresponding to the edges of the irreducible prism. The calculations have been made for $m_A = 39.948$ amu. Along the symmetry directions the branches can be classified as longitudinal and transverse; elsewhere this is not possible. The heavy lines near the origin show the elastic behavior based on the measured elastic constants.[78] Also shown in the figure are recent experimental data[79] obtained by neutron scattering (Sec. 6.1).† The results for the direction $\Gamma P$ have been included to illustrate how the accidental degeneracy along [110] is removed when one moves away from that direction.§

### 2.9.2 Critical points

We shall now examine the critical points in some detail. The majority of these can be located using simple symmetry arguments. A useful rule is to look for them in mirror planes and in particular along the symmetry directions and at symmetry points in such planes. For example, at the point $X$ it is evident from Figure 2.20 that the components of grad $\omega_j(\mathbf{q})$ vanish for all the three branches along three orthogonal directions, two of the $XW$ type and one of the $X\Gamma$ type (see also Figure 2.19). Thus $X$ is indeed a CP. Similar remarks can be made concerning the other symmetry points.

It is interesting to examine the constant-frequency surfaces around a few representative CP's. In Figure 2.21 are illustrated the sections of such surfaces in a set of orthogonal planes through $X$. Taking the top branch first, we note that the point $X$ is clearly a maximum for that branch. In

---

*On the other hand, if one works with a LJ potential for which the observed spacing equals $1.542\sigma$, then the potential can be regarded as central in the sense of Sec. 2.8.1. In this case, it is necessary in the interests of consistency to include interactions up to as many neighbors as are required for fulfilling the sum rule (2.229) to a reasonable accuracy.
†Batchelder et al.[80] have made measurements for $A^{36}$ at 4.2 K. Their results could also be plotted on our curves by multiplying their frequencies by $(35.968/39.948)^{1/2}$.
§See also Figure 2.5 and the remarks on ordered labeling in Sec. 2.3.3. The present results should give an idea of the rationale behind such a labeling scheme.

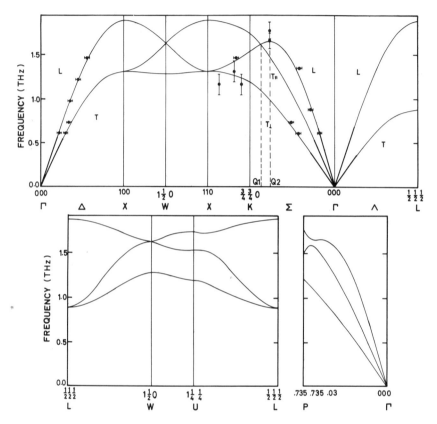

**Figure 2.20**
Dispersion curves for argon obtained using the model discussed in the text. The heavy
lines through the origin are computed using measured elastic constants (from ref. 78). The
points denote experimental results (from ref. 79).

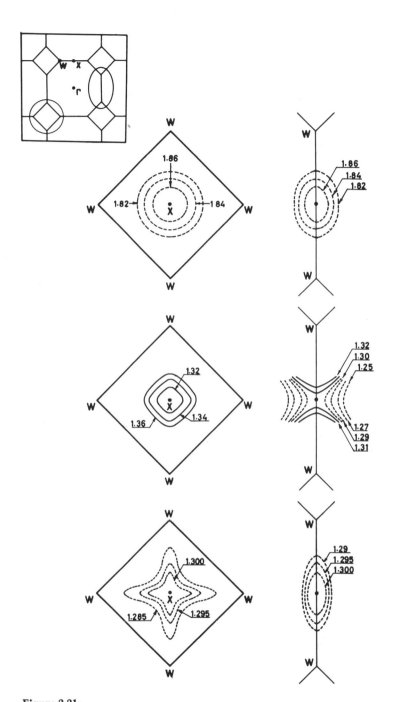

**Figure 2.21**
Frequency contours around $X$ for the three branches. For each branch the contours are displayed in two orthogonal planes through $X$, whose disposition relative to the BZ is indicated at the top left-hand corner (see also Figure 2.19). The third plane through $X$ and orthogonal to the above two is identical with the plane $\Gamma XW$. The nature of the contours is discussed in the text.

addition there is no degeneracy and, since the gradient vanishes, the point is an ACP of type $P_3$, a fact that is reflected in the constant-frequency contours which, as expected, are sections of ellipsoids (see Figure 2.6). In the middle and lower branches, $X$ corresponds to an FCP since a degeneracy exists and the gradient vanishes completely. The classification for the middle branch is $P_1$(FCP) while that for the lowest branch is $P_3$(FCP). The degeneracy produces warping of the contours, particularly striking in the case of the lower branch.

Some further examples of constant-frequency contours are given in Figure 2.22. Shown here are the contours around $W$ for the middle and the lower branches. For the former, $W$ is evidently a maximum since the frequency decreases in every direction away from it. Further $W$ cannot be an ACP for that branch owing to the degeneracy. Consider now three orthogonal directions, two along $WM$ and the third along $WX$. Along $WM$ the components of grad $\omega_2(\mathbf{q} \rightarrow W)$ must vanish since the planes normal to $MWM$ and passing through $W$ (that is, the planes $q_x = (2\pi/a)$ or $q_z = 0$) have reflection symmetry. On the other hand, the component of the gradient along $WX$ changes sign discontinuously as shown by the dispersion curves themselves. Thus for the middle branch, $W$ is a CP of type $P_3$(SCP 1). The sharp angularity of the BZ at $W$ causes considerable distortion in the contours as the frequency difference increases. Care is therefore required in making a proper assessment of the surfaces relative to the CP. For the lowest branch, $W$ is an ACP of type $P_2$. It may be observed that the calculations in the plane $q_z = 0$ have been made for much larger frequency intervals than in the case of the middle branch; the shapes observed are therefore not "pure."

We now turn to the systematic identification and assignment of the CP's. The first step is to hunt carefully in the mirror planes with the help of the dispersion curves and if necessary also constant-frequency contours and look for points with the required behavior for the slope components. Such an examination easily reveals CP's at $X$, $L$, and $W$, and these are cataloged in Table 2.7. In addition there are also CP's at $Q_1$ and $Q_2$ along $\Gamma K$ as inspection of Figure 2.20 shows. Furthermore, because the top branch has a maximum both at $X$ and $L$, one might expect an additional CP for the top branch in the (110) plane. Indeed a detailed study of the dispersion relations in this plane does reveal such a point designated $J$ with coordinates $(2\pi/a)(0.1, 0.1, 0.8)$.

Having located the CP's and examined the degeneracies as well as the slope behavior, the sector numbers $(P, N)$ and the topological weights must then be assigned. The former involves a careful examination of the shapes of the constant-frequency surfaces in the neighborhood of the CP.

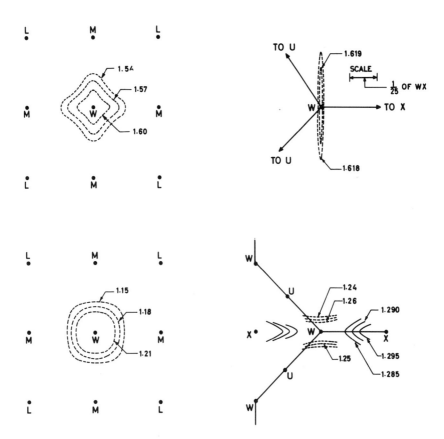

**Figure 2.22**
Frequency contours around *W* for the middle and lower branches.

**Table 2.7**
Critical points in Argon.

| Branch | Point | Orthogonal directions | Behavior of slope component | $(P,N)$ | Index $i$ | Weight $w$ | Multiplicity | Classification | Mult. × $w$ |
|---|---|---|---|---|---|---|---|---|---|
| Bottom | $\Gamma(000)$ | $\Gamma X$ $\Gamma X$ $\Gamma X$ | Disc. Disc. Disc. | $(1,0)$ | 0 | 1 | 1 | $P_0(\text{SCP 3})$ | 1 |
| | $X(100)$ | $X\Gamma$ $XW$ $XW$ | 0 0 0 | $(0,1)$ | 3 | 1 | 3 | $P_3(\text{FCP})$ | 3 |
| | $W(1,\frac{1}{2},0)$ | $WM$ $WM$ $WX$ | 0 0 0 | $(2,1)$ | 2 | 1 | 6 | $P_2$ | 6 |
| | $L(\frac{1}{2},\frac{1}{2},\frac{1}{2})$ | $L\Gamma$ $LU$ $LW$ | 0 0 0 | $(1,2)$ | 1 | 1 | 4 | $P_1(\text{FCP})$ | 4 |
| Middle | $\Gamma(000)$ | $\Gamma X$ $\Gamma X$ $\Gamma X$ | Disc. Disc. Disc. | $(1,0)$ | 0 | 1 | 1 | $P_0(\text{SCP 3})$ | 1 |
| | $X(100)$ | $X\Gamma$ $XW$ $XW$ | 0 0 0 | $(1,2)$ | 1 | 1 | 3 | $P_1(\text{FCP})$ | 3 |
| | $W(1,\frac{1}{2},0)$ | $WM$ $WM$ $WX$ | 0 0 Disc. | $(0,1)$ | 3 | 1 | 6 | $P_3(\text{SCP 1})$ | 6 |
| | $L(\frac{1}{2},\frac{1}{2},\frac{1}{2})$ | $L\Gamma$ $LU$ $LW$ | 0 0 0 | $(1,2)$ | 1 | 1 | 4 | $P_1(\text{FCP})$ | 4 |
| | $Q_1(0.65,0.65,0)$ | $A$ $B$ $C$ | Disc. 0 0 | $(2,1)$ | 2 | 1 | 12 | $P_2(\text{SCP 1})$ | 12 |
| Top | $\Gamma(000)$ | $\Gamma X$ $\Gamma X$ $\Gamma X$ | Disc. Disc. Disc. | $(1,0)$ | 0 | 1 | 1 | $P_0(\text{SCP 3})$ | 1 |
| | $X(100)$ | $X\Gamma$ $XW$ $XW$ | 0 0 0 | $(0,1)$ | 3 | 1 | 3 | $P_3$ | 3 |
| | $W(1,\frac{1}{2},0)$ | $WM$ $WM$ $WX$ | 0 0 Disc. | $(1,4)$ | 1 | 3 | 6 | $P_1(\text{SCP 1})$ | 18 |
| | $L(\frac{1}{2},\frac{1}{2},\frac{1}{2})$ | $L\Gamma$ $LU$ $LW$ | 0 0 0 | $(0,1)$ | 3 | 1 | 4 | $P_3$ | 4 |

| | | | | | | | |
|---|---|---|---|---|---|---|---|
| $Q_1(0.65,0.65,0)$ $\begin{cases} A \\ B \\ C \end{cases}$ | Disc.<br>0<br>0 | $(1,0)$ | 0 | 1 | 12 | $P_0$(SCP 1) | 12 |
| $Q_2(0.56,0.56,0)$ $\begin{cases} D \\ E \\ F \end{cases}$ | 0<br>0<br>0 | $(1,2)$ | 1 | 1 | 12 | $P_1$ | 12 |
| $J(0.1,0.1,0.8)$ $\begin{cases} G \\ H \\ I \end{cases}$ | 0<br>0<br>0 | $(2,1)$ | 2 | 1 | 24 | $P_2$ | 24 |

| | | |
|---|---|---|
| A: $Q_1K$ | D: $Q_2K$ | G: $JL$ |
| B: Through $Q_1$ & in | E: Through $Q_2$ & in | H: $\perp$ to $JL$ and in (110) |
| $q_z = 0$ plane | $q_z = 0$ plane | plane |
| C: Through $Q_1$ & $\parallel$ to $Z$ axis | F: Through $Q_2$ & $\parallel$ to $Z$ axis | I: $\perp$ to G and H |

In most instances this is straightforward but care is required. The top branch at $W$ provides an interesting illustration. A cursory inspection of the frequencies near $W$ suggests that they increase in all directions away from $W$. However, this is not true in the immediate vicinity of the "line of contact" shown dashed in Figure 2.19. This is a line along which one encounters accidental degeneracy like that at $Q_1$ (the degeneracy at $W$ is symmetry required and not accidental). If one moves away from $W$ along such directions, the frequency decreases from 1.620 THz to 1.609 THz at $Q_1$. Since there are four such directions around every $W$ point, there are four cones within which $\omega_3(\mathbf{q}) < \omega_3(\mathbf{q} \to W)$. Along all other directions the frequency increases. The top branch at $W$ thus has sector numbers $(1, 4)$ instead of $(1, 0)$ as a casual inspection might suggest.

Once the sector numbers are known, it is straightforward to assign the topological weights using the prescriptions of Sec. 2.3.4. The task remains of ensuring that no wrong classification has been made nor have obvious CP's been missed. The Morse relations (2.98) prove useful at this point. Using the numbers given in the last column of Table 2.7, it is readily verified that the CP's listed in that table do indeed satisfy the Morse relations. For example, for the top branch we have from the last column of that table, $N_0 = 13$, $N_1 = 30$, $N_2 = 24$, and $N_3 = 7$ which clearly fulfill (2.98). We may thus claim to have found a set of CP's which contains those required by symmetry considerations and which also satisfies the Morse relations. While this does not guarantee that other CP's in off-symmetry directions might not have been missed, it at least ensures that a minimal set has been located and properly assigned. In general, unless the dispersion curves have complicated behavior (an infrequent occurrence), the CP's determined as above may be expected to exhaust all those possible.

### 2.9.3 Frequency distribution

We next consider singularities in the frequency distribution. This may be done by reference to Figure 2.23, which shows the frequency spectrum for our model computed using the method of Gilat and Raubenheimer.[33]* Exact diagonalization was carried out in the present case at 5947 mesh points in the irreducible prism. Since frequencies in the prism are known, those in the rest of the BZ can also be determined by Eq. (2.69). In this context it is important to remember that the multiplicity governed by Eq. (2.69) depends on the value of $\mathbf{q}$. For example, corresponding to the point $(\eta, 0, 0)$ $(\eta < 2\pi/a)$ in the prism the multiplicity is 6, while for a general point it is 48. Thus in computing $g(\omega)$ by sorting the results obtained from the prism, proper weights must be associated with the eigenvalues.† In the present calculations, 2000 channels were employed to sort the frequencies and construct the histogram.

The frequency distribution shown in Figure 2.23 clearly exhibits several singular features. It can be seen (at least in the case of the prominent singularities) that the shapes obtained here do indeed correspond to those predicted earlier in Sec. 2.3. Attention is drawn to the ACP of type $P_2$ at $J$ which contributes a big peak on account of its multiplicity. For further elucidation, we have plotted in Figure 2.24 the derivative spectra around some of the critical frequencies. These are even more revealing and the reader will find it useful to compare the actual shapes with those predicted earlier.

### 2.9.4 Dynamics using the cubic unit cell

In the example just discussed, the primitive cell was used as the fundamental cell because of considerations of mathematical and computational labor. Nothing a priori would prevent our using the cubic unit cell, and it in fact is interesting to consider the problem in this context.‡ The results thus obtained for the [100] direction are shown in Figure 2.25. The

---

*The calculations were kindly done by P. R. Vijayaraghavan using a computer program developed by Raubenheimer and Gilat and described in the Oak Ridge National Laboratory Report $TM$-1425.

†For a further discussion of this point, see the ORNL report of Raubenheimer and Gilat cited above. See also B. Dayal and S. P. Singh, Proc. Roy. Soc. (London) **76,** 777 (1960).

‡It is worth pointing out that if we take the three edges of the cubic cell as the basis vectors (as is usually done), then the lattice is not generated by integral translations alone. Rather the translational group includes elements of the form $\{E|(\mathbf{i}+\mathbf{j})a/2\}$, $\mathbf{i}, \mathbf{j}$ unit vectors along $x, y$ axes. It would appear from this that the space group is nonsymmorphic. However, the above fractional translations are not essential since they can be eliminated by a different choice of basis. On the other hand, for the diamond lattice, which consists of two interpenetrating fcc lattices, fractional translations are essential and the space group is truly nonsymmorphic.

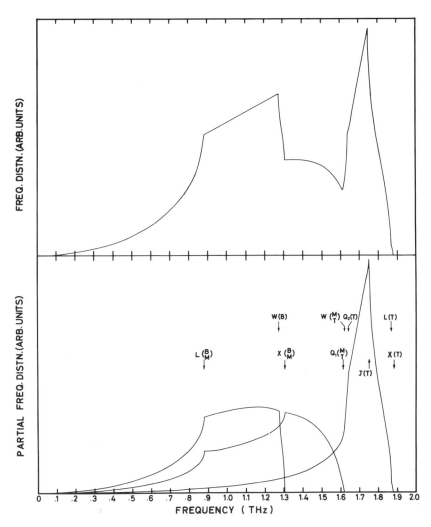

**Figure 2.23**
Frequency distributions $g_j(\omega)$ and $g(\omega)$ for solid argon. The critical features are identified by indicating the point in the BZ, the branch being shown in parentheses; for example, $W(B)$ denotes the critical frequency for the bottom branch at $W$. Compare the critical features with those of Figures 2.7 and 2.9. (See also Table 2.7). Notice that SCP show up less prominently than do ACP.

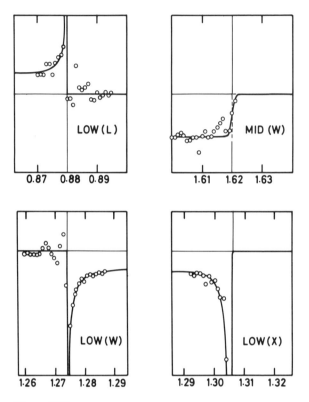

**Figure 2.24**
Representative plots of $g_j'(\omega)$. The points are based on computed data while the lines are guides to the eye. Compare with Figures 2.7 and 2.9. The finite slope at the critical frequency for the middle branch at $W$ is due to the finite resolution in the calculations, that is, the finite mesh used.

presence of four atoms in the cell results in twelve branches which clearly must represent a remapping of those presented earlier. Consider Figure 2.26 which shows the (100) plane of the reciprocal lattice. Superposed on this is the corresponding section of the lattice generated by the primitive cell. Also shown are the sections of the BZ's appropriate to the two lattices. It is evident from the figure that the [100] direction in the present problem is equivalent to the segments $AB$, $BC$, $DE$, and $EF$ of the previous BZ. The present results therefore represent just a folding of the earlier results for $\Gamma X$ and $XWX$ as is evident from a comparison of Figures 2.20 and 2.25.

The normal-mode frequencies are thus independent of the choice of the fundamental cell, as they should be. The algebra of their calculation and the representation of the final results will differ, however, depending on how the cell is chosen. In this context the primitive cell is always to be preferred.

In this chapter we have concentrated mainly on setting up the force-

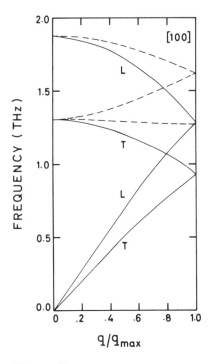

**Figure 2.25**
Dispersion curves for argon along [100] based on the cubic cell. The dispersion curves here represent essentially a remapping of the curves along $\Gamma X$, $XWX$ of Figure 2.20. The curves for $XWX$ of the earlier figure are shown here with dashed lines.

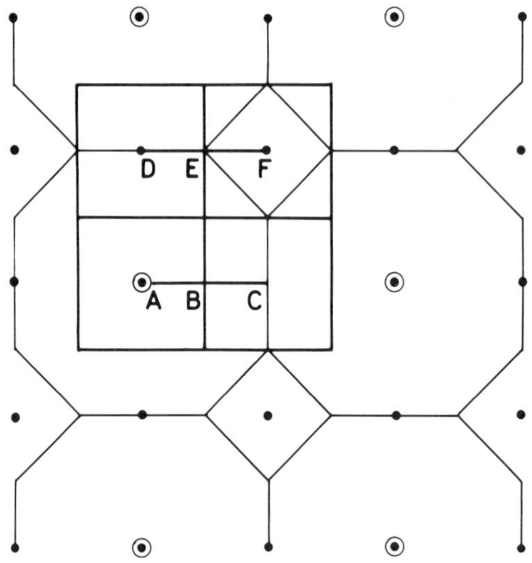


**Figure 2.26**
(100) plane of the reciprocal lattice of argon. The open circles and the dots denote respectively the lattice points based on the primitive and cubic cells. Also shown are sections of the BZ in the two representations. The [100] direction of the cubic BZ is equivalent to the segments *AB*, *BC*, *DE*, and *EF* of the conventional BZ.

constant formalism for lattice dynamics. The force constants were considered as formal parameters though some indications were given of the forms acquired by them under certain special circumstances (Sec. 2.8). In Chapter 4 we shall see how the force constants of insulating crystals may be expressed in greater detail by using suitable phenomenological models, while in Chapter 6 we shall review in general the success of the force-constant models in explaining experimental data. In the last chapter we shall take up the question of calculating the force constants and the dispersion curves from first principles.

## References

1. M. Born and J. R. Oppenheimer, Ann. Physik **84**, 457 (1927).

2. BH, Chapter IV and Appendix VII.

3. J. C. Slater, *Quantum Theory of Molecules and Solids* (McGraw-Hill, New York, 1963), Vol. I.

4. Seitz, p. 470.

5. J. Ziman, *Principles of the Theory of Solids* (Cambridge University Press, London, 1965), p. 169.

6. G. V. Chester, Advan. Phys. **10**, 357 (1961).

7. E. G. Brovman and Yu. Kagan, Zh. Eksp. Theor. Fiz. **52**, 557 (1967) [Sov. Phys.— JETP **25**, 365 (1967)].

8. M. Born and Th. von Kármán, Z. Physik **13**, 297 (1912).

9. G. F. Koster, Solid State Phys. **5**, 173 (1957).

10. Kittel, p. 20.

11. Standards Committee of IRE, Proc. IRE **37**, 1378 (1949).

12. K. Londsdale, *International Tables of X-ray Crystallography* (Kynoch, Birmingham, 1965), Vol. I.

13. H. Goldstein, *Classical Mechanics* (Addison-Wesley, Cambridge, Mass., 1950) p. 126.

14. BH, p. 222.

15. H. Margenau and G. M. Murphy, *Mathematics of Physics and Chemistry* (Van Nostrand, Princeton, 1943); F. B. Hildebrand, *Methods of Applied Mathematics* (Prentice-Hall, New York, 1952).

16. Goldstein, reference 13 above, p. 154.

17. BH, p. 153.

18. BH, p. 295.

19. L. Brillouin, *Propagation of Waves in Periodic Structures* (Dover, New York, 1953).

20. N. F. Mott and H. Jones, *Theory of Metals and Alloys* (Dover, New York, 1958), p. 152.

21. Seitz, p. 292.

22. C. V. Raman, Proc. Indian Acad. Sci. **14A**, 317 (1941).

23. BH, Appendix IV.

24. J. R. Hardy, Phil. Mag. **7**, 315 (1962).

25. G. H. Begbie and M. Born, Proc. Roy. Soc. (London) **A188**, 179 (1947); G. H. Begbie, ibid., 189 (1947).

26. H. M. J. Smith, Phil. Trans. Roy. Soc. (London) **241**, 105 (1948).

27. W. Cochran, Rept. Progr. Phys. **26**, 1 (1963).

28. G. Venkataraman and V. C. Sahni, Rev. Mod. Phys. **42**, 409 (1970), Appendix I.

29. J. E. Lynch Jr., Thesis, University of Michigan (1968).

30. P. Debye, Ann. Physik **39**, 789 (1912).

31. A. A. Maradudin, E. W. Montroll, G. H. Weiss, and I. P. Ipatova, *Theory of Lattice Dynamics in the Harmonic Approximation* (Academic Press, New York, 1971), second edition.

32. G. Gilat and G. Dolling, Phys. Letters **8**, 304 (1964).

33. G. Gilat and L. J. Raubenheimer, Phys. Rev. **144**, 390 (1966); see also G. Gilat, J. Comput. Phys. **10**, 432 (1972). This article contains references to developments in this area.

34. L. van Hove, Phys. Rev. **89**, 1189 (1953).

35. J. C. Phillips, Phys. Rev. **104**, 1263 (1956).

36. J. F. Nye, *Physical Properties of Crystals* (Clarendon Press, Oxford, 1957).

37. BH, Chapter III.

38. BH, p. 279.

39. BH, Chapter V.

40. R. Courant and D. Hilbert, *Methods of Mathematical Physics* (Interscience, New York, 1953), Vol. I, p. 6.

41. BH, p. 240 ff.

42. G. Leibfried and W. Ludwig, Z. Physik **160**, 80 (1960).

43. L. T. Hedin, Ark. Fysik **18**, 369 (1960).

44. M. Lax, in Wallis, p. 583.

45. R. Srinivasan, Phys. Rev. **144**, 620 (1966).

46. P. N. Keating, Phys. Rev. **145**, 637, 674 (1966).

47. BH, Ch. VI.

48. K. Fuchs, Proc. Roy. Soc. (London) **A153**, 622 (1936).

49. Kittel, p. 245.

50. R. E. Peierls, *Quantum Theory of Solids* (Clarendon Press, Oxford, 1955) Chapter 2.

51. BH, Ch. VI.

52. G. Leibfried and W. Ludwig, Solid State Phys. **12**, 275 (1961).

53. H. Hahn, in Chalk River Proceedings, Vol. I, p. 37.

54. G. Leibfried, in Wallis, p. 237.

55. BH, p. 297.

56. F. R. Gantmacher, *The Theory of Matrices* (Chelsea, New York, 1960), Chapter X.

57. Courant and Hilbert, reference 40 above.

58. Peierls, reference 50 above, Chapter 2.

59. A. Ghatak and L. S. Kothari, *Introduction to Lattice Dynamics* (Addison-Wesley, Reading, Mass., 1972).

60. H. H. Jensen, in Bak, p. 1.

61. R. A. Cowley, Advan. Phys. **12**, 421 (1963).

62. K. Parlinski, Acta Phys. Polon. **35**, 223 (1969).

63. Goldstein, reference 13 above, p. 130.

64. J. van Kranendonk and V. F. Sears, Can. J. Phys. **44**, 313 (1966).

65. G. Venkataraman and V. C. Sahni, Rev. Mod. Phys. **42**, 409 (1970).

66. G. Herzberg, *Infrared and Raman Spectra of Polyatomic Molecules* (Van Nostrand, New York, 1945).

67. R. S. Krishnan, Proc. Indian Acad. Sci. **A26**, 432 (1947).

68. E. L. Wagner and D. F. Hornig, J. Chem. Phys. **18**, 296 (1950).

69. Seitz, p. 303.

70. Mott and Jones, reference 20 above, p. 65.

71. A. S. Davydov, *Theory of Molecular Excitons* (McGraw-Hill, New York, 1962).

72. BH, p. 246; see also N. K. Pope, Acta Cryst. **2**, 325 (1949).

73. B. N. Brockhouse, E. D. Hallman, and S. C. Ng, in *Magnetic and Inelastic Scattering of Neutrons by Metals*, edited by T. J. Rowland and P. A. Beck (Gordon and Breach, New York, 1970).

74. T. Wolfram, G. W. Lehman, and R. E. DeWames, Phys. Rev. **129**, 2483 (1963).

75. E. B. Wilson Jr., J. C. Decius, and P. C. Cross, *Molecular Vibrations* (McGraw-Hill, New York, 1955).

76. S. J. Cyvin, *Molecular Vibrations and Mean Square Amplitudes* (Elsevier, Amsterdam, 1968).

77. A. Czachor, Phys. Stat. Sol. **53**, K65 (1972); ICTP Trieste, Report IC/71/109.

78. M. Gsänger, H. Egger, and E. Lüscher, Phys. Letters **27A**, 695 (1968).

79. H. Egger, M. Gsänger, E. Lüscher, and B. Dorner, Phys. Letters **28A**, 433 (1968).

80. D. N. Batchelder, B. C. G. Haywood, and D. H. Saunderson, J. Phys. C **4**, 910 (1971).

# 3 Group Theory in Lattice Dynamics

In this chapter we discuss the role of group theory in the study of lattice vibrations. Although not mandatory, it is desirable to apply group-theoretical methods to the calculation of phonon spectra in order to force a simplification of the eigenvalue problem and to gain greater insight into the transformation properties of eigenvectors, in turn facilitating the calculation of such quantities as optical selection rules. In presenting this chapter we presume the reader has a basic knowledge of finite groups and their representations, the application of representation theory in quantum mechanics, and preferably also in the study of molecular vibrations. The reader unfamiliar with these ideas should consult the standard texts.[1-8]

## 3.1 Introduction

Our present interest lies in simplifying the eigenvalue problem (2.51) by using group-theoretic techniques. Now the basic strategy underlying the application of group theory to the eigenvalue problem

$$\mathbf{AX} = \lambda \mathbf{X},  \tag{3.1}$$

where $\lambda$ and $\mathbf{X}$ are the eigenvalue and the eigenvector, respectively, of the Hermitian matrix $\mathbf{A}$, is to discover a group of unitary transformations which commute with $\mathbf{A}$. For the generalized eigenvalue problem

$$\mathbf{AX} = \lambda \mathbf{BX}  \tag{3.2}$$

(see (2.82)), the transformations sought must commute with both $\mathbf{A}$ and $\mathbf{B}$. If such a group of transformations can be found, then the possibility of simplifying the eigenvalue problem exists.

Consider next a cyclic crystal. The task of finding its normal-mode frequencies amounts to solving the equation

$$\mathbf{dE}(\eta) = \omega_\eta^2 \mathbf{E}(\eta),  \tag{3.3}$$

where $\mathbf{d}$ is a $3nN$-dimensional matrix with elements

$$d_{\alpha\beta}\begin{pmatrix} l & l' \\ k & k' \end{pmatrix} = (m_k m_{k'})^{-1/2} \phi_{\alpha\beta}\begin{pmatrix} l & l' \\ k & k' \end{pmatrix}$$

and $\mathbf{E}(\eta)$ is a $3nN$-component column matrix. Equation (3.3) is written in the spirit of a molecular vibration problem, the cyclic crystal being visualized as a giant molecule. Recognizing that the crystal is invariant under the space group,* one could in principle apply group theory to the prob-

---

*The space group is here regarded as *finite* since, for a cyclic crystal, the number of elements in the translational group is finite. This enables the ready use of the theory of finite groups.

lem (3.3) exactly as in the study of molecular vibrations. However, in practice, the eigenvalue equation which one seeks to solve is not (3.3) but

$$\mathbf{D}(\mathbf{q})\mathbf{e}\begin{pmatrix}\mathbf{q}\\j\end{pmatrix} = \omega_j^2(\mathbf{q})\mathbf{e}\begin{pmatrix}\mathbf{q}\\j\end{pmatrix}, \tag{3.4}$$

obtained from (3.3) by making explicit use of the translational symmetry of the crystal.

With regard to the latter problem, it was pointed out by Bouckaert et al. in a classic paper[9] that whenever attention is focused on a particular vector $\mathbf{q}$ in the BZ (be it in lattice dynamics or band structure), it is sufficient to consider a group $G(\mathbf{q})$ called the *group of the wave vector*. This group is comprised of those elements $\mathcal{R}_m = \{\mathbf{R}|\mathbf{x}(m) + \mathbf{v}(R)\}$ of the space group which have the property

$$\mathbf{R}\mathbf{q} = \mathbf{q} \text{ or } \mathbf{q} + \mathbf{G}, \tag{3.5}$$

where $\mathbf{G}$ is a reciprocal lattice vector. In other words, the elements of $G(\mathbf{q})$ are such that their *rotational parts* acting on $\mathbf{q}$ either leave it invariant or change it by $\mathbf{G}$. Clearly $G(\mathbf{q})$ is a subgroup of the space group $G$ and itself contains the translational group $\mathcal{T}$ as a subgroup. The physical reason for restricting attention to $G(\mathbf{q})$ is simply that the elements of $G(\mathbf{q})$ *acting on the crystal* leave unaltered the direction of a wave traveling along $\mathbf{q}$.

In the literature,[7,8] $G(\mathbf{q})$ is sometimes referred to as a *little group*.* For symmorphic groups, the problem of constructing the irreducible representations (IR) of $G(\mathbf{q})$ is quite straightforward. In the case of nonsymmorphic groups, however, it turns out that some methods based on the factor-group approach[10] lead to representations not all of which are IR's of $G(\mathbf{q})$. Only some meet the requirements of being an irreducible representation, and these "relevant" ones are often termed *allowable representations* (also *small representations*).† The allowable representations (AR) of $G(\mathbf{q})$ may be employed in the problem (3.4) in much the same way as point-group representations are used in the molecular vibration problem. This may be called the allowable or small-representation approach to lattice dynamics. Rather than following this method, we adopt here the more recent *multiplier-representation approach*.[11] This involves the use of the point group that underlies $G(\mathbf{q})$, namely $G_0(\mathbf{q})$, and its *irreducible multiplier representations* (IMR). The elements of $G_0(\mathbf{q})$ are made up of only the rotational parts of $G(\mathbf{q})$, that is, $G_0(\mathbf{q}) = \{\mathbf{R}\}$. The two approaches mentioned here are, of course, related.

---

*More specifically as the little group of the second kind relative to $G$, $\mathcal{T}$, and $\mathbf{q}$.[7]
†Technically, an IR of $G(\mathbf{q})$ is called an allowable representation (relative to $\mathcal{T}$ and $\mathbf{q}$) if it subduces a multiple of $\mathbf{q}$ on $\mathcal{T}$, that is, if each pure translational element $\{\mathbf{E}|\mathbf{x}(m)\}$ is represented as $\exp[-i\mathbf{q}\cdot\mathbf{x}(m)]$ times a unit matrix. See ref. 7 for further details.

### 3.1.1 Concerning multiplier representations in general

A few preliminaries concerning multiplier representations are in order. Consider a group $G$ with elements $\{R\}$. Let us associate with every element $R$ a matrix $\tau(R)$ such that the set $\tau = \{\tau(R)\}$ obeys the multiplication rule

$$\tau(R_i)\tau(R_j) = \phi(R_i, R_j)\tau(R_iR_j) \qquad \text{for all } R_i, R_j \in G. \tag{3.6}$$

The matrices $\{\tau(R)\}$ are said to form a multiplier representation of $G$ corresponding to the *factor system* $\{\phi(R_i, R_j)\}$. In the literature this representation is sometimes referred to as a *ray, projective,* or *weighted* representation. If the *multiplier* or *weight factor* $\phi(R_i, R_j)$ is unity for every $R_i, R_j \in G$, then $\{\tau(R)\}$ is identical to the ordinary representation of the group $G$. The representation $\{\tau(R)\}$ is said to be an IMR if the matrices $\tau(R)$ are irreducible.

If two IMR's $\tau^\sigma$ and $\tau^s$ of a group belong to the same factor system, then they obey the orthogonality relation

$$\sum_{R \in G} [\phi(R, R^{-1})]^{-1} \tau^s_{im}(R) \tau^\sigma_{lj}(R^{-1}) = \left(\frac{g}{n_s}\right)\delta_{s\sigma}\delta_{ij}\delta_{ml}, \tag{3.7a}$$

where $g$ is the order of the group and $n_s$ is the dimensionality of $\tau^s$. If in addition the representations $\tau^\sigma$ and $\tau^s$ are unitary, then the orthogonality relation acquires the familiar form[11]

$$\sum_{R \in G} \tau^s_{im}(R) \tau^{\sigma*}_{jl}(R) = \left(\frac{g}{n_s}\right)\delta_{s\sigma}\delta_{ij}\delta_{ml}. \tag{3.7b}$$

### 3.1.2 IMR's required in lattice dynamics

We next turn our attention to the IMR's of $G_0(\mathbf{q})$ and their relationship to the allowable representations of $G(\mathbf{q})$. For this purpose, consider first an allowable representation of $G(\mathbf{q})$, $\gamma^s(\mathbf{q})$ say, in which $\gamma^s(\mathbf{q}; \mathcal{R}_m)$ is the matrix representative of $\mathcal{R}_m$. Define now a matrix $\tau^s(\mathbf{q}; \mathbf{R})$ related to $\gamma^s(\mathbf{q}; \mathcal{R}_m)$ by

$$\tau^s(\mathbf{q}; \mathbf{R}) = \exp\{i\mathbf{q}\cdot[\mathbf{v}(R) + \mathbf{x}(m)]\}\gamma^s(\mathbf{q}; \mathcal{R}_m). \tag{3.8}$$

A consideration of the explicit form of $\gamma^s(\mathbf{q}; \mathcal{R}_m)$ reveals that the same matrix $\tau^s(\mathbf{q}; \mathbf{R})$ is associated with all the $N$ elements $\mathcal{R}_m$ corresponding to a fixed $\mathbf{R}$ but different $\mathbf{x}(m)$. Hence a *many-to-one correspondence* exists between the elements of $G(\mathbf{q})$ and the matrices $\tau^s$. However, there is a one-to-one correspondence between the elements of $G_0(\mathbf{q})$ and the matrices $\tau^s$. Even so, the matrices $\tau^s$ do not provide an IR of $G_0(\mathbf{q})$, but rather they provide an IMR as will now be shown.

Consider the following:

$$\tau^s(\mathbf{q}; \mathbf{R}_i)\tau^s(\mathbf{q}; \mathbf{R}_j)$$
$$= \exp\{i\mathbf{q}\cdot[\mathbf{v}(R_i) + \mathbf{x}(m_i)]\}\exp\{i\mathbf{q}\cdot[\mathbf{v}(R_j) + \mathbf{x}(m_j)]\}\,\gamma^s(\mathbf{q}; \mathscr{R}_{m_i}\mathscr{R}_{m_j})$$
$$= \exp\{i\mathbf{q}\cdot[\mathbf{v}(R_i) + \mathbf{x}(m_i)]\}\exp\{i\mathbf{q}\cdot[\mathbf{v}(R_j) + \mathbf{x}(m_j)]\}$$
$$\quad \times \exp\{-i\mathbf{q}\cdot[\mathbf{R}_i\mathbf{v}(R_j) + \mathbf{R}_i\mathbf{x}(m_j) + \mathbf{v}(R_i) + \mathbf{x}(m_i)]\}\,\tau^s(\mathbf{q}; \mathbf{R}_i\mathbf{R}_j)$$
$$= \exp\{i(\mathbf{q} - \mathbf{R}_i^{-1}\mathbf{q})\cdot[\mathbf{v}(R_j) + \mathbf{x}(m_j)]\}\tau^s(\mathbf{q}; \mathbf{R}_i\mathbf{R}_j). \tag{3.9}$$

The first step makes use of the result $\gamma^s(\mathbf{q}; \mathscr{R}_{m_i})\gamma^s(\mathbf{q}; \mathscr{R}_{m_j}) = \gamma^s(\mathbf{q}; \mathscr{R}_{m_i}\mathscr{R}_{m_j})$ appropriate to allowable representations, which follows from the group multiplication property. The last step uses the relation $\mathbf{A}\cdot\mathbf{R}\mathbf{B} = \mathbf{R}^{-1}\mathbf{A}\cdot\mathbf{B}$, where $\mathbf{A}$ and $\mathbf{B}$ are two vectors and $\mathbf{R}$ is a rotation in three dimensions. From (3.5), the factor $(\mathbf{q} - \mathbf{R}_i^{-1}\mathbf{q})$ occurring in (3.9) can be equal to zero or to some reciprocal lattice vector. Accordingly we may write

$$(\mathbf{q} - \mathbf{R}_i^{-1}\mathbf{q}) = \mathbf{G}(\mathbf{q}, R_i), \tag{3.10}$$

making explicit the dependence of $\mathbf{G}$ on both $\mathbf{q}$ and $R_i$. $\mathbf{G}$ can be nonzero only if $\mathbf{q}$ is on the boundary of the BZ. Writing

$$\exp\{i[(\mathbf{q} - \mathbf{R}_i^{-1}\mathbf{q})\cdot\mathbf{v}(R_j)]\} = \phi(\mathbf{q}; \mathbf{R}_i, \mathbf{R}_j) \tag{3.11}$$

and remembering that $\mathbf{G}\cdot\mathbf{x}(m) = 2\pi \times$ integer, we can express (3.9) as

$$\tau^s(\mathbf{q}; \mathbf{R}_i)\tau^s(\mathbf{q}; \mathbf{R}_j) = \phi(\mathbf{q}; \mathbf{R}_i, \mathbf{R}_j)\tau^s(\mathbf{q}; \mathbf{R}_i\mathbf{R}_j). \tag{3.12}$$

Comparing (3.12) and (3.6), we see that indeed the set of matrices $\tau^s(\mathbf{q}) = \{\tau^s(\mathbf{q}; \mathbf{R})\}$, $\mathbf{R} \in G_0(\mathbf{q})$, furnish a multiplier representation of $G_0(\mathbf{q})$ corresponding to the factor system $\{\phi(\mathbf{q}; \mathbf{R}_i, \mathbf{R}_j)\}$. Further it may be argued by *reductio ad absurdum* that since $\gamma^s(\mathbf{q})$ is irreducible $\tau^s(\mathbf{q})$ is also. Combining (3.10) and (3.11) we obtain for the multiplier

$$\phi(\mathbf{q}; \mathbf{R}_i, \mathbf{R}_j) = \exp\{i\mathbf{G}(\mathbf{q}, R_i)\cdot\mathbf{v}(R_j)\}. \tag{3.13}$$

It is clear from (3.13) that—
1. All the multipliers are unity if the space group is symmorphic (in which case $\mathbf{v}(R) = 0$ for every $R$), and consequently the representations of $G_0(\mathbf{q})$ are always ordinary representations.
2. All the multipliers are unity even for nonsymmorphic groups provided $\mathbf{q}$ is within the BZ (whence $\mathbf{G}(\mathbf{q}, R_i) = 0$); once again the multiplier representations reduce to the ordinary representations.
3. Some of the multipliers can be complex numbers of modulus unity if the space group is nonsymmorphic and $\mathbf{q}$ is on the zone boundary. In this case the multiplier representations are different from the ordinary representations.

This means that, whether or not the space group $G$ is symmorphic, the IMR's in most situations (1 and 2 above) reduce to the ordinary representations of the point group $G_0(\mathbf{q})$ which can be readily obtained from books on molecular physics.* For case 3, however, the IMR's must be obtained by special techniques available for that purpose.[12] In passing it may be mentioned that several compilations of tables of IMR's have recently appeared.[13–15]

## 3.2 Role of $G_0(\mathbf{q})$

We are now prepared to discuss the actual details of applying group theory to the eigenvalue equation (3.4). One may identify four steps in the process:

1. The first step is to discover a set of unitary matrices which commute with $\mathbf{D}(\mathbf{q})$. These matrices will be obtained by considering the transformation of vectors in the space of the eigenvectors of $\mathbf{D}(\mathbf{q})$. The matrices obtained will be found to furnish a reducible multiplier representation of $G_0(\mathbf{q})$.

2. Utilizing the latter property, the eigenvectors of $\mathbf{D}(\mathbf{q})$ are next given group-theoretic labels.

3. The next stage is to construct *symmetry-adapted* vectors which have the same transformation properties as the eigenvectors. For this purpose it is not necessary to know the eigenvectors (which, of course, would amount to begging the question).

4. As the final step, $\mathbf{D}(\mathbf{q})$ is transformed to a more convenient form $\mathscr{D}(\mathbf{q})$ by referring the matrix to the vectors obtained in step 3. This transformation of the dynamical matrix provides the essential simplification of the eigenvalue problem.

### 3.2.1 Construction of unitary matrices that commute with $\mathbf{D}(\mathbf{q})$

In order to construct the unitary matrices mentioned above, we must examine the transformations induced by space-group operations in the $3n$-dimensional space spanned by $\{\mathbf{e}\binom{\mathbf{q}}{j}\}$. The reason for this may be appreciated by recalling the arguments leading to (2.26). There, by starting from the transformation of displacements and subsequently imposing invariance conditions on the potential energy, we arrived at a commutation relation involving force-constant and transformation matrices. By pursuing an analogous approach we can arrive at a commutation rule for

---

*Care is required, however, since books on molecular physics often combine pairs of complex one-dimensional representations related by time-reversal symmetry into two-dimensional representations.

$D(\mathbf{q})$; this clearly must involve matrices describing the transformation of Fourier components of the displacements, that is, of the Bloch vectors.

We recall from (2.160) that atomic displacements can be expressed in the form

$$u_\alpha \binom{l}{k} = \frac{1}{\sqrt{Nm_k}} \sum_{\mathbf{q}} \exp[i\mathbf{q}\cdot\mathbf{x}(l)]\zeta_\alpha(k|\mathbf{q}).$$

Thus, corresponding to a given $\mathbf{q}$,

$$u_\alpha \binom{l}{k} = \frac{1}{\sqrt{Nm_k}} \{\zeta_\alpha(k|\mathbf{q}) \exp[i\mathbf{q}\cdot\mathbf{x}(l)] + \text{c.c.}\}. \tag{3.14}$$

Here c.c. denotes the complex conjugate of the preceding quantity, and must be included since $\mathbf{u}$ must be real. We are seeking the transformation of the Bloch vectors $\zeta(\mathbf{q})$, which are $3n$-component vectors existing in the same space as the $\mathbf{e}\binom{\mathbf{q}}{j}$'s.

Next, as is done in the study of molecular vibrations, we consider two displacement configurations CI and CII. Of these, CI has displacements as in (3.14) while CII is a configuration derived from CI such that if

$$\binom{\lambda}{\kappa} \xrightarrow{\mathscr{R}_m} \binom{l}{k}, \text{ that is, } \mathbf{x}\binom{l}{k} = \mathscr{R}_m\mathbf{x}\binom{\lambda}{\kappa}, \tag{3.15}$$

then we associate with every site $\binom{l}{k}$ a displacement $\mathbf{Ru}\binom{\lambda}{\kappa}$. Formally let us write the displacements in CII as

$$\mathbf{u}'\binom{l}{k} = \frac{1}{\sqrt{Nm_k}} \{\zeta'(k|\mathbf{q}) \exp[i\mathbf{q}\cdot\mathbf{x}(l)] + \text{c.c.}\}. \tag{3.16}$$

On the other hand, from our definition,

$$\mathbf{u}'\binom{l}{k} = \mathbf{Ru}\binom{\lambda}{\kappa} = \frac{1}{\sqrt{Nm_\kappa}} \{\mathbf{R}\zeta(\kappa|\mathbf{q}) \exp[i\mathbf{q}\cdot\mathbf{x}(\lambda)] + \text{c.c.}\}. \tag{3.17}$$

Now from (3.15)

$$\mathbf{x}(\lambda) = \mathscr{R}_m^{-1}\mathbf{x}\binom{l}{k} - \mathbf{x}(\kappa)$$
$$= \mathbf{R}^{-1}(\mathbf{x}(l) + \mathbf{x}(k)) - \mathbf{x}(\kappa) - \mathbf{R}^{-1}[\mathbf{v}(R) + \mathbf{x}(m)].$$

Therefore

$$\mathbf{q}\cdot\mathbf{x}(\lambda) = \mathbf{Rq}\cdot\mathbf{x}(l) + \mathbf{Rq}\cdot\mathbf{x}(k) - \mathbf{Rq}\cdot\mathscr{R}_m\mathbf{x}(\kappa).$$

Since the sublattices $\kappa$ and $k$ must be occupied by the same species, $m_\kappa = m_k$ and (3.17) becomes

$$
\mathbf{u}'\binom{l}{k} = \frac{1}{\sqrt{Nm_k}} (\mathbf{R}\zeta(\kappa|\mathbf{q}) \exp\{i\mathbf{Rq}\cdot[\mathbf{x}(k) - \mathscr{R}_m\mathbf{x}(\kappa)]\}
$$
$$
\times \exp[i\mathbf{Rq}\cdot\mathbf{x}(l)] + \text{c.c.}). \tag{3.18}
$$

Remembering (3.5) and that $\mathbf{G}\cdot\mathbf{x}(m) = 2\pi \times$ integer, we can write (3.18) as

$$
\mathbf{u}'\binom{l}{k} = \frac{1}{\sqrt{Nm_k}} (\mathbf{R}\zeta(\kappa|\mathbf{q}) \exp\{i\mathbf{Rq}\cdot[\mathbf{x}(k) - \mathscr{R}_m\mathbf{x}(\kappa)]\}
$$
$$
\times \exp[i\mathbf{q}\cdot\mathbf{x}(l)] + \text{c.c.}). \tag{3.19}
$$

Comparing (3.16) and (3.19), we obtain

$$
\zeta(k|\mathbf{q}) \xrightarrow{\mathscr{R}_m} \zeta'(k|\mathbf{q})
$$
$$
= \mathbf{R}\zeta(\kappa|\mathbf{q}) \exp\{i\mathbf{Rq}\cdot[\mathbf{x}(k) - \mathscr{R}_m\mathbf{x}(\kappa)]\}. \tag{3.20}
$$

This result may be expressed compactly in matrix notation by

$$
\zeta'(\mathbf{q}) = \Gamma(\mathbf{q}; \mathscr{R}_m)\zeta(\mathbf{q}), \tag{3.21}
$$

where $\zeta$ and $\zeta'$ are $3n$-component column matrices and $\Gamma(\mathbf{q}; \mathscr{R}_m)$ is a $3n$-dimensional square matrix with elements

$$
\Gamma_{\alpha\beta}(kk'|\mathbf{q}; \mathscr{R}_m) = R_{\alpha\beta}\delta(k, F(k', R)) \exp\{i\mathbf{Rq}\cdot[\mathbf{x}(k) - \mathscr{R}_m\mathbf{x}(k')]\}. \tag{3.22}
$$

The Kronecker delta $\delta(k, F(k', R))$ describes the interchange of sublattices, if any, vanishing unless $k$ corresponds to the sublattice $F(k', R)$ reached from $k'$ via $\mathscr{R}_m$.*

The matrix $\Gamma(\mathbf{q}; \mathscr{R}_m)$ describes the transformation of vectors in the space of the Bloch vectors associated with a particular $\mathbf{q}$. It is easily proved[11] that $\Gamma$ is unitary. This is also to be expected intuitively, since $\Gamma$ merely "rotates" $\zeta(\mathbf{q})$ to $\zeta'(\mathbf{q})$ and therefore involves no change of length.

We now argue that the matrices $\Gamma(\mathbf{q}; \mathscr{R}_m)$ corresponding to every element $\mathscr{R}_m$ of $G(\mathbf{q})$ commute with $\mathbf{D}(\mathbf{q})$. We first observe from Eq. (2.172) that the eigenvalue equation (3.4) is essentially associated with a potential function $\Phi(\mathbf{q}) = \zeta^\dagger(\mathbf{q})\mathbf{D}(\mathbf{q})\zeta(\mathbf{q})$. Let us now examine the effect on $\Phi(\mathbf{q})$ consequent to the operation $\mathscr{R}_m$ on the crystal. This causes a transformation of displacements as in (3.17), and the crystal goes from configuration CI to CII. Since $\mathbf{q}$ is unchanged, the potential function associated with CII may be written as $\Phi'(\mathbf{q}) = [\zeta'(\mathbf{q})]^\dagger\mathbf{D}(\mathbf{q})\zeta'(\mathbf{q})$. Furthermore, since the potential itself must remain invariant, we have $\Phi(\mathbf{q}) = \Phi'(\mathbf{q})$. Hence†

---

*Note that the sublattice interchanges do not depend on the lattice translational part of $\mathscr{R}_m$, since the same sublattice is reached from $k$ for all $\mathbf{x}(m)$. For this reason we write $F(k, R)$ instead of $F(k, \mathscr{R}_m)$.

†In the present discussion, we treat $\zeta_\alpha(k|\mathbf{q})$ and $\zeta'_\alpha(k|\mathbf{q})$ as "coordinates" pertaining to CI and CII in the same spirit as in molecular physics. A more elaborate derivation of (3.23) using explicitly the Cartesian displacements as the coordinates is also possible.

$$\Phi(\mathbf{q}) = \zeta^\dagger(\mathbf{q})\mathbf{D}(\mathbf{q})\zeta(\mathbf{q})$$
$$= [\zeta'(\mathbf{q})]^\dagger \mathbf{D}(\mathbf{q})\zeta'(\mathbf{q})$$
$$= \zeta^\dagger(\mathbf{q})\Gamma^\dagger(\mathbf{q}; \mathscr{R}_m)\mathbf{D}(\mathbf{q})\Gamma(\mathbf{q}; \mathscr{R}_m)\zeta(\mathbf{q}) \quad \text{(using (3.21))} \qquad (3.23)$$

which implies

$$\mathbf{D}(\mathbf{q}) = \Gamma^\dagger(\mathbf{q}; \mathscr{R}_m)\mathbf{D}(\mathbf{q})\Gamma(\mathbf{q}; \mathscr{R}_m). \qquad (3.24)$$

This shows that for every element $\mathscr{R}_m$ in $G(\mathbf{q})$ exists a unitary matrix $\Gamma(\mathbf{q}; \mathscr{R}_m)$ which commutes with $\mathbf{D}(\mathbf{q})$. It is further possible to show that the set of matrices $\{\Gamma(\mathbf{q}; \mathscr{R}_m)\}$, $\mathscr{R}_m \in G(\mathbf{q})$, provides a $3n$-dimensional representation of $G(\mathbf{q})$; this forms the starting point of the allowable-representation approach to the lattice dynamical problem.

The matrices pertaining to the multiplier-representation approach are related to the $\Gamma$ matrices through[11] (see also (3.8))

$$\mathbf{T}(\mathbf{q}; \mathbf{R}) = \Gamma(\mathbf{q}; \mathscr{R}_m) \exp\{i\mathbf{q}\cdot[\mathbf{v}(R) + \mathbf{x}(m)]\}. \qquad (3.25)$$

Using the result

$$\mathbf{R}\mathbf{q}\cdot[\mathbf{x}(k) - \mathscr{R}_m\mathbf{x}(k')] + \mathbf{q}\cdot[\mathbf{v}(R) + \mathbf{x}(m)]$$
$$= \mathbf{q}\cdot\mathbf{x}(k) + \mathbf{G}\cdot\mathbf{x}(k) - \mathbf{q}\cdot\mathscr{R}_m\mathbf{x}(k') - \mathbf{G}\cdot\mathscr{R}_m\mathbf{x}(k') + \mathbf{q}\cdot[\mathbf{v}(R) + \mathbf{x}(m)]$$
$$= \mathbf{q}\cdot[\mathbf{x}(k) - \mathbf{R}\mathbf{x}(k')] + \mathbf{G}\cdot[\mathbf{x}(k) - \mathscr{R}_m\mathbf{x}(k')]$$

and Eqs. (3.22) and (3.25), we may write the elements of $\mathbf{T}(\mathbf{q}; \mathbf{R})$ as

$$T_{\alpha\beta}(kk'|\mathbf{q}; \mathscr{R}_m) = R_{\alpha\beta}\delta(k, F(k', R)) \exp\{i\mathbf{q}\cdot[\mathbf{x}(k) - \mathbf{R}\mathbf{x}(k')]\}$$
$$\times \exp\{i\mathbf{G}\cdot[\mathbf{x}(k) - \mathscr{R}_m\mathbf{x}(k')]\}.$$

Now the elements of $\mathbf{T}$ involving $k$ and $k'$ are nonzero only if $k = F(k', R)$. This means

$$\mathscr{R}_m\mathbf{x}(k') = \mathscr{R}_m\mathbf{x}\binom{0}{k'} = \mathbf{x}(L) + \mathbf{x}(k),$$

where $\binom{L}{k}$ is the site to which $\binom{0}{k'}$ is carried by $\mathscr{R}_m$. Therefore

$$\mathbf{G}\cdot[\mathbf{x}(k) - \mathscr{R}_m\mathbf{x}(k')] = -\mathbf{G}\cdot\mathbf{x}(L),$$

and so

$$T_{\alpha\beta}(kk'|\mathbf{q}; \mathbf{R}) = R_{\alpha\beta}\delta(k, F(k', R)) \exp\{i\mathbf{q}\cdot[\mathbf{x}(k) - \mathbf{R}\mathbf{x}(k')]\}. \qquad (3.26)$$

It may be noted that, like (3.8), Eq. (3.25) establishes a many-to-one correspondence. In other words, the same matrix $\mathbf{T}(\mathbf{q}; \mathbf{R})$ is associated with each of the $N$ matrices $\Gamma(\mathbf{q}; \mathscr{R}_m)$ corresponding to the $N$ elements $\mathscr{R}_m$ with a fixed rotational part $\mathbf{R}$ but variable translational part $\mathbf{x}(m)$. Further, there is a one-to-one correspondence between the matrices

$\mathbf{T}(\mathbf{q}; \mathbf{R})$ and the elements of $G_0(\mathbf{q})$, and it is this set $T(\mathbf{q}) = \{\mathbf{T}(\mathbf{q}; \mathbf{R})\}$, $\mathbf{R} \in G_0(\mathbf{q})$, which is of present interest.

The matrices $\mathbf{T}(\mathbf{q}; \mathbf{R})$ have the following properties:

1. They are unitary; this follows immediately from (3.25) upon remembering that $\Gamma$ is unitary.

2. $\mathbf{T}(\mathbf{q}; \mathbf{R})\mathbf{D}(\mathbf{q})\mathbf{T}^{\dagger}(\mathbf{q}; \mathbf{R}) = \mathbf{D}(\mathbf{q})$, \hfill (3.27)

that is, the matrices $\mathbf{T}(\mathbf{q}; \mathbf{R})$ commute with $\mathbf{D}(\mathbf{q})$ for all $R$; this follows from (3.24).

3. They provide a $3n$-dimensional multiplier representation of $G_0(\mathbf{q})$ corresponding to the factor system defined in (3.13)*; this is straightforward to demonstrate using the form of the $\mathbf{T}$ matrix given above.

### 3.2.2 Group-theoretical labeling of eigenvectors

Finding the $\mathbf{T}$ matrices completes the first phase of the program. Given the existence of the set $T(\mathbf{q}) = \{\mathbf{T}(\mathbf{q}, \mathbf{R})\}$, $\mathbf{R} \in G_0(\mathbf{q})$, we next assert that the eigenvectors $\mathbf{e}\binom{\mathbf{q}}{j}$ can be classified in terms of the IMR's of $G_0(\mathbf{q})$. To see this, consider the equation

$$\mathbf{D}(\mathbf{q})\mathbf{e}\binom{\mathbf{q}}{j} = \omega_j^2(\mathbf{q})\mathbf{e}\binom{\mathbf{q}}{j}.$$

Since $\mathbf{T}(\mathbf{q}, \mathbf{R})$ commutes with $\mathbf{D}(\mathbf{q})$, this equation may be written

$$\mathbf{D}(\mathbf{q})\left[\mathbf{T}(\mathbf{q}; \mathbf{R})\mathbf{e}\binom{\mathbf{q}}{j}\right] = \omega_j^2(\mathbf{q})\left[\mathbf{T}(\mathbf{q}; \mathbf{R})\mathbf{e}\binom{\mathbf{q}}{j}\right].$$

This means that $\mathbf{T}(\mathbf{q}; \mathbf{R})\mathbf{e}\binom{\mathbf{q}}{j}$ must be some linear combination of the eigenvectors of $\mathbf{D}(\mathbf{q})$ corresponding to the eigenvalue $\omega_j^2(\mathbf{q})$. To express this, we introduce the double index $\sigma\lambda$ in place of $j$ to label the eigenvectors. Here $\sigma$ refers to the *distinct* eigenvalues $\omega_\sigma^2(\mathbf{q})$ of degeneracy $f_\sigma$, and $\lambda = 1, 2, \ldots, f_\sigma$ labels the $f_\sigma$ orthonormal eigenvectors belonging to $\omega_\sigma^2(\mathbf{q})$. In this notation,

$$\mathbf{D}(\mathbf{q})\mathbf{e}\binom{\mathbf{q}}{\sigma\lambda} = \omega_\sigma^2(\mathbf{q})\mathbf{e}\binom{\mathbf{q}}{\sigma\lambda}$$

and

$$\mathbf{T}(\mathbf{q}; \mathbf{R})\mathbf{e}\binom{\mathbf{q}}{\sigma\lambda} = \sum_{\lambda'=1}^{f_\sigma} A_{\lambda'\lambda}^\sigma(\mathbf{q}; \mathbf{R})\mathbf{e}\binom{\mathbf{q}}{\sigma\lambda'} \hfill (3.28a)$$

for every $\mathbf{R}$ in $G_0(\mathbf{q})$. Attention is called to the fact that the summation on

---

*The multiplier representations we shall be considering from now on will invariably pertain to the factor system defined in (3.13).

the right-hand side is over the first index. This is in keeping with standard group-theoretic practice and is adopted to guarantee that the matrices $\mathbf{A}^\sigma(\mathbf{q}; \mathbf{R})$ with elements $A^\sigma_{\lambda'\lambda}(\mathbf{q}; \mathbf{R})$ have the properties of a representation. Indeed, by using the fact that $\{\mathbf{T}(\mathbf{q}; \mathbf{R})\}$ provides a multiplier representation of $G_0(\mathbf{q})$, it may be shown easily that the set of matrices $A^\sigma(\mathbf{q}) = \{\mathbf{A}^\sigma(\mathbf{q}; \mathbf{R})\}$ also furnish a $f_\sigma$-dimensional multiplier representation of $G_0(\mathbf{q})$.[11] One can in fact go further and assert that, except in the rare event of an accidental degeneracy, $A^\sigma(\mathbf{q})$ is an irreducible multiplier representation,[11] so that using the earlier notation we may write (3.28a) as

$$\mathbf{T}(\mathbf{q}; \mathbf{R})\mathbf{e}\begin{pmatrix}\mathbf{q}\\\sigma\lambda\end{pmatrix} = \sum_{\lambda'=1}^{f_\sigma} \tau^\sigma_{\lambda'\lambda}(\mathbf{q}; \mathbf{R})\mathbf{e}\begin{pmatrix}\mathbf{q}\\\sigma\lambda'\end{pmatrix}. \tag{3.28b}$$

Thus the transformation properties of $\mathbf{e}\begin{pmatrix}\mathbf{q}\\j\end{pmatrix}$ can be discussed in terms of the IMR's of $G_0(\mathbf{q})$. This in turn implies that the eigenvectors and the eigenvalues may be given the same labels as the IMR's according to which the eigenvectors transform.

### 3.2.3 Construction of symmetry-adapted vectors
Besides providing convenient labels, the transformation property of eigenvectors mentioned above suggests a way of simplifying the determination of eigenvalues and the eigenvectors. The idea is to express the dynamical matrix with respect to a set of $3n$ symmetry vectors which have the same transformation properties as the eigenvectors. Since an eigenvector involves a linear combination of symmetry vectors having the *same* transformation properties, this change of basis renders simpler the determination of eigenvectors and eigenvalues. Presently we shall discuss the details, but first we must determine the number of occurrences of the various IMR's in the reducible representation $T(\mathbf{q})$. This is given by the formula†

$$c_s = \frac{1}{h}\sum_{R \in G_0(\mathbf{q})} [\chi^s(\mathbf{q}; \mathbf{R})]^*\chi(\mathbf{q}; \mathbf{R}), \tag{3.29}$$

where $h$ is the order of $G_0(\mathbf{q})$ and

$$\chi(\mathbf{q}; \mathbf{R}) = \mathrm{Tr}\,\mathbf{T}(\mathbf{q}; \mathbf{R}), \quad \chi^s(\mathbf{q}; \mathbf{R}) = \mathrm{Tr}\,\tau^s(\mathbf{q}; \mathbf{R}). \tag{3.30}$$

---

†Formula (3.29) assumes that the IMR's are unitary. In general the number of times an IMR $\{\tau^s(R)\}$ of a group of order $h$ occurs in a reducible multiplier representation $\{\mathbf{T}(R)\}$ of the same group is given by[11]

$$c_s = \frac{1}{h}\sum_{R \in G} \mathrm{Tr}[\mathbf{T}(R)]\,\mathrm{Tr}[\tau^s(R^{-1})][\phi(R, R^{-1})]^{-1}.$$

It is pertinent to note in this context that the matrices $\{\mathbf{A}^\sigma(\mathbf{q}; \mathbf{R})\}$ introduced in (3.28a) constitute a unitary representation. For proof, see ref. 11.

Here $c_s$ is the number of times the IMR $\tau^s(\mathbf{q})$ of dimensionality $f_s$ occurs in the reducible multiplier representation $T(\mathbf{q})$. Correspondingly there are $c_s$ (generally distinct) eigenvalues $\omega_{s,1}^2(\mathbf{q})$, $\omega_{s,2}^2(\mathbf{q})$, ..., $\omega_{s,c_s}^2(\mathbf{q})$, each of which is $f_s$-fold degenerate. The associated eigenvectors are labeled $\mathbf{e}\binom{\mathbf{q}}{sa\lambda}$, where $s$ labels the IMR, $a$ ($= 1, 2, \ldots, c_s$) indexes the occurrence and $\lambda$ ($= 1, 2, \ldots, f_s$) labels the different eigenvectors corresponding to a given squared frequency $\omega_{sa}^2(\mathbf{q})$. Thus the eigenvectors for the above set of eigenvalues are

$$\left\{ \mathbf{e}\binom{\mathbf{q}}{s11}, \mathbf{e}\binom{\mathbf{q}}{s12}, \ldots, \mathbf{e}\binom{\mathbf{q}}{s1f_s} \right\},$$

$$\left\{ \mathbf{e}\binom{\mathbf{q}}{s21}, \mathbf{e}\binom{\mathbf{q}}{s22}, \ldots, \mathbf{e}\binom{\mathbf{q}}{s2f_s} \right\},$$

$$\cdots\cdots\cdots\cdots\cdots\cdots\cdots\cdots\cdots$$

$$\left\{ \mathbf{e}\binom{\mathbf{q}}{sc_s1}, \mathbf{e}\binom{\mathbf{q}}{sc_s2}, \ldots, \mathbf{e}\binom{\mathbf{q}}{sc_sf_s} \right\}. \tag{3.31}$$

In this way, *even without solving the problem numerically, one can determine the number of distinct eigenvalues, their degeneracies, and the symmetry properties of the associated eigenvectors.*

Having determined the occurrences of the various IMR's, the next step is to construct the requisite symmetry-adapted vectors. For this purpose, one forms the projection operator $P_{\lambda\lambda}^s(\mathbf{q})$ defined by

$$P_{\lambda\lambda}^s(\mathbf{q}) = \frac{f_s}{h} \sum_{\mathbf{R}\in\bar{G}_0(\mathbf{q})} [\tau_{\lambda\lambda}^s(\mathbf{q}; \mathbf{R})]^* \mathbf{T}(\mathbf{q}; \mathbf{R}). \tag{3.32}$$

By applying this operator systematically to the following orthogonal unit vectors of the $3n$-dimensional space of $\{\mathbf{e}\binom{\mathbf{q}}{j}\}$:

$$\begin{pmatrix}1\\0\\0\\\vdots\\0\end{pmatrix}, \begin{pmatrix}0\\1\\0\\\vdots\\0\end{pmatrix}, \ldots, \begin{pmatrix}0\\0\\0\\\vdots\\1\end{pmatrix} \quad \updownarrow \; 3n \text{ elements,} \tag{3.33}$$

one projects out $c_s$ mutually orthogonal vectors. After normalization these may be labeled $\boldsymbol{\Sigma}\binom{\mathbf{q}}{s1\lambda}, \boldsymbol{\Sigma}\binom{\mathbf{q}}{s2\lambda}, \ldots, \boldsymbol{\Sigma}\binom{\mathbf{q}}{sc_s\lambda}$. Each of these vectors transforms according to the $\lambda$th row of the IMR $\tau^s(\mathbf{q})$; further, corresponding to each are $(f_s - 1)$ partners which transform according to the remaining rows of the IMR $\tau^s(\mathbf{q})$. The partners of $\boldsymbol{\Sigma}\binom{\mathbf{q}}{sa\lambda}$ (for each $a$) may be generated by applying to that vector the following $(f_s - 1)$ operators:

$$P_{\mu\lambda}^s(\mathbf{q}) = \frac{f_s}{h} \sum_{\mathbf{R}\in G_0(\mathbf{q})} [\tau_{\mu\lambda}^s(\mathbf{q};\mathbf{R})]^* \dot{\mathbf{T}}(\mathbf{q};\mathbf{R}),$$

$$(\mu = 1, 2, \ldots, \lambda - 1, \lambda + 1, \ldots, f_s). \tag{3.34}$$

In this fashion one can obtain $c_s$ orthonormal sets

$$\left\{ \mathbf{\Sigma}\begin{pmatrix}\mathbf{q}\\s11\end{pmatrix}, \mathbf{\Sigma}\begin{pmatrix}\mathbf{q}\\s12\end{pmatrix}, \ldots, \mathbf{\Sigma}\begin{pmatrix}\mathbf{q}\\s1f_s\end{pmatrix} \right\},$$

$$\left\{ \mathbf{\Sigma}\begin{pmatrix}\mathbf{q}\\s21\end{pmatrix}, \mathbf{\Sigma}\begin{pmatrix}\mathbf{q}\\s22\end{pmatrix}, \ldots, \mathbf{\Sigma}\begin{pmatrix}\mathbf{q}\\s2f_s\end{pmatrix} \right\},$$

$$\cdots\cdots\cdots\cdots\cdots\cdots\cdots\cdots$$

$$\left\{ \mathbf{\Sigma}\begin{pmatrix}\mathbf{q}\\sc_s1\end{pmatrix}, \mathbf{\Sigma}\begin{pmatrix}\mathbf{q}\\sc_s2\end{pmatrix}, \ldots, \mathbf{\Sigma}\begin{pmatrix}\mathbf{q}\\sc_sf_s\end{pmatrix} \right\}, \tag{3.35}$$

each containing $f_s$ vectors and corresponding to the $c_s$ occurrences of $\tau^s(\mathbf{q})$ in $T(\mathbf{q})$; these have the same transformation properties as the eigenvectors. The process can be repeated for all the IMR's of $G_0(\mathbf{q})$ which occur in $T(\mathbf{q})$, so that a total of $3n$ symmetry vectors $\{\mathbf{\Sigma}(^{\mathbf{q}}_{sa\lambda})\}$ may be constructed.

It is useful to note that the $\mathbf{\Sigma}$ vectors each correspond to the motion of one type of atom only. This follows from a consideration of the effect of the $\mathbf{T}$ operators on the vectors (3.33).

### 3.2.4 Block diagonalization of the dynamical matrix
The symmetry-adapted vectors $\{\mathbf{\Sigma}(^{\mathbf{q}}_{sa\lambda})\}$ have the property that

$$\mathbf{\Sigma}^\dagger\begin{pmatrix}\mathbf{q}\\s'a'\lambda'\end{pmatrix}\mathbf{D}(\mathbf{q})\mathbf{\Sigma}\begin{pmatrix}\mathbf{q}\\sa\lambda\end{pmatrix} = 0 \text{ unless } s = s' \text{ and } \lambda = \lambda'. \tag{3.36}$$

This is the statement of the usual "matrix-element theorem" of group theory. From (3.36) it follows that if we form a $3n$-dimensional matrix $\mathbf{\Sigma}(\mathbf{q})$ by employing the following grouping of the vectors

$$\left\{ \cdots, \mathbf{\Sigma}\begin{pmatrix}\mathbf{q}\\s11\end{pmatrix}, \mathbf{\Sigma}\begin{pmatrix}\mathbf{q}\\s21\end{pmatrix}, \cdots, \mathbf{\Sigma}\begin{pmatrix}\mathbf{q}\\sc_s1\end{pmatrix}, \right.$$

$$\mathbf{\Sigma}\begin{pmatrix}\mathbf{q}\\s12\end{pmatrix}, \mathbf{\Sigma}\begin{pmatrix}\mathbf{q}\\s22\end{pmatrix}, \cdots, \mathbf{\Sigma}\begin{pmatrix}\mathbf{q}\\sc_s2\end{pmatrix},$$

$$\cdots\cdots\cdots\cdots\cdots\cdots\cdots\cdots$$

$$\mathbf{\Sigma}\begin{pmatrix}\mathbf{q}\\s1f_s\end{pmatrix}, \mathbf{\Sigma}\begin{pmatrix}\mathbf{q}\\s2f_s\end{pmatrix}, \cdots, \mathbf{\Sigma}\begin{pmatrix}\mathbf{q}\\sc_sf_s\end{pmatrix},$$

$$\left.\cdots\cdots\cdots\cdots\cdots\cdots\cdots\cdots \right\}, \tag{3.37}$$

that is, if we form $\boldsymbol{\Sigma}(\mathbf{q})$ by grouping the vectors belonging to the same $\lambda$ and same $s$, then

$$\mathscr{D}(\mathbf{q}) = \boldsymbol{\Sigma}^{\dagger}(\mathbf{q})\,\mathbf{D}(\mathbf{q})\boldsymbol{\Sigma}(\mathbf{q}) \tag{3.38}$$

will be block diagonal. In particular, there are $f_s$ diagonal blocks $\mathscr{D}^{s\lambda}(\mathbf{q})$ $(\lambda = 1, \ldots, f_s)$, each of dimensionality $c_s$. $\mathscr{D}(\mathbf{q})$ thus has the following structure (focusing attention on a particular value of $s$):

$$\mathscr{D}(\mathbf{q}) = \begin{pmatrix} \ddots & & & \\ & \mathscr{D}^{s1} & & \\ & & \ddots & \\ & & & \mathscr{D}^{sf_s} \\ & & & & \ddots \end{pmatrix}. \tag{3.39}$$

This form of the dynamical matrix facilitates considerably the problem of determining the eigenvalues and the eigenvectors by enabling us to deal separately with blocks pertaining to different IMR's. For example, to find the $c_s$ distinct eigenvalues $\omega^2_{s,1}(\mathbf{q})$, $\omega^2_{s,2}(\mathbf{q})$, $\ldots$, $\omega^2_{s,c_s}(\mathbf{q})$ (each $f_s$-fold degenerate) and their associated eigenvectors corresponding to $\tau^s(\mathbf{q})$, we need consider only one of the $c_s$-dimensional matrices $\mathscr{D}^{s\lambda}(\mathbf{q})$. The explicit equations to be solved are obtained as follows:

Express first the eigenvector $\mathbf{e}\begin{pmatrix} \mathbf{q} \\ sa\lambda \end{pmatrix}$ as a linear combination of all symmetry vectors belonging to the same $s$ and same $\lambda$, that is, write

$$\mathbf{e}\begin{pmatrix} \mathbf{q} \\ sa\lambda \end{pmatrix} = \sum_{b=1}^{c_s} \boldsymbol{\Sigma}\begin{pmatrix} \mathbf{q} \\ sb\lambda \end{pmatrix} C\begin{pmatrix} \mathbf{q} \\ sb\lambda, sa\lambda \end{pmatrix}, \tag{3.40}$$

where the $C$'s are complex coefficients. Substitution of (3.40) into (3.4) gives

$$\sum_{b=1}^{c_s} \mathbf{D}(\mathbf{q})\boldsymbol{\Sigma}\begin{pmatrix} \mathbf{q} \\ sb\lambda \end{pmatrix} C\begin{pmatrix} \mathbf{q} \\ sb\lambda, sa\lambda \end{pmatrix} = \omega^2_{sa}(\mathbf{q}) \sum_{b'=1}^{c_s} \boldsymbol{\Sigma}\begin{pmatrix} \mathbf{q} \\ sb'\lambda \end{pmatrix} C\begin{pmatrix} \mathbf{q} \\ sb'\lambda, sa\lambda \end{pmatrix}. \tag{3.41}$$

Multiplying both sides from the left by $\boldsymbol{\Sigma}^{\dagger}\begin{pmatrix} \mathbf{q} \\ sb''\lambda \end{pmatrix}$, where $b''$ is one of the set $1, 2, \ldots, c_s$, and subsequently rearranging, we obtain

$$\sum_{b=1}^{c_s} \left\{ \boldsymbol{\Sigma}^{\dagger}\begin{pmatrix} \mathbf{q} \\ sb''\lambda \end{pmatrix} [\mathbf{D}(\mathbf{q}) - \omega^2_{sa}(\mathbf{q})\mathbb{1}_{3n}] \boldsymbol{\Sigma}\begin{pmatrix} \mathbf{q} \\ sb\lambda \end{pmatrix} \right\} C\begin{pmatrix} \mathbf{q} \\ sb\lambda, sa\lambda \end{pmatrix} = 0,$$

$$(b'' = 1, 2, \ldots, c_s). \tag{3.42}$$

For (3.42) to have nontrivial solutions for the coefficients $C\begin{pmatrix} \mathbf{q} \\ sb\lambda, sa\lambda \end{pmatrix}$, $\omega^2_{sa}$ should coincide with the eigenvalues of the matrix whose $b''b$th element is

$$\boldsymbol{\Sigma}^{\dagger}\begin{pmatrix} \mathbf{q} \\ sb''\lambda \end{pmatrix} \mathbf{D}(\mathbf{q})\boldsymbol{\Sigma}\begin{pmatrix} \mathbf{q} \\ sb\lambda \end{pmatrix}, \tag{3.43a}$$

that is, with the roots of

$$\|\mathscr{D}^{s\lambda}(\mathbf{q}) - \omega^2(\mathbf{q})\mathbb{1}_{c_s}\| = 0. \tag{3.43b}$$

This equation must be solved by brute force, no further simplification due to symmetry being possible. However, it is a much simpler equation to solve then the usual $3n$-degree equation. Solving (3.43) gives the $c_s$ distinct eigenvalues $\omega^2_{s,1}(\mathbf{q}), \ldots, \omega^2_{s,c_s}(\mathbf{q})$. The same eigenvalues are obtained from each of the matrices $\mathscr{D}^{s\lambda}(\mathbf{q})$. To determine the eigenvectors one must find the coefficients $C\binom{\mathbf{q}}{sb\lambda,sa\lambda}$ (see Eq. (3.40)). These can be solved for by introducing the eigenvalues successively into (3.41). Once the various eigenvectors pertaining to the $\lambda$th row are known, their partners may be generated by application of the operator defined in (3.34). The process may then be repeated for the other occurring IMR's.

As already indicated (see Sec. 2.2.6), the calculation of eigenvalues and eigenvectors is nowadays generally performed on a computer. The problem fed to the machine is

$$\mathscr{D}(\mathbf{q})\mathbf{C}(\mathbf{q}) = \mathbf{C}(\mathbf{q})\mathbf{\Omega}(\mathbf{q}), \tag{3.44}$$

where $\mathbf{C}(\mathbf{q})$ is the matrix of eigenvectors of the (transformed) dynamical matrix $\mathscr{D}(\mathbf{q})$. Indeed even $\mathscr{D}$ itself may be assembled in a computer since it is now feasible to carry out the construction of projection operators and the generation of $\Sigma(\mathbf{q})$ in a machine. It is easily proved that (3.44) is equivalent to (2.52). From the latter equation we have, remembering $\Sigma$ is unitary,

$$\Sigma^\dagger \mathbf{D}\Sigma\Sigma^\dagger \mathbf{e} = (\Sigma^\dagger \mathbf{e})\mathbf{\Omega},$$

from which (3.44) immediately follows upon using (3.38) and making the identification

$$\mathbf{C}(\mathbf{q}) = \Sigma^\dagger(\mathbf{q})\mathbf{e}(\mathbf{q}). \tag{3.45}$$

Since $\mathscr{D}(\mathbf{q})$ is block diagonal, $\mathbf{C}(\mathbf{q})$ is also.

Computer diagonalization routines commonly available not only deliver the eigenvalues but also the eigenvectors, which in the present case are the vectors that make up $\mathbf{C}(\mathbf{q})$. The eigenvectors of the original problem, namely (2.52), can then be obtained from the inverse of (3.45):

$$\mathbf{e}(\mathbf{q}) = \Sigma(\mathbf{q})\mathbf{C}(\mathbf{q}), \tag{3.46}$$

which is the matrix form of (3.40).

In the special case in which $c_s = 1$, (3.41) simplifies to

$$\Sigma^\dagger\binom{\mathbf{q}}{s\lambda}\mathbf{D}(\mathbf{q})\Sigma\binom{\mathbf{q}}{s\lambda} = \omega_s^2(\mathbf{q})\Sigma^\dagger\binom{\mathbf{q}}{s\lambda}\Sigma\binom{\mathbf{q}}{s\lambda} = \omega_s^2(\mathbf{q}). \tag{3.47}$$

In other words, the frequency corresponding to an IMR which occurs *only once* is obtainable purely from symmetry considerations, no further diagonalization being necessary. Additionally, $\Sigma\binom{q}{s\lambda}$ can itself be chosen as the eigenvector.

An interesting corollary is the observation of Elliott and Thorpe[16] that in a mode corresponding to an IMR which occurs only once, only a single (chemical) species of atoms vibrates, all other species being stationary. One example is provided by the eigenvector for ZnS considered in the last chapter:

$$\frac{1}{\sqrt{2}}\begin{pmatrix} 0 \\ 0 \\ 0 \\ 1 \\ 0 \\ e^{-i\pi/2} \end{pmatrix}$$

(see Sec. 2.2.12). This pertains to the point $W$ in Figure 2.19. The IMR corresponding to this eigenvector occurs only once;[17,18] therefore only one species of atoms is expected to move, as borne out by the eigenvector.

Summarizing, group-theoretic applications in lattice dynamics exploit the existence of a set of matrices $T(\mathbf{q}) = \{\mathbf{T}(\mathbf{q};\mathbf{R})\}$, $\mathbf{R} \in G_0(\mathbf{q})$, that commute with $\mathbf{D}(\mathbf{q})$. These matrices furnish a $3n$-dimensional reducible multiplier representation of $G_0(\mathbf{q})$, and as a result the eigenvectors $\mathbf{e}\binom{q}{j}$ may be labeled in terms of the IMR's of $G_0(\mathbf{q})$ occurring in $T(\mathbf{q})$, as $\mathbf{e}\binom{q}{sa\lambda}$. Furthermore, using the projection-operator technique, one can construct symmetry vectors $\Sigma\binom{q}{sa\lambda}$ which enable the dynamical matrix to be brought into a block-diagonal form through the transformation $\Sigma^\dagger(\mathbf{q})\mathbf{D}(\mathbf{q})\Sigma(\mathbf{q})$. In passing we may note that the utility of group theory is greatest at symmetry points and along symmetry directions in the BZ.

### 3.3 Compatibility relations

We have seen how group-theoretic considerations may be invoked in order to classify the modes corresponding to a given vector $\mathbf{q}$ in the BZ. The question arises, How do the classifications and the degeneracies change as one moves from $\mathbf{q}$ to a neighboring point $\mathbf{q}'$? The answer depends on how the symmetry of the point group of the wave vector is altered. The particular case where the group $G_0(\mathbf{q}')$ is a subgroup of $G_0(\mathbf{q})$ is of special interest, for here the above question can be answered by examining the *compatibilities* between the IMR's of $G_0(\mathbf{q})$ and $G_0(\mathbf{q}')$. Since we have assumed the latter group to be a subgroup of the former, the IMR's $\tau^s(\mathbf{q})$ of $G_0(\mathbf{q})$ are in general reducible with respect to $G_0(\mathbf{q}')$ and can be expressed as a direct sum as follows:

$$\tau^s(\mathbf{q}) = \sum_{\oplus} c_\sigma \tau^\sigma(\mathbf{q}'). \tag{3.48}$$

The number $c_\sigma$ of times the IMR $\tau^\sigma(\mathbf{q}')$ occurs in the reducible representation $\{\tau^s(\mathbf{q}; \mathbf{R})\}$, $\mathbf{R} \in G_0(\mathbf{q}')$, [it is to be remembered that $\mathbf{R}$ now corresponds to elements common to $G_0(\mathbf{q})$ and $G_0(\mathbf{q}')$] can be obtained by the standard decomposition formula (see (3.29)). Equation (3.48) is the key for finding the changes in classifications and degeneracies, and some examples of its application will be considered later. It is worth noting that, formally, the present problem is similar to that of changes in the vibration spectra of molecules brought about by a lowering of symmetry of the molecule and to the problem of crystal-field splitting of atomic levels.

### 3.4 On degeneracies

Basically two categories of degeneracies are encountered in the study of normal vibrations of a crystal. The first arises from symmetry, both spatial and time-reversal; those due to spatial symmetry have already been studied while those due to time-reversal symmetry will be examined in a later section. Degeneracies caused by symmetry are termed *essential degeneracies*. On the other hand, degeneracies not required by symmetry may also occur and are called *accidental degeneracies* (AD). They manifest themselves as contacts or crossings between two or more branches belonging to nonequivalent or equivalent IMR's.

The various types of AD's have been investigated in detail by Herring[19] in connection with electron energy bands in crystals. These considerations can be extended to normal vibrations[20] as noted by Herring himself. Among the AD's, he identifies a type that arises due to rather special circumstances, namely to a special form of the potential of the problem, the degeneracy disappearing upon making an infinitesimal change in the potential consistent with the symmetry of the crystal. In lattice dynamics, an analogous situation would be caused by a special set of force constants, with the degeneracy disappearing when the values are changed by an infinitesimal amount in a way consistent with the crystal symmetry. As noted by Herring, such degeneracies are rarely encountered and may be formally classified as "highly improbable."*

---

*Herring uses the terminology "vanishingly improbable." What is implied is that the probability for such a degeneracy is vanishingly small. To avoid confusion we prefer the jargon introduced above.

Accidental degeneracies other than those of the "highly improbable" variety can also occur. As an illustration of such an AD involving two branches belonging to nonequivalent IMR's, let us consider two points $\mathbf{q}_1$ and $\mathbf{q}_2$ within the BZ, lying along the same direction $\hat{\mathbf{q}}$ and having the same wave-vector group. The IMR's associated with these points are therefore identical. Suppose that numerical calculations show that at $\mathbf{q}_1$ a branch belonging to a certain IMR $\tau^b(\mathbf{q})$ has a higher frequency than another which belongs to a nonequivalent IMR $\tau^a(\mathbf{q})$. Suppose further that at $\mathbf{q}_2$ the situation is reversed, as indicated in Figure 3.1. The question is, What are the forms of the dispersion curves for these branches in the region between $\mathbf{q}_1$ and $\mathbf{q}_2$? Since the $\tau$ labels for these curves cannot change suddenly, the two branches evidently must cross at some point $\mathbf{q}_3$. This contact is not symmetry determined and is therefore an AD. However, it is not of the "highly improbable" variety since, in general, any infinitesimal change of force constants (consistent with symmetry) still produces an intersection, though probably shifted. This kind of AD has already been encountered in the argon example (see Sec. 2.9).

In a similar fashion, AD's other than the "highly improbable" variety but associated with branches belonging to equivalent IMR's are also possible, though under special circumstances. Detailed criteria for their occurrence are available in the literature[19,20].

A degeneracy that all crystals exhibit is the threefold degeneracy of the

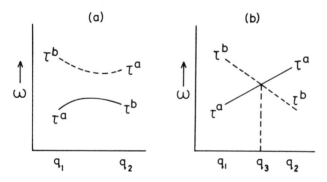

**Figure 3.1**
Illustration of an AD involving two branches corresponding to nonequivalent IMR's. At all points from $q_1$ to $q_2$ the two relevant IMR's are $\tau^a$ and $\tau^b$. Further, the ordering of the frequencies at $q_1$ and $q_2$ are as indicated. Two possibilities for the behavior of the branches in the region between $q_1$ and $q_2$ are shown. In (a) the group theoretical labels change for the *same* branch but this is not permitted. Hence the branches must cross at some point $q_3$, leading to an AD as shown in (b). (Note, however, that if the branches in (b) are numbered in an ordered sequence as discussed in Sec. 2.3, the group-theoretic labels for the branches change suddenly at $q_3$.)

acoustic branches at $\mathbf{q} = 0$. This arises from the invariance of the vibrational Hamiltonian under an infinitesimal translation.

## 3.5 Independent elements of $\mathbf{D}(\mathbf{q})$

In the last chapter we saw how symmetry considerations permit the determination of the structure of the force-constant matrices and the number of independent force constants. Analogously, by systematically exploiting the commutation relation (3.27)

$$\mathbf{D}(\mathbf{q}) = \mathbf{T}(\mathbf{q}; \mathbf{R})\mathbf{D}(\mathbf{q})\mathbf{T}^\dagger(\mathbf{q}; \mathbf{R}), \qquad \mathbf{R} \in G_0(\mathbf{q}),$$

it is possible to determine the structure of $\mathbf{D}(\mathbf{q})$ and the number of independent elements in it. An estimate of the latter may also be made by a different method. We use the fact that the transformation

$$\mathbf{D}(\mathbf{q}) \to \mathcal{D}(\mathbf{q}) = \Sigma^\dagger(\mathbf{q})\mathbf{D}(\mathbf{q})\Sigma(\mathbf{q}),$$

being based on symmetry considerations alone, preserves the number of independent elements. This is similar to the transformation

$$\phi\begin{pmatrix} L & L' \\ K & K' \end{pmatrix} = \mathbf{S}\phi\begin{pmatrix} l & l' \\ k & k' \end{pmatrix}\tilde{\mathbf{S}},$$

where the force-constant matrices on both sides have the same number of independent constants. Referring now to the structure of $\mathcal{D}(\mathbf{q})$ as given in (3.39) and further noting that $\mathcal{D}^{s\lambda}(\mathbf{q})$ can have a maximum of $c_s(c_s + 1)/2$ independent elements (since $\mathcal{D}(\mathbf{q})$ is Hermitian), it follows that the maximum number of independent elements in $\mathcal{D}(\mathbf{q})$ and therefore in $\mathbf{D}(\mathbf{q})$ is given by

$$\sum_s c_s(c_s + 1)/2, \tag{3.49}$$

where $s$ indexes the nonequivalent IMR's of $G_0(\mathbf{q})$ which occur in $T(\mathbf{q})$. Result (3.49) takes cognizance of the fact that the elements of $\mathcal{D}^{s\lambda}(\mathbf{q})$ corresponding to different $\lambda$ but the same $s$ are not independent of each other. Time-reversal symmetry, which we have not considered yet, could lead to a reduction of the number specified by (3.49).

## 3.6 Equivalence of results for different members of a star

We now consider how the above results pertain to vectors in the BZ which are related to $\mathbf{q}$ by symmetry. This requires first the concept of the *star of* $\mathbf{q}$.

Let $\mathbf{S}_1 = \mathbf{E}, \mathbf{S}_2, \ldots, \mathbf{S}_{g_0}$ denote the elements of the point group of the crystal. Given $\mathbf{q}$, compute

$$\mathbf{q}_i = \mathbf{S}_i\mathbf{q} \tag{3.50}$$

for all $i \, (= 1, 2, \ldots, g_0)$. Three possibilities arise:

1. $\mathbf{q}_i = \mathbf{q}$, \hfill (3.51a)

2. $\mathbf{q}_i = \mathbf{q} + \mathbf{G}(S_i), \mathbf{G}(S_i) \neq 0$, \hfill (3.51b)

3. $\mathbf{q}_i \neq \mathbf{q}$ or $\mathbf{q} + \mathbf{G}$. \hfill (3.51c)

The second possibility can occur if $\mathbf{q}$ lies on the surface of the BZ. We include $S_i$ in parentheses to emphasize that the reciprocal lattice vector $\mathbf{G}$ depends on $S_i$. Suppose that through the operations (3.50) we arrive at $(v - 1)$ vectors corresponding to (3.51c). Label these as $\mathbf{q}_2, \mathbf{q}_3, \ldots, \mathbf{q}_v$. The $v$ vectors $\mathbf{q}_1 = \mathbf{q}, \mathbf{q}_2, \ldots, \mathbf{q}_v$ together form a star. Stated in words, the star of $\mathbf{q}$ is the set of vectors comprised of $\mathbf{q}$ and of *nonequivalent vectors* (not equivalent in the sense of (3.51a or b)) generated from $\mathbf{q}$ through the application of point-group operations.

Intuitively one expects the frequencies corresponding to different members of a star to be the same. For example, in argon (see Sec. 2.9) the points $(\eta, \eta, \eta)$ and $(\eta, \bar{\eta}, \eta)$ in the BZ both belong to the family of $[111]$ directions and as such the frequencies at the two points are expected to be the same. This result can be established rigorously as will be done presently.

Now as a result of the application of the space-group operation $\mathscr{S}_m$ to the crystal, a wave initially propagating along $\mathbf{q}$ changes its direction of propagation to $\mathbf{Sq}$. Correspondingly, the Bloch vector $\boldsymbol{\zeta}(\mathbf{q})$ is transformed to $\boldsymbol{\zeta}'(\mathbf{Sq})$, and this transformation may be described in terms of a $3n$-dimensional unitary matrix $\boldsymbol{\Gamma}(\mathbf{q}; \mathscr{S}_m)$ similar to that defined in (3.21–3.22), that is,

$$\boldsymbol{\zeta}(\mathbf{q}) \xrightarrow{\mathscr{S}_m} \boldsymbol{\zeta}'(\mathbf{Sq}) = \boldsymbol{\Gamma}(\mathbf{q}; \mathscr{S}_m)\boldsymbol{\zeta}(\mathbf{q}). \tag{3.52}$$

The elements of $\boldsymbol{\Gamma}(\mathbf{q}; \mathscr{S}_m)$ may be deduced in the same fashion as in (3.22) and are found to be[11]

$$\Gamma_{\alpha\beta}(kk'|\mathbf{q}; \mathscr{S}_m) = S_{\alpha\beta}\delta(k, F(k', S))\exp\{i\mathbf{q}\cdot[\mathscr{S}_m^{-1}\mathbf{x}(k) - \mathbf{x}(k')]\}. \tag{3.53}$$

Next, as in the derivation of (3.23) we demand that the potential energies $\Phi(\mathbf{q})$ and $\Phi'(\mathbf{Sq})$ before and after the application of $\mathscr{S}_m$ be identical. This then gives[11]

$$\mathbf{D}(\mathbf{Sq}) = \boldsymbol{\Gamma}(\mathbf{q}; \mathscr{S}_m)\mathbf{D}(\mathbf{q})\boldsymbol{\Gamma}^\dagger(\mathbf{q}; \mathscr{S}_m). \tag{3.54}$$

It follows immediately that $\mathbf{D}(\mathbf{Sq})$ has the same set of eigenvalues as $\mathbf{D}(\mathbf{q})$, that is, $\{\omega_j^2(\mathbf{q})\} = \{\omega_{j'}^2(\mathbf{Sq})\}$. With a suitable labeling convention we can write

$$\omega_j^2(\mathbf{q}) = \omega_j^2(\mathbf{Sq}), \tag{3.55}$$

the result we expected intuitively.

The eigenvectors associated with different members of a star are related, for a nondegenerate mode $j$, by

$$\mathbf{e}\begin{pmatrix} \mathbf{Sq} \\ j \end{pmatrix} = e^{i\phi}\boldsymbol{\Gamma}(\mathbf{q}; \mathscr{S}_m)\,\mathbf{e}\begin{pmatrix} \mathbf{q} \\ j \end{pmatrix}, \tag{3.56}$$

where $\phi$ is a phase factor. Following Maradudin and Vosko[11] we choose the phase factor such that*

$$\mathbf{e}\begin{pmatrix} \mathbf{Sq} \\ j \end{pmatrix} = \boldsymbol{\Gamma}(\mathbf{q}; \{\mathbf{S}|\mathbf{v}(S)\})\,\mathbf{e}\begin{pmatrix} \mathbf{q} \\ j \end{pmatrix}. \tag{3.57}$$

If $\omega_j^2(\mathbf{q})$ is degenerate, then $\mathbf{e}\binom{\mathbf{Sq}}{\sigma\lambda}$ is a linear combination of $\mathbf{e}\binom{\mathbf{q}}{\sigma\lambda'}$, ($\lambda' = 1, 2, \ldots, f_\sigma$). However, with no loss of generality we can choose the linear combination such that (3.57) still holds[11].

Thus we see that if the eigenvalue problem is solved for one member of the star, the results for the other members can be derived through (3.55) and (3.57). This fact enables us to confine attention to the irreducible prism of the BZ while making dispersion curve calculations.

It is worth pointing out that in the approach involving (3.3) and the complete machinery of the space group, the set of eigenvalues associated with different members of a star and related to each other as in (3.55) are classified in terms of the IR's *of the space group*. On the other hand, the present approach concentrates on one member of a star and then derives the results for the other members via (3.55) and (3.57).

## 3.7 Application to other forms of the dynamical equation

In Sec. 2.2.14 it was noted that the dynamical equation may be written in forms other than given in (3.4). The application of group theory in such cases requires a slight modification of the procedure discussed above. Consider, for example, the form

$$\omega^2(\mathbf{q})\mathbf{mU}(\mathbf{q}) = \mathbf{B}(\mathbf{q})\mathbf{U}(\mathbf{q}) \tag{3.58}$$

(see (2.76)). As remarked in Sec. 3.1, this case requires that we construct

---

*Note that (3.57) applies only if $\mathbf{q}$ and $\mathbf{Sq}$ are distinct members of a star, that is, $\mathbf{Sq}$ corresponds to (3.51c).

matrices which commute with both $\mathbf{B}(\mathbf{q})$ and $\mathbf{m}$. It is easily verified that the $\mathbf{T}$ matrices defined in (3.26) have this property so that group theory may be applied to Eq. (3.58) in essentially the same fashion as before. A minor difference is that the matrix-element theorems are

$$\left.\begin{array}{l} \Sigma^\dagger\begin{pmatrix}\mathbf{q}\\s'b'\lambda'\end{pmatrix}\mathbf{B}(\mathbf{q})\Sigma\begin{pmatrix}\mathbf{q}\\sa\lambda\end{pmatrix} \\[2ex] \Sigma^\dagger\begin{pmatrix}\mathbf{q}\\s'b'\lambda'\end{pmatrix}\mathbf{m}\Sigma\begin{pmatrix}\mathbf{q}\\sa\lambda\end{pmatrix} \end{array}\right\} \text{vanish unless } s = s' \text{ and } \lambda = \lambda'. \tag{3.59}$$

Correspondingly, the determinantal equation (3.43b) becomes

$$\|\Sigma^\dagger(\mathbf{q})[\mathbf{B}(\mathbf{q}) - \omega^2(\mathbf{q})\mathbf{m}]\Sigma(\mathbf{q})\| = 0. \tag{3.60}$$

In the case of the dynamical equation

$$\omega^2(\mathbf{q})\mathbf{m}\mathbf{U}(\mathbf{q}) = \mathbf{M}(\mathbf{q})\mathbf{U}(\mathbf{q}), \tag{3.61}$$

we must alter the definition of the matrix $\Gamma(\mathbf{q}; \mathscr{R}_m)$ to accomodate differences in the phase factors. The appropriately commuting matrix now is

$$\Gamma_{\alpha\beta}(kk'|\mathbf{q}; \mathscr{R}_m) = R_{\alpha\beta}\delta(k, F(k', R))$$
$$\times \exp\{-i\mathbf{q}\cdot[\mathbf{v}(R) + \mathbf{x}(m)]\} \exp\{i\mathbf{G}\cdot[\mathbf{x}(k) - \mathbf{v}(R)]\}. \tag{3.62}$$

This form is particularly attractive since, for $\mathbf{q}$ inside the BZ, the phase factors for the $\mathbf{T}$ matrices based on (3.62) vanish for symmorphic as well as nonsymmorphic groups. Only for some surface points associated with non-symmorphic groups does the phase factor become important.

Like Eqs. (3.58) and (3.61), the eigenvalue equation pertaining to external modes in complex crystals has the generalized form (3.2), and similar considerations apply to that problem. The details are adequately described in the literature[21] and will not be repeated here.

### 3.8 Time-reversal symmetry

In classical physics, time-reversal symmetry implies that, given a closed system of interacting particles evolving according to a set of equations, if at any instant the motions of all the particles are reversed, then the subsequent evolution of the system is governed by the same dynamical equations. Since reversing the motions is equivalent to going back in time, one says that for a closed classical system the time-reversed sequence is also a possible solution of the dynamical equations.

In quantum mechanics, one asserts in a similar fashion that if the wave function $\Psi(r, t)$ (here $r$ stands collectively for all the particles in the system) evolves according to the Schrödinger equation

$$i\hbar\frac{\partial\Psi}{\partial t} = H\Psi, \tag{3.63}$$

then the time-reversed wave function denoted formally by $\hat{T}\Psi$ evolves according to the equation

$$i\hbar\frac{\partial}{\partial t}[\hat{T}\Psi] = H[\hat{T}\Psi].\dagger \tag{3.64}$$

The time-reversed state is defined by

$$\hat{T}\Psi(r, t) = \Psi^*(r, -t). \tag{3.65}$$

Formally, the necessary condition for (3.64) to hold is

$$\hat{T}H\hat{T}^{-1} = H, \tag{3.66}$$

that is, the commutation of the time-reversal operator with the Hamiltonian.

The operator $\hat{T}$ has the important properties of being antilinear‡ and antiunitary.§

The importance of time-reversal symmetry is that, in some instances, it leads to degeneracies in the eigenfunctions of $H$. This topic is well discussed in group theory texts. To recall briefly:

Let $\Psi_n$ denote a stationary state of $H$, that is,

$$\Psi_n(r, t) = \psi_n(r)e^{-iE_n t/\hbar}. \tag{3.67}$$

Then

$$\hat{T}\Psi_n(r, t) = \psi_n^*(r)e^{-iE_n t/\hbar}. \tag{3.68}$$

---

†In writing (3.64) it is assumed that $H(t) = H(-t)$ and that $H$ is real. The latter is true as long as no magnetic fields are present.
‡An operator $\hat{O}$ is said to be linear if

$$\hat{O}(a\psi + b\phi) = a\hat{O}\psi + b\hat{O}\phi,$$

where $a$ and $b$ are complex and $\psi$ and $\phi$ are functions on which $\hat{O}$ can act. An antilinear operator $\hat{A}$ is defined by

$$\hat{A}(a\psi + b\phi) = a^*\hat{A}\psi + b^*\hat{A}\phi.$$

§If $\hat{O}$ is unitary, then it preserves inner products, that is,

$$(\hat{O}\psi, \hat{O}\phi) = (\psi, \phi).$$

On the other hand, an antiunitary operator $\hat{A}$ has the property

$$(\hat{A}\psi, \hat{A}\phi) = (\psi, \phi)^* = (\phi, \psi).$$

From (3.63–3.64) it is seen that $\psi_n(r)$ and $\psi_n^*(r)$ satisfy the eigenvalue equations

$$H\psi_n(r) = E_n\psi_n(r), \tag{3.69}$$
$$H\psi_n^*(r) = E_n\psi_n^*(r). \tag{3.70}$$

$\psi_n^*(r)$ may be expressed in terms of the complex conjugation operator $K$ as $K\psi_n(r)$ and is often referred to as the time-reversed eigenfunction. Equations (3.69–3.70) reveal that since $K\psi_n(r)$ belongs to the same eigenvalue $E_n$ as does $\psi_n(r)$, an extra degeneracy may exist provided the degenerate sets of eigenfunctions $\{\psi_n(r)\}$ and $\{K\psi_n(r)\}$ are linearly independent. This problem can be examined group theoretically, but requires an extension of the customary ideas pertaining to unitary groups. To be more explicit, suppose $H$ is invariant under a group $G$ of unitary operations that correspond to the spatial symmetry of the system. This then enables the investigation of the degeneracies in the eigenvalues consequent to spatial symmetry. To examine the consequences of time-reversal symmetry, one enlarges $G$ to a new group $\mathscr{G}$ by adding elements $\hat{T}G$ so that, symbolically,

$$\mathscr{G} = G + \hat{T}G. \tag{3.71}$$

An important feature of this extension is that the elements of coset $\hat{T}G$ are *antiunitary*. The question of additional degeneracies can now be discussed in terms of the representations of the enlarged group $\mathscr{G}$. These have somewhat different properties than do the representations of unitary groups and are referred to as *corepresentations*. It turns out, however, that in order to examine the occurrence of additional degeneracies, it is not necessary to deal explicitly with the corepresentations of $\mathscr{G}$. Rather it is sufficient to note that $\{\psi_n(r)\}$ and $\{K\psi_n(r)\}$ transform according to IR's of $G$ which are complex conjugates of each other. One then appeals to the well-known Frobenius-Schur criteria[1–8] to see whether these complex conjugate representations are equivalent and sorts the problem out.

Directing attention now specifically to the role of time-reversal symmetry in lattice dynamics, recall first that the substitution of the wavelike solution

$$u_\alpha\binom{l}{k} = U_\alpha(k|\mathbf{q}) \exp\{i[\mathbf{q}\cdot\mathbf{x}(l) - \omega t]\} \tag{3.72}$$

into the equations of motion (2.34) led to the eigenvalue equation

$$\mathbf{D}(\mathbf{q})\mathbf{e}\binom{\mathbf{q}}{j} = \omega_j^2(\mathbf{q})\mathbf{e}\binom{\mathbf{q}}{j}.$$

Since the crystal is an isolated system, we can invoke the principle of time-reversal symmetry and use the same equations of motion to calculate the displacements for the time-reversed motion. The latter corresponds to a wave propagating in the opposite direction:

$$u_\alpha\binom{l}{k} = U_\alpha(k|-\mathbf{q}) \exp\{i[-\mathbf{q}\cdot\mathbf{x}(l) - \omega(-\mathbf{q})t]\}. \tag{3.73}$$

When (3.73) is substituted into the equations of motion, we get

$$\mathbf{D}(-\mathbf{q})\mathbf{e}\binom{-\mathbf{q}}{j} = \omega_j^2(-\mathbf{q})\mathbf{e}\binom{-\mathbf{q}}{j}. \tag{3.74}$$

We know, however, that

$$\mathbf{D}(-\mathbf{q}) = \mathbf{D}^*(\mathbf{q}) \tag{3.75}$$

(see (2.63)), whence (3.74) becomes

$$\mathbf{D}^*(\mathbf{q})\mathbf{e}\binom{-\mathbf{q}}{j} = \omega_j^2(-\mathbf{q})\mathbf{e}\binom{-\mathbf{q}}{j}. \tag{3.76}$$

On the other hand, taking the complex conjugate of (3.4) gives

$$\mathbf{D}^*(\mathbf{q})\mathbf{e}^*\binom{\mathbf{q}}{j} = \omega_j^2(\mathbf{q})\mathbf{e}^*\binom{\mathbf{q}}{j}. \tag{3.77}$$

From the last two equations it is seen that the set of eigenvalues $\{\omega_j^2(\mathbf{q})\}$ and $\{\omega_j^2(-\mathbf{q})\}$ must be identical, a result familiar from the last chapter. As remarked there, the branches associated with $-\mathbf{q}$ may be labeled such that

$$\omega_j^2(\mathbf{q}) = \omega_j^2(-\mathbf{q}) \tag{3.78a}$$

and

$$\mathbf{e}^*\binom{\mathbf{q}}{j} = \mathbf{e}\binom{-\mathbf{q}}{j}, \tag{3.78b}$$

The latter† requires the phase convention mentioned in Sec. 2.2.13.

We see that the equality of the normal-mode frequencies associated with wave vectors $\mathbf{q}$ and $-\mathbf{q}$ is really a consequence of time-reversal symmetry. The significance of this remark is readily appreciated by considering a crystal belonging to the triclinic space group $C_1^1$. Since this

---

†Labeling of the branches according to (3.78b) is always possible in the absence of conflicting constraints. However, if we insist upon (3.57) and if $\omega_j(\mathbf{q})$ is degenerate because of time-reversal symmetry, then it turns out that we may not also demand (3.78b). See ref. 11.

group has no rotational symmetry, Eq. (3.55) cannot be invoked and the frequencies associated with two different $\mathbf{q}$ vectors (in particular $\mathbf{q}$ and $-\mathbf{q}$) would be expected to be equal only accidentally. However, if explicit numerical calculations are made, it is found that the results obtained satisfy (3.78a). Consideration of time-reversal symmetry shows that this degeneracy is expected. Numerical calculations always yield this result because time-reversal symmetry is an intrinsic part of the equations of motion.

Time-reversal symmetry can cause degeneracies beyond that of (3.78a) if $\mathbf{q}$ and $-\mathbf{q}$ belong to the same star. To consider this point, we enlarge $G_0(\mathbf{q})$ to a new group $G_0(\mathbf{q}, -\mathbf{q})$ defined† by

$$G_0(\mathbf{q}, -\mathbf{q}) = G_0(\mathbf{q}) + \hat{T}\mathbf{S}_-G_0(\mathbf{q}), \qquad (3.79)$$

where $\mathbf{S}_-$ is an element of the point group of the crystal which sends $\mathbf{q}$ to $-\mathbf{q}$, that is, $\mathbf{S}_-\mathbf{q} = -\mathbf{q}$. With the elements of $G_0(\mathbf{q})$ we associate the usual unitary matrices $\mathbf{T}(\mathbf{q}; \mathbf{R})$. With the elements of the coset $\hat{T}\mathbf{S}_-G_0(\mathbf{q})$ we associate matrices[11] defined by

$$\hat{T}_{\alpha\beta}(kk'|\mathbf{q}; \mathbf{S}_-\mathbf{R}) = \exp\{i\mathbf{q}\cdot[\mathbf{v}(S_-R) + \mathbf{x}(m)]\}$$
$$\times K\Gamma_{\alpha\beta}(kk'|\mathbf{q};\{\mathbf{S}_-\mathbf{R}|\mathbf{v}(S_-R) + \mathbf{x}(m)\}). \qquad (3.80)$$

Maradudin and Vosko[11] have shown that
1. $\hat{T}(\mathbf{q}; \mathbf{S}_-\mathbf{R})$ is antiunitary, that is,

$$(\hat{T}(\mathbf{q}; \mathbf{S}_-\mathbf{R})\psi, \hat{T}(\mathbf{q}; \mathbf{S}_-\mathbf{R})\phi) = (\phi, \psi),$$

where $\psi$ and $\phi$ are vectors in the $3n$-dimensional space of $\{\mathbf{e}(^{\mathbf{q}}_j)\}$;

2. $\hat{T}\mathbf{D}(\mathbf{q})\hat{T}^{-1} = \mathbf{D}(\mathbf{q})$; $\qquad\qquad\qquad\qquad\qquad\qquad$ (3.81a)

3. $\mathbf{D}(\mathbf{q})\hat{T}(\mathbf{q}; \mathbf{S}_-\mathbf{R})\mathbf{e}\left(\begin{matrix}\mathbf{q}\\sa\lambda\end{matrix}\right) = \omega_{sa}^2(\mathbf{q})\hat{T}(\mathbf{q}; \mathbf{S}_-\mathbf{R})\,\mathbf{e}\left(\begin{matrix}\mathbf{q}\\sa\lambda\end{matrix}\right).$ $\quad$ (3.81b)

Property 2 is referred to as the time-reversal invariance of the dynamical matrix, in the same spirit as (3.66). The corresponding commutation relation (3.27) is known as invariance under spatial symmetry. Property 3 shows that $\hat{T}(\mathbf{q}; \mathbf{S}_-\mathbf{R})\mathbf{e}(^{\mathbf{q}}_{sa\lambda})$ is itself an eigenvector of $\mathbf{D}(\mathbf{q})$.

We now have the kind of situation reviewed earlier, with $\mathbf{D}(\mathbf{q})$ and $\hat{T}(\mathbf{q}; \mathbf{S}_-\mathbf{R})$ the analogs of $H$ and $\hat{T}$. Whether degeneracy beyond that dictated by spatial symmetry exists depends on whether $\{\mathbf{e}(^{\mathbf{q}}_{sa\lambda})\}$ and $\{\hat{T}(\mathbf{q}; \mathbf{S}_-\mathbf{R})\mathbf{e}(^{\mathbf{q}}_{sa\lambda})\}$ are linearly independent. This question can be

---

†The enlarging of $G_0(\mathbf{q})$ to $G_0(\mathbf{q}, -\mathbf{q})$ is here done in the spirit of (3.71). This makes our $G_0(\mathbf{q}, -\mathbf{q})$ slightly different from the point group $G_0(\mathbf{q}, -\mathbf{q}) = G_0(\mathbf{q}) + \mathbf{S}_-G_0(\mathbf{q})$ defined in ref. 11. The final analysis is unaffected, however, since the matrix operators associated with corresponding elements are the same.

analyzed via the Frobenius-Schur criteria, first considered in regard to solid-state problems by Herring[22] and stated in terms of the IR's of $G(\mathbf{q})$. Here they will be stated in terms of the IMR's of $G_0(\mathbf{q})$.[11]

Central to these criteria is the sum

$$Y(\mathbf{q}) = \sum_{R \in \bar{G}_0(\mathbf{q})} \phi(\mathbf{q}; \mathbf{S}_-\mathbf{R}, \mathbf{S}_-\mathbf{R}) \chi^s(\mathbf{q}; [\mathbf{S}_-\mathbf{R}]^2), \tag{3.82}$$

where $\phi$ is a phase factor given by

$$\phi = \exp\{-i[\mathbf{q} + (\mathbf{S}_-\mathbf{R})^{-1}\mathbf{q}] \cdot \mathbf{v}(S_-R)\}. \tag{3.83}$$

Note that since $\mathbf{S}_-\mathbf{R}$ sends $\mathbf{q}$ to $-\mathbf{q}$, the element $(\mathbf{S}_-\mathbf{R})^2$ is contained in $G_0(\mathbf{q})$. Consequently the character $\chi^s(\mathbf{q}; [\mathbf{S}_-\mathbf{R}]^2)$ merely represents the character of the element $[\mathbf{S}_-\mathbf{R}]^2$ of $G_0(\mathbf{q})$ in the IMR $\tau^s(\mathbf{q})$.

We now consider three cases:

Case 1. $Y(\mathbf{q}) = h$. \hfill (3.84a)

    In this case there is no additional degeneracy.

Case 2. $Y(\mathbf{q}) = -h$. \hfill (3.84b)

    There is an additional degeneracy of the form $\omega_{sa}^2(\mathbf{q}) = \omega_{sa'}^2(\mathbf{q})$, $(a \neq a')$; that is, two frequencies corresponding to different occurrences of the same IMR $\tau^s(\mathbf{q})$ are degenerate. $\tau^s(\mathbf{q})$ must occur an even number of times.

Case 3. $Y(\mathbf{q}) = 0$. \hfill (3.84c)

    There is an additional degeneracy of the form $\omega_{sa}^2(\mathbf{q}) = \omega_{s'a'}^2(\mathbf{q})$, $(s \neq s')$; that is, two frequencies belonging to different IMR's $\tau^s(\mathbf{q})$ and $\tau^{s'}(\mathbf{q})$ are rendered equal by time-reversal symmetry. $\tau^s(\mathbf{q})$ and $\tau^{s'}(\mathbf{q})$ must occur in pairs.

The possibility of extra degeneracy if $\mathbf{q}$ and $-\mathbf{q}$ belong to the same star arises because one can envisage additional (antiunitary) operators of the type defined in (3.80) which commute with $\mathbf{D}(\mathbf{q})$. There are, however, two special situations wherein one can have additional antiunitary operators without the need for elements of the type $\mathbf{S}_-$. One of these corresponds to the case of points on the surface of the BZ which lie halfway to a reciprocal lattice point $\mathbf{G}$. In this case, since $\mathbf{q} = \mathbf{G}/2$ we have

$$\mathbf{D}\left(\frac{\mathbf{G}}{2}\right) = \mathbf{D}\left(\frac{\mathbf{G}}{2} - \mathbf{G}\right) \qquad \text{(using (2.64))}$$

$$= \mathbf{D}\left(-\frac{\mathbf{G}}{2}\right) = \mathbf{D}^*\left(\frac{\mathbf{G}}{2}\right). \qquad \text{(from (3.75)).}$$

Therefore,

$$\mathbf{D}^*(\mathbf{G}/2) = K\mathbf{D}(\mathbf{G}/2)K \qquad \text{(by definition)}$$
$$= \mathbf{D}(\mathbf{G}/2). \tag{3.85}$$

In other words, if $\mathbf{q} = \mathbf{G}/2$ then $K$ itself commutes with $\mathbf{D}(\mathbf{q})$. This additional symmetry then presents the possibility of extra degeneracy which may be examined via the criteria of (3.84), with $Y(\mathbf{q})$ now given by[11,20]

$$Y(\mathbf{q}) = \sum_{R\in\bar{G}_0(\mathbf{q})} \exp\tfrac{1}{2}\{-i[\mathbf{G} + \mathbf{R}^{-1}\mathbf{G}]\cdot\mathbf{v}(R)\}\chi^s(\mathbf{G}/2, \mathbf{R}^2). \tag{3.86}$$

A similar situation obtains at $\mathbf{q} = 0$ and here $Y$ is given by

$$Y(0) = \sum_{R\in\bar{G}_0} \chi^s(0, \mathbf{R}^2). \tag{3.87}$$

Two points are worth emphasizing. Firstly, the criteria (3.84) merely predict whether or not extra degeneracies occur. The actual origin of the degeneracies lies in the commutation rule (3.81) or (3.85). The second remark is that these commutation relations are also important in simplifying the structure of $\mathbf{D}(\mathbf{q})$ and in determining its number of independent elements (see Sec. 3.5).

The above discussion of time-reversal symmetry can be extended in a straightforward manner to external modes.[21]

### 3.9 Application of group theory—an example

We discuss now the application of group theory to the dynamics of Te, to illustrate the ideas discussed in the preceding sections.

Tellurium belongs to the trigonal class and crystallizes in two forms, namely, $D_3^4$ ($P3_1\,21$) and $D_3^6$ ($P3_2 21$). Here we shall discuss the application of group theory to the dynamics of the former structure. The atoms are arranged in parallel helical chains that spiral about the $c$ axis. The helices contain three atoms per turn and are arranged in a hexagonal array, so that each primitive cell contains three atoms. The relevant crystallographic information pertaining to this space group is displayed in Table 3.1, which is reproduced from the International Tables of X-ray Crystallography.[23] The atomic arrangement is illustrated in Figure 3.2, while the Brillouin Zone[24] is shown in Figure 3.3. The Cartesian axes are chosen in accord with conventions prescribed by the Standards Committee of the IRE.[25] It may be seen by reference to Table 3.1 that of all the possible positions that could be occupied by atoms in the space group $D_3^4$, only those corresponding to the $a$ positions are occupied in Te. It is also worth noting that atom 1 of the primitive cell (corresponding to the $a$ site

**Table 3.1**
Crystallographic information for space group $D_3^4$ from ref. 23.

| Trigonal | 32 | | $P\,3_1\,2\,1$ | No. 152 | $P3_121$ |
|---|---|---|---|---|---|
| | | | | | $D_3^4$ |

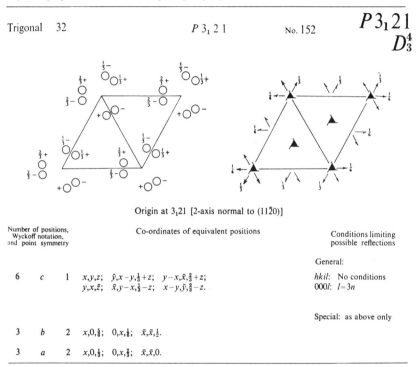

Origin at $3_121$ [2-axis normal to $(11\bar{2}0)$]

| Number of positions, Wyckoff notation, and point symmetry | | | Co-ordinates of equivalent positions | Conditions limiting possible reflections |
|---|---|---|---|---|
| | | | | General: |
| 6 | $c$ | 1 | $x,y,z;\quad \bar{y},x-y,\tfrac{1}{3}+z;\quad y-x,\bar{x},\tfrac{2}{3}+z;$ $y,x,\bar{z};\quad \bar{x},y-x,\tfrac{1}{3}-z;\quad x-y,\bar{y},\tfrac{2}{3}-z.$ | $hkil$: No conditions $000l$: $l=3n$ |
| | | | | Special: as above only |
| 3 | $b$ | 2 | $x,0,\tfrac{5}{6};\quad 0,x,\tfrac{1}{6};\quad \bar{x},\bar{x},\tfrac{1}{2}.$ | |
| 3 | $a$ | 2 | $x,0,\tfrac{1}{3};\quad 0,x,\tfrac{2}{3};\quad \bar{x},\bar{x},0.$ | |

position $\bar{x}\ \bar{x}\ 0$) is not only not at the origin (as is usually assumed), but is physically outside the cell! This, however, does not cause any complication.

We shall now

1. classify the normal modes corresponding to the point $Z$ of the Brillouin zone;

2. construct symmetry-adapted vectors for $Z$ using projection operators;

3. discuss the restrictions imposed by symmetry on the structure of the dynamical matrix for $Z$, and

4. block diagonalize the dynamical matrix so constructed. In addition we shall discuss compatibility relations along $\Gamma\Delta Z$.

The elements corresponding to any given space group are easily identified from the information given in the International Tables[23] corresponding to the transformation of a general point $(x, y, z)$. For instance, consider the transformation $(x, y, z) \rightarrow (\bar{y}, x - y, z + \tfrac{1}{3})$ in Table 3.1. By successively taking $(x, y, z)$ to be $(1, 0, 0)$, $(0, 1, 0)$ and $(0, 0, 1)$, it is easily

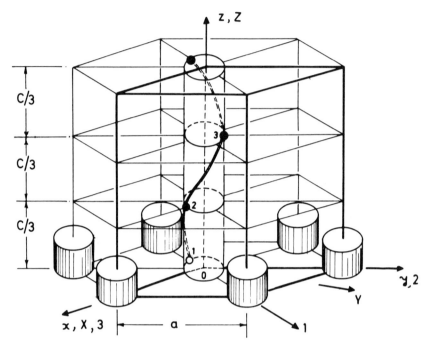

**Figure 3.2**
Atomic arrangement in Te. This consists of a hexagonal array of spirals, each having three atoms per turn. A typical spiral is shown, while others are indicated as truncated cylinders. The bold lines mark the boundaries of the primitive cell, while 1, 2, 3 label the atoms associated with the cell. $(xyz)$ refer to the (oblique) crystallographic axes. The $(XYZ)$ refer to the Cartesian axes and $(01, 02, 03)$ the twofold axes.

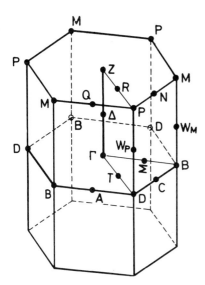

**Figure 3.3**
Brillouin zone for Tellurium. (After ref. 24.)

established that the transformation corresponds to $\{\mathbf{C}_3(Z)|\tau\}$, where $\mathbf{C}_3(Z)$ denotes a counterclockwise rotation through $(2\pi/3)$ about the $Z$ axis, and $\tau = \frac{1}{3}c\mathbf{k}$ denotes a fractional translation of $(c/3)$ along the $z$ axis. Further, referred to the axes $x\,y\,z$ of the International Tables, $\mathbf{C}_3(Z)$ has the matrix representation

$$\begin{pmatrix} 0 & -1 & 0 \\ 1 & -1 & 0 \\ 0 & 0 & 1 \end{pmatrix}.$$

In this manner, it is straightforward to deduce all transformation matrices corresponding to the point group underlying $D_3^4$ and also to establish that the symmetry elements of $D_3^4$ are given by

$\{\mathbf{E}|0\}$, $\{\mathbf{C}_3(Z)|\tau\}$, $\{\mathbf{C}_3^2(Z)|2\tau\}$,
$\{\mathbf{C}_2(1)|0\}$, $\{\mathbf{C}_2(2)|\tau\}$, $\{\mathbf{C}_2(3)|2\tau\}$,

and the products of these with the elements of the translation group $\mathscr{T}$. In the above, $\mathbf{C}_2(i)$ $(i = 1, 2, 3)$ represents a $180°$ rotation about the direction $i$ identified in Figure 3.2.

Now the group transformations listed in the International Tables refer to oblique axes. On the other hand, we require transformation matrices specified with reference to orthogonal axes. The latter are easily obtained from the former through the transformation

$$\mathbf{O}(\text{orthogonal}) = \mathbf{A}^{-1}\mathbf{O}(\text{oblique})\,\mathbf{A},$$

where $\mathbf{A}$ for the present case is given by

$$\begin{pmatrix} 1 & \dfrac{1}{\sqrt{3}} & 0 \\[2ex] 0 & \dfrac{2}{\sqrt{3}} & 0 \\[2ex] 0 & 0 & 1 \end{pmatrix}.$$

Thus for $\mathbf{C}_3(Z)$ we obtain

$$\begin{pmatrix} 1 & -\dfrac{1}{2} & 0 \\[1.5ex] 0 & \dfrac{\sqrt{3}}{2} & 0 \\[1.5ex] 0 & 0 & 1 \end{pmatrix} \begin{pmatrix} 0 & -1 & 0 \\ 1 & -1 & 0 \\ 0 & 0 & 1 \end{pmatrix} \begin{pmatrix} 1 & \dfrac{1}{\sqrt{3}} & 0 \\[1.5ex] 0 & \dfrac{2}{\sqrt{3}} & 0 \\[1.5ex] 0 & 0 & 1 \end{pmatrix} = \begin{pmatrix} -\dfrac{1}{2} & -\dfrac{\sqrt{3}}{2} & 0 \\[1.5ex] \dfrac{\sqrt{3}}{2} & -\dfrac{1}{2} & 0 \\[1.5ex] 0 & 0 & 1 \end{pmatrix}.$$

The three-dimensional matrices (referred to Cartesian axes) corresponding to the various point-group operations obtained thus are shown in
Table 3.2.

Turning next to the application of group theory to the point $Z$, we note
first that the point group corresponding to this wave vector is the same as
the point group of the crystal, namely $D_3$. The factor system corresponding
to $Z$ is easily worked out using Eq. (3.13) and the information provided

**Table 3.2**
Matrices for the point-group operations, referred to Cartesian axes.

| | | | |
|---|---|---|---|
| $E$ | $\begin{pmatrix} 1 & & \\ & 1 & \\ & & 1 \end{pmatrix}$ | $C_2(1)$ | $\begin{pmatrix} -\dfrac{1}{2} & \dfrac{\sqrt{3}}{2} & 0 \\[1.5ex] \dfrac{\sqrt{3}}{2} & \dfrac{1}{2} & 0 \\[1.5ex] 0 & 0 & -1 \end{pmatrix}$ |
| $C_3(Z)$ | $\begin{pmatrix} -\dfrac{1}{2} & -\dfrac{\sqrt{3}}{2} & 0 \\[1.5ex] \dfrac{\sqrt{3}}{2} & -\dfrac{1}{2} & 0 \\[1.5ex] 0 & 0 & 1 \end{pmatrix}$ | $C_2(2)$ | $\begin{pmatrix} -\dfrac{1}{2} & -\dfrac{\sqrt{3}}{2} & 0 \\[1.5ex] -\dfrac{\sqrt{3}}{2} & \dfrac{1}{2} & 0 \\[1.5ex] 0 & 0 & -1 \end{pmatrix}$ |
| $C_3^2(Z)$ | $\begin{pmatrix} -\dfrac{1}{2} & \dfrac{\sqrt{3}}{2} & 0 \\[1.5ex] -\dfrac{\sqrt{3}}{2} & -\dfrac{1}{2} & 0 \\[1.5ex] 0 & 0 & 1 \end{pmatrix}$ | $C_2(3)$ | $\begin{pmatrix} 1 & & \\ & -1 & \\ & & -1 \end{pmatrix}$ |

above and is given in Table 3.3. The IMR's corresponding to this factor system are listed in Table 3.4. These have been obtained using the method discussed in reference 12. In this context it is important to remember that the axes employed in generating the IMR's must be the same as those employed in the dynamical calculations so as to be consistent with regard to the factor system and various other phase factors.

The next step is to construct the set of matrices

$$T(Z) = \{\mathbf{T}(Z; \mathbf{R})\}, \qquad \mathbf{R} \in G_0(Z).$$

The exponential factors $\exp\{i\mathbf{q}\cdot[\mathbf{x}(k) - \mathbf{R}\mathbf{x}(k')]\}$ necessary for this purpose (see Eq. (3.26)) are summarized in Table 3.5. Using this table, the $\mathbf{T}$ matrices are found to be

$$\mathbf{T}(Z; \mathbf{E}) = \begin{bmatrix} \mathbf{E} & & \\ & \mathbf{E} & \\ & & \mathbf{E} \end{bmatrix};$$

$$\mathbf{T}(Z; \mathbf{C}_3(Z)) = \mu \begin{bmatrix} \cdot & \cdot & -\mathbf{C}_3(Z) \\ \mathbf{C}_3(Z) & \cdot & \cdot \\ \cdot & \mathbf{C}_3(Z) & \cdot \end{bmatrix};$$

$$\mathbf{T}(Z; \mathbf{C}_3^2(Z)) = \lambda \begin{bmatrix} \cdot & \mathbf{C}_3^2(Z) & \\ \cdot & \cdot & \mathbf{C}_3^2(Z) \\ -\mathbf{C}_3^2(Z) & \cdot & \cdot \end{bmatrix};$$

$$\mathbf{T}(Z; \mathbf{C}_2(1)) = \begin{bmatrix} \mathbf{C}_2(1) & \cdot & \cdot \\ \cdot & \cdot & -\mathbf{C}_2(1) \\ \cdot & -\mathbf{C}_2(1) & \cdot \end{bmatrix};$$

$$\mathbf{T}(Z; \mathbf{C}_2(2)) = \mu \begin{bmatrix} \cdot & \mathbf{C}_2(2) & \cdot \\ \mathbf{C}_2(2) & \cdot & \cdot \\ \cdot & \cdot & -\mathbf{C}_2(2) \end{bmatrix};$$

$$\mathbf{T}(Z; \mathbf{C}_2(3)) = -\lambda \begin{bmatrix} \cdot & \cdot & \mathbf{C}_2(3) \\ \cdot & \mathbf{C}_2(3) & \cdot \\ \mathbf{C}_2(3) & \cdot & \cdot \end{bmatrix};$$

where $\mu = \exp(i\pi/3)$, $\lambda = \exp(-i\pi/3)$.                    (3.88)

The traces of these matrices are given in the last row of Table 3.4, using which the decomposition of $T(Z)$ is easily effected through an application of (3.29). The results are summarized in the last column of Table 3.4.

In discussing the transformation properties of eigenvectors, Maradudin and Vosko[11] have pointed out that for IMR's of dimensionality $\geqslant 2$ it is desirable that: (1) the matrices $\tau^s(\mathbf{q}; \mathbf{R})$ be real for those group elements $\mathbf{R}$ for which $\mathbf{T}(\mathbf{q}; \mathbf{R})$ commutes with the antiunitary operator $\hat{\mathbf{T}}(\mathbf{q}; \mathbf{S}_-)$; (2) as many of these as possible be in diagonal form. This leads to the useful

**Table 3.3**

Factor system for the point $Z$. Values of $\phi(Z; \mathbf{R}_i, \mathbf{R}_j)$ are given in the table.

| $\mathbf{R}_i$ | $\mathbf{R}_J$ | | | | | |
|---|---|---|---|---|---|---|
|  | E | $\mathbf{C}_3(Z)$ | $\mathbf{C}_3^2(Z)$ | $\mathbf{C}_2(1)$ | $\mathbf{C}_2(2)$ | $\mathbf{C}_2(3)$ |
| E | 1 | 1 | 1 | 1 | 1 | 1 |
| $\mathbf{C}_3(Z)$ | 1 | 1 | 1 | 1 | 1 | 1 |
| $\mathbf{C}_3^2(Z)$ | 1 | 1 | 1 | 1 | 1 | 1 |
| $\mathbf{C}_2(1)$ | 1 | $\omega$ | $\omega^2$ | 1 | $\omega$ | $\omega^2$ |
| $\mathbf{C}_2(2)$ | 1 | $\omega$ | $\omega^2$ | 1 | $\omega$ | $\omega^2$ |
| $\mathbf{C}_2(3)$ | 1 | $\omega$ | $\omega^2$ | 1 | $\omega$ | $\omega^2$ |

$\omega = \exp(2\pi i/3)$

**Table 3.4**

IMR's of $G_0(Z)$.

| IMR | E | $\mathbf{C}_3(Z)$ | $\mathbf{C}_3^2(Z)$ | $\mathbf{C}_2(1)$ | $\mathbf{C}_2(2)$ | $\mathbf{C}_2(3)$ | $C_{z_i}$ |
|---|---|---|---|---|---|---|---|
| $Z_1$ | 1 | $\omega^2$ | $\omega$ | 1 | $\omega^2$ | $\omega$ | 1 |
| $Z_2$ | 1 | $\omega^2$ | $\omega$ | $-1$ | $-\omega^2$ | $-\omega$ | 2 |
| $Z_3$ | $\begin{pmatrix} 1 & 0 \\ 0 & 1 \end{pmatrix}$ | $\begin{pmatrix} \omega & 0 \\ 0 & 1 \end{pmatrix}$ | $\begin{pmatrix} \omega^2 & 0 \\ 0 & 1 \end{pmatrix}$ | $\begin{pmatrix} 0 & 1 \\ 1 & 0 \end{pmatrix}$ | $\begin{pmatrix} 0 & \omega \\ 1 & 0 \end{pmatrix}$ | $\begin{pmatrix} 0 & \omega^2 \\ 1 & 0 \end{pmatrix}$ | 3 |
| $Z_3'$ | $\begin{pmatrix} 1 & 0 \\ 0 & 1 \end{pmatrix}$ | $\frac{\mu}{2}\begin{pmatrix} 1 & -\sqrt{3} \\ \sqrt{3} & 1 \end{pmatrix}$ | $\frac{\lambda}{2}\begin{pmatrix} 1 & \sqrt{3} \\ -\sqrt{3} & 1 \end{pmatrix}$ | $\begin{pmatrix} 1 & 0 \\ 0 & -1 \end{pmatrix}$ | $\frac{\mu}{2}\begin{pmatrix} 1 & \sqrt{3} \\ \sqrt{3} & -1 \end{pmatrix}$ | $\frac{\lambda}{2}\begin{pmatrix} 1 & -\sqrt{3} \\ -\sqrt{3} & -1 \end{pmatrix}$ | |
| Tr $\mathbf{T}(Z;\mathbf{R})$ | 9 | 0 | 0 | $-1$ | $\mu$ | $\lambda$ | |

$\tau(Z_3'; \mathbf{R}) = \mathbf{U}\tau(Z_3; \mathbf{R})\mathbf{U}^{-1}; \mathbf{U} = \dfrac{1}{\sqrt{2}}\begin{pmatrix} 1 & 1 \\ -i & i \end{pmatrix}$

$\mu = \exp(i\pi/3), \lambda = \exp(-i\pi/3)$

**Table 3.5**

Exponential factors for $Z$. Only the factors for those pairs of sublattices which are interchanged consequent to symmetry operation are listed.

|  | E | $\mathbf{C}_3(Z)$ | $\mathbf{C}_3^2(Z)$ | $\mathbf{C}_2(1)$ | $\mathbf{C}_2(2)$ | $\mathbf{C}_2(3)$ |
|---|---|---|---|---|---|---|
| 11 | 1 | $\cdot$ | $\cdot$ | 1 | $\cdot$ | $\cdot$ |
| 12 | $\cdot$ | $\cdot$ | $\lambda$ | $\cdot$ | $\mu$ | $\cdot$ |
| 13 | $\cdot$ | $-\mu$ | $\cdot$ | $\cdot$ | $\cdot$ | $-\lambda$ |
| 21 | $\cdot$ | $\mu$ | $\cdot$ | $\cdot$ | $\mu$ | $\cdot$ |
| 22 | 1 | $\cdot$ | $\cdot$ | $\cdot$ | $\cdot$ | $-\lambda$ |
| 23 | $\cdot$ | $\cdot$ | $\lambda$ | $-1$ | $\cdot$ | $\cdot$ |
| 31 | $\cdot$ | $\cdot$ | $-\lambda$ | $\cdot$ | $\cdot$ | $-\lambda$ |
| 32 | $\cdot$ | $\mu$ | $\cdot$ | $-1$ | $\cdot$ | $\cdot$ |
| 33 | 1 | $\cdot$ | $\cdot$ | $\cdot$ | $-\mu$ | $\cdot$ |

property that the eigenvectors resulting from the use of $\tau^s(\mathbf{q})$ are also eigenvectors of $\hat{\mathbf{T}}(\mathbf{q};\mathbf{S}_-)$. Since $Z$ is on the zone boundary, the operator $K$ plays the role of $\hat{\mathbf{T}}(\mathbf{q};\mathbf{S}_-)$, and all we require is that $\tau^s(Z;\mathbf{R})$ be real for those $\mathbf{R}$ for which $\mathbf{T}(Z;\mathbf{R})$ commutes with $K$, that is, for $\mathbf{R} = \mathbf{E}$ and $\mathbf{C}_2(1)$. Inspection of Table 3.4 shows that though the matrices of $\tau(Z_3)$ meet the reality requirement (1), they are deficient with respect to the requirement (2), since a unitary transformation through the matrix

$$\mathbf{U} = \frac{1}{\sqrt{2}}\begin{pmatrix} 1 & 1 \\ -i & i \end{pmatrix}$$

delivers an equivalent representation (labeled $Z_3'$ in Table 3.4) in which the representative of $\mathbf{C}_2(1)$ is also diagonal. From now on whenever we talk of the representation $\tau(Z_3)$, one may assume that we have the latter form of matrices in mind. This ensures that the eigenvectors transforming according to $Z_3$ are also eigenvectors of $K$ and hence real. (Eigenvectors transforming according to the one-dimensional IMR's $Z_1$ and $Z_2$ are automatically eigenvectors of $K$). Thus all eigenvectors of $D(Z)$ are guaranteed to be real, which can also be demanded on the basis that the eigenvectors of a real symmetric matrix (see (3.85)) can always be chosen to be real.

Next we consider the construction of symmetry-adapted vectors appropriate to $Z$. This is done through the use of the projection operator technique. Consider, for instance, the operator

$$P_{11}(Z_3) = \sum_{\mathbf{R}\in \bar{G}_0(Z)} \tau^*_{11}(Z_3;\mathbf{R})\mathbf{T}(Z;\mathbf{R}). \tag{3.89}$$

In writing the above expression we have omitted the group-theoretic normalization factor since we shall be normalizing the symmetry vectors anyway. Using Table 3.4 and the matrices $\mathbf{T}(Z;\mathbf{R})$ given in (3.88), it is easily seen that

$$P_{11}(Z_3) = \begin{pmatrix} \mathbf{E}+\mathbf{C}_2(1) & \frac{1}{2}(\mathbf{C}_3^2(Z)+\mathbf{C}_2(2)) & -\frac{1}{2}(\mathbf{C}_3(Z)+\mathbf{C}_2(3)) \\ \frac{1}{2}(\mathbf{C}_3(Z)+\mathbf{C}_2(2)) & \mathbf{E}-\frac{1}{2}\mathbf{C}_2(3) & \frac{1}{2}\mathbf{C}_3^2(Z)-\mathbf{C}_2(1) \\ -\frac{1}{2}(\mathbf{C}_3^2(Z)+\mathbf{C}_2(3)) & \frac{1}{2}\mathbf{C}_3(Z)-\mathbf{C}_2(1) & \mathbf{E}-\frac{1}{2}\mathbf{C}_2(2) \end{pmatrix}.$$

By applying this operator systematically to the 9 orthonormal unit vectors

$$\begin{pmatrix} 1 \\ 0 \\ 0 \\ \vdots \\ 0 \end{pmatrix} \begin{pmatrix} 0 \\ 1 \\ 0 \\ \vdots \\ 0 \end{pmatrix}, \ldots, \begin{pmatrix} 0 \\ 0 \\ 0 \\ \vdots \\ 1 \end{pmatrix} \left. \begin{array}{c} \uparrow \\ \\ 9 \text{ elements} \\ \\ \downarrow \end{array} \right.$$

spanning the 9-dimensional space of the present problem, we can project three nonzero linearly independent vectors. After normalization these are

$$\frac{1}{2\sqrt{6}}\begin{pmatrix}2\\2\sqrt{3}\\0\\-2\\0\\0\\-1\\\sqrt{3}\\0\end{pmatrix},\quad \frac{1}{2\sqrt{2}}\begin{pmatrix}0\\0\\0\\0\\2\\0\\-\sqrt{3}\\-1\\0\end{pmatrix},\quad \frac{1}{\sqrt{2}}\begin{pmatrix}0\\0\\0\\0\\0\\1\\0\\0\\1\end{pmatrix}.$$

The partners of these vectors are found by acting on them with the projection operator $P_{21}(Z_3)$. The symmetry-adapted vectors appropriate to the IMR's $Z_1$ and $Z_2$ are found in a similar manner. The problem is simpler in this case since the question of partners does not arise, $Z_1$ and $Z_2$ being one dimensional. In this way we obtain 9 symmetry-adapted vectors, which are then assembled into the transformation matrix $\Sigma(Z)$ as shown below.

$$\Sigma(Z)=\begin{pmatrix}
\frac{1}{2\sqrt{3}} & \frac{1}{2} & 0 & \frac{1}{\sqrt{6}} & 0 & 0 & 0 & -\frac{1}{\sqrt{2}} & 0\\[4pt]
\frac{1}{2} & -\frac{1}{2\sqrt{3}} & 0 & \frac{1}{\sqrt{2}} & 0 & 0 & 0 & \frac{1}{\sqrt{6}} & 0\\[4pt]
0 & 0 & \frac{1}{\sqrt{3}} & 0 & 0 & 0 & 0 & 0 & \frac{-2}{\sqrt{6}}\\[4pt]
\frac{1}{\sqrt{3}} & 0 & 0 & -\frac{1}{\sqrt{6}} & 0 & 0 & -\frac{1}{\sqrt{2}} & 0 & 0\\[4pt]
0 & -\frac{1}{\sqrt{3}} & 0 & 0 & \frac{1}{\sqrt{2}} & 0 & 0 & -\frac{1}{\sqrt{6}} & 0\\[4pt]
0 & 0 & -\frac{1}{\sqrt{3}} & 0 & 0 & \frac{1}{\sqrt{2}} & 0 & 0 & -\frac{1}{\sqrt{6}}\\[4pt]
\frac{1}{2\sqrt{3}} & -\frac{1}{2} & 0 & -\frac{1}{2\sqrt{6}} & \frac{-\sqrt{3}}{2\sqrt{2}} & 0 & \frac{1}{2\sqrt{2}} & -\frac{1}{2\sqrt{2}} & 0\\[4pt]
-\frac{1}{2} & -\frac{1}{2\sqrt{3}} & 0 & \frac{1}{2\sqrt{2}} & -\frac{1}{2\sqrt{2}} & 0 & \frac{-\sqrt{3}}{2\sqrt{2}} & -\frac{1}{2\sqrt{6}} & 0\\[4pt]
0 & 0 & \frac{1}{\sqrt{3}} & 0 & 0 & \frac{1}{\sqrt{2}} & 0 & 0 & \frac{1}{\sqrt{6}}
\end{pmatrix}$$

$$\underbrace{\phantom{Z_1}}_{Z_1}\ \underbrace{\phantom{Z_2}}_{Z_2}\ \underbrace{\phantom{Z_3Z_3}}_{Z_3}\ \underbrace{\phantom{Z_3Z_3}}_{Z_3}$$

$$(3.90)$$

We now consider the symmetry-imposed restrictions on the structure of the dynamical matrix $\mathbf{D}(Z)$. This requires a consideration of the commutation relation (3.27), involving the unitary operators, and the relation (3.81) (or 3.85)), involving antiunitary operators. With regard to (3.27), it is sufficient to restrict consideration to the generating elements of $G_0(\mathbf{q})$, since these exhaust all the consequences of *spatial* symmetry. Thus for the point $Z$ we need only consider the invariances

$$\mathbf{D}(Z) = \mathbf{T}(Z;\mathbf{R})\mathbf{D}(Z)\mathbf{T}^{\dagger}(Z;\mathbf{R}), \qquad \mathbf{R} = \mathbf{C}_3(Z),\, \mathbf{C}_2(3). \tag{3.91}$$

By assuming the most general structure possible for the Hermitian matrix $\mathbf{D}(Z)$ and carrying through the matrix multiplication, one can arrive at the *symmetry-determined* structure of the dynamical matrix exactly as one does in the case of force constants (see Sec. 2.9). Relation (3.91) must, of course, be supplemented by restrictions imposed by time-reversal symmetry which, in the present case, reduces to a use of (3.85) which demands $\mathbf{D}(\mathbf{q})$ to be real.

Taking all these restrictions together, the symmetry-determined form of $\mathbf{D}(Z)$ is found to be as follows:

$$\mathbf{D}\begin{pmatrix}Z\\11\end{pmatrix} = \begin{vmatrix} \alpha(11) & \frac{\sqrt{3}}{2}[\gamma(11)-\alpha(11)] & -\sqrt{3}\beta(11) \\ \frac{\sqrt{3}}{2}[\gamma(11)-\alpha(11)] & \gamma(11) & \beta(11) \\ -\sqrt{3}\beta(11) & \beta(11) & \delta(11) \end{vmatrix},$$

$$\mathbf{D}\begin{pmatrix}Z\\12\end{pmatrix} = \begin{vmatrix} \alpha(12) & \beta(12) & \gamma(12) \\ -\beta(12)+\sqrt{3}[\alpha(12)-\delta(12)] & \delta(12) & \varepsilon(12) \\ \frac{1}{2}[\sqrt{3}\varepsilon(12)+\gamma(12)] & \frac{1}{2}[\sqrt{3}\gamma(12)-\varepsilon(12)] & \eta(12) \end{vmatrix},$$

$$\mathbf{D}\begin{pmatrix}Z\\12\end{pmatrix} = \tilde{\mathbf{D}}\begin{pmatrix}Z\\21\end{pmatrix},$$

$$\mathbf{D}\begin{pmatrix}Z\\13\end{pmatrix} = \begin{vmatrix} \frac{1}{2}[\alpha(12)-3\delta(12)] & \beta(12)+\frac{\sqrt{3}}{2}[\delta(12)-\alpha(12)] & \frac{\sqrt{3}}{2}\varepsilon(12)-\frac{\gamma(12)}{2} \\ -\beta(12)-\frac{\sqrt{3}}{2}[\delta(12)-\alpha(12)] & \frac{1}{2}[\delta(12)-3\alpha(12)] & \frac{1}{2}[\sqrt{3}\gamma(12)+\varepsilon(12)] \\ \frac{1}{2}[\gamma(12)-\sqrt{3}\varepsilon(12)] & \frac{1}{2}[\sqrt{3}\gamma(12)+\varepsilon(12)] & -\eta(12) \end{vmatrix},$$

$$\mathbf{D}\begin{pmatrix}Z\\22\end{pmatrix} = \begin{vmatrix} \frac{1}{2}[3\gamma(11)-\alpha(11)] & 0 & 0 \\ 0 & \frac{1}{2}[3\alpha(11)-\gamma(11)] & -2\beta(11) \\ 0 & -2\beta(11) & \delta(11) \end{vmatrix},$$

$$\mathbf{D}\begin{pmatrix} Z \\ 23 \end{pmatrix}$$

$$= \begin{vmatrix} \alpha(12) & \beta(12) + \sqrt{3}[\delta(12) - \alpha(12)] & -\dfrac{1}{2}[\sqrt{3}\varepsilon(12) + \gamma(12)] \\ -\beta(12) & \delta(12) & \dfrac{1}{2}[\sqrt{3}\gamma(12) - \varepsilon(12)] \\ -\gamma(12) & \varepsilon(12) & \eta(12) \end{vmatrix},$$

$$\mathbf{D}\begin{pmatrix} Z \\ 31 \end{pmatrix} = \tilde{\mathbf{D}}\begin{pmatrix} Z \\ 13 \end{pmatrix},$$

$$\mathbf{D}\begin{pmatrix} Z \\ 32 \end{pmatrix} = \tilde{\mathbf{D}}\begin{pmatrix} Z \\ 23 \end{pmatrix},$$

$$\mathbf{D}\begin{pmatrix} Z \\ 33 \end{pmatrix} = \begin{vmatrix} \alpha(11) & \dfrac{\sqrt{3}}{2}[\alpha(11) - \gamma(11)] & \sqrt{3}\beta(11) \\ \dfrac{\sqrt{3}}{2}[\alpha(11) - \gamma(11)] & \gamma(11) & \beta(11) \\ \sqrt{3}\beta(11) & \beta(11) & \delta(11) \end{vmatrix}.$$

Here we have used Greek symbols to denote the distinct elements of the dynamical matrix.

The task remains of bringing $\mathbf{D}(Z)$ to the block-diagonal form $\mathscr{D}(Z)$. This is done through the transformation

$$\mathscr{D}(Z) = \Sigma^{\dagger}(Z)\mathbf{D}(Z)\Sigma(Z)$$

(see Eq. (3.38)). The resulting $\mathscr{D}(Z)$ has the form

$$\mathscr{D}(Z) = \begin{pmatrix} \mathscr{D}(Z_1) & & & \\ & \mathscr{D}(Z_2) & & \\ & & \mathscr{D}^{(1)}(Z_3) & \\ & & & \mathscr{D}^{(2)}(Z_3) \end{pmatrix},$$

where $\mathscr{D}(Z_1)$ is one-dimensional, $\mathscr{D}(Z_2)$ is two-dimensional, and the remaining two matrices are three-dimensional. Explicitly,

$$\mathscr{D}(Z_1) = \tfrac{1}{2}[3\gamma(11) - \alpha(11)] + 4\alpha(12) - 3\delta(12) - \sqrt{3}\beta(12),$$

$$\mathscr{D}(Z_2) = \begin{bmatrix} \tfrac{1}{2}[3\alpha(11) - \gamma(11)] + \delta(12) - \sqrt{3}\beta(12) & -2\beta(11) - \sqrt{3}\gamma(12) + \varepsilon(12) \\ -2\beta(11) - \sqrt{3}\gamma(12) + \varepsilon(12) & \delta(11) - 2\eta(12) \end{bmatrix}.$$

Similar expressions may be written for $\mathscr{D}^{(1)}(Z_3)$ and $\mathscr{D}^{(2)}(Z_3)$ which, incidentally, lead to identical eigenvalues (see remarks following (3.39)). It is worth emphasizing that the block-diagonal form $\mathscr{D}(Z)$ results only when $\Sigma(Z)$ is organized according to the sequence (3.37) (see also (3.90)).

Finally we consider the compatibility relations along $\Gamma\Delta Z$. The additional information necessary for this purpose (the character tables for $\Gamma$ and $\Delta$) is given in Table 3.6. Now $G_0(\Delta)$ is a subgroup of both $G_0(\Gamma)$ and $G_0(Z)$. The representations of $\Gamma$ and $Z$ are therefore both reducible with respect to $G_0(\Delta)$. The reduction is easily achieved using standard procedure in conjunction with the tables given and leads to the following compatibilities.

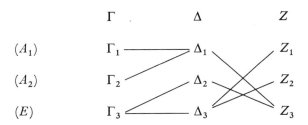

The use of this information is illustrated in Figure 3.4 which displays the measured dispersion curves[26] for Te along $\Gamma\Delta Z$. The solid lines are guides to the eye. The group-theoretical assignments made by the authors of ref. 26 are also shown, and these are in accord with the compatibility relations given above.*

**Table 3.6**
Character tables for $\Gamma$ and $\Delta$.

| $\Gamma$ | $E$ | $2C_3$ | $3C_2$ |
|---|---|---|---|
| $\Gamma_1(A_1)$ | 1 | 1 | 1 |
| $\Gamma_2(A_2)$ | 1 | 1 | $-1$ |
| $\Gamma_3(E)$ | 2 | $-1$ | 0 |
| $\Delta$ | $E$ | $C_3$ | $C_3^2$ |
| $\Delta_1$ | 1 | 1 | 1 |
| $\Delta_2$ | 1 | $\omega$ | $\omega^2$ |
| $\Delta_3$ | 1 | $\omega^2$ | $\omega$ |

$\omega = \exp(2\pi i/3)$
Note: The symbols $A_1$, $A_2$, and $E$ refer to the notation used commonly by spectroscopists and molecular physicists.

*Tellurium exhibits infrared activity. This results in the splitting of the modes belonging to $\Gamma_3$ for $\hat{q}$ along directions other than $\Gamma Z$. This poses problems in labeling the long-wavelength modes, as discussed in Sec. 4.4.

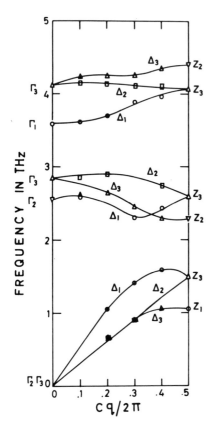

**Figure 3.4**
Dispersion curves for Te along $\Gamma\Delta Z$ obtained from experiments. The solid lines are guides
to the eye. The various branches are labeled group-theoretically and in a manner consistent with compatibility relations. (After ref. 26.)

## References

1. V. Heine, *Group Theory in Quantum Mechanics* (Pergamon, New York, 1960).

2. M. Tinkham, *Group Theory and Quantum Mechanics* (McGraw-Hill, New York, 1964).

3. G. Ya. Lyubarskii, *The Application of Group Theory in Physics* (Pergamon, New York, 1960).

4. R. McWeeny, *Symmetry* (Pergamon, New York, 1963).

5. M. Hamermesh, *Group Theory and its Application to Physical Problems* (Addison-Wesley, Reading, Mass., 1962).

6. L. M. Falicov, *Group Theory and its Physical Applications* (University of Chicago Press, Chicago, 1966).

7. J. S. Lomont, *Applications of Finite Groups* (Academic Press, New York, 1959).

8. H. W. Streitwolf, *Group Theory in Solid State Physics* (Macdonald, London, 1971).

9. L. P. Bouckert, R. Smoluchowski, and E. Wigner, Phys. Rev. **50**, 58 (1936).

10. See reference 6 above, p. 174.

11. A. A. Maradudin and S. H. Vosko, Rev. Mod. Phys. **40**, 1 (1968).

12. V. C. Sahni and G. Venkataraman, Phys. kondens Materie **11**, 199, 212 (1970).

13. O. V. Kovalev, *Irreducible Representations of Space Groups* (Gordon and Breach, New York, 1965).

14. J. Zak, A. Casher, M. Gluck, and Y. Gur, *Irreducible Representations of Space Groups* (Benjamin, New York, 1969).

15. S. C. Miller and W. F. Love, *Tables of Irreducible Representations of Space Groups and Corepresentations of Magnetic Space Groups* (Pruett, Boulder, Colorado, 1967).

16. R. J. Elliott and M. F. Thorpe, Proc. Phys. Soc. (London) **91**, 903 (1967).

17. G. Venkataraman, L. A. Feldkamp, and J. S. King, University of Michigan report, 01358-1-T (1968).

18. H. Montgomery, Proc. Roy. Soc. (London) **A309**, 521 (1969).

19. C. Herring, Phys. Rev. **52**, 365 (1937). Reprinted in R. S. Knox and A. Gold, *Symmetry in the Solid State* (Benjamin, New York, 1964).

20. J. L. Warren, Rev. Mod. Phys. **40**, 38 (1968).

21. G. Venkataraman and V. C. Sahni, Rev. Mod. Phys. **42**, 409 (1970).

22. C. Herring, Phys. Rev. **52**, 361 (1937). (Reprinted in Knox and Gold, see reference 19 above).

23. K. Londsdale, *International Tables of X-ray Crystallography* (Kynoch, Birmingham, 1965), Vol. I.

24. M. Hulin, Ann. Phys. (Paris) **8**, 647 (1963).

25. Standards Committee of IRE, Proc. IRE **37**, 1378 (1949).

26. B. M. Powell and P. Martel, in *Proceedings of the Tenth International Conference on the Physics of Semiconductors* (1970), p. 851.

# 4  Electric Field and Polarization Effects in Insulating Crystals

The Born–von Kármán formalism introduced in Chapter 2 can be and has been applied to a wide class of crystals—metals, insulators, and semiconductors. In its practical application, the force constants are often regarded as disposable parameters whose values are to be suitably fixed. While adopting such a procedure one must of course guard against its indiscriminate application. Especially in the case of insulating and semiconducting crystals, it is necessary to link the force constants explicitly to certain important physical processes so as to obtain realistic results. More specifically, we must adapt the force-constant formalism to include the interaction of the vibrations with the electric field induced by the vibrations themselves. This is the task we undertake in this chapter.

It is convenient to treat the problem in two related parts:
1. The production and consequences of the macroscopic electric field, and
2. the effects of polarization, induced by the ionic displacements and by the distortion of the electron cloud.

We split off the macroscopic field, itself a consequence of polarization, for special treatment because the primary results may be obtained, in the long-wavelength limit, without recourse to a specific model. Other effects of electrical polarization may be examined in detail only by using models, some of which will be discussed later.

We note that electronic polarization is a feature common to both ionic and homopolar insulating crystals (diamond, for example). To this extent, much of the discussion of this chapter will be relevant to crystals of either category.

## 4.1 Influence of the macroscopic field on the q = 0 optic modes— an example

An overall appreciation of the role of the macroscopic field vis-à-vis the long-wavelength vibrations is best obtained by considering a specific example. Accordingly we consider the simplest among the ionic crystals, namely cubic diatomic crystals such as those of the NaCl, CsCl, and (cubic) ZnS structures.

### 4.1.1 Splitting of the q = 0 optic modes
Figure 4.1 shows dispersion curves for ZnS computed as in the example of Sec. 2.9, assuming only a short-range interaction extending to nearest neighbors. Such a model is clearly unrealistic since it ignores the long-range Coulomb interactions between the ions. In the curves of Figure 4.2 this deficiency has been corrected by assuming that the force field has a short-range part, as above, plus a Coulombic part associated with electro-

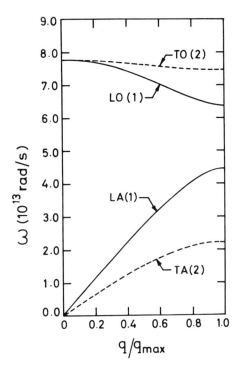

**Figure 4.1**
Dispersion curves along [100] for zinc blende, assuming nearest-neighbor interactions only. The numbers in parentheses indicate the degeneracies of the concerned branches. The labels TA, LA, TO, LO signify transverse acoustic, longitudinal acoustic, transverse optic and, longitudinal optic. Observe the triple degeneracy of the $\mathbf{q} = 0$ modes.

static interactions between the various ions, the latter being regarded as point charges. Two versions of the calculations are shown; that in Figure 4.2a includes the entire Coulombic effect while that in Figure 4.2b retains only the "short-range part" of the Coulomb interactions.* It is evident that the residual "long-range part" of the Coulomb contribution produces an important modification: the splitting of the threefold degeneracy of the $\mathbf{q} = 0$ optic modes by raising the frequency of the LO mode above that of the TO mode. Later it will be seen that this "long-range part" corresponds to the macroscopic field arising out of the vibrations. (See also Appendix 2).

---

*The manner in which these calculations can be made is discussed in Sec. 4.2 and in Appendix 2, where the definition of the terms "short-range part" and "long-range part" of the Coulomb interaction is made more precise.

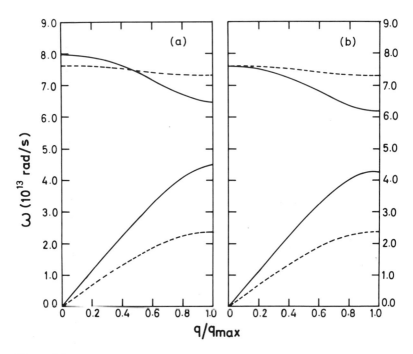

**Figure 4.2**
Dispersion curves similar to those in Figure 4.1 but with Coulomb interactions between the
ions allowed for. In (a) the full Coulomb contribution is included while in (b) only the
"short-range part" of the Coulomb interaction is included.

The origin of the macroscopic field can be understood physically as
follows: Consider a long-wavelength (that is, $\mathbf{q} \to 0$) optic vibration and
suppose that as a result there is a *slowly varying* polarization density
$\mathbf{P}(\mathbf{r}) = \mathbf{P}_0 \exp(i\mathbf{q} \cdot \mathbf{r})$.* The latter arises from the movement of the elementary charges in the crystal and is often described in terms of ele-
mentary dipoles. For example, considering displacement dipoles only one
could write

$$\mathbf{P}_0 = \frac{1}{v} Ze[\mathbf{U}_0(1) - \mathbf{U}_0(2)]. \tag{4.1}$$

In this chapter and elsewhere we shall take $e = +4.8 \times 10^{-10}$ esu. The
charge on the positive ion is denoted by $Ze$ while $\mathbf{U}_0(1)$ and $\mathbf{U}_0(2)$ are the
displacement amplitudes of the ions in the two sublattices in the limit

---

*In general, not all long-wavelength optic vibrations need be accompanied by polariza-
tion waves. However, in cubic diatomic crystals polarizations are associated with all the
optic modes.

$\mathbf{q} \to 0$. Therefore $Ze[\mathbf{U}_0(1) - \mathbf{U}_0(2)]$ represents the amplitude of the *displacement* dipole moment per cell, whence $\mathbf{P}_0$ is the amplitude of the dipole-moment density. The polarization $\mathbf{P}(\mathbf{r})$ gives rise to a slowly varying macroscopic field $\mathbf{E}(\mathbf{r})$ of the form $\mathbf{E}(\mathbf{r}) = \mathbf{E}_0(\hat{\mathbf{q}}) \exp(i\mathbf{q} \cdot \mathbf{r})$, where $\mathbf{E}$ and $\mathbf{P}$ are related by the well known equations of electrostatics

$$\mathbf{V} \cdot (\mathbf{E} + 4\pi\mathbf{P}) = 0, \mathbf{V} \times \mathbf{E} = 0. \tag{4.2}$$

When solved, Eqs. (4.2) lead to the result[1]

$$\mathbf{E} = -4\pi\hat{\mathbf{q}} \, (\hat{\mathbf{q}} \cdot \mathbf{P}), \tag{4.3}$$

where $\hat{\mathbf{q}}$ is a unit vector in the direction of $\mathbf{q}$ (See (A2.33)). Equation (4.3) shows that the polarization developed during optic vibrations results in a macroscopic field directed along $\mathbf{q}$ and therefore *longitudinal*.

Associated with the field $\mathbf{E}$ and the polarization $\mathbf{P}$ is an energy density $-\mathbf{P} \cdot \mathbf{E}$. Remembering that $\mathbf{P} \| [\mathbf{U}(1) - \mathbf{U}(2)]$, it follows that for the TO mode the contribution from this source vanishes since $\mathbf{P} \perp \mathbf{q}$ whereas $\mathbf{E} \| \mathbf{q}$. On the other hand, in the LO mode $[\mathbf{U}(1) - \mathbf{U}(2)] \| \mathbf{P} \| \mathbf{E} \| \mathbf{q}$, and this interaction energy term produces an additional restoring force which raises the frequency of the LO mode above that of the TO mode.

### 4.1.2 Electronic polarization effects—LST relation
We next consider the effects of electronic polarization. These were ignored in writing (4.1) by attributing $\mathbf{P}_0$ entirely to the displacement of *point* charges. One way to improve upon this unrealistic hypothesis would be to write

$$\mathbf{P} = \frac{1}{v} \{Ze[\mathbf{U}(1) - \mathbf{U}(2)] + \mathscr{E}(\alpha_+ + \alpha_-)\}, \tag{4.4}$$

where $\mathscr{E}$ is the local field acting on the ions and $\alpha_\pm$ are their electronic polarizabilities. In contrast to (4.1), (4.4) includes the effects of electronic polarization arising from the *local* field $\mathscr{E}$. The latter refers to the field acting on a particular ion due to the elementary dipoles residing on all the others and is discussed further in Appendix 2. Use of (4.4) in (4.3) gives a new value for the macroscopic field amplitude which now includes the effects of electronic polarization.

Actually the long-wavelength frequencies themselves demonstrate the presence of electronic polarization effects. Lyddane, Sachs, and Teller (LST)[2] have shown that in cubic diatomic crystals

$$\frac{\omega_{LO}^2(\mathbf{q} \approx 0)}{\omega_{TO}^2(\mathbf{q} \approx 0)} = \frac{\varepsilon(0)}{\varepsilon(\infty)}, \tag{4.5}$$

where $\varepsilon(0)$ is the static dielectric constant and $\varepsilon(\infty)$ is the so-called high-frequency dielectric constant. The latter term signifies the dielectric constant at a frequency much greater than vibrational frequencies but well below electronic transition frequencies. An applied electric field of such a frequency will therefore not be able to produce any lattice displacements; only the electrons can respond and thus $\varepsilon(\infty)$ includes only electronic polarization effects.* Since $\varepsilon(\infty)$ is a measure of electronic distortion, it follows from the LST relation that the vibrational frequencies are indeed affected by the electronic polarization.

### 4.1.3 Phenomenological treatment of the long-wavelength optic modes—electrostatic approximation

We shall now give a brief phenomenological discussion of the behavior of the long-wavelength optic modes in cubic diatomic crystals, with suitable allowance for electronic polarization effects. This will also be a forerunner to a generalized treatment applicable to insulating crystals of arbitrary structure, to be given later.

We take as the two basic variables (1) $\mathbf{W}$ defined by

$$\mathbf{W} = (\bar{m}/v)^{1/2}[\mathbf{U}(1) - \mathbf{U}(2)], \tag{4.6}$$

where $\bar{m}$ is the reduced mass of the two atoms in the primitive cell; and (2) $\mathbf{E}$ the macroscopic field. The use of $\mathbf{W}$ is suggested by the center-of-mass-preserving motion of the two sublattices in the $\mathbf{q} \approx 0$ optic modes. We postulate an energy density function[3]

$$\phi = -\tfrac{1}{2}\{b_{11}\mathbf{W}^2 + 2b_{12}\mathbf{W}\cdot\mathbf{E} + b_{22}\mathbf{E}^2\} \tag{4.7}$$

which, we observe, also includes the energy density associated with the polarization produced by the field $\mathbf{E}$. By adopting a specific model for the lattice, the coefficients $b_{11}$, etc. may be deduced in terms of lattice quantities.[4] For the present we take (4.7) for granted and deduce from it the equations of motion. These are

$$-\frac{\partial\phi}{\partial\mathbf{W}} = \text{force conjugate to } \mathbf{W}$$

$$= b_{11}\mathbf{W} + b_{12}\mathbf{E} = \ddot{\mathbf{W}}, \tag{4.8}$$

$$-\frac{\partial\phi}{\partial\mathbf{E}} = \mathbf{P}$$

$$= b_{12}\mathbf{W} + b_{22}\mathbf{E}. \tag{4.9}$$

---

*Note that $\varepsilon(\infty)$ does *not* refer to the dielectric constant at infinite frequency (which in fact must be unity). An alternate way to characterize $\varepsilon(\infty)$ is as the dielectric constant at zero frequency with the lattice held rigid.

The macroscopic field $\mathbf{E}$ set up during the vibrations is related to $\mathbf{P}$ as in (4.3). From (4.8) we note that the equation of motion for the displacement is affected by the macroscopic field, while from (4.9) we see that the polarization is not solely due to displacement effects but also includes the effects of electronic polarization through the phenomenological parameter $b_{22}$.

Specializing the above equations to the TO mode, we get $\mathbf{E} = 0$ since $\mathbf{P} \perp \mathbf{q}$, whence

$$\ddot{\mathbf{W}} = b_{11}\mathbf{W}.$$

By considering a solution of the form $\mathbf{W} = \mathbf{W}_0 \exp[i(\mathbf{q} \cdot \mathbf{r} - \omega_{TO}t)]$, we obtain

$$\omega_{TO}^2 = -b_{11}. \tag{4.10}$$

On the other hand, for the LO mode $\mathbf{E} = -4\pi\mathbf{P}$ since $\mathbf{P} \parallel \hat{\mathbf{q}}$. Hence from (4.8) and (4.9) we deduce

$$\begin{aligned}
\omega_{LO}^2 &= -b_{11} + \frac{4\pi b_{12}^2}{1 + 4\pi b_{22}} \\
&= \omega_{TO}^2 + \frac{4\pi b_{12}^2}{1 + 4\pi b_{22}}, \tag{4.11}
\end{aligned}$$

which shows that the LO frequency is indeed increased by the macroscopic field. It is worth observing that both the LO and TO frequencies are independent of $\mathbf{q}$ in the long-wavelength limit.

### 4.1.4 Phenomenological treatment of the dielectric properties

Equations (4.8) and (4.9) may also be used to determine the dielectric behavior of cubic diatomic crystals for frequencies in the region of the vibrational frequencies. Prior to doing so, we recall that the dielectric properties of a small system, such as an assembly of molecules, are deduced by calculating the polarization via the equation $\mathbf{P} = \chi\mathbf{E}$, where $\chi$ is the dielectric susceptibility, and subsequently using the relation $\mathbf{D} = \varepsilon\mathbf{E}$ $= \mathbf{E} + 4\pi\mathbf{P}$. Here it is assumed that the field $\mathbf{E}$ is uniform over the entire system. A similar strategy may be employed for crystals by first considering a region comprising several unit cells which is large enough to be regarded as macroscopic, yet small compared to the crystal as a whole, and then evaluating the polarization produced by a field $\mathbf{E}$ which is sufficiently uniform over this region. It must be remembered, however, that the field $\mathbf{E}$, though viewed as "externally imposed" over a local portion (of macroscopic dimensions) of the crystal, includes the contribution arising from the polarization of the crystal itself. In other words, suppose for purposes of computing $\chi$ one introduces initially a field $\mathbf{E}'$ due to an agency not associated with the crystal.* The field $\mathbf{E}$ actually driving a typical local region is then $\mathbf{E}'$ plus that due to polarization (initially triggered by $\mathbf{E}'$ itself). Thus $\mathbf{E}$ is the total self-consistently determined macroscopic field acting on a local region. In the formal calculations, however, it is not necessary to identify the components of $\mathbf{E}$ since what is required is $\chi = P/E$. For purposes of calculating $\mathbf{P}$, therefore, it is adequate to regard $\mathbf{E}$ itself as "externally imposed" and then use (4.8) and (4.9) in the spirit of a driver-driven problem as in the molecular case.[5] In identifying a local region and studying its polarization we effectively assume that different local regions behave independently and are dynamically decoupled.[5]

Let $\mathbf{E} = \mathbf{E}_0 e^{i\omega t}$ be the "external field" in the sense explained above, $\omega$ being arbitrary. The applied field forces both $\mathbf{W}$ and $\mathbf{P}$ to oscillate at the same frequency $\omega$, and the forced oscillations may be studied by introducing the solutions

$$\begin{pmatrix} \mathbf{W} \\ \mathbf{P} \\ \mathbf{E} \end{pmatrix} = \begin{pmatrix} \mathbf{W}_0 \\ \mathbf{P}_0 \\ \mathbf{E}_0 \end{pmatrix} \exp[i\omega t] \tag{4.12}$$

into (4.8) and (4.9) which yield, after a few manipulations,

$$\mathbf{P} = \left\{ b_{22} + \frac{b_{12}^2}{-b_{11} - \omega^2} \right\} \mathbf{E}. \tag{4.13}$$

Therefore

$$\begin{aligned} \varepsilon(\omega) &= 1 + 4\pi P/E \\ &= 1 + 4\pi b_{22} + 4\pi b_{12}^2/(-b_{11} - \omega^2) \\ &= \varepsilon(\infty) + \frac{4\pi b_{12}^2}{\omega_{TO}^2 - \omega^2}, \end{aligned} \tag{4.14}$$

where $\varepsilon(\infty) = 1 + 4\pi b_{22}$ denotes the high-frequency dielectric constant. The LST relation follows immediately from (4.10), (4.11) and (4.14).

### 4.1.5 Phenomenological treatment of long-wavelength modes—retardation effects included

Returning to the problem of free vibrations, we recall that the results deduced in (4.10) and (4.11) for the transverse and longitudinal frequencies were obtained within the electrostatic approximation, which as-

---

*For finite crystals, $\mathbf{E}'$ is readily visualized as that due to an externally applied field, and the value of $\mathbf{E}'$ may be deduced by appropriate boundary conditions. For infinite crystals $\mathbf{E}'$ may be regarded as produced by Maxwell's demon!

sumes that electrical interactions are instantaneous. However, this is not strictly true since electrical interactions are retarded and propagate with the velocity of light. Accordingly it is necessary to modify the earlier discussion by solving the lattice equations in conjunction with Maxwell's equations[6]

$$\nabla \cdot \mathbf{E} = -4\pi \nabla \cdot \mathbf{P}, \text{ (a)} \qquad \nabla \cdot \mathbf{H} = 0, \text{ (b)}$$

$$\nabla \times \mathbf{E} = -\frac{1}{c}\dot{\mathbf{H}}, \text{ (c)} \qquad \nabla \times \mathbf{H} = \frac{1}{c}(\dot{\mathbf{E}} + 4\pi\dot{\mathbf{P}}), \text{ (d)}, \qquad (4.15)$$

rather than with (4.2). Use of the plane-wave solutions

$$\begin{pmatrix} \mathbf{W} \\ \mathbf{P} \\ \mathbf{E} \\ \mathbf{H} \end{pmatrix} = \begin{pmatrix} \mathbf{W}_0 \\ \mathbf{P}_0 \\ \mathbf{E}_0 \\ \mathbf{H}_0 \end{pmatrix} \exp i[\mathbf{q}\cdot\mathbf{r} - \omega t] \qquad (4.16)$$

in the lattice and Maxwell's equations then leads to

$$-\omega^2 \mathbf{W} = b_{11}\mathbf{W} + b_{12}\mathbf{E}, \qquad (4.17a)$$

$$\mathbf{P} = b_{12}\mathbf{W} + b_{22}\mathbf{E}, \qquad (4.17b)$$

$$\mathbf{q}\cdot(\mathbf{E} + 4\pi\mathbf{P}) = 0, \qquad (4.17c)$$

$$\mathbf{q}\cdot\mathbf{H} = 0, \qquad (4.17d)$$

$$\mathbf{q} \times \mathbf{E} = \frac{\omega}{c}\mathbf{H}, \qquad (4.17e)$$

$$\mathbf{q} \times \mathbf{H} = -\frac{\omega}{c}(\mathbf{E} + 4\pi\mathbf{P}). \qquad (4.17f)$$

Taking the cross product of $\mathbf{q}$ with (4.17e) and remembering the rule for triple cross product of vectors, we obtain

$$(\mathbf{q}\cdot\mathbf{E})\mathbf{q} - q^2\mathbf{E} = \frac{\omega}{c}\mathbf{q} \times \mathbf{H} = -\frac{\omega^2}{c^2}(\mathbf{E} + 4\pi\mathbf{P})$$

or

$$\mathbf{E}(n^2 - 1) = 4\pi\mathbf{P} + (\mathbf{q}\cdot\mathbf{E})\mathbf{q}\frac{c^2}{\omega^2} = 4\pi\mathbf{P} - 4\pi n^2(\hat{\mathbf{q}}\cdot\mathbf{P})\hat{\mathbf{q}}, \qquad (4.18)$$

where $n = cq/\omega$. Comparison with (4.3) shows that the inclusion of retardation frees $\mathbf{E}$ from being purely longitudinal by introducing transverse components. These are readily obtained by writing $\mathbf{E} = \mathbf{E}_\mathrm{l} + \mathbf{E}_\mathrm{t}$ and $\mathbf{P} = \mathbf{P}_\mathrm{l} + \mathbf{P}_\mathrm{t}$. We then get

$$\mathbf{E}_\mathrm{l} = -4\pi\mathbf{P}_\mathrm{l} \quad \text{(a)}; \qquad \mathbf{E}_\mathrm{t} = \frac{4\pi}{(n^2 - 1)}\mathbf{P}_\mathrm{t}. \quad \text{(b)} \qquad (4.19)$$

The result for the longitudinal field is the same as before while that for the transverse field is new. Observe that when $c$ is set equal to infinity, $\mathbf{E}_t$ vanishes (although $\mathbf{P}_t$ may not), and $\mathbf{E}$ becomes purely longitudinal, the electrostatic result.

To obtain the longitudinal frequency in the present case, we take (4.19a) in conjunction with (4.17a and b). This yields

$$\omega^2 = -b_{11} + \frac{4\pi b_{12}^2}{1 + 4\pi b_{22}} = \omega_{LO}^2, \tag{4.20}$$

showing that the longitudinal solution is unaffected by retardation effects.

On the other hand, for transverse waves we obtain from (4.17a and b),

$$\mathbf{P}_t = \left\{ b_{22} + \frac{b_{12}^2}{-b_{11} - \omega^2} \right\} \mathbf{E}_t. \tag{4.21}$$

Combining this with (4.19b) and demanding $\mathbf{E}_t \neq 0$ then leads to

$$\frac{q^2 c^2}{\omega^2} = n^2 = \left\{ 1 + 4\pi b_{22} + \frac{4\pi b_{12}^2}{-b_{11} - \omega^2} \right\} = \varepsilon(\omega). \tag{4.22}$$

When solved, (4.22) yields two doubly degenerate solutions for $\omega^2$ as a function of $q$. The transverse solutions are thus affected by retardation and show dispersion.

The physical implication of the foregoing result becomes apparent when we recall that waves in (isotropic or cubic) media accompanied by *transverse* electric fields really represent photons propagating in the medium.* In the present instance, waves accompanied by transverse electric fields also involve the vibrations of atoms. In this sense, they have a dual charracter, being both photonlike and phononlike. One often speaks of this phenomenon as a manifestation of *photon-phonon coupling* and the quanta of such coupled photon-phonon modes are referred to as *polaritons*. Equation (4.22) gives in fact the dispersion relation for polaritons in cubic diatomic crystals. Attention is called to the fact that the doubly degenerate solutions of (4.22) really involve four degrees of freedom corresponding to every $q$. Two of these are "mechanical" degrees of freedom associated with the transverse mechanical vibrations, and the remaining two are "electromagnetic" degrees of freedom associated with $\mathbf{E}_t$. It is interesting to observe that had we written the dispersion relations for photons in the usual spirit of optics, we would have had

$$\omega = \frac{cq}{n} = \frac{cq}{[\varepsilon(\infty)]^{1/2}}, \tag{4.23}$$

where $n = [\varepsilon(\infty)]^{1/2}$ is the optical refractive index. This result, which

---

*In this context recall Maxwell's famous remark, "Light consists of transverse undulations of the same medium which is the cause of electric and magnetic phenomena."

implies that **P** has no component from vibrations, is generalized in (4.22) since, in the frequency range under consideration, waves accompanied by transverse electric fields have not only photonlike but also phononlike character.

It is important to bear in mind that the photon-phonon interaction we are considering is *not* the interaction of phonons with photons coming from outside. It is rather the coupling between the transverse vibrations and the transverse electric field orginating from the transverse current densities set up by the *vibrations themselves*.[7] It is because of this that the vibrational and field equations were solved together to obtain self-con-sistency. The transverse-field effects are important only in the context of retardation.

We must, however, also note the following. Suppose photons with frequency $\sim \omega_{TO}$ are allowed to fall on a crystal from outside. Once in-side, the electromagnetic radiation is mixed with the lattice vibrations and its propagation is governed by the same rules as for polaritons. This im-plies that the optical properties of ionic crystals in the infrared (that is, $\omega \sim \omega_{TO} \sim 10^{13}$ rad/s) are closely related to the polariton spectrum. The study of the polariton part of the phonon spectrum thus acquires special importance.

Figure 4.3 illustrates schematically the dispersion relations under various conditions. The polariton spectrum of ZnS computed from (4.22) using measured values for the various parameters is shown in Figure 4.4. It is seen that the electrostatic approximation (in which both $\omega_{TO}(q)$ and $\omega_{LO}(q)$ are dispersionless on the scale of $q$ shown) breaks down when $q \sim \omega/c$, that is, $\sim 5 \times 10^3$ cm$^{-1}$, and that for these and smaller $q$ values retardation effects become important.

Summarizing, the macroscopic field primarily influences the long-wave-length ($\mathbf{q} \rightarrow 0$) frequencies of the optic modes. If $\mathbf{q} \rightarrow 0$ is interpreted in the sense that $q$ is small but still $\gg \omega/c$ then the electrostatic approximation is adequate. For smaller values of $q$, retardation effects must be considered and these lead to polariton formation. Because of the field-vibration inter-action, the dielectric properties of the crystal are also dependent on the optic modes, the dependence being expressed succintly through (4.14) and the LST relation. In the next few sections we shall extend these considera-tions to ionic crystals of arbitrary structure.

## 4.2 Rigid-ion treatment of long-wavelength phenomena in ionic crystals of arbitrary structure

We now wish to extend the considerations of the preceding section by first examining in some detail the origin of the macroscopic field and then

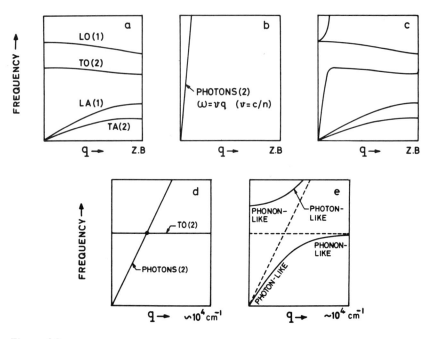

**Figure 4.3**

Schematic drawings of the dispersion curves along a symmetry direction such as [100] in a cubic diatomic crystal. (a) Curves appropriate to the electrostatic approximation. (b) Photon dispersion curves in the absence of coupling with vibrations. In this case, the velocity $v = (c/n)$, where $n = [\varepsilon(\infty)]^{1/2}$ is the refractive index. If photon-phonon mixing is allowed for in the sense discussed in the text, the dispersion curves become modified as in (c). For purposes of visualization, the photonlike parts in both (b) and (c) have been shown with a smaller slope than is appropriate. Otherwise it would be impossible to sketch the acoustic branches in the same figure. In (d) and (e) are shown just the photon and the TO branches without and with mixing, respectively. The $q$ range of (a), (b), and (c) extends to the zone boundary (Z.B.), while that of (d) and (e) is less by about a factor of $10^4$.

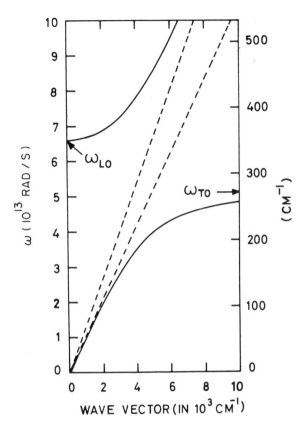

**Figure 4.4**

Polariton spectrum for ZnS computed using Eq. (4.22) and the measured values of $\omega_{TO}$, $\omega_{LO}$, $\varepsilon(0)$, and $\varepsilon(\infty)$. The dashed lines are based on the velocity of light in ZnS appropriate to the high- and the low-frequency refractive indices. This figure is to be compared with Figure 4.3e. Observe that the lower polariton branch approaches $\omega_{TO}$ asymptotically.

studying its influence on the long-wavelength optic modes of ionic crystals of arbitrary structure. For convenience we shall initially assume a simple model in which, for purposes of describing the electrostatic interactions, the ions are viewed as point charges. Since electronic polarization is clearly excluded, the present model is sometimes called the *rigid-ion model* or the *point-ion model*.* It must be pointed out that, though introduced here in the context of a study of the long-wavelength modes, the rigid-ion model can be used for a complete calculation of the dispersion curves, as has been done in arriving at those shown in Figure 4.2.

### 4.2.1 Long-wavelength equations and the origin of the macroscopic field

In the rigid-ion model, the force constants are regarded as made up of $\phi_{\alpha\beta}^{SR}\left(\begin{smallmatrix} l & l' \\ k & k' \end{smallmatrix}\right)$ arising from short-range forces and $\phi_{\alpha\beta}^{C}\left(\begin{smallmatrix} l & l' \\ k & k' \end{smallmatrix}\right)$ arising from Coulomb interactions. Through the use of (2.225), the latter constant is determined to be

$$\phi_{\alpha\beta}^{C}\left(\begin{matrix} l & l' \\ k & k' \end{matrix}\right) = Z_k Z_{k'} e^2 \left(\frac{\delta_{\alpha\beta}}{x^3} - \frac{3x_\alpha x_\beta}{x^5}\right)_{x=|x\left(\begin{smallmatrix} l & l' \\ k & k' \end{smallmatrix}\right)|}, \quad (lk) \neq (l'k'). \qquad (4.24)$$

The self term is defined through

$$\phi_{\alpha\beta}^{C}\left(\begin{matrix} l & l \\ k & k \end{matrix}\right) = -\sum_{l'k'}{}' \phi_{\alpha\beta}^{C}\left(\begin{matrix} l & l' \\ k & k' \end{matrix}\right). \qquad (4.25)$$

In the above $Z_k e$ denotes the charge on the ions in the $k$th sublattice. The assignment of the numerical value of $Z_k$ is a tricky problem and depends very much on the type of crystal one is dealing with. In the case of a clear-cut ionic crystal there is no conceptual difficulty, and $Z_k e$ may be regarded as the ionic charge or the charge appropriate to a simple chemical picture of an ionic crystal. For example, in NaCl we would have $Z_+ = 1$ and $Z_- = -1$. On the other hand, for some substances such as the III-V compounds, which exhibit considerable covalency effects, there is difficulty since not only is the ionic value for $Z_k$ inappropriate, but also one does not know how to associate a net charge with each atom. Furthermore, as we shall see later (Sec. 4.6), even if one manages to assign a value for $Z_k$, it turns out that owing to the effects of electronic polarization, the charge one must associate with the ions during vibrations is different from $Z_k e$. This is true of all insulating crystals, particularly those in which the valence electrons are loosely bound and hence susceptible to readjustments during vibrations. For the present we ignore all these complications and

*In the literature it is also sometimes referred to as the Born model.

merely suppose that an assignment of $Z_k e$ is possible and that this value remains unchanged during vibrations. Note, incidentally, that in the context of present assumptions $Z_k \equiv 0$ for all $k$ in homopolar crystals.

Reverting to a discussion of the dynamics, we shall, for purposes of treating the long-wavelength aspects, find it convenient to consider the dynamical equation in the form

$$\omega^2(\mathbf{q}) m_k U_\alpha(k|\mathbf{q}) = \sum_{k'\beta} M_{\alpha\beta}\begin{pmatrix} \mathbf{q} \\ k\ k' \end{pmatrix} U_\beta(k'|\mathbf{q}) \tag{4.26}$$

(see (2.80)). Since $\phi_{\alpha\beta} = \phi_{\alpha\beta}^C + \phi_{\alpha\beta}^{SR}$, we may write

$$\omega^2(\mathbf{q}) m_k U_\alpha(k|\mathbf{q}) = \sum_{k'\beta} \left[ M_{\alpha\beta}^{SR}\begin{pmatrix} \mathbf{q} \\ k\ k' \end{pmatrix} + M_{\alpha\beta}^C\begin{pmatrix} \mathbf{q} \\ k\ k' \end{pmatrix} \right] U_\beta(k'|\mathbf{q}). \tag{4.27}$$

The evaluation of $M_{\alpha\beta}^C$, the Coulomb contribution to the dynamical matrix, is a nontrivial problem discussed in Appendix 2. Here we note that $M_{\alpha\beta}^C(\begin{smallmatrix} \mathbf{q} \\ kk' \end{smallmatrix})$ does not have a unique limit as $\mathbf{q} \to 0$, and that the limiting value depends on the direction by which $\mathbf{q} = 0$ is approached. However, it is possible to write $M_{\alpha\beta}^C$ as (see (A2.19))

$$M_{\alpha\beta}^C\begin{pmatrix} \mathbf{q} \\ k\ k' \end{pmatrix} = \overline{M}_{\alpha\beta}^C\begin{pmatrix} \mathbf{q} \\ k\ k' \end{pmatrix} + \frac{4\pi e^2}{v} Z_k Z_{k'} \hat{q}_\alpha \hat{q}_\beta, \tag{4.28}$$

where $\overline{M}_{\alpha\beta}^C$ is regular* at $q = 0$.

In the literature, the Coulomb matrix $M_{\alpha\beta}^C$ sometimes appears as

$$M_{\alpha\beta}^C\begin{pmatrix} \mathbf{q} \\ k\ k' \end{pmatrix} = Z_k e C_{\alpha\beta}\begin{pmatrix} \mathbf{q} \\ k\ k' \end{pmatrix} Z_{k'} e, \tag{4.29}$$

where $C_{\alpha\beta}(\begin{smallmatrix} \mathbf{q} \\ kk' \end{smallmatrix})$ is defined in (A2.24). On occasion, the Coulombic contribution is also given in terms of the local field, and Eq. (4.27) is written as follows (for details see Appendix 2):

$$\omega^2(\mathbf{q}) m_k U_\alpha(k|\mathbf{q}) = \sum_{k'\beta} M_{\alpha\beta}^{SR}\begin{pmatrix} \mathbf{q} \\ k\ k' \end{pmatrix} U_\beta(k'|\mathbf{q}) - Z_k e \mathscr{E}_\alpha(k|\mathbf{q}). \tag{4.30}$$

Here $\mathscr{E}(k|\mathbf{q})$ is the amplitude of the local field acting on the ions in the $k$th sublattice due to the dipoles created by ionic displacements.

Irrespective of which description one uses, the Coulombic contribution to the dynamical matrix may be explicitly calculated. Upon adopting suitable models for the short-range forces, one can then completely calcu-

---

*A necessary condition for a function $f(x, y, z, \ldots)$ of several variables to be regular at the point $(x_0, y_0, z_0, \ldots)$ is that it tends to a unique limit no matter how that point is approached from the neighborhood. The second term on the right-hand side of (4.28) clearly does not fulfill this requirement.

late the dispersion curves in the rigid-ion model. Later we shall have occasion to consider such calculations, especially for alkali halides. Our present interest lies, however, in adapting (4.27) to the long-wavelength modes to facilitate the examination of specific effects associated with the macroscopic field. For this purpose we use (4.28) whence (4.27) becomes

$$\omega^2(\mathbf{q})m_k U_\alpha(k|\mathbf{q}) = \sum_{k'\beta} M_{\alpha\beta}\binom{\mathbf{q}}{k\,k'} U_\beta(k'|\mathbf{q})$$
$$- Z_k e\left\{-4\pi\hat{q}_\alpha \sum_\beta \hat{q}_\beta\left[\frac{1}{v}\sum_{k'} Z_{k'}\cdot e U_\beta(k'|\mathbf{q})\right]\right\}, \qquad (4.31)$$

where

$$\overline{\mathbf{M}} = \overline{\mathbf{M}}^C + \mathbf{M}^{SR} \qquad (4.32)$$

is regular at $\mathbf{q} = 0$ since $\mathbf{M}^{SR}$, like $\overline{\mathbf{M}}^C$, is always regular at $\mathbf{q} = 0$. In the long-wavelength limit the quantity in the square brackets on the right-hand side of (4.31) can be identified as the amplitude of the polarization component $P_\beta$ associated with displacements (see (4.1)). This enables the second term to be recast as $-Z_k e E_\alpha$, where $E_\alpha$ is the $\alpha$th component of the macroscopic field $\mathbf{E}$, related to $\mathbf{P}$ via (4.3). Equation (4.31) may thus be expressed in matrix notation as

$$\omega^2(\mathbf{q})\mathbf{m}\mathbf{U}(\mathbf{q}) = \overline{\mathbf{M}}(\mathbf{q})\mathbf{U}(\mathbf{q}) - \check{\mathbf{Z}}\mathbf{E}. \qquad (4.33)$$

The definitions of $\mathbf{m}$, $\overline{\mathbf{M}}(\mathbf{q})$ and $\mathbf{U}(\mathbf{q})$ appear in Sec. 2.2; $\check{\mathbf{Z}}$ is the transpose of a $3 \times 3n$ array of the following structure:

$$\mathbf{Z} = e\begin{pmatrix} Z_1 & 0 & 0 & Z_2 & 0 & 0 & \cdots & Z_n & 0 & 0 \\ 0 & Z_1 & 0 & 0 & Z_2 & 0 & \cdots & 0 & Z_n & 0 \\ 0 & 0 & Z_1 & 0 & 0 & Z_2 & \cdots & 0 & 0 & Z_n \end{pmatrix}; \qquad (4.34)$$

$\mathbf{E}$ is a column matrix with components $E_x$, $E_y$, $E_z$. Supplementing (4.33) are the polarization equation

$$\mathbf{P} = \frac{1}{v}\mathbf{Z}\mathbf{U}(\mathbf{q}) \qquad (4.35)$$

and the field equation

$$\mathbf{E} = -4\pi\hat{\mathbf{Q}}\mathbf{P}, \qquad (4.36)$$

where $\hat{\mathbf{Q}}$ is a symmetric tensor with elements $\hat{q}_\alpha \hat{q}_\beta$, $(\alpha, \beta = x, y, z)$. It may be verified that, like $\mathbf{E}$, $\mathbf{P}$ is a three-component column matrix.

Equations (4.33), (4.35), and (4.36) provide the starting point of our present discussion. Such trios will be frequently encountered in this chapter. In each case, the polarization and the field equations will first be employed to express the dynamical equation entirely in terms of displacements; the resulting equation will then be solved as usual.

Since we are primarily interested in the long-wavelength optic modes, we might as well take the $\mathbf{q} \to 0$ limit of $\overline{\mathbf{M}}$ and write (4.33) as*

$$\omega^2(\hat{\mathbf{q}})\mathbf{m}\mathbf{U}(\hat{\mathbf{q}}) = \overline{\mathbf{M}}(0)\mathbf{U}(\hat{\mathbf{q}}) - \check{\mathbf{Z}}\mathbf{E}(\hat{\mathbf{q}}),\tag{4.37}$$

where $\hat{\mathbf{q}}$ in parentheses signifies that the concerned quantity pertains to the $\mathbf{q} \to 0$ limit taken in the direction $\hat{\mathbf{q}}$, the limit being understood in the sense discussed previously (Sec. 4.1.5).

## 4.2.2 Solution of the long-wavelength equations under different electrical conditions

We now wish to discuss the solutions of (4.37) under various electrical conditions: (a) $\mathbf{E} = 0$, (b) $\mathbf{E} \neq 0$ but $\mathbf{D} = 0$, (c) both $\mathbf{E}$ and $\mathbf{D} \neq 0$. The significance of these various solutions will emerge later.

**(a) $\mathbf{E} = 0$**

Here (4.37) simplifies to

$$\omega^2\mathbf{m}\mathbf{U} = \overline{\mathbf{M}}(0)\mathbf{U}.\tag{4.38}$$

Since the dynamical matrix does not depend upon $\hat{\mathbf{q}}$, neither do the eigenvalues and the eigenvectors which may therefore be designated as $\omega_j^2(0)$ and $\mathbf{e}\binom{0}{j}$, $(j = 1, 2, \ldots, 3n)$. Of the $3n$ eigenvalues of $\overline{\mathbf{M}}(0)$, three vanish corresponding to the acoustic solutions. For convenience we divide the remaining $(3n - 3)$ optic solutions into two categories:

1. $\mathbf{Z}\mathbf{e}\binom{0}{j} = 0,$ \hfill (4.39)

2. $\mathbf{Z}\mathbf{e}\binom{0}{j} \neq 0.$ \hfill (4.40)

From (4.35) we see that if (4.39) is true then $\mathbf{P}$ vanishes for the corresponding mode.† Hence the physical significance of this categorization is that modes of type 1 do not have displacive dipole moments associated with them while modes of type 2 do. Assume now that there are $r$ modes in the latter category. The corresponding frequencies are often referred to as

---

*The $\mathbf{q}$-dependence of $\overline{\mathbf{M}}$ is important in the discussion of the acoustic modes. See Sec. 4.3.
†For the acoustic modes, $\mathbf{Z}\mathbf{e}\binom{0}{j} = \text{const} \times \sum_k Z_k e = 0$ owing to the neutrality of the primitive cell.

*dispersion frequencies,* while the modes themselves are termed *infrared active* (IR active), since their dipole moments enable them to interact with electromagnetic radiation of the same frequency. Indeed the IR-active modes can interact not only with external fields but also with the macroscopic field generated by the vibrations themselves. The latter interaction, suppressed in the present discussion, will be treated in case (c).

It must be emphasized that the division of optic solutions into IR active and IR inactive, as per (4.39) and (4.40), is presently based on the rigid-ion model. Later we shall amend this division to make it model independent. While (4.39) gives guidance to the actual IR-active modes in ionic crystals, it suggests (erroneously) that IR activity is not possible in homopolar crystals since $Z$ vanishes for such crystals, as noted previously. The possible existence of IR-active modes in homopolar crystals will be commented upon later.

**(b) $E \neq 0$ but $D = 0$**

Since $D = E + 4\pi P$, it follows that in the present case $E = -4\pi P$. We may thus write (4.37) as

$$\omega^2 \mathbf{m} \mathbf{U} = \left[ \overline{\mathbf{M}}(0) + \frac{4\pi}{v} \tilde{\mathbf{Z}} \mathbf{Z} \right] \mathbf{U}. \tag{4.41}$$

Once again the dynamical matrix and consequently the eigenvalues and eigenvectors are independent of $\hat{\mathbf{q}}$. The characteristic equation which delivers the eigenvalues is

$$\left\| \left[ \overline{\mathbf{M}}(0) + \frac{4\pi}{v} \tilde{\mathbf{Z}} \mathbf{Z} \right] - \mathbf{m} \omega^2 \right\| = 0. \tag{4.42}$$

Among the $(3n - 3)$ optic solutions, only $r$ differ from the corresponding solutions of (4.38). These $r$ frequencies will be designated $\omega_{\lambda,l}$, $(\lambda = 1, 2, \ldots, r)$; they are called *longitudinal frequencies* because $E \| P$. It will be argued later that, in crystals of orthorhombic or higher symmetry, to every dispersion frequency $\omega_\lambda(0)$, $(\lambda = 1, \ldots, r)$, corresponds a longitudinal counterpart $\omega_{\lambda,l}$.

---

*The dispersion frequency $\omega_{TO}$ in cubic diatomic crystals is sometimes called the *restrahlen* (residual ray) frequency. This nomenclature arose in the early part of the century when it was found that upon multiply reflecting a polychromatic infrared beam by mirrors of the material concerned, there resulted a narrow band beam with an edge frequency characteristic of the reflecting material. These rays were called residual rays, and their frequency was linked to that of the oscillations of the positive ions with respect to negative ions.

**(c) E ≠ 0, D ≠ 0**

This is the most interesting of the three cases since here the macroscopic field is allowed full play. The equations to be solved are (4.33), (4.35), and (4.36). The latter two yield

$$E = -4\pi \hat{Q}P = -\frac{4\pi}{v} \hat{Q}ZU. \tag{4.43}$$

Used in (4.33), this gives

$$\omega^2(\hat{q})mU(\hat{q}) = \left[\bar{M}(0) + \frac{4\pi}{v}\check{Z}\hat{Q}Z\right]U(\hat{q}). \tag{4.44}$$

The limiting frequencies $\omega_\lambda(\hat{q})$ are thus obtained by solving the equation

$$\left\|\left[\bar{M}(0) + \frac{4\pi}{v}\check{Z}\hat{Q}Z\right] - m\omega^2(\hat{q})\right\| = 0. \tag{4.45}$$

Except in the case of the IR-active modes, (4.44) and (4.45) yield the same solutions as (4.38). For the IR-active modes, however, the additional contribution $(4\pi/v)\check{Z}\hat{Q}Z$ to the dynamical matrix in general causes the limiting frequencies to be direction dependent.

### 4.2.3 Long-wavelength equations pertaining to the IR-active modes only

We have noted that in cases (b) and (c) only $r$ out of the $3n$ solutions are different from those of case (a). It is often convenient to focus attention on these by projecting out of the equations of motion just the part pertaining to the IR-active modes. We shall illustrate this by reconsidering case (b). The equations of the problem are

$$\bar{M}(0)e'(J)A(J) - \check{Z}E = \omega_J^2 me'(J)A(J), \tag{4.46}$$

$$P = \frac{1}{v}Ze'(J)A(J), \tag{4.47}$$

$$E = -4\pi P. \tag{4.48}$$

Here $e'(J)$, $(J = 1, \ldots, 3n)$, denotes the eigenvector of the present problem and can be expressed formally as a linear combination of the vectors $e\binom{0}{j}$, $(j = 1, \ldots, 3n)$, of case (a), since the latter form a complete set (see remarks in Sec. 2.2.7). The constant $A(J) = A\binom{0}{J}$ has dimensions of length and must be included to maintain dimensional consistency since the eigenvector $e'(J)$, like $e\binom{0}{j}$, is dimensionless (see Sec. 2.2.14). We thus write

$$\mathbf{e}'(J) = \sum_{j=1}^{3n} a_{jJ}\mathbf{e}\begin{pmatrix} 0 \\ j \end{pmatrix}. \tag{4.49}$$

In matrix form this is

$$[\cdots, \mathbf{e}'(J), \cdots] = \left[\cdots, \mathbf{e}\begin{pmatrix} 0 \\ j \end{pmatrix}, \cdots\right]\mathbf{a}, \tag{4.50}$$

where $\mathbf{a}$ is a $3n$-dimensional matrix with elements $a_{jJ}$. In fact, $\mathbf{a}$ can be regarded as made up by stacking column vectors $\mathbf{a}(J)$ with elements $a_{jJ}$, $(j = 1, \ldots, 3n)$. Substituting (4.49) in (4.46) and (4.47), we get

$$\sum_j \left[\overline{\mathbf{M}}(0)\mathbf{e}\begin{pmatrix} 0 \\ j \end{pmatrix}\right]a_{jJ}A(J) - \check{\mathbf{Z}}\mathbf{E} = \omega_J^2 \sum_j \left[m\mathbf{e}\begin{pmatrix} 0 \\ j \end{pmatrix}\right]a_{jJ}A(J) \tag{4.51}$$

$$\mathbf{P} = \frac{1}{v}\sum_j \left[\mathbf{Z}\mathbf{e}\begin{pmatrix} 0 \\ j \end{pmatrix}\right]a_{jJ}A(J). \tag{4.52}$$

Next we multiply by $\check{\mathbf{e}}\begin{pmatrix} 0 \\ j' \end{pmatrix}$ from the left* and make use of the ortho-normalization (see Sec. 2.2.14)

$$\check{\mathbf{e}}\begin{pmatrix} 0 \\ j' \end{pmatrix}m\mathbf{e}\begin{pmatrix} 0 \\ j \end{pmatrix} = M\delta_{j'j} \tag{4.53}$$

and the related property

$$\check{\mathbf{e}}\begin{pmatrix} 0 \\ j' \end{pmatrix}\overline{\mathbf{M}}(0)\mathbf{e}\begin{pmatrix} 0 \\ j \end{pmatrix} = \delta_{j'j}M\Omega_{jj}(0) = \delta_{j'j}M\omega_j^2(0), \tag{4.54}$$

where $M$ is the mass of the primitive cell. This then leads to

$$\sum_j [M\Omega_{jj}(0)\delta_{j'j} - \omega_J^2 M\delta_{j'j}]a_{jJ}A(J) - \left[\check{\mathbf{e}}\begin{pmatrix} 0 \\ j' \end{pmatrix}\check{\mathbf{Z}}\right]\mathbf{E} = 0,$$
$$j' = 1, 2, \ldots, 3n. \tag{4.55}$$

In matrix form this is

$$M[\Omega(0) - \omega_J^2\mathbb{1}_{3n}]\mathbf{a}(J)A(J) - \check{\boldsymbol{\zeta}}\mathbf{E} = 0, \tag{4.56}$$

where $\boldsymbol{\zeta}$ is the transformed $(3 \times 3n)$ charge array defined by

$$\boldsymbol{\zeta} = \mathbf{Z}\left[\cdots, \mathbf{e}\begin{pmatrix} 0 \\ j \end{pmatrix}, \cdots\right]. \tag{4.57}$$

In this notation,

$$\mathbf{P} = \frac{1}{v}\boldsymbol{\zeta}\mathbf{a}(J)A(J). \tag{4.58}$$

Let us now arrange the matrix $\left[\cdots, \mathbf{e}\begin{pmatrix} 0 \\ j \end{pmatrix}, \cdots\right]$ such that the $r$ eigen-

*Remember $\mathbf{e}\begin{pmatrix} 0 \\ j \end{pmatrix}$ can be chosen real since $\mathbf{q} = 0$. See Sec. 2.2.

vectors corresponding to the $r$ IR-active modes are to the left. Correspondingly we also arrange $\mathbf{\Omega}(0)$ as

$$\mathbf{\Omega}(0) = \begin{pmatrix} \mathbf{\Omega}(\text{IR}) & 0 \\ 0 & \mathbf{\Omega}(\text{NIR}) \end{pmatrix}, \tag{4.59}$$

where $\mathbf{\Omega}(\text{IR})$ is the $r$-dimensional diagonal matrix of the squares of the dispersion frequencies and $\mathbf{\Omega}(\text{NIR})$ is the corresponding $(3n - r)$-dimensional matrix of eigenvalues corresponding to the non-IR-active modes. Now in view of the arrangement adopted for the eigenvector matrix, it is clear that $\zeta$ has the structure

$$\zeta = \begin{matrix} \xleftarrow{\quad r \quad} \xleftarrow{\quad\quad (3n-r) \quad\quad} \\ \left[\begin{array}{c|c} \mathscr{L} & 0 \end{array}\right] \end{matrix} \updownarrow 3 \,, \tag{4.60}$$

since $\mathbf{Ze}\binom{0}{j} = 0$ for the non-IR-active modes. Furthermore, since $\mathbf{e}' = \mathbf{e}$ for these modes, the matrix $\mathbf{a}$ must have the structure

$$\mathbf{a} = \left(\begin{array}{c|c} & 0 \\ \hline 0 & \mathbb{1}_{(3n-r)} \end{array}\right) \begin{matrix} \updownarrow r \\ \\ \updownarrow (3n-r) \end{matrix} \tag{4.61}$$

$$\xleftarrow{\quad r \quad} \xleftarrow{\quad\quad (3n-r) \quad\quad}$$

Bearing in mind the structures of $\mathbf{a}$, $\zeta$ and $\mathbf{\Omega}(0)$, the parts of (4.56) and (4.58) pertaining to the IR-active modes can be written as

$$M[\mathbf{\Omega}(\text{IR}) - \omega_j^2 \mathbb{1}_r]\mathbf{a}'(J)A(J) - \mathscr{Z}\mathbf{E} = 0, \tag{4.62}$$

$$\mathbf{P} = \frac{1}{v} \mathscr{L}\mathbf{a}'(J)A(J), \tag{4.63}$$

where $\mathbf{a}'(J)$ is an $r$-component column matrix, identifiable as the appropriate column of the upper left submatrix of Eq. (4.61). Writing (4.62) and (4.63) without the label $J$ and denoting $M\mathbf{a}'A$ by $\mathbf{b}$, we get

$$[\mathbf{\Omega}(\text{IR}) - \omega^2 \mathbb{1}_r]\mathbf{b} - \mathscr{Z}\mathbf{E} = 0, \tag{4.64}$$

$$\mathbf{P} = \frac{1}{Mv} \mathscr{L}\mathbf{b}. \tag{4.65}$$

Eliminating $\mathbf{E}$ as usual, we have

$$\left[ (\mathbf{\Omega}(\text{IR}) - \omega^2 \mathbb{1}_r) + \frac{4\pi}{Mv} \mathscr{Z}\mathscr{Z} \right] \mathbf{b} = 0 \tag{4.66}$$

which is to be compared with (4.41). The determinantal equation resulting is

$$\left\| [\mathbf{\Omega}(\text{IR}) - \omega^2 \mathbb{1}_r] + \frac{4\pi}{Mv} \mathscr{Z}\mathscr{Z} \right\| = 0, \tag{4.67}$$

which is the part of (4.42) pertaining to only those modes which are affected by the macroscopic field. Solution of (4.67) leads directly to the longitudinal frequencies.

An analogous treatment applied to case (c) yields the basic equations

$$[\mathbf{\Omega}(\text{IR}) - \omega^2(\hat{\mathbf{q}})\mathbb{1}_r]\mathbf{b} - \mathscr{Z}\mathbf{E} = 0, \tag{4.68}$$

$$\frac{4\pi}{Mv} \hat{\mathbf{Q}} \mathscr{Z}\mathbf{b} + \mathbf{E} = 0. \tag{4.69}$$

(Compare (4.69) with (4.43).) Eliminating $\mathbf{E}$ leads to

$$\left\| [\mathbf{\Omega}(\text{IR}) - \omega^2(\hat{\mathbf{q}})\mathbb{1}_r] + \frac{4\pi}{Mv} \mathscr{Z}\hat{\mathbf{Q}}\mathscr{Z} \right\| = 0. \tag{4.70}$$

### 4.2.4 Consequences of the macroscopic field—some examples

The important result which emerges from the foregoing discussion is that the macroscopic field essentially *produces a mixing of the IR-active modes* in the sense of (4.49). A related consequence, which will be justified later, is that in crystals of *orthorhombic or higher symmetry* one can find at least three directions along which the limiting frequencies $\omega_\lambda(\hat{\mathbf{q}})$ reduce to either the dispersion frequencies or the longitudinal frequencies. It is in this context that cases (a) and (b) are interesting.

Let us amplify these remarks with two brief illustrations.

**(a) Cubic diatomic crystals.** We are already aware that the dispersion frequency (found by setting $\mathbf{E} = 0$) and the longitudinal frequency (computed by taking $\mathbf{E} \| \mathbf{P}$) are respectively $\omega_{\text{TO}}$ and $\omega_{\text{LO}}$. Consider now the limiting optic frequencies $\omega_j(\hat{\mathbf{q}})$ for an arbitrary direction $\hat{\mathbf{q}}$. It is well established (for example by numerous reported calculations) that for all directions $\hat{\mathbf{q}}$ one of the three optic frequencies is $\omega_{\text{LO}}$ and the others are $\omega_{\text{TO}}$. (See also Figure 4.2). Thus, in cubic crystals, every direction is such that the limiting frequencies fall into the longitudinal and dispersion categories.

**(b) Wurtzite.** This is a hexagonal crystal with four atoms in the cell. There are accordingly 9 optic frequencies of which only 3 are IR active. These can be classified into a doubly degenerate mode with vibrations perpendicular to the $c$ axis and a nondegenerate mode with vibrations parallel to that axis. The corresponding dispersion and longitudinal frequencies may be labeled $\omega_\perp(0)$, $\omega_\parallel(0)$, $\omega_{\perp,l}$, and $\omega_{\parallel,l}$, respectively. Let $\theta$ denote the angle between $\hat{\mathbf{q}}$ and the $c$ axis. For $\theta = 0°$, the limiting frequencies are $\omega_\perp(0)$, $\omega_\perp(0)$, and $\omega_{\parallel,l}$. For $\theta = 90°$, on the other hand, the limiting frequencies are $\omega_\perp(0)$, $\omega_\parallel(0)$, and $\omega_{\perp,l}$. Hence, for these two directions the limiting frequencies fall entirely into the dispersion or the longitudinal categories. For intermediate angles, one of the frequencies is always $\omega_\perp(0)$ while the other two are combinations of $\omega_\perp(0)$, $\omega_\parallel(0)$, $\omega_{\perp,l}$, and $\omega_{\parallel,l}$. The $\theta$-variations of these frequencies have in fact been experimentally observed[8] and are sketched in Figure 4.5.

The case of a nondegenerate IR-active mode whose dispersion frequency is far removed from any other deserves special mention. In this case, one could to a first approximation ignore the effects of mixing with the other

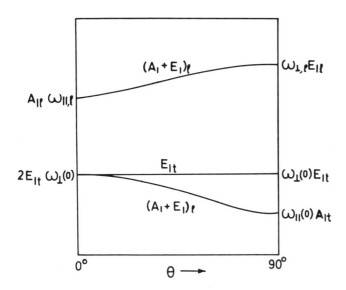

**Figure 4.5**
Angular variation of the limiting frequencies in a wurtzite-type crystal in the electrostatic approximation. Here $\theta$ denotes the angle between $\hat{\mathbf{q}}$ and the $c$ axis; $A_1$ and $E_1$ are group-theoretic symbols pertaining to the dispersion frequencies $\omega_\parallel(0)$ and $\omega_\perp(0)$, respectively. With electrostatic effects present, the $\mathbf{q} \approx 0$ modes can be labeled only in a hybrid fashion using group-theoretic labels as illustrated. The problem of proper classification is discussed in Sec. 4.4.

IR-active modes and treat the problem via perturbation theory. This gives*

$$\omega_\lambda^2(\hat{\mathbf{q}}) = \omega_\lambda^2(0) + \frac{1}{M}\tilde{\mathbf{e}}\binom{0}{\lambda}\left[\frac{4\pi}{v}\tilde{\mathbf{Z}}\hat{\mathbf{Q}}\mathbf{Z}\right]\mathbf{e}\binom{0}{\lambda}. \tag{4.71}$$

Remembering (4.35), this result can be written

$$\omega_\lambda^2(\hat{\mathbf{q}}) = \omega_\lambda^2(0) + \frac{4\pi v}{M}[\mathbf{P}(\lambda)\cdot\hat{\mathbf{q}}]^2\frac{1}{A^2\binom{0}{\lambda}}$$

$$= \omega_\lambda^2(0) + \frac{4\pi}{\rho A^2\binom{0}{\lambda}}P^2(\lambda)\cos^2\phi, \tag{4.72}$$

where $\rho$ is the density and $\phi$ is the angle between $\hat{\mathbf{q}}$ and $\mathbf{P}(\lambda)$.

Several examples of such $\cos^2\phi$ variations are known; one of these is illustrated in Figure 4.6.[9] In some situations there is a slight departure from a pure $\cos^2\phi$ dependence due to admixture from a neighboring IR-active mode, easily allowed for by a suitable modification of the perturba-

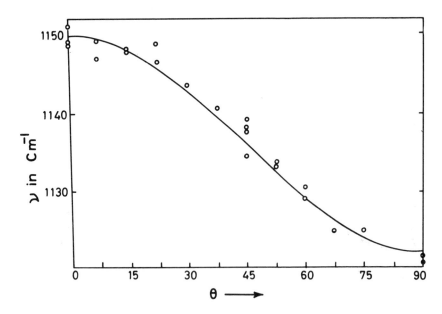

**Figure 4.6**
Angular variation of one of the long-wavelength modes in LiClO$_4\cdot$3H$_2$O. The solid line is a $\cos^2\phi$ fit. (After Mathieu and Couture-Mathieu, ref. 9.)

*Henceforth the frequency of an IR-active mode will be denoted by a subscript $\lambda$ rather than by $j$.

tion treatment. Hence in favorable circumstances one need not consider the mixing with all the modes but may confine attention to an appropriate subset. Such a treatment, though approximate, may often be quite adequate.

It is necessary to emphasize here that even though we take the limit $\mathbf{q} \to 0$, the limiting frequencies are valid only for $q \gg (\omega/c)$, since the electrostatic approximation has been employed. Hence there is no contradiction in the fact that the limiting frequencies for different $\mathbf{q}$ are different. For smaller values of $q$ retardation effects must be included, upon which the limiting frequencies become regular (Sec. 4.3.8).

### 4.2.5 Macroscopic-field effects viewed as a plasma contribution

It is illuminating to examine briefly the role of the macroscopic field in a different fashion. Let us suppress the $\bar{\mathbf{M}}(0)$ term in (4.44) and write the eigenvalue equation as

$$\omega^2(\hat{\mathbf{q}})\,m\xi(\hat{\mathbf{q}}) = \frac{4\pi}{v}[\check{\mathbf{Z}}\hat{\mathbf{Q}}\mathbf{Z}]\xi(\hat{\mathbf{q}}). \tag{4.73}$$

This is the equation for a plasma with the same charge composition as our ionic crystal. The equation has one nonzero eigenvalue

$$\omega^2(\hat{\mathbf{q}}) = \frac{4\pi}{v}\sum_k (Z_k e)^2/m_k = \Omega_p^2 \quad \text{(say)} \tag{4.74}$$

which, we observe, is direction independent. The corresponding eigenvector $\xi$ has components

$$\xi_\alpha(k|\hat{\mathbf{q}}) = \frac{Z_k e}{m_k}\hat{q}_\alpha \left[ M^{-1}\sum_k \frac{(Z_k e)^2}{m_k} \right]^{-1/2}. \tag{4.75}$$

This result suggests that the macroscopic-field effects make a net contribution of $\Omega_p^2$, where $\Omega_p$ has the character of a plasma frequency for the ionic system. This contribution is shared between the $r$ IR-active modes, the sharing pattern being direction dependent.

### 4.2.6 Field quenching by free carriers

It was remarked previously that the macroscopic field is a consequence of the macroscopic polarization resulting from the long-wavelength vibrations. If, however, there are mobile carriers in the crystal, then they will suitably redistribute themselves so as to suppress this polarization and hence the macroscopic field itself. This is precisely what happens in metals and alloys. One often refers to this phenomenon by saying that in such systems the long-range Coulomb interactions are screened by the conduction electrons. In ionic crystals, on the other hand, the screening is incomplete and consequently the fields are not fully suppressed.

It might be asked whether even in ionic crystals it is possible to quench the fields by introducing free carriers. Indeed it is, at least in semiconducting compounds where the carrier concentration may be fairly easily altered. The results of recent Raman-scattering experiments (see Sec. 5.3) on GaAs support this observation.[10] It is known that in normal GaAs (that with low carrier concentration), the $\mathbf{q} \approx 0$ optic modes are split by the macroscopic field. The Raman-scattering experiments show that as the carrier concentration is increased (by doping), two longwavelength modes are observed in addition to the TO mode. One of these has a frequency lower than the TO mode and the other a higher frequency. The observed variation of these frequencies with carrier concentration is shown in Figure 4.7. For large carrier concentration, only a single optic

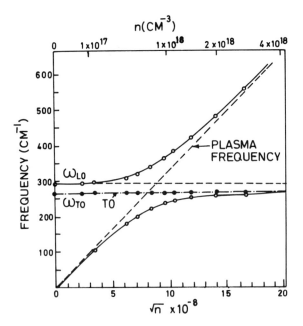

**Figure 4.7**
Frequencies of the long-wavelength modes in GaAs as a function of carrier concentration. For every concentration, three frequencies are seen; one of these always has the value $\omega_{TO}$ while the other two occur with variable frequencies. The latter arise from the mixing of plasmons with the LO phonon. In the absence of such mixing, the concentration variation of the plasmon and the LO frequencies are as shown by the dashed lines. (After Mooradian and McWhorter, ref. 10.)

vibrational frequency is seen, namely that corresponding to the TO mode which is now triply degenerate, characteristic of a suppressed macroscopic field. Besides this frequency, there is one higher up which is in fact the plasma frequency[11] $[4\pi n e^2/m^* \varepsilon(\infty)]^{1/2}$ of the carriers, where $n$ is the carrier concentration and $m^*$ is the effective mass. The results shown in Figure 4.7 illustrate the mixing between the plasmon (the name given to the quantum associated with plasma oscillations) and the LO phonon in degenerate semiconductors. In the zeroth approximation these do not interact and their behavior is as sketched by the dotted lines. When the plasmon frequency approaches the LO frequency, they begin to interact strongly since both involve similar types of charge fluctuations, one being associated with the ionic system and the other with the electronic system. This coupling causes the frequencies to "repel" and behave as shown. The mixing under discussion is analogous to the photon-phonon interaction considered earlier, the parameter $n^{1/2}$ playing a role similar to that of **q**.

Perspective on the results of Figure 4.7 may be gained from Figure 4.8 which shows qualitatively the dispersion curves in specimens of various carrier concentrations.[12] When the concentration is finite, a plasmon branch accompanies the usual phonon branches. The dashed and solid

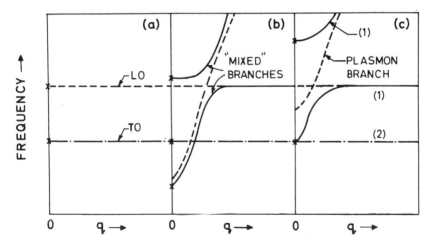

**Figure 4.8**
Schematic drawing of the long-wavelength portion of the optic branches in GaAs for various carrier concentrations. (a), (b), and (c) depict, respectively, the situation in a pure crystal, a partially doped crystal, and a heavily doped crystal. In the latter two cases the LO and the plasmon branches (dashed lines) mix and appear as shown by the solid lines. The crosses at $q = 0$ correspond to frequencies observed in the Raman-scattering experiments. (See previous figure). Observe in (c) a triple degeneracy at $q = 0$ (at the frequency $\omega_{TO}$), characteristic of the absence of macroscopic fields.

lines indicate, respectively, the nature of this plasmon branch in the absence and presence of interactions with phonons. The crosses denote the frequencies observed in the Raman-scattering experiment. Note that for high carrier concentrations the LO and TO branches merge at $\mathbf{q} = 0$ as in Figure 4.2b. Besides the plasmon–LO phonon coupling, mixing is also possible between plasmons and polaritons leading to new coupled modes called *plasmaritons*.[13] These arise due to the interaction between the transverse fields associated with the plasmons and polaritons.

Examples of measured dispersion curves in degenerate semiconductors will be presented in Chapter 6.

## 4.3 Long-wavelength phenomena in model-independent form for crystals of arbitrary structure

### 4.3.1 Basic equations
Having studied the origin of the macroscopic field and its effect on the optic modes within a specific model, we now wish to investigate the influence of the field in a model-independent form in relation to both the optic and the acoustic modes.

It can be shown[14] that, irrespective of the model assumed for electronic polarization, the long-wavelength aspects of lattice dynamics of insulating crystals are described by the equation

$$\omega^2(\mathbf{q})\mathbf{m}\mathbf{U}(\mathbf{q}) = \bar{\mathbf{M}}(\mathbf{q})\mathbf{U}(\mathbf{q}) - \mathbf{Z}^\dagger(\mathbf{q})\mathbf{E}(\mathbf{q}), \tag{4.76}$$

supplemented by the polarization equation

$$\mathbf{P}(\mathbf{q}) = \frac{1}{v}\mathbf{Z}(\mathbf{q})\mathbf{U}(\mathbf{q}) + \chi(\infty)\mathbf{E}(\mathbf{q}) \tag{4.77}$$

and the field equation

$$\mathbf{E}(\mathbf{q}) = -4\pi\hat{\mathbf{Q}}\mathbf{P}(\mathbf{q}) \tag{4.78a}$$

or

$$\mathbf{E}(\mathbf{q}) = \frac{4\pi}{n^2 - 1}[\mathbf{P}(\mathbf{q}) - n^2\hat{\mathbf{Q}}\mathbf{P}(\mathbf{q})]. \tag{4.78b}$$

Equation (4.78a) applies in the electrostatic approximation, while (4.78b) must be used if retardation is to be included.‡ We recognize in Eqs. (4.76–4.78) the generalization of the trio (4.33), (4.35), and (4.36). For later

‡The field amplitude $\mathbf{E}(\mathbf{q})$ now depends both on the magnitude and direction of $\mathbf{q}$ since $\mathbf{P}$ depends likewise. In the limit $\mathbf{q} \to 0$ and for optic modes, the field merely becomes direction dependent. This will be seen later.

convenience we note that (4.78b) may also be written as

$$\hat{\mathbf{q}}^{\text{tr}}\mathbf{E}(\mathbf{q}) = -4\pi\hat{\mathbf{q}}^{\text{tr}}\mathbf{P}(\mathbf{q}), \tag{4.79}$$

where $\hat{\mathbf{q}}^{\text{tr}}$ is the transpose of the column matrix $\hat{\mathbf{q}}$ with components $\hat{q}_\alpha$, $(\alpha = x, y, z)$.

All the quantities appearing above are familiar except for $\chi(\infty)$ and $\mathbf{Z}(\mathbf{q})$. The former is the high-frequency susceptibility, related to the symmetric dielectric tensor $\varepsilon(\infty)$ by

$$\varepsilon(\infty) = \mathbb{1}_3 + 4\pi\chi(\infty).\S \tag{4.80}$$

$\mathbf{Z}(\mathbf{q})$ is a $(3 \times 3n)$ matrix with elements $Z_{\alpha\beta}(k|\mathbf{q})$ and may be regarded as representing the charge interacting with the macroscopic field. Following Cochran and Cowley,[15] we shall refer to $\mathbf{Z}(\mathbf{q})$ as the *apparent-charge matrix*. By virtue of its transformation properties, the quantity $\mathbf{Z}(k|\mathbf{q})$ with nine elements (corresponding to a fixed $k$), can be viewed as the wave-vector-dependent *apparent-charge tensor* for atoms in the $k$th sublattice. Unlike the corresponding quantity $\mathbf{Z}_k = Z_k e \mathbb{1}_3$ of the rigid-ion model, $\mathbf{Z}(k|\mathbf{q})$ may have both diagonal and off-diagonal elements, the values of which depend on the details of the electronic polarization (see Sec. 4.6). $\mathbf{Z}(\mathbf{q})$ has the property[14]

$$Z_{\alpha\beta}(k|\mathbf{q}) = Z_{\alpha\beta}^*(k|-\mathbf{q}) \tag{4.81}$$

(see (2.63)). Furthermore, the $\mathbf{q} \to 0$ limit of $\mathbf{Z}(\mathbf{q})$ is well defined, and the condition that a uniform translation must produce no macroscopic fields leads to the sum rule[15]

$$\sum_k Z_{\alpha\beta}(k|0) = 0 \text{ for all } \alpha, \beta. \tag{4.82}$$

It can be shown (as will be discussed in Chapter 7) that (4.82) arises basically from the overall electrical neutrality of the primitive cell. Crystal symmetry imposes the requirement

$$\mathbf{R}\mathbf{Z}(k|\mathbf{q}) = \mathbf{Z}(F(k, R)|\mathbf{q})\mathbf{R}, \tag{4.83}$$

where $\mathbf{R}$ is an element of $G_0(\mathbf{q})$ and $F(k, R)$ is as defined in Chapter 3. Equation (4.83) may be obtained from results deduced in Sec. 4.4. The corresponding equation obeyed by $\mathbf{Z}(k|0)$ is

$$\mathbf{S}\mathbf{Z}(k|0) = \mathbf{Z}(F(k, S)|0)\mathbf{S}, \tag{4.84}$$

where $\mathbf{S}$ is an element of the point group of the crystal.

§Strictly speaking one should use the frequency- and wave-vector-dependent dielectric function $\varepsilon(\mathbf{q}, \omega \to \infty)$ where the limit $\omega \to \infty$ is to be understood in the sense discussed earlier. However, for simplicity we use $\varepsilon(\infty) = \varepsilon(\mathbf{q} \to 0, \omega \to \infty)$, since we are concerned with phenomena at $\mathbf{q} \to 0$.

It is worth emphasizing that both $\mathbf{Z}(k|0)$ and $\mathbf{Z}(k|\mathbf{q})$ can be nonvanishing even though $\mathbf{Z}_k = 0$. Further comments on this will be offered later.

Equations (4.76) and (4.77) may be derived from a suitably constructed phenomenological potential function:

$$
\begin{aligned}
\Phi(\mathbf{q}) = & \sum_{\alpha\beta} \sum_{kk'} u_\alpha^*(k|\mathbf{q}) M_{\alpha\beta}\!\begin{pmatrix}\mathbf{q}\\k\ k'\end{pmatrix} u_\beta(k'|\mathbf{q}) \\
& - \sum_{\alpha\beta}\sum_k [e_\alpha^*(\mathbf{q}) Z_{\alpha\beta}(k|\mathbf{q}) u_\beta(k|\mathbf{q}) + u_\beta^*(k|\mathbf{q}) Z_{\alpha\beta}^*(k|\mathbf{q}) e_\alpha(\mathbf{q})] \\
& - v \sum_{\alpha\beta} e_\alpha^*(\mathbf{q}) \chi_{\alpha\beta}(\infty) e_\beta(\mathbf{q}),
\end{aligned}
\tag{4.85}
$$

where $u_\alpha(k|\mathbf{q})$ are the displacement components and $e_\alpha(\mathbf{q})$ the field components. (Compare with the potential function introduced near (3.23). See also (4.7).) Proceeding as we did in Sec. 4.1,

$$
\begin{aligned}
-\frac{\partial \Phi}{\partial u_\alpha^*(k|\mathbf{q})} &= -\sum_{k'\beta} M_{\alpha\beta}\!\begin{pmatrix}\mathbf{q}\\k\ k'\end{pmatrix} u_\beta(k'|\mathbf{q}) + \sum_\beta Z_{\alpha\beta}^*(k|\mathbf{q}) e_\beta(\mathbf{q}) \\
&= m_k \ddot{u}_\alpha(k|\mathbf{q}).
\end{aligned}
\tag{4.86}
$$

Writing

$$
\begin{pmatrix} u_\alpha(k|\mathbf{q}) \\ e_\alpha(\mathbf{q}) \end{pmatrix} = \begin{pmatrix} U_\alpha(k|\mathbf{q}) \\ E_\alpha(\mathbf{q}) \end{pmatrix} e^{-i\omega t},
\tag{4.87}
$$

the preceding equation becomes

$$
m_k \omega^2(\mathbf{q}) U_\alpha(k|\mathbf{q}) = \sum_{k'\beta} M_{\alpha\beta}\!\begin{pmatrix}\mathbf{q}\\k\ k'\end{pmatrix} U_\beta(k'|\mathbf{q}) - \sum_\beta Z_{\alpha\beta}^*(k|\mathbf{q}) E_\beta(\mathbf{q}).
\tag{4.88}
$$

Similarly

$$
\begin{aligned}
-\frac{1}{v}\frac{\partial \Phi}{\partial e_\alpha^*(\mathbf{q})} &= \left[\frac{1}{v}\sum_{\beta k} Z_{\alpha\beta}(k|\mathbf{q}) U_\beta(k|\mathbf{q}) + \sum_\beta \chi_{\alpha\beta}(\infty) E_\beta(\mathbf{q})\right] e^{-i\omega t} \\
&= P_\alpha(\mathbf{q}) e^{-i\omega t}.
\end{aligned}
\tag{4.89}
$$

Equations (4.88) and (4.89) are in agreement with (4.76) and (4.77), respectively. These equations could be solved in a comprehensive form and the various features pertaining to the optic and acoustic modes could then be projected out individually.[14] Rather than adopt this approach, we shall here consider each aspect separately.

### 4.3.2 Sound waves in ionic and piezoelectric crystals

In Chapter 2 we remarked that the mechanical and electrical effects are not separable in ionic crystals and must be discussed together. We shall now direct attention to this problem in the context of acoustic waves. In other words, we shall first study the influence of the macroscopic field on the frequency and the velocity of sound waves, and later deduce lattice-theoretic expressions for the elastic and piezoelectric constants of the crystal by comparing the macroscopic and the lattice-theoretic expressions, as we did in Sec. 2.4. To facilitate our discussion we shall make use of the diagram in Figure 4.9. Here the stress **S** and the electric field **E** represent the macroscopic "forces" and the strain **s** and the dielectric displacement **D** denote the "results" produced by these forces.* The figure also illustrates the "principal effects" (the purely mechanical or purely electrical effects) and the "coupled effects."

Based on the diagram, the macroscopic electromechanical phenomena may be described via the equations†

$$dS_{\alpha\gamma} = \sum_{\beta\lambda}\left(\frac{\partial S_{\alpha\gamma}}{\partial s_{\beta\lambda}}\right)_E ds_{\beta\lambda} + \sum_{\beta}\left(\frac{\partial S_{\alpha\gamma}}{\partial E_\beta}\right)_s dE_\beta, \tag{4.90}$$

$$dD_\alpha = \sum_{\beta\gamma}\left(\frac{\partial D_\alpha}{\partial s_{\beta\gamma}}\right)_E ds_{\beta\gamma} + \sum_{\beta}\left(\frac{\partial D_\alpha}{\partial E_\beta}\right)_s dE_\beta. \tag{4.91}$$

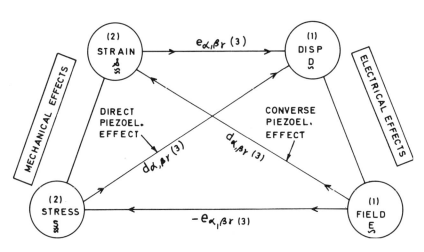

**Figure 4.9**
Schematic drawing illustrating the interplay of macroscopic electrical and mechanical effects in piezoelectric crystals at 0 K. (After Nye, ref. 16.)

*We are assuming the crystal to be at 0 K and ignoring thermal effects. This is in the same spirit as in Sec. 2.4. If thermal effects are included then Figure 4.9 must be enlarged.
†We use **s** and **E** as the independent variables rather than **S** and **E** to facilitate comparison with the lattice-theoretic treatment.

The partial derivatives appearing here can be related to the constants pertaining to the principal and the coupled effects. Thus (for details, see Nye[16]; note, however, that Nye uses a different system of units)

$$\left(\frac{\partial S_{\alpha\gamma}}{\partial s_{\beta\lambda}}\right)_{\mathbf{E}} = C_{\alpha\gamma,\beta\lambda}^{E}, \quad \left(\frac{\partial D_{\alpha}}{\partial E_{\beta}}\right)_{s} = \varepsilon_{\alpha\beta}^{s}(0),$$

$$\left(\frac{\partial D_{\alpha}}{\partial s_{\beta\gamma}}\right)_{\mathbf{E}} = 4\pi e_{\alpha,\beta\gamma}, \quad \left(\frac{\partial S_{\alpha\gamma}}{\partial E_{\beta}}\right)_{s} = -\frac{1}{4\pi}\left(\frac{\partial D_{\beta}}{\partial s_{\alpha\gamma}}\right)_{\mathbf{E}} = -e_{\beta,\alpha\gamma}. \tag{4.92}$$

Here $C_{\alpha\gamma,\beta\lambda}^{E}$ and $\varepsilon_{\alpha\beta}^{s}(0)$ are the elastic-constant and dielectric tensors,* respectively. They are already familiar to us but have been superscripted to indicate the conditions of their measurement. Such specifications in general become necessary when there are several interconnected variables and one is exploring the relations between a pair of them. In this circumstance, the variables not involved could, in principle, be allowed to vary in an arbitrary fashion. However, one is most often interested in the situation where either the "force" variable or the "result" variable from among the noninvolved set is held equal to zero. In the former case one has *free* constants and in the latter *clamped* constants. Thus $C_{\alpha\gamma,\beta\lambda}^{E}$ refer to the free elastic constants, since they describe the stress-strain relation when the field **E** is absent. Similarly, $\varepsilon_{\alpha\beta}^{s}(0)$ is the clamped dielectric tensor. It might be noted here that the distinction between clamped and free dielectric constants is irrelevant for $\varepsilon(\infty)$ since the lattice is regarded as undisturbed in the definition of this quantity. Furthermore, a distinction is necessary in respect to the static dielectric constant only for piezoelectric crystals.

The coefficients $e_{\alpha,\beta\gamma}$, which form a third-rank tensor, are referred to as the *piezoelectric constants*, and are typically $\sim \pm 10^{4}$–$10^{5}$ esu/cm². In Voigt's notation (Sec. 2.4) they can be expressed as $e_{\alpha i}$ ($\alpha = 1, 2, 3$; $i = 1, \ldots, 6$).

The 18 constants $e_{\alpha i}$ are not all independent if the crystal has nontrivial symmetry. In NaCl, for example, all the $e_{\alpha i}$ vanish, while in cubic ZnS $e_{14} = e_{25} = e_{36} \neq 0$ while all others vanish.†

The other modulus pertaining to the coupled effects in Figure 4.9 is also a third-rank tensor and is denoted by $d_{\alpha,\beta\gamma}$. It is referred to as the *piezoelectric modulus* and is related to $e_{\alpha,\beta\gamma}$ via[16]

$$e_{\alpha,\mu\nu} = \sum_{\beta\gamma} d_{\alpha,\beta\gamma} C_{\beta\gamma,\mu\nu}^{E}. \tag{4.93}$$

With the definitions (4.92), the integrated forms of (4.90) and (4.91) are

---

*A zero has been included in parentheses to make clear that we are here dealing with the *static* or low-frequency dielectric constant.
†The effect of crystal symmetry on the $e_{\alpha i}$ is discussed in several books.[16,17]

$$S_{\alpha\gamma} = \sum_{\beta\lambda} C^E_{\alpha\gamma,\beta\lambda} S_{\beta\lambda} - \sum_\beta e_{\beta,\alpha\gamma} E_\beta, \tag{4.94}$$

$$D_\alpha = 4\pi \sum_{\beta\gamma} e_{\alpha,\beta\gamma} S_{\beta\gamma} + \sum_\beta \varepsilon^s_{\alpha\beta}(0) E_\beta. \tag{4.95}$$

Remembering $\mathbf{D} = \mathbf{E} + 4\pi\mathbf{P}$, (4.95) can be written

$$P_\alpha = \sum_{\beta\gamma} e_{\alpha,\beta\gamma} S_{\beta\gamma} + \sum_\beta \chi^s_{\alpha\beta}(0) E_\beta, \tag{4.96}$$

where $\chi^s(0)$ is the clamped, static susceptibility.

Equations (4.94–4.96) are the basic equations governing macroscopic phenomena in a piezoelectric crystal (at 0 K). Based on (4.94), we may write the equation governing the propagation of acoustic waves:

$$\rho\ddot{u}_\alpha = \sum_\gamma \frac{\partial S_{\alpha\gamma}}{\partial x_\gamma}$$
$$= \sum_\gamma \sum_{\beta\lambda} C^E_{\alpha\gamma,\beta\lambda} \frac{\partial^2 u_\beta}{\partial x_\gamma \partial x_\lambda} - \sum_\gamma \sum_\beta e_{\beta,\alpha\gamma} \frac{\partial E_\beta}{\partial x_\gamma}. \tag{4.97}$$

Upon adopting the trial solutions

$$\begin{pmatrix} \mathbf{u}(\mathbf{x},t) \\ \mathbf{E}(\mathbf{x},t) \end{pmatrix} = \begin{pmatrix} \mathbf{u}(\mathbf{q}) \\ \mathbf{E}(\mathbf{q}) \end{pmatrix} \exp i\{\mathbf{q}\cdot\mathbf{x} - \omega(\mathbf{q})t\}, \tag{4.98}$$

Eq. (4.97) becomes

$$\rho\omega^2(\mathbf{q})u_\alpha(\mathbf{q}) = \sum_\beta \left[\sum_{\gamma\lambda} C^E_{\alpha\gamma,\beta\lambda} q_\gamma q_\lambda\right] u_\beta(\mathbf{q}) + i\sum_\beta \left(\sum_\gamma e_{\beta,\alpha\gamma} q_\gamma\right) E_\beta(\mathbf{q}), \tag{4.99a}$$

or, in matrix notation,

$$\rho\omega^2(\mathbf{q})\mathbf{u}(\mathbf{q}) = \mathbf{A}(\mathbf{q})\mathbf{u}(\mathbf{q}) + i\tilde{\mathbf{B}}(\mathbf{q})\mathbf{E}(\mathbf{q}), \tag{4.99b}$$

where $\mathbf{A}(\mathbf{q})$ is defined in terms of the (electrically free)elastic constants as in (2.115) and $\mathbf{B}(\mathbf{q})$ is defined by

$$B_{\beta\alpha}(\mathbf{q}) = [\tilde{\mathbf{B}}(\mathbf{q})]_{\alpha\beta} = \sum_\gamma e_{\beta,\alpha\gamma} q_\gamma. \tag{4.100}$$

Using the Voigt notation, the elements of $\mathbf{B}(\mathbf{q})$ are given explicitly as follows:

$$B_{\alpha 1}(\mathbf{q}) = e_{\alpha 1} q_1 + e_{\alpha 6} q_2 + e_{\alpha 5} q_3,$$
$$B_{\alpha 2}(\mathbf{q}) = e_{\alpha 6} q_1 + e_{\alpha 2} q_2 + e_{\alpha 4} q_3,$$
$$B_{\alpha 3}(\mathbf{q}) = e_{\alpha 5} q_1 + e_{\alpha 4} q_2 + e_{\alpha 3} q_3, \quad (\alpha = 1, 2, 3). \tag{4.101}$$

Similarly, by considering the polarization wave

$$\mathbf{P}(\mathbf{x},t) = \mathbf{P}(\mathbf{q}) \exp\{i(\mathbf{q}\cdot\mathbf{x} - \omega t)\},$$

we obtain from (4.96)

$$\mathbf{P}(\mathbf{q}) = i\mathbf{B}(\mathbf{q})\mathbf{u}(\mathbf{q}) + \chi^s(0)\mathbf{E}(\mathbf{q}), \tag{4.102}$$

in addition to which we have the relation

$$\mathbf{E}(\mathbf{q}) = -4\pi\hat{\mathbf{Q}}\mathbf{P}(\mathbf{q}) \qquad (4.103)$$

defining the macroscopic field.*

Equations (4.99), (4.102), and (4.103) are the macroscopic analogues of (4.76–4.78) and pertain explicitly to the sound waves. They may be combined in the form

$$\rho\omega^2(\mathbf{q})\mathbf{u}(\mathbf{q}) = [\mathbf{A}(\mathbf{q}) + 4\pi\tilde{\mathbf{B}}(\mathbf{q})(\mathbb{1}_3 + 4\pi\hat{\mathbf{Q}}\chi^s(0))^{-1}\hat{\mathbf{Q}}\mathbf{B}(\mathbf{q})]\mathbf{u}(\mathbf{q})$$
$$= \mathscr{A}(\mathbf{q})\mathbf{u}(\mathbf{q}) \qquad (4.104)$$

which then leads to the equation

$$\left\|\frac{1}{\rho}\mathscr{A}(\mathbf{q}) - \omega^2(\mathbf{q})\mathbb{1}_3\right\| = 0 \qquad (4.105)$$

for the sound-wave frequencies $\omega_j(\mathbf{q})$, $(j = 1, 2, 3)$. Equation (4.104) is to be compared with Eq. (2.114) for nonionic crystals. The matrix $\mathscr{A}(\mathbf{q})$ replaces $\mathbf{A}(\mathbf{q})$ and can be expressed in terms of "effective elastic constants" $C^*_{\alpha\gamma,\beta\lambda}(\hat{\mathbf{q}})$ via the same formula as earlier (see (2.115)). However, the effective elastic constants are now direction dependent as a consequence of the macroscopic field. An expression for these constants may be obtained as follows: Observe first

$$(\mathbb{1}_3 + 4\pi\hat{\mathbf{Q}}\chi)^{-1}\hat{\mathbf{Q}} = \hat{\mathbf{Q}} - \hat{\mathbf{Q}}4\pi\chi\hat{\mathbf{Q}} + \hat{\mathbf{Q}}4\pi\chi\hat{\mathbf{Q}}4\pi\chi\hat{\mathbf{Q}} - \cdots \qquad (4.106)$$

Furthermore,

$$\hat{\mathbf{Q}}\chi\hat{\mathbf{Q}} = \begin{pmatrix}\hat{q}_x \\ \hat{q}_y \\ \hat{q}_z\end{pmatrix}\left(\sum_{\alpha\beta}\hat{q}_\alpha\chi_{\alpha\beta}\hat{q}_\beta\right)(\hat{q}_x\hat{q}_y\hat{q}_z)$$
$$= \left(\sum_{\alpha\beta}\hat{q}_\alpha\chi_{\alpha\beta}\hat{q}_\beta\right)\hat{\mathbf{Q}}. \qquad (4.107)$$

Therefore, with $K = \sum_{\alpha\beta}\hat{q}_\alpha\chi^s_{\alpha\beta}(0)\hat{q}_\beta$,

$$(\mathbb{1}_3 + 4\pi\hat{\mathbf{Q}}\chi^s(0))^{-1}\hat{\mathbf{Q}} = \hat{\mathbf{Q}}\{1 - 4\pi K + (4\pi K)^2 - \cdots\}$$
$$= \frac{\hat{\mathbf{Q}}}{1 + 4\pi\sum_{\alpha\beta}\hat{q}_\alpha\chi^s_{\alpha\beta}(0)\hat{q}_\beta}$$
$$= \frac{\hat{\mathbf{Q}}}{\hat{\mathbf{q}}^{tr}(\mathbb{1}_3 + 4\pi\chi^s(0))\hat{\mathbf{q}}}$$
$$= [\hat{\mathbf{Q}}/\varepsilon^s_0(\hat{\mathbf{q}})], \text{ say.} \qquad (4.108)$$

*The electrostatic description is adequate for fields associated with sound waves since $(\omega/c) \ll q$. Retardation effects need not be considered.

Employing this notation, we may write

$$\mathscr{A}_{\alpha\beta}(\mathbf{q}) = A_{\alpha\beta}(\mathbf{q}) + \frac{4\pi}{\varepsilon_0^s(\hat{\mathbf{q}})} \sum_{\gamma\lambda} \left\{ \sum_{\mu\nu} [e_{\mu,\alpha\gamma}\hat{q}_\mu \hat{q}_\nu e_{\nu,\beta\lambda}] q_\gamma q_\lambda \right\}.$$

Comparison with (2.115) then gives

$$C^*_{\alpha\gamma,\beta\lambda}(\hat{\mathbf{q}}) = C^E_{\alpha\gamma,\beta\lambda} + 4\pi \sum_{\mu\nu} \frac{e_{\mu,\alpha\gamma}\hat{q}_\mu \hat{q}_\nu e_{\nu,\beta\lambda}}{\varepsilon_0^s(\hat{\mathbf{q}})}. \tag{4.109}$$

An interesting feature of the constants $C^*_{ij}(\hat{\mathbf{q}})$ is that for special directions they reduce to either the electrically free constants $C^E_{ij}$ or the electrically clamped constants $C^D_{ij}$. An expression for the latter may be obtained from (4.99), (4.102), and (4.103) by noting that $\mathbf{D} = 0$ with $\mathbf{E} \neq 0$ implies $\mathbf{E} = -4\pi\mathbf{P}$. This gives

$$\begin{aligned}\rho\omega^2(\mathbf{q})\mathbf{u}(\mathbf{q}) &= \mathscr{A}^D(\mathbf{q})\mathbf{u}(\mathbf{q}) \\ &= [\mathbf{A}(\mathbf{q}) + 4\pi\tilde{\mathbf{B}}(\mathbf{q})[\varepsilon^s(0)]^{-1}\mathbf{B}(\mathbf{q})]\mathbf{u}(\mathbf{q}),\end{aligned}$$

whence by comparison with (2.115)

$$C^D_{\alpha\gamma,\beta\lambda} = C^E_{\alpha\gamma,\beta\lambda} + 4\pi \sum_{\mu\nu} e_{\mu,\alpha\gamma}[\varepsilon^s(0)]^{-1}_{\mu\nu} e_{\nu,\beta\lambda}. \tag{4.110}$$

The similarities in the behavior of $C^*_{ij}(\hat{\mathbf{q}})$ and of the limiting optic frequencies $\omega_\lambda(\hat{\mathbf{q}})$ could not have escaped the reader's notice and are traceable in both cases to the influence of the macroscopic field.

As an illustration, let us consider the effective constants in the zinc blende structure, the simplest which exhibits piezoelectric properties. Since the crystal class is cubic, there are only three independent elastic constants, namely, $C_{11}$, $C_{12}$ and $C_{44}$,[16] and one independent piezoelectric constant, $e_{14}$. Table 4.1 summarizes the $C^*_{ij}$'s or their combinations appropriate to the various modes along the three principal directions. Transverse waves in the (100) plane with displacements along [001] are characterized by $C^*_{44}(\hat{\mathbf{q}})$, which becomes $C^E_{44}$ for $\hat{\mathbf{q}}\|[100]$ and $C^D_{44}$ for $\hat{\mathbf{q}}\|[110]$ owing to the different electrical conditions in the two situations. Even more striking is the fact that whereas $C^D_{44}$ enters the expression for the T2 mode ($\hat{\mathbf{q}}\|[110]$), $C^E_{44}$ appears in the expression for the longitudinal wave corresponding to the same $\hat{\mathbf{q}}$.

### 4.3.3 Long-wavelength expansion of lattice dynamical equations and expressions for the elastic and piezoelectric constants

Our next task is to obtain lattice-theoretic expressions for $C^E_{ij}$ and $e_{\alpha i}$. To begin, an expansion of (4.76) is made similar to that in Sec. 2.4. We replace $\mathbf{q}$ by $\eta\mathbf{q}$ and expand each $\mathbf{q}$-dependent quantity in a power series

**Table 4.1**
Sound velocities in the principal directions in zinc-blende-structure crystals.

| Direction of propagation | Polarization of sound wave | $C'$ |
|---|---|---|
| [100] | L | $C_{11}$ |
|  | T | $C_{44}^E$ |
| [111] | L | $\frac{1}{3}(C_{11} + 2C_{12} + 4C_{44}^D)$ |
|  | T | $\frac{1}{3}(C_{11} - C_{12} + C_{44}^E)$ |
| [110] | L | $\frac{1}{2}(C_{11} + C_{12} + 2C_{44}^E)$ |
|  | T1 | $\frac{1}{2}(C_{11} - C_{12})$ |
|  | T2 | $C_{44}^D$ |

Note: The velocity of sound is given by $(C'/\rho)^{1/2}$, where $\rho$ is the density. For the direction [110], mode T2 refers to transverse waves with displacements along [001], while mode T1 refers to transverse waves with displacements in the plane (001). Further, using (4.109) and (4.110) it can be shown that $C_{11}^*(\hat{\mathbf{q}}) = C_{11}^D = C_{11}^E$ and likewise for $C_{12}^*(\hat{\mathbf{q}})$. For this reason, $C_{11}$ and $C_{12}$ have no superscripts.

in $\eta$, the parameter eventually being allowed to take the value unity. The resulting series for $\bar{\mathbf{M}}(\eta\mathbf{q})$, $\mathbf{U}\binom{\eta\mathbf{q}}{j}$ and $\omega_j(\eta\mathbf{q})$ are of the same form as in (2.120) and (2.132). In addition we have

$$\mathbf{Z}(\eta\mathbf{q}) = \mathbf{Z}(0) + \eta \sum_\gamma \frac{\partial \mathbf{Z}(\mathbf{q})}{\partial q_\gamma}\bigg|_{\mathbf{q}=0} q_\gamma + \frac{\eta^2}{2} \sum_{\gamma\lambda} \frac{\partial^2 \mathbf{Z}(\mathbf{q})}{\partial q_\gamma \partial q_\lambda}\bigg|_{\mathbf{q}=0} q_\gamma q_\lambda + \cdots$$

$$(4.111\text{a})$$

$$= \mathbf{Z}(0) + i\eta \sum_\gamma \mathbf{Z}_\gamma^{(1)} q_\gamma + \frac{\eta^2}{2} \sum_{\gamma\lambda} \mathbf{Z}_{\gamma\lambda}^{(2)} q_\gamma q_\lambda + \cdots, \qquad (4.111\text{b})$$

$$\mathbf{E}\binom{\eta\mathbf{q}}{j} = \mathbf{E}^{(0)}\binom{\mathbf{q}}{j} + i\eta \mathbf{E}^{(1)}\binom{\mathbf{q}}{j} + \frac{\eta^2}{2} \mathbf{E}^{(2)}\binom{\mathbf{q}}{j} + \cdots. \qquad (4.112)$$

Some of the properties of the expansion matrices to be used later are (see Sec. 2.4):*

$$\bar{\mathbf{M}}^{(0)} = \bar{\mathbf{M}}^{(0)\text{tr}}, \qquad (4.113\text{a})$$

$$\bar{\mathbf{M}}_\gamma^{(1)} = -\bar{\mathbf{M}}_\gamma^{(1)\text{tr}}, \qquad (4.113\text{b})$$

$$\bar{\mathbf{M}}_{\gamma\lambda}^{(2)} = \bar{\mathbf{M}}_{\gamma\lambda}^{(2)\text{tr}}, \qquad (4.113\text{c})$$

$$\sum_{k'} M_{\alpha\beta}^{(0)}(kk') = 0 \qquad \text{for all } \alpha, \beta \text{ and } k, \qquad (4.114\text{a})$$

$$\sum_k M_{\alpha\beta}^{(0)}(kk') = 0 \qquad \text{for all } \alpha, \beta \text{ and } k', \qquad (4.114\text{b})$$

*In general these properties can be deduced as in Sec. 2.4 by noting first that $\mathbf{M}(\mathbf{q}) = \bar{\mathbf{M}}(\mathbf{q}) + (4\pi/v\varepsilon_\infty (\hat{\mathbf{q}}))\mathbf{Z}^\dagger(\mathbf{q})\hat{\mathbf{Q}}\mathbf{Z}(\mathbf{q})$ (see (4.139)). All requirements on $\mathbf{M}(\mathbf{q})$ are now separately insisted upon for the two parts whence (4.113–4.117) result.

$$\sum_{k'} M^{(1)}_{\alpha\beta,\gamma}(kk') = \sum_{k'} M^{(1)}_{\alpha\gamma,\beta}(kk'), \tag{4.115}$$

$$\sum_{kk'} M^{(1)}_{\alpha\beta,\gamma}(kk') = 0, \tag{4.116}$$

$$\sum_{k} Z^{(1)}_{\alpha\beta,\gamma}(k) = \sum_{k} Z^{(1)}_{\alpha\gamma,\beta}(k). \tag{4.117}$$

Introducing the various expansions into (4.76) and equating powers of $\eta$, we get to second order*

$$\overline{\mathbf{M}}^{(0)}\mathbf{U}^{(0)} - [\mathbf{Z}^{(0)}]^\dagger\mathbf{E}^{(0)} = 0, \tag{4.118a}$$

$$\overline{\mathbf{M}}^{(0)}\mathbf{U}^{(1)} + \sum_{\gamma} \overline{\mathbf{M}}^{(1)}_\gamma q_\gamma \mathbf{U}^{(0)} = [\mathbf{Z}^{(0)}]^\dagger\mathbf{E}^{(1)} + \sum_{\gamma} [\mathbf{Z}^{(1)}_\gamma]^\dagger q_\gamma \mathbf{E}^{(0)}, \tag{4.118b}$$

$$\overline{\mathbf{M}}^{(0)}\mathbf{U}^{(2)} - [\mathbf{Z}^{(0)}]^\dagger\mathbf{E}^{(2)} = 2[\omega^{(1)}_j(\mathbf{q})]^2 m\mathbf{U}^{(0)}$$
$$- \sum_{\gamma\lambda} \overline{\mathbf{M}}^{(2)}_{\gamma\lambda} q_\gamma q_\lambda \mathbf{U}^{(0)} + 2\sum_{\gamma} \overline{\mathbf{M}}^{(1)}_\gamma q_\gamma \mathbf{U}^{(1)}$$
$$+ \sum_{\gamma\lambda} [\mathbf{Z}^{(2)}_{\gamma\lambda}]^\dagger q_\gamma q_\lambda \mathbf{E}^{(0)} - 2\sum_{\gamma} [\mathbf{Z}^{(1)}_\gamma]^\dagger q_\gamma \mathbf{E}^{(1)}. \tag{4.118c}$$

These equations are now solved by the procedure used in Sec. 2.4. If the crystal does not have any spontaneous polarization,† then $\mathbf{E}^{(0)} = 0$ and the zero-order equation reduces to

$$\overline{\mathbf{M}}^{(0)}\mathbf{U}^{(0)} = 0$$

which, being of the same form as Eq. (2.133a), has the three nontrivial solutions $\mathbf{U}^{(0)}\binom{\mathbf{q}}{j}$, $(j = 1, 2, 3)$, given in (2.134). The solubility condition for the first-order equation is then

$$\tilde{\mathbf{U}}^{(0)H}\binom{\mathbf{q}}{j'}\left\{-\left(\sum_{\gamma} \overline{\mathbf{M}}^{(1)}_\gamma q_\gamma \mathbf{U}^{(0)}\binom{\mathbf{q}}{j}\right) - [\mathbf{Z}^{(0)}]^\dagger \mathbf{E}^{(1)}\binom{\mathbf{q}}{j}\right)\right\} = 0, \quad (j,j'=1,2,3),$$

which may be demonstrated to be satisfied. Solving the first-order equation then yields

$$\mathbf{U}^{(1)}\binom{\mathbf{q}}{j} = -\boldsymbol{\Gamma}\left(\sum_{\gamma} \overline{\mathbf{M}}^{(1)}_\gamma q_\gamma \mathbf{U}^{(0)}\binom{\mathbf{q}}{j} - [\mathbf{Z}^{(0)}]^\dagger\mathbf{E}^{(1)}\binom{\mathbf{q}}{j}\right),$$

with

---

*As in Sec. 2.4 we temporarily suspend indicating the $\mathbf{q}$ dependence.
†Certain crystals have the property of developing an electric polarization when their temperature is changed. This phenomenon is called *pyroelectricity*. In such crystals $\mathbf{E}^{(0)} \neq 0$ if $T > 0$ K, and hence if we are using the present formalism to discuss (in a quasi-harmonic sense) elastic properties of pyroelectric crystals at finite temperatures, then we must take note of the nonvanishing of the zero-order field $\mathbf{E}^{(0)}$.

$$\boldsymbol{\Gamma} = \begin{pmatrix} 0 & 0 & 0 & \cdots & 0 \\ 0 & 0 & 0 & \cdots & 0 \\ 0 & 0 & 0 & \cdots & 0 \\ & \cdots & & & \\ & \cdots & & \boldsymbol{\Gamma}' & \\ & \cdots & & & \\ 0 & 0 & 0 & & \end{pmatrix}; \quad \boldsymbol{\Gamma}' = \begin{pmatrix} \overline{\mathbf{M}}^{(0)}(22) & \cdots & \overline{\mathbf{M}}^{(0)}(2n) \\ \cdots\cdots\cdots\cdots\cdots\cdots\cdots \\ \cdots\cdots\cdots\cdots\cdots\cdots\cdots \\ \overline{\mathbf{M}}^{(0)}(n2) & \cdots & \overline{\mathbf{M}}^{(0)}(nn) \end{pmatrix}^{-1}$$

We now rewrite the second-order equation as

$$\overline{\mathbf{M}}^{(0)}\mathbf{U}^{(2)} = \Big( 2m[\omega_j^{(1)}]^2 \mathbf{U}^{(0)} - \sum_{\gamma\lambda} \overline{\mathbf{M}}^{(2)}_{\gamma\lambda} q_\gamma q_\lambda \mathbf{U}^{(0)}$$
$$+ 2\sum_\gamma \overline{\mathbf{M}}^{(1)}_\gamma q_\gamma \mathbf{U}^{(1)} + \sum_{\gamma\lambda} [\mathbf{Z}^{(2)}_{\gamma\lambda}]^\dagger q_\gamma q_\lambda \mathbf{E}^{(0)}$$
$$- 2\sum_\gamma [\mathbf{Z}^{(1)}_\gamma]^\dagger q_\gamma \mathbf{E}^{(1)} + [\mathbf{Z}^{(0)}]^\dagger \mathbf{E}^{(2)} \Big). \tag{4.119}$$

Remembering $\mathbf{E}^{(0)} = 0$ (by assumption) and that

$$\tilde{\mathbf{U}}^{(0)H}[\mathbf{Z}^{(0)}]^\dagger \mathbf{E}^{(2)} = 0$$

which follows from the sum rule (4.82) (note that $\mathbf{Z}^{(0)}$, $\mathbf{Z}^{(1)}_\gamma$ are all real quantities), we may express the solubility condition for the second-order equation as

$$\tilde{\mathbf{U}}^{(0)H}\binom{\mathbf{q}}{j'} \Big\{ 2m[\omega_j^{(1)}]^2 \mathbf{U}^{(0)}\binom{\mathbf{q}}{j}$$
$$- \Big[ \sum_{\gamma\lambda} \overline{\mathbf{M}}^{(2)}_{\gamma\lambda} q_\gamma q_\lambda + 2\sum_{\gamma\lambda} \overline{\mathbf{M}}^{(1)}_\gamma \boldsymbol{\Gamma} \overline{\mathbf{M}}^{(1)}_\lambda q_\gamma q_\lambda \Big] \mathbf{U}^{(0)}\binom{\mathbf{q}}{j}$$
$$+ \Big( 2\sum_\gamma \overline{\mathbf{M}}^{(1)}_\gamma \boldsymbol{\Gamma}[\mathbf{Z}^{(0)}]^\dagger q_\gamma - 2\sum_\gamma [\mathbf{Z}^{(1)}_\gamma]^\dagger q_\gamma \Big) \mathbf{E}^{(1)}\binom{\mathbf{q}}{j} \Big\} = 0, \ (j, j' = 1, 2, 3). \tag{4.120}$$

As in (2.141) we obtain from this last equation

$$\rho[\omega_j^{(1)}(\mathbf{q})]^2 u_\alpha\binom{\mathbf{q}}{j} = \sum_\beta \Big\{ \sum_{\gamma\lambda} [\alpha\beta, \gamma\lambda] q_\gamma q_\lambda + \sum_{\gamma\lambda} (\alpha\gamma, \beta\lambda) q_\gamma q_\lambda \Big\} u_\beta\binom{\mathbf{q}}{j}$$
$$- \sum_\beta \Big( \sum_\gamma [\beta, \alpha\gamma] q_\gamma \Big) E_\beta^{(1)}\binom{\mathbf{q}}{j}, \ (j = 1, 2, 3), \tag{4.121}$$

where

$$[\alpha\beta, \gamma\lambda] = \frac{1}{2v} \Big( \sum_{kk'} M^{(2)}_{\alpha\beta, \gamma\lambda}(kk') \Big), \tag{4.122}$$

$$(\alpha\gamma, \beta\lambda) = -\frac{1}{v} \sum_{\mu\nu} \sum_{k'k''} \Big( \sum_k M^{(1)}_{\mu\alpha, \gamma}(k'k) \Big) \Gamma_{\mu\nu}(k'k'') \Big( \sum_{k'''} M^{(1)}_{\nu\beta, \lambda}(k''k''') \Big), \tag{4.123}$$

$$[\beta, \alpha\gamma] = \frac{1}{v} \sum_{\mu\nu} \sum_{k'k''} \left( \sum_{k} M^{(1)}_{\alpha\mu,\gamma}(kk') \right)$$

$$\times \ \Gamma_{\mu\nu}(k'k'') Z_{\beta\nu}(k''|0) \ - \frac{1}{v} \sum_{k} Z^{(1)}_{\beta\alpha,\gamma}(k). \tag{4.124}$$

The square and the round brackets above have the same properties as in (2.144) and (2.145). In addition,

$$[\beta, \alpha\gamma] = [\beta, \gamma\alpha]. \tag{4.125}$$

Comparing (4.121) with (4.99a) then leads to

$$C^{E}_{\alpha\gamma,\beta\lambda} = [\alpha\beta, \gamma\lambda] + [\beta\gamma, \alpha\lambda] - [\beta\lambda, \alpha\gamma] + (\alpha\gamma, \beta\lambda), \tag{4.126}$$

$$e_{\beta,\alpha\gamma} = [\beta, \alpha\gamma]. \tag{4.127}$$

Equation (4.126) shows that the elastic constants of electrically free ionic and piezoelectric crystals are given by the same expression as in (2.146), the square and the round brackets having additional contributions arising from the "short-range part" of the Coulomb interactions. (Remember that the "long-range part" gives rise to the macroscopic field and has been separated out.)

Expressions (4.122–4.124) are rather formal but may be rendered amenable to numerical evaluation by considering explicit models. An example will be given later (Sec. 4.6). As an exercise, the reader could specialize these results to the rigid-ion model and verify that they yield results previously obtained by Born and Huang.[18]

### 4.3.4 Limiting optic frequencies in the electrostatic approximation

We next direct attention to the optic modes and consider their behavior in the electrostatic approximation. The discussion here is simply an extension of that presented earlier for the rigid-ion model. The basic equations as deduced from (4.76–4.78) are

$$\omega^2(\hat{\mathbf{q}})m\mathbf{U}(\hat{\mathbf{q}}) = \bar{\mathbf{M}}(0)\mathbf{U}(\hat{\mathbf{q}}) - \tilde{\mathbf{Z}}(0)\mathbf{E}(\hat{\mathbf{q}}), \tag{4.128}$$

$$\mathbf{P}(\hat{\mathbf{q}}) = \frac{1}{v}\mathbf{Z}(0)\mathbf{U}(\hat{\mathbf{q}}) + \chi(\infty)\mathbf{E}(\hat{\mathbf{q}}), \tag{4.129}$$

$$\mathbf{E}(\hat{\mathbf{q}}) = -4\pi\hat{\mathbf{Q}}\mathbf{P}(\hat{\mathbf{q}}). \tag{4.130}$$

A comparison with the corresponding equations (4.37), (4.35), and (4.36) of the rigid-ion model reveals that $\mathbf{Z}(0)$ replaces $\mathbf{Z}$ and that the total polarization contains an electronic contribution. We now consider three cases as in Sec. 4.2.

**(a) E = 0.** This case defines the "unperturbed" frequencies via the equation

$$m\omega^2 \mathbf{U} = \overline{\mathbf{M}}(0)\mathbf{U}. \tag{4.131}$$

Once again the $(3n - 3)$ optic frequencies can be classified as IR active or otherwise according as

$$\mathbf{Z}(0)\mathbf{e}\begin{pmatrix} 0 \\ \cdot \\ j \end{pmatrix} \tag{4.132}$$

is nonvanishing or not. While for ionic crystals IR activity as deduced from (4.40) yields the same results as (4.132), a difference is possible in the case of homopolar crystals. If for the latter symmetry leads to the result

$$\mathbf{Z}(k|0) = \mathbf{Z}(k'|0) \text{ for every } (kk'), \tag{4.133}$$

then, on account of (4.82), we have $\mathbf{Z}(k|0) = 0$ for every $k$. This is precisely what happens in crystals of the diamond structure and explains the absence of the macroscopic field and hence that of LO-TO splitting. On the other hand, in crystals of the tellurium structure (see Sec. 3.9), symmetry considerations[19] show that $\mathbf{Z}(0)$ has the structure

$$\mathbf{Z}(0) = \begin{vmatrix} A & 0 & 0 & -\dfrac{1}{2}A & -\dfrac{\sqrt{3}}{2}A & \sqrt{\dfrac{3}{2}}B & -\dfrac{1}{2}A & \dfrac{\sqrt{3}}{2}A & -\sqrt{\dfrac{3}{2}}B \\[2mm] 0 & -A & -\sqrt{2}B & -\dfrac{\sqrt{3}}{2}A & \dfrac{1}{2}A & \dfrac{1}{\sqrt{2}}B & \dfrac{\sqrt{3}}{2}A & \dfrac{1}{2}A & \dfrac{1}{\sqrt{2}}B \\[2mm] 0 & C & 0 & -\dfrac{\sqrt{3}}{2}C & -\dfrac{1}{2}C & 0 & \dfrac{\sqrt{3}}{2}C & -\dfrac{1}{2}C & 0 \end{vmatrix},$$

$A$, $B$, $C$ being constants. The above structure clearly does not meet the requirement (4.133) and the possibility therefore exists of tellurium exhibiting infrared activity. A more detailed analysis[19] in fact shows that there are 5 IR-active modes, and these correspond to the $\Gamma_2(A_2)$ and $\Gamma_3(E)$ representations (see Sec. 3.9). We shall later discuss how $\mathbf{Z}(0)$ may be explicitly calculated with reference to specific models for electronic polarization (see Sec. 4.6).

**(b) E ≠ 0 but D = 0.** This case leads to the longitudinal frequencies. Noting that here $\mathbf{E} = -4\pi\mathbf{P}$, we obtain from (4.129) the result

$$\mathbf{E} = -\frac{4\pi}{v}\boldsymbol{\varepsilon}^{-1}(\infty)\mathbf{Z}(0)\mathbf{U}. \tag{4.134}$$

Substituting in (4.128), we obtain

$$m\omega^2\mathbf{U} = \left(\overline{\mathbf{M}}(0) + \frac{4\pi}{v}\tilde{\mathbf{Z}}(0)\,\boldsymbol{\varepsilon}^{-1}(\infty)\mathbf{Z}(0)\right)\mathbf{U}; \tag{4.135}$$

the requirement of nontrivial solutions for $\mathbf{U}$ then gives

$$\left\| \left[ \overline{\mathbf{M}}(0) + \frac{4\pi}{v} \check{\mathbf{Z}}(0) \, \varepsilon^{-1}(\infty) \mathbf{Z}(0) \right] - m\omega^2 \right\| = 0. \tag{4.136}$$

Only for the $r$ IR-active modes are the solutions of (4.136) different from those obtained from (4.131).

(c) $\mathbf{E} \neq \mathbf{0}$ and $\mathbf{D} \neq \mathbf{0}$. We follow a procedure similar to that in case (b) but with $\mathbf{E}$ given by (4.130). From Eqs. (4.129) and (4.130) we then obtain

$$\frac{4\pi}{v} \hat{\mathbf{Q}} \mathbf{Z}(0) \mathbf{U}(\hat{\mathbf{q}}) + (\mathbb{1}_3 + 4\pi \hat{\mathbf{Q}} \chi(\infty)) \mathbf{E}(\hat{\mathbf{q}}) = 0. \tag{4.137}$$

Solving for $\mathbf{E}$ and substituting in (4.128), we obtain

$$m\omega^2(\hat{\mathbf{q}}) \mathbf{U}(\hat{\mathbf{q}}) = \left[ \overline{\mathbf{M}}(0) + \frac{4\pi}{v} \check{\mathbf{Z}}(0) [\mathbb{1}_3 + 4\pi \hat{\mathbf{Q}} \chi(\infty)]^{-1} \hat{\mathbf{Q}} \mathbf{Z}(0) \right] \mathbf{U}(\hat{\mathbf{q}}). \tag{4.138}$$

(Observe the structural similarity to (4.104)). As in (4.107) we now use the result

$$[\mathbb{1}_3 + 4\pi \hat{\mathbf{Q}} \chi(\infty)]^{-1} \hat{\mathbf{Q}} = \frac{\hat{\mathbf{Q}}}{\hat{\mathbf{q}}^{\mathrm{tr}} \varepsilon(\infty) \hat{\mathbf{q}}}$$

$$\equiv \frac{\hat{\mathbf{Q}}}{\varepsilon_\infty(\hat{\mathbf{q}})}. \tag{4.139}$$

Equation (4.138) then becomes

$$m\omega^2(\hat{\mathbf{q}}) \mathbf{U}(\hat{\mathbf{q}}) = \left[ \overline{\mathbf{M}}(0) + \frac{4\pi}{v \varepsilon_\infty(\hat{\mathbf{q}})} \check{\mathbf{Z}}(0) \hat{\mathbf{Q}} \mathbf{Z}(0) \right] \mathbf{U}(\hat{\mathbf{q}}), \tag{4.140}$$

whence the limiting frequencies are obtained from

$$\left\| \left[ \overline{\mathbf{M}}(0) + \frac{4\pi}{v \varepsilon_\infty(\hat{\mathbf{q}})} \check{\mathbf{Z}}(0) \hat{\mathbf{Q}} \mathbf{Z}(0) \right] - m\omega^2(\hat{\mathbf{q}}) \right\| = 0. \tag{4.141}$$

Once again, the roots of the above equation are different from the "unperturbed" eigenvalues only in the case of IR-active modes.

If attention is restricted to these modes alone, then the relevant equations are (see (4.68) and (4.69)):

$$[\boldsymbol{\Omega}(\mathrm{IR}) - \omega^2(\hat{\mathbf{q}}) \mathbb{1}_r] \mathbf{b} - \mathscr{Z}(0) \mathbf{E} = 0, \tag{4.142}$$

$$\frac{4\pi}{Mv} \hat{\mathbf{Q}} \mathscr{Z}(0) \mathbf{b} + [\mathbb{1}_3 + 4\pi \hat{\mathbf{Q}} \chi(\infty)] \mathbf{E} = 0. \tag{4.143}$$

Here $\mathscr{L}(0)$ is the analogue of $\mathscr{L}$ in (4.60) and is defined through

$$
\mathbf{Z}(0)\left[\cdots, \mathbf{e}\binom{0}{j}, \cdots\right] = \left[\begin{array}{c|c} \mathscr{L}(0) & 0 \end{array}\right] \updownarrow 3 . \tag{4.144}
$$
$$
\xleftarrow{\ \ r\ \ } \xleftarrow{\ \ (3n-r)\ \ }
$$

The determinantal equation resulting from (4.142) and (4.143) is

$$
\left\| [\mathbf{\Omega}(\text{IR}) - \omega^2(\hat{\mathbf{q}})\mathbb{1}_r] + \frac{4\pi}{Mv\varepsilon_\infty(\hat{\mathbf{q}})}\, \tilde{\mathscr{L}}(0)\hat{\mathbf{Q}}\mathscr{L}(0) \right\| = 0, \tag{4.145}
$$

which when solved gives the limiting frequencies for the IR-active modes.*
Similarly, the longitudinal frequencies are obtainable from

$$
\left\| [\mathbf{\Omega}(\text{IR}) - \omega^2\mathbb{1}_r] + \frac{4\pi}{Mv}\, \tilde{\mathscr{L}}(0)\boldsymbol{\varepsilon}^{-1}(\infty)\mathscr{L}(0) \right\| = 0. \tag{4.146}
$$

It is easily verified that these expressions reduce to those given earlier for the rigid-ion model under appropriate conditions. As in the rigid-ion model, the solutions of (4.145) are distributed between the dispersion and longitudinal categories for special directions in crystals of orthorhombic or higher symmetry.

The relationship between the macroscopic field and the long-wavelength limit of the apparent charge matrix (see also (4.43)) may be obtained by combining (4.137) and (4.139):

$$
\mathbf{E}(\hat{\mathbf{q}}) = -\frac{4\pi}{v\varepsilon_\infty(\hat{\mathbf{q}})}\, \hat{\mathbf{Q}}\mathbf{Z}(0)\mathbf{U}(\hat{\mathbf{q}}). \tag{4.147}
$$

### 4.3.5 Generalized LST relation

We turn next to the task of generalizing the LST relation. We shall first evaluate a ratio which is the analogue of the quantity $(\omega_{L0}^2/\omega_{T0}^2)$ in (4.5). The dielectric response of the crystal will then be examined, and by combining the two results the desired formula will be obtained.

Consider Eq. (4.145). For convenience we shall regard the $(3 \times r)$ array $\mathscr{L}(0)$ occurring therein as being made up of $r$ three-component column matrices $\mathscr{L}_\lambda$ in the following manner:

$$
\mathscr{L}(0) = [\cdots, \mathscr{L}_\lambda, \cdots]. \tag{4.148}
$$

Here

$$
\mathscr{L}_\lambda = \mathbf{Z}(0)\mathbf{e}\binom{0}{\lambda} \tag{4.149}
$$

*This result has been deduced independently by S. M. Shapiro and J. D. Axe, Phys. Rev. **B6**, 2420 (1972).

and, when multiplied by the amplitude $A\binom{0}{\lambda}$ (see Sec. 2.2.14), defines the dipole moment arising from the displacements in the $\lambda$th IR-active mode.

We now recall two results of matrix algebra.[20]

1. If **A** is an $n$-dimensional matrix with a characteristic equation

$$\| \mathbf{A} - \mathbb{1}_n x \| = c(x) = (x - \lambda_1) \cdots (x - \lambda_n) = 0,$$

then

$$\| \mathbf{A} \| = \prod_{i=1}^{n} \lambda_i. \tag{4.150a}$$

2. Given the matrices **A** and **B** defined by $A_{ij} = \delta_{ij} A_{ii}$ and $B_{ij} = b_i b_j$, $b_i$ and $b_j$ being numbers, the determinant of their sum can be expressed as

$$\| \mathbf{A} + \mathbf{B} \| = \| \mathbf{A} \| \left( 1 + \sum_{i=1}^{n} \frac{b_i b_i}{A_{ii}} \right). \tag{4.150b}$$

Making use of these results, we obtain

$$\left\| \boldsymbol{\Omega}(\mathrm{IR}) + \frac{4\pi}{Mv\varepsilon_\infty(\hat{\mathbf{q}})} \mathscr{L}(0) \hat{\mathbf{Q}} \mathscr{L}(0) \right\|$$

$$= \| \boldsymbol{\Omega}(\mathrm{IR}) \| \left( 1 + \frac{4\pi}{Mv\varepsilon_\infty(\hat{\mathbf{q}})} \sum_{\lambda=1}^{r} [\hat{\mathbf{q}}^{\,\mathrm{tr}} \mathscr{L}(0)]_\lambda [\mathscr{L}(0)\hat{\mathbf{q}}]_\lambda / \Omega_{\lambda\lambda}(0) \right)$$

$$= \prod_{\lambda=1}^{r} \omega_\lambda^2(\hat{\mathbf{q}}). \tag{4.151}$$

In this derivation, the dummy index $\lambda$ refers to the IR-active modes. Furthermore, the elements of the $r$-dimensional matrix $[\mathscr{L}(0)\hat{\mathbf{Q}}\mathscr{L}(0)]$ are viewed as $[\hat{\mathbf{q}}^{\,\mathrm{tr}} \mathscr{L}(0)]_{\lambda'}[\mathscr{L}(0)\hat{\mathbf{q}}]_\lambda$, $(\lambda, \lambda' = 1, 2, \ldots, r)$. Remembering now that $\| \boldsymbol{\Omega}(\mathrm{IR}) \| = \prod_\lambda \omega_\lambda^2(0)$, we get

$$\frac{\displaystyle\prod_{\lambda=1}^{r} \omega_\lambda^2(\hat{\mathbf{q}})}{\displaystyle\prod_{\lambda=1}^{r} \omega_\lambda^2(0)} = \frac{\left\{ \varepsilon_\infty(\hat{\mathbf{q}}) + \dfrac{4\pi}{Mv} \hat{\mathbf{q}}^{\,\mathrm{tr}} \mathscr{L}(0)\, \boldsymbol{\Omega}^{-1}(\mathrm{IR}) \mathscr{L}(0)\hat{\mathbf{q}} \right\}}{\varepsilon_\infty(\hat{\mathbf{q}})}$$

$$= \frac{\hat{\mathbf{q}}^{\,\mathrm{tr}} \left\{ \boldsymbol{\varepsilon}(\infty) + \dfrac{4\pi}{Mv} \mathscr{L}(0)\, \boldsymbol{\Omega}^{-1}(\mathrm{IR}) \mathscr{L}(0) \right\} \hat{\mathbf{q}}}{\hat{\mathbf{q}}^{\,\mathrm{tr}} \boldsymbol{\varepsilon}(\infty) \hat{\mathbf{q}}}. \tag{4.152}$$

The quantity on the left-hand side is the analogue of $(\omega_{\mathrm{LO}}^2/\omega_{\mathrm{TO}}^2)$; that on the right-hand side will presently be related to the dielectric function $\varepsilon(\omega)$.

The determination of $\boldsymbol{\varepsilon}(\omega)$ is accomplished as in the diatomic case by considering the *forced vibrations* produced by a field $\mathbf{E} = \mathbf{E}_0 e^{i\omega t}$ ($\omega$ arbitrary). Noting that the field can interact with the IR-active modes only, we may set up the following equations for the forced motions:

$$\Omega(\text{IR})\mathbf{b} - \mathscr{Z}(0)\mathbf{E} = \omega^2\mathbf{b}, \tag{4.153}$$

$$\frac{1}{Mv}\mathscr{Z}(0)\mathbf{b} + \chi(\infty)\mathbf{E} = \mathbf{P}. \tag{4.154}$$

Here $\mathbf{b}$ plays the same role as the corresponding quantity in (4.64–4.65) except that it now pertains to forced oscillations. Eliminating $\mathbf{b}$,

$$\mathbf{P} = \chi(\omega)\mathbf{E}$$

$$= \left\{\chi(\infty) + \frac{1}{Mv}\mathscr{Z}(0)\left[\Omega(\text{IR}) - \omega^2\mathbb{1}_r\right]^{-1}\mathscr{Z}(0)\right\}\mathbf{E}.$$

Therefore

$$\varepsilon(\omega) = \mathbb{1}_3 + 4\pi\chi(\omega)$$

$$= \varepsilon(\infty) + \frac{4\pi}{Mv}\mathscr{Z}(0)[\Omega(\text{IR}) - \omega^2\mathbb{1}_r]^{-1}\mathscr{Z}(0). \tag{4.155}$$

In view of this result,

$$\frac{\prod\limits_{\lambda=1}^{r}\omega_\lambda^2(\hat{\mathbf{q}})}{\prod\limits_{\lambda=1}^{r}\omega_\lambda^2(0)} = \frac{\hat{\mathbf{q}}^{\text{tr}}\varepsilon(0)\hat{\mathbf{q}}}{\hat{\mathbf{q}}^{\text{tr}}\varepsilon(\infty)\hat{\mathbf{q}}}, \tag{4.156}$$

which is the generalization of the LST formula we sought.[15]

At this point the question arises whether $\varepsilon(0)$ in (4.156) refers to the clamped or the free dielectric constant. The answer is not obvious from the derivation just given. However, a deeper analysis starting with an electric field of the form $\mathbf{E} = \mathbf{E}_0 e^{i(\mathbf{q}\cdot\mathbf{r}-\omega t)}$ followed by a systematic expansion of various quantities in powers of $\mathbf{q}$ (as in Sec. 4.3.3) shows that the polarization equation can be manipulated to a form comparable to (4.102), leading to the result[21]

$$\chi^s(0) = \chi^s(\infty) + \frac{1}{Mv}\mathscr{Z}(0)[\Omega(\text{IR})]^{-1}\mathscr{Z}(0). \tag{4.157}$$

The static dielectric constant occurring in (4.156) is thus to be identified with the clamped constant, a result which appears to have been first noted by Cochran.[22]

### 4.3.6 Some remarks on $\varepsilon(\omega)$

It is necessary to supplement the result (4.155) with some remarks.

1. $\varepsilon(\omega)$ is actually a complex quantity and includes both one-phonon as

well as multiphonon contributions. However, as determined in (4.155) the tensor $\boldsymbol{\varepsilon}(\omega)$ is real and contains only the one-phonon contribution. Further comments on this matter will be made in the next chapter. Confining attention here to (4.155), we note that the structure of the tensor is dictated by crystal symmetry as summarized in Table 4.2.

2. Next let us write (4.155) in component form as

$$\varepsilon_{\alpha\beta}(\omega) = \varepsilon_{\alpha\beta}(\infty) + \frac{4\pi}{Mv} \sum_{\lambda=1}^{r} \frac{\mathscr{L}_{\alpha\lambda}(0)\mathscr{L}_{\beta\lambda}(0)}{\omega_{\lambda}^2(0) - \omega^2}. \tag{4.158}$$

In the study of the optical properties of crystals, a formula of this kind is often simply written down from heuristic considerations and is referred to as the *oscillator formula*. This practice goes back to the days when a similar formula was derived for the refractive index $n(\omega)$ $(n^2(\omega) = \varepsilon(\omega))$ by treating the bound electrons as harmonic oscillators capable of interacting with electromagnetic radiation.[23] The quantity $[4\pi\mathscr{L}_{\alpha\lambda}(0)\mathscr{L}_{\beta\lambda}(0)/Mv\omega_{\lambda}^2(0)]$ is called the *oscillator strength* and is seen to be related primarily to $\mathscr{L}_{\lambda}$ and hence to the dipole moment associated with the IR-active mode of frequency $\omega_{\lambda}(0)$.

3. Being related to a dipole moment, $\mathscr{L}_{\lambda}$ must transform essentially like the vector $\mathbf{r}$. On the other hand, from the definition in (4.149) it can be

**Table 4.2**
Effect of crystal symmetry on the dielectric tensor $\varepsilon(\omega)$. The Cartesian axes are chosen according to conventions specified by Nye (ref. 16).

| System | $\varepsilon(\omega)$ |
|---|---|
| Cubic (1) | $\begin{pmatrix} \varepsilon_{xx}(\omega) & & \\ & \varepsilon_{xx}(\omega) & \\ & & \varepsilon_{xx}(\omega) \end{pmatrix}$ |
| Tetragonal $\left.\begin{array}{l}\\ \text{Hexagonal}\\ \text{Trigonal}\end{array}\right\}$ (2) | $\begin{pmatrix} \varepsilon_{xx}(\omega) & & \\ & \varepsilon_{xx}(\omega) & \\ & & \varepsilon_{zz}(\omega) \end{pmatrix}$ |
| Orthorhombic (3) | $\begin{pmatrix} \varepsilon_{xx}(\omega) & & \\ & \varepsilon_{yy}(\omega) & \\ & & \varepsilon_{zz}(\omega) \end{pmatrix}$ |
| Monoclinic (4) | $\begin{pmatrix} \varepsilon_{xx}(\omega) & 0 & \varepsilon_{xz}(\omega) \\ 0 & \varepsilon_{yy}(\omega) & 0 \\ \varepsilon_{xz}(\omega) & 0 & \varepsilon_{zz}(\omega) \end{pmatrix}$ |
| Triclinic (6) | $\begin{pmatrix} \varepsilon_{xx}(\omega) & \varepsilon_{xy}(\omega) & \varepsilon_{xz}(\omega) \\ \varepsilon_{xy}(\omega) & \varepsilon_{yy}(\omega) & \varepsilon_{yz}(\omega) \\ \varepsilon_{xz}(\omega) & \varepsilon_{yz}(\omega) & \varepsilon_{zz}(\omega) \end{pmatrix}$ |

Note: For uniaxial crystals (that is, tetragonal, etc.) one usually writes $\varepsilon_{xx}$ as $\varepsilon_{\perp}$ and $\varepsilon_{zz}$ as $\varepsilon_{\parallel}$. The structures given above apply separately to the real and imaginary parts if $\boldsymbol{\varepsilon}$ is complex, provided the crystal does not belong to the monoclinic or triclinic systems. The numbers in parentheses indicate the number of independent elements.

argued that $\mathscr{L}_\lambda$ must transform according to the same IMR of $G_0(\mathbf{q} = 0)$ as does $\mathbf{e}\binom{0}{\lambda}$.* It follows that if $\mathscr{L}_\lambda$ is nonvanishing, $\mathbf{e}\binom{0}{\lambda}$ must belong to that IMR $G_0(\mathbf{q} = 0)$ according to which the components of $\mathbf{r}$ also transform. This forms the basis for the customary identification of the IR-active modes, a practice well known in molecular spectroscopy. In this, one first picks out the IMR's of $G_0(\mathbf{q} = 0)$ according to which $x$, $y$, $z$ transform. This may be done quite independently of other considerations from information often listed in the character tables for the point groups. Next, from a decomposition of the reducible representation $T(\mathbf{q} = 0)$ one determines the IMR's corresponding to the $\mathbf{q} = 0$ modes. If any of these are the same as those appropriate to $x$, $y$, or $z$, then the corresponding mode is deemed to be IR active, since its transformation properties are similar to $\mathbf{r}$ and therefore by implication similar to the dipole moment.

It is worth emphasizing that this method of identifying the IR-active modes involves essentially a *converse argument*. In other words, we pick out modes which have the same transformation properties as $x, y, z$ and then identify them as IR active since they transform the same way as do the components of the dipole-moment vector. However, this does not guarantee that the mode in question has a nonvanishing dipole moment. The $\mathbf{q} = 0$ acoustic modes provide an example. Hence for a mode to possess a dipole moment it is necessary but not sufficient that it transform like $x$, $y$, or $z$.

4. In crystals of orthorhombic or higher symmetry, the vectors $\mathscr{L}_\lambda$ must necessarily be directed along the principal axes of the tensor $\varepsilon(\omega)$, that is, along the Cartesian axes, provided these are chosen according to the convention implied in Table 4.2. This remark can be substantiated not only by an examination of the IMR's of point groups corresponding to the higher-symmetry classes, but is also seen to be plausible from (4.158), where we note that $\varepsilon(\omega)$ involves a sum of the form $\mathscr{L}_\lambda\mathscr{L}_\lambda$. Evidently if some $\mathscr{L}_\lambda$ had more than one nonzero component, $\varepsilon$ would not be diagonal with respect to the crystallographic axes.

The significance of the foregoing remarks may be better appreciated by the specific example of uniaxial crystals. Here the vibrations of the atoms (in the absence of the macroscopic field) are either parallel or perpen-

*This can be proved formally by using the result
$$\mathbf{R}\mathbf{Z}(0) = \mathbf{Z}(0)\mathbf{T}(0; \mathbf{R})$$
obtainable from Eq. (4.218). Therefore
$$\mathbf{R}\mathscr{L}_\lambda = \mathbf{R}\mathbf{Z}(0)\mathbf{e}\binom{0}{\lambda} = \mathbf{Z}(0)\mathbf{T}(0; \mathbf{R})\mathbf{e}\binom{0}{\lambda} \quad (\lambda \to sa\mu)$$
$$= \mathbf{Z}(0) \sum_\nu \tau^s_{\nu\mu}(0; \mathbf{R})\mathbf{e}\binom{0}{sa\nu}$$
$$= \sum_\nu \tau^s_{\nu\mu}(0; \mathbf{R})\mathscr{L}_{sa\nu}.$$

dicular to the $c$ axis (that is, the $z$ axis), and the same is therefore true of the vectors $\mathscr{L}_\lambda$. The parallel vibrations are nondegenerate while the perpendicular vibrations are doubly degenerate. (Recall the wurtzite example). Therefore $\mathscr{L}(0)$ has the following structure:

$$\mathscr{L}(0) = [\mathscr{L}_\perp(0) | \mathscr{L}'_\perp(0) | \mathscr{L}_\parallel(0)]. \tag{4.159}$$

$\mathscr{L}_\parallel(0)$ is the array made up of the dipole vectors $\mathscr{L}_\lambda$ pertaining to the parallel vibrations while similar meanings apply to $\mathscr{L}_\perp(0)$ and $\mathscr{L}'_\perp(0)$. The double degeneracy of the perpendicular vibrations results in two orthogonal vectors $\mathscr{L}_\lambda$ and $\mathscr{L}'_\lambda$ for each mode, and these have been grouped separately in (4.159). It is clear that if $\mathbf{q}$ is along the $c$ axis (say), then

$$\hat{\mathbf{q}}^{\mathrm{tr}}\mathscr{L}(0) = (0, 0, 1)\mathscr{L}(0) = [0|0|(0, 0, 1)\mathscr{L}_\parallel(0)].$$

Thus the matrix

$$\mathbf{\Omega}(\mathrm{IR}) + \frac{4\pi}{Mv\varepsilon_\infty(\hat{\mathbf{q}})}\,\mathscr{L}(0)\hat{\mathbf{Q}}\mathscr{L}(0)$$

has the structure

$$\begin{bmatrix} \mathbf{\Omega}_\perp(\mathrm{IR}) & & \\ & \mathbf{\Omega}_\perp(\mathrm{IR}) & \\ & & \mathbf{\Omega}_\parallel(\mathrm{IR}) \end{bmatrix} + \frac{4\pi}{Mv\varepsilon_\parallel(\infty)}\begin{bmatrix} 0 & & \\ & 0 & \\ & & \mathscr{L}_\parallel(0)\mathscr{L}_\parallel(0) \end{bmatrix}.$$

The limiting frequencies for $\mathbf{q}$ along the $c$ axis are therefore given by

$$\|\mathbf{\Omega}_\perp(\mathrm{IR}) - \omega^2\mathbb{1}\| = 0 \tag{4.160}$$

and

$$\left\|\mathbf{\Omega}_\parallel(\mathrm{IR}) + \frac{4\pi}{Mv\varepsilon_\parallel(\infty)}\,\mathscr{L}_\parallel(0)\mathscr{L}_\parallel(0) - \omega^2\mathbb{1}\right\| = 0, \tag{4.161}$$

where

$$\varepsilon_\infty(\hat{\mathbf{q}}\|z) = \varepsilon_{zz}(\infty) = \varepsilon_\parallel(\infty).$$

Equation (4.160) yields the dispersion frequencies appropriate to the perpendicular vibrations, while (4.161) gives the longitudinal frequencies for the parallel vibrations. The physical reasons for this should also be clear. Since $\hat{\mathbf{q}}\|z$ axis, the macroscopic field (which is $\|\mathbf{q}$) is inoperative for the perpendicular vibrations; for the parallel vibrations $\mathbf{E}\|\mathbf{q}\|\mathbf{P}$, whence the frequencies are longitudinal. In brief, if $\hat{\mathbf{q}}$ is along the principal axes of $\mathbf{\varepsilon}(\omega)$ for crystals of orthorhombic or higher symmetry, then the limiting frequencies are distributed entirely between the longitudinal and dispersion categories. This substantiates remarks made earlier in Sec. 4.2. Note, incidentally, that since in *cubic* crystals every direction qualifies to be a principal axis, the limiting frequencies are *always* of the longitudinal or dispersion types. In general, however, every limiting frequency $\omega_\lambda(\hat{\mathbf{q}})$ arises from a mixing of the $r$ IR-active modes, and correspondingly every eigenvector $\mathbf{e}\binom{\mathbf{q}}{\lambda}$ is a linear combination of the set $\{\mathbf{e}\binom{0}{\nu}\}$, $(\nu = 1, \ldots, r)$, of "unperturbed" eigenvectors.

5. The determinantal equation (4.145) can be cast in an alternate form which emphasizes the viewpoint of the spectroscopist. To achieve this we assume that the inverse of the operator $[\mathbf{\Omega}(\text{IR}) - \omega^2 \mathbb{1}_r]$ exists and eliminate $\mathbf{b}$ in (4.142–4.143), giving

$$\left\{\mathbb{1}_3 + 4\pi\hat{\mathbf{Q}}\chi(\infty) + \frac{4\pi}{Mv}\hat{\mathbf{Q}}\mathcal{Z}(0)\left[\mathbf{\Omega}(\text{IR}) - \omega^2\mathbb{1}_r\right]^{-1}\tilde{\mathcal{Z}}(0)\right\}\mathbf{E} = 0. \quad (4.162)$$

Now, in the electrostatic approximation $\hat{\mathbf{Q}}\mathbf{E} = \mathbf{E}$. Therefore Eq. (4.162) becomes*

$$\{\hat{\mathbf{q}}^{\text{tr}}\mathbf{\varepsilon}(\omega)\hat{\mathbf{q}}\}\mathbf{E} = 0. \quad (4.163)$$

Upon demanding nontrivial solutions for $\mathbf{E}$ this leads to

$$\hat{\mathbf{q}}^{\text{tr}}\mathbf{\varepsilon}(\omega)\hat{\mathbf{q}} = 0, \quad (4.164)$$

a form readily amenable to numerical work using spectroscopically determined $\mathbf{\varepsilon}(\omega)$. Observe that while for $\hat{\mathbf{q}}$ along arbitrary directions (4.164) can deliver all the limiting frequencies $\omega_\lambda(\hat{\mathbf{q}})$, $(\lambda = 1, 2, \ldots, r)$, the situation is different for $\hat{\mathbf{q}}$ along a principal-axis direction (in crystals of orthorhombic or higher symmetry). In this case, as noted earlier, the limiting frequencies are entirely in the longitudinal and dispersive categories; since the modes of the latter type do not have electric fields associated with them (in the electrostatic approximation), Eq. (4.164), which now reduces to

$$\varepsilon_{\alpha\alpha}(\omega) \equiv \varepsilon_\alpha(\omega) = 0, \quad (4.165)$$

delivers only the *longitudinal* frequencies. Equation (4.165) is frequently employed by spectroscopists to deduce the longitudinal frequencies from the measured dispersion formula.

6. We earlier remarked that, in crystals of orthorhombic or higher symmetry, $\mathcal{Z}_\lambda$ has only one nonvanishing component and further that $\mathbf{\varepsilon}$ is diagonal. Consequently the dynamical matrix

$$\mathbf{\Omega}(\text{IR}) + \frac{4\pi}{Mv}\mathcal{Z}(0)\mathbf{\varepsilon}^{-1}(\infty)\mathcal{Z}(0),$$

---

*A manipulation as in (4.107) is used here.

which determines the longitudinal frequencies, can for such crystals be written as

$$\left\{ \omega_\lambda^2(0) + \frac{4\pi}{Mv} \frac{\mathscr{L}_{\alpha\lambda}^2(0)}{\varepsilon_{\alpha\alpha}(\infty)} \right\} \delta_{\lambda\lambda'}, \tag{4.166}$$

where $\alpha$ denotes the Cartesian component corresponding to which $\mathscr{L}_\lambda$ has a nonvanishing value. From this we see that corresponding to every dispersion frequency $\omega_\lambda(0)$ in such crystals is a longitudinal counterpart $\omega_{\lambda,l}$ whose square has the value given in (4.166). Once again this substantiates a remark made in Sec. 4.2.

7. Now the dielectric constant is defined via Maxwell's constitutive equation

$$\mathbf{D} = \varepsilon\mathbf{E}. \tag{4.167}$$

Here it is assumed that both $\mathbf{D}$ and $\mathbf{E}$ pertain to the same space point and the same time. A possible generalization of (4.167) is

$$\mathbf{D}(\mathbf{r}, t) = \int_{-\infty}^{t} dt' \, \varepsilon(t, t') \mathbf{E}(\mathbf{r}, t') \tag{4.168}$$

which relates $\mathbf{D}$ to fields $\mathbf{E}$ acting at *earlier* times. Assuming invariance under time translation, (4.168) can be written as

$$\mathbf{D}(\mathbf{r}, t) = \int_{-\infty}^{t} dt' \, \varepsilon(t - t') \mathbf{E}(\mathbf{r}, t'). \tag{4.169}$$

The quantity $\varepsilon(\omega)$ deduced in (4.155) is simply the Fourier transform of $\varepsilon(t)$. A still further generalization gives

$$\mathbf{D}(\mathbf{r}, t) = \int d\mathbf{r}' \int_{-\infty}^{t} dt' \, \varepsilon(\mathbf{r}, \mathbf{r}', t - t') \mathbf{E}(\mathbf{r}', t'). \tag{4.170}$$

To the extent the medium can be regarded as homogeneous (this is possible if $\mathbf{E}(\mathbf{r}, t)$ is slowly varying in space), (4.170) may be written as[24]

$$\mathbf{D}(\mathbf{r}, t) = \int d\mathbf{r}' \int_{-\infty}^{t} dt' \, \varepsilon(\mathbf{r} - \mathbf{r}', t - t') \mathbf{E}(\mathbf{r}', t'). \tag{4.171}$$

As earlier, we could define a Fourier transform $\varepsilon(\mathbf{q}, \omega)$ which would include the effects of spatial dispersion as well.[24] An expression for $\varepsilon(\mathbf{q}, \omega)$ may be deduced[25] as in Sec. 4.3.5 by starting with a field $\mathbf{E} = \mathbf{E}_0 e^{i(\mathbf{q}\cdot\mathbf{r} - \omega t)}$. As is to be expected

$$\varepsilon(\omega) = \lim_{\mathbf{q} \to 0} \varepsilon(\mathbf{q}, \omega). \tag{4.172}$$

The $\mathbf{q}$ dependence of $\varepsilon$ is generally unimportant except while discussing optical activity.[26]

### 4.3.7 An example of the application of the LST result

We shall illustrate the application of (4.156) by considering the example of the uniaxial crystal $CaWO_4$ (space group $C_{4h}^6$—$I4_1/a$) in which the dispersion frequencies are all in the parallel or perpendicular categories. Group-theoretical analysis[27] shows that there are 8 IR-active modes, 4 of the parallel type and 4 of the (doubly degenerate) perpendicular type. They carry respectively the labels $A_u$ and $E_u$ and refer to the IMR's of the point group $C_{4h}$ underlying the space group of $CaWO_4$.

The experimental information[28] concerning the IR-active modes in $CaWO_4$ is summarized in Table 4.3. Suppose $\hat{\mathbf{q}} \| c$ axis. As noted in our discussion on wurtzite (see Sec. 4.2), the limiting frequencies are of the types $\omega_{\perp,\lambda}(0)$ and $\omega_{\|,\lambda',l}$. Therefore, from (4.156)

$$
\frac{\prod_\lambda \omega_\lambda^2(\hat{\mathbf{q}})}{\prod_\lambda \omega_\lambda^2(0)} = \frac{[\omega_{\perp,1}^2(0)\omega_{\perp,2}^2(0)\omega_{\perp,3}^2(0)\omega_{\perp,4}^2(0)]^2}{[\omega_{\perp,1}^2(0)\omega_{\perp,2}^2(0)\omega_{\perp,3}^2(0)\omega_{\perp,4}^2(0)]^2}
$$

$$
\times \frac{\omega_{\|,1,l}^2\omega_{\|,2,l}^2\omega_{\|,3,l}^2\omega_{\|,4,l}^2}{\omega_{\|,1}^2(0)\omega_{\|,2}^2(0)\omega_{\|,3}^2(0)\omega_{\|,4}^2(0)}
$$

$$
= \frac{\varepsilon_{zz}(\omega = 0)}{\varepsilon_{zz}(\omega = \infty)}
$$

$$
= \frac{\varepsilon_\|(0)}{\varepsilon_\|(\infty)}. \tag{4.173}
$$

Similarly, for $\hat{\mathbf{q}} \| y$ axis the limiting frequencies are $\omega_{\perp,i}(0)$, $\omega_{\perp,i,l}$ and $\omega_{\|,i}(0)$, $(i = 1, \ldots, 4)$. Therefore,

$$
\frac{\prod_{\lambda=1}^r \omega_\lambda^2(\hat{\mathbf{q}})}{\prod_{\lambda=1}^r \omega_\lambda^2(0)} \quad \frac{\prod_{i=1}^4 \omega_{\perp,i,l}^2}{\prod_{i=1}^4 \omega_{\perp,i}^2(0)} = \frac{\varepsilon_{yy}(0)}{\varepsilon_{yy}(\infty)} = \frac{\varepsilon_\perp(0)}{\varepsilon_\perp(\infty)}. \tag{4.174}
$$

Substitution of experimental values gives the following results:

Eq. (4.173), left-hand side: 2.72; right-hand side: 2.77,

Eq. (4.174) left-hand side: 2.88; right-hand side: 3.06,

The agreement may be judged to be fair considering the experimental accuracies. It must be remembered that result (4.156) has been obtained within the harmonic approximation. Anharmonic effects could destroy the agreement. However, Cochran and Cowley[15] estimate that (4.156) is accurate to at least 5%.

As an exercise, the reader may evaluate the LST relation for $\hat{\mathbf{q}}$ making an angle $\theta$ with the $c$ axis such that $0° < \theta < 90°$. The limiting frequencies may be obtained from Eq. (4.145) by explicitly setting up the

**Table 4.3**
Experimental information concerning the IR-active modes in CaWO$_4$ (after Barker, ref. 28).

Parallel vibrations (nondegenerate)

| Dispersion frequency $\omega_{\parallel,\lambda}(0)$ (cm$^{-1}$) | Longitudinal frequency $\omega_{\parallel,\lambda,l}$ (cm$^{-1}$) | Oscillator strength $S_{\parallel,\lambda}$ |
|---|---|---|
| 180 | 182 | 0.30 |
| 237 | 327 | 4.65 |
| 435 | 448 | 0.22 |
| 778 | 893 | 0.97 |

$\varepsilon_{\parallel}(\omega) = \varepsilon_{\parallel}(\infty) + \sum_{\lambda=1}^{4} \frac{S_{\parallel,\lambda}\omega^2_{\parallel,\lambda}(0)}{\omega^2_{\parallel,\lambda}(0) - \omega^2}$ ; $\varepsilon_{\parallel}(\infty) = 3.5$, $\varepsilon_{\parallel}(0) = 9.7$.

Perpendicular vibrations (doubly degenerate)

| Dispersion frequency $\omega_{\perp,\lambda}(0)$ (cm$^{-1}$) | Longitudinal frequency $\omega_{\perp,\lambda,l}$ (cm$^{-1}$) | Oscillator strength $S_{\perp,\lambda}$ |
|---|---|---|
| 143 | 147.5 | 1.5 |
| 202 | 248 | 3.5 |
| 309 | 363 | 1.06 |
| 793 | 905 | 0.93 |

$\varepsilon_{\perp}(\omega) = \varepsilon_{\perp}(\infty) + \sum_{\lambda=1}^{4} \frac{S_{\perp,\lambda}\omega^2_{\perp,\lambda}(0)}{\omega^2_{\perp,\lambda}(0) - \omega^2}$ ; $\varepsilon_{\perp}(\infty) = 3.4$, $\varepsilon_{\perp}(0) = 10.4$.

secular determinant using the observed dispersion frequencies to construct $\Omega(\text{IR})$ and the measured oscillator strengths $S_{\lambda}$ to obtain $\mathscr{L}(0)$ via the relation

$$\frac{4\pi\mathscr{L}^2_{\lambda}}{Mv\omega^2_{\lambda}(0)} = S_{\lambda}. \tag{4.175}$$

Alternatively one may use (4.164).

An interesting corollary may be deduced from (4.175). Suppose we consider the plasma problem corresponding to (4.73) for the present case. This is defined by the equation

$$\omega^2(\hat{\mathbf{q}})\xi(\hat{\mathbf{q}}) = \left[ \frac{4\pi}{Mv\varepsilon_{\infty}(\hat{\mathbf{q}})} \mathscr{L}(0)\,\hat{\mathbf{Q}}\mathscr{L}(0) \right]\xi(\hat{\mathbf{q}}).$$

Taking $\hat{\mathbf{q}}$ to be along the $\alpha$th principal axis* (that is, along the Cartesian axis $\alpha$), the plasma frequency $\Omega_{p\alpha}$ is found to be

$$\Omega^2_{p\alpha} = \sum_{\lambda} \Omega^2_{p\alpha,\lambda} = \frac{4\pi}{Mv} \sum_{\lambda} \frac{\mathscr{L}^2_{\alpha\lambda}(0)}{\varepsilon_{\alpha\alpha}(\infty)}. \tag{4.176}$$

*We are here considering the case of crystals of orthorhombic or higher symmetry.

In contrast to (4.73), we note that here the frequency is direction dependent due to electronic polarization effects.

Equation (4.176) suggests that associated with each dispersion frequency $\omega_\lambda(0)$ is a plasma contribution $[4\pi \mathscr{L}^2_{\alpha\lambda}(0)/Mv\varepsilon_{\alpha\alpha}(\infty)]^{1/2}$ when $\hat{\mathbf{q}}$ is along the $\alpha$th principal axis. This observation can be corroborated by considering the case of CaWO$_4$. From (4.173)

$$\frac{\prod_{i=1}^{4} \omega^2_{\|,i,l}}{\prod_{i=1}^{4} \omega^2_{\|,i}(0)} = \prod_{i=1}^{4}\left(1 + \frac{\omega^2_{\|,i,l} - \omega^2_{\|,i}(0)}{\omega^2_{\|,i}(0)}\right)$$

$$\approx 1 + \sum_{i=1}^{4} \frac{\omega^2_{\|,i,l} - \omega^2_{\|,i}(0)}{\omega^2_{\|,i}(0)}$$

$$= 1 + \sum_{i=1}^{4} \frac{\Omega^2_{p,\|,i}}{\omega^2_{\|,i}(0)} \quad \text{(say)}$$

$$= 1 + \frac{4\pi}{Mv}\sum_{i=1}^{4} \frac{\mathscr{L}^2_{\|,i}}{\varepsilon_\|(\infty)\omega^2_{\|,i}(0)} \quad \text{(recalling (4.166)).}$$

Therefore, the net plasma contribution $\Omega^2_{p,\|}$ is shared as

$$\Omega^2_{p,\|,i} = \frac{4\pi\mathscr{L}^2_{\|,i}}{Mv\varepsilon_\|(\infty)} = \frac{S_{\|,i}\omega^2_{\|,i}(0)}{\varepsilon_\|(\infty)}. \tag{4.177}$$

The prescription[29] (4.176) for the sharing of the "plasma contribution" appears to work reasonably well, as the reader may confirm for CaWO$_4$ by numerical calculations.

### 4.3.8 Polaritons and plasmaritons

We now consider the modifications to the results of Sec. 4.3.4 that arise when retardation effects are included. Later the effects of free carriers will be examined. Since only the IR-active modes are affected, the pertinent starting equations are:

$$[\mathbf{\Omega}(\text{IR}) - \omega^2(\mathbf{q})\mathbb{1}_r]\mathbf{b} - \mathscr{L}(0)\mathbf{E} = 0, \tag{4.178a}$$

$$-\frac{1}{Mv}\mathscr{L}(0)\mathbf{b} + \frac{1}{4\pi}\left[\frac{c^2q^2}{\omega^2}(\mathbb{1}_3 - \hat{\mathbf{Q}}) - \boldsymbol{\varepsilon}(\infty)\right]\mathbf{E} = 0. \tag{4.178b}$$

Of these, (4.178a) is the same as Eq. (4.142) except that we have now permitted $\omega$ to be a function of $|\mathbf{q}|$ as well.* Equation (4.178b), however, differs from its electrostatic counterpart (4.143) and may be derived by

---

*Strictly speaking, $\Omega$ (IR) in (4.178) is slightly different from that in (4.142) in that it is now obtained in the diagonalization of a matrix $\overline{\mathbf{M}}(0)$ which includes the effects of retardation. However, as Born and Huang[30] have pointed out, for practical purposes this distinction may be ignored.

using the property

$$\hat{\mathbf{Q}}\mathbf{E} = -4\pi\hat{\mathbf{Q}}\mathbf{P} \tag{4.179}$$

(see (4.79)) in conjunction with the polarization equation (4.154)

$$\mathbf{P} = \frac{1}{Mv}\mathscr{L}(0)\mathbf{b} + \chi(\infty)\mathbf{E}.$$

The solution of Eqs. (4.178) leads to the polariton dispersion curves. This solution can be obtained in various ways which, though equivalent, appear superficially different. Since the various forms occur in the literature, it is useful to consider them here.

One approach would be to demand that $\mathbf{b}$ and $\mathbf{E}$ have nontrivial solutions and thus write the characteristic equation as

$$\left\| \begin{matrix} \Omega(\text{IR}) - \omega^2(\mathbf{q})\mathbb{1}_r, & -\tilde{\mathscr{L}}(0) \\ -\dfrac{1}{Mv}\mathscr{L}(0) & \dfrac{1}{4\pi}\mathscr{L} \end{matrix} \right\| = 0, \tag{4.180}$$

where

$$\mathscr{L} = \frac{c^2 q^2}{\omega^2}(\mathbb{1}_3 - \hat{\mathbf{Q}}) - \varepsilon(\infty). \tag{4.181}$$

Alternatively, by eliminating $\mathbf{E}$ from Eqs. (4.178) we obtain

$$\left\| [\Omega(\text{IR}) - \omega^2(\mathbf{q})\mathbb{1}_r] - \frac{4\pi}{Mv}\tilde{\mathscr{L}}(0)\mathscr{L}^{-1}\mathscr{L}(0) \right\| = 0. \tag{4.182}$$

This equation is to be compared with (4.145). On the basis of the example discussed in Sec. 4.1, one would expect that in the limit $q \to \infty$ all but the two solutions of (4.182) which correspond to pure photon modes must go over those of (4.145). This is indeed so, for in the limit $q \to \infty$, $n^2$ also $\to \infty$, whence the field equation (4.78b) simplifies to (4.78a). Correspondingly, one has in place of Eqs. (4.178) their electrostatic counterparts (4.142–4.143), which then lead to (4.145).*

A third approach to discussing the polariton problem would be to write from (4.178a)

$$\mathbf{b} = [\Omega(\text{IR}) - \omega^2(\mathbf{q})\mathbb{1}_r]^{-1}\tilde{\mathscr{L}}(0)\mathbf{E}, \tag{4.183}$$

where it is supposed that the inverse of the operator $[\Omega(\text{IR}) - \omega^2\mathbb{1}_r]$

---

*Alternatively, one may make use of the property

$$\lim_{n\to\infty}(\mathscr{L}^{-1})_{\alpha\beta} = \lim_{n\to\infty}\left\{ -\frac{\hat{q}_\alpha\hat{q}_\beta}{\varepsilon_\infty(\hat{q})} + O\left(\frac{1}{n^2}\right) \right\} = -\frac{\hat{q}_\alpha\hat{q}_\beta}{\varepsilon_\infty(\hat{q})}.$$

For further details see ref. 14.

exists. Using (4.183) in (4.178b) then leads to an equation in **E** from which we obtain

$$\left\|\frac{\omega^2}{c^2}\boldsymbol{\varepsilon}(\omega) - q^2(\mathbb{1}_3 - \hat{\mathbf{Q}})\right\| = 0, \tag{4.184}$$

upon remembering the definition of $\boldsymbol{\varepsilon}(\omega)$ given in (4.155). Equation (4.184) is the well-known *Fresnel's equation* and is popular with spectroscopists since it is amenable for calculations using $\boldsymbol{\varepsilon}(\omega)$ as measured from optical experiments. It is often written as

$$\|\boldsymbol{\varepsilon}(\omega) - n^2(\mathbb{1}_3 - \hat{\mathbf{Q}})\| = 0, \tag{4.185}$$

a form more suited to the calculation of the two allowed values of the refractive index corresponding to waves of frequency $\omega$ and with electromagnetic fields propagating in a given direction $\hat{\mathbf{q}}$. On the other hand, when (4.184), a generalization of (4.164) to include retardation effects, is expanded, it yields a polynomial in $\omega^2$ and is appropriate for computing the $\omega$ vs **q** relations. We shall give further consideration to (4.184) in the following chapter.

To obtain yet another form in which to write the polariton equation, we express (4.178) in terms of **b** and $\mathbf{E}_t$, the transverse field, rather than in terms of **b** and **E**. Let us first write

$$\mathbf{E} = \mathbf{E}_l + \mathbf{E}_t \tag{4.186}$$

where the subscripts l and t denote longitudinal and transverse (with respect to **q**), respectively. By definition,

$$\mathbf{E}_l = \hat{\mathbf{Q}}\mathbf{E} = -4\pi\hat{\mathbf{Q}}\mathbf{P} \quad \text{(from (4.78b))}.$$

Using (4.154),

$$[\mathbb{1}_3 + 4\pi\hat{\mathbf{Q}}\chi(\infty)]\mathbf{E}_l = -\frac{4\pi}{Mv}\hat{\mathbf{Q}}\mathcal{Z}(0)\mathbf{b} - 4\pi\hat{\mathbf{Q}}\chi(\infty)\mathbf{E}_t. \tag{4.187}$$

Exploiting the property $\hat{\mathbf{Q}}\mathbf{E}_l = \mathbf{E}_l$, the left-hand side may be written

$$[\hat{\mathbf{Q}} + 4\pi\hat{\mathbf{Q}}\chi(\infty)\hat{\mathbf{Q}}]\mathbf{E}_l = \hat{\mathbf{Q}}[\mathbb{1}_3 + 4\pi\chi(\infty)]\hat{\mathbf{Q}}\mathbf{E}_l$$
$$= \varepsilon_\infty(\hat{\mathbf{q}})\mathbf{E}_l.$$

(see Eq. (4.139)). Hence

$$\mathbf{E}_l = -\frac{4\pi}{Mv\varepsilon_\infty(\hat{\mathbf{q}})}\hat{\mathbf{Q}}\mathcal{Z}(0)\mathbf{b} - \frac{4\pi}{\varepsilon_\infty(\hat{\mathbf{q}})}\hat{\mathbf{Q}}\chi(\infty)\mathbf{E}_t. \tag{4.188}$$

Substituting (4.186) and (4.188) in (4.178a), we get

$$\left[ \boldsymbol{\Omega}(\text{IR}) - \omega^2(\mathbf{q})\mathbb{1}_r + \frac{4\pi}{Mv\varepsilon_\infty(\hat{\mathbf{q}})} \mathscr{Z}(0)\,\hat{\mathbf{Q}}\mathscr{Z}(0) \right] \mathbf{b}$$

$$- \left[ \mathscr{Z}(0) - \frac{4\pi}{\varepsilon_\infty(\hat{\mathbf{q}})}\,\mathscr{Z}(0)\,\hat{\mathbf{Q}}\chi(\infty) \right] \mathbf{E}_t = 0. \qquad (4.189)$$

For convenience, let us introduce two unit vectors $\hat{\boldsymbol{\eta}}_1$ and $\hat{\boldsymbol{\eta}}_2$ which are mutually perpendicular and also normal to $\hat{\mathbf{q}}$. We may then write

$$\mathbf{E}_t = \hat{\boldsymbol{\eta}}_1\mathfrak{E}_1 + \hat{\boldsymbol{\eta}}_2\mathfrak{E}_2 = [\hat{\boldsymbol{\eta}}_1, \hat{\boldsymbol{\eta}}_2]\begin{bmatrix} \mathfrak{E}_1 \\ \mathfrak{E}_2 \end{bmatrix} = \mathbf{N}\mathfrak{E}, \qquad (4.190)$$

where $\mathbf{N}$ is the following $(3 \times 2)$ matrix:

$$\mathbf{N} = \begin{bmatrix} \hat{\eta}_{1x} & \hat{\eta}_{2x} \\ \hat{\eta}_{1y} & \hat{\eta}_{2y} \\ \hat{\eta}_{1z} & \hat{\eta}_{2z} \end{bmatrix}. \qquad (4.191)$$

In terms of (4.190), Eq. (4.189) becomes

$$\left[ \boldsymbol{\Omega}(\text{IR}) - \omega^2(\mathbf{q})\mathbb{1}_r + \frac{4\pi}{Mv\varepsilon_\infty(\hat{\mathbf{q}})} \mathscr{Z}(0)\,\hat{\mathbf{Q}}\mathscr{Z}(0) \right] \mathbf{b} - \tilde{\psi}(0)\mathfrak{E} = 0, \quad (4.192)$$

where

$$\tilde{\psi}(0) = \left[ \mathscr{Z}(0) - \frac{4\pi}{\varepsilon_\infty(\hat{\mathbf{q}})}\,\mathscr{Z}(0)\,\hat{\mathbf{Q}}\chi(\infty) \right]\mathbf{N} \qquad (4.193)$$

is an $(r \times 2)$ matrix. We shall also express Eq. (4.178b) in terms of $\mathbf{b}$ and $\mathfrak{E}$. Substitute from (4.186), (4.188), and (4.190) into (4.178b) and use

$$[\mathbb{1}_3 - \hat{\mathbf{Q}}]\mathbf{E}_t = \mathbf{E}_t, \ \hat{\mathbf{Q}}\mathbf{E}_l = \mathbf{E}_l, \ [\mathbb{1}_3 - \hat{\mathbf{Q}}]\hat{\mathbf{Q}} = 0.$$

Then

$$\left[ -\frac{1}{Mv}\mathscr{Z}(0) + \frac{1}{Mv\varepsilon_\infty(\hat{\mathbf{q}})}\,\varepsilon(\infty)\hat{\mathbf{Q}}\mathscr{Z}(0) \right]\mathbf{b}$$

$$+ \left[ \frac{n^2}{4\pi} - \frac{\varepsilon(\infty)}{4\pi} + \frac{1}{\varepsilon_\infty(\hat{\mathbf{q}})}(\mathbb{1}_3 + 4\pi\chi(\infty))\hat{\mathbf{Q}}\chi(\infty) \right]\mathbf{N}\mathfrak{E} = 0.$$

Multiplying from the left by $\tilde{\mathbf{N}}$ and noting that $\tilde{\mathbf{N}}\hat{\mathbf{Q}} = 0$ (since $\hat{\boldsymbol{\eta}}_1, \hat{\boldsymbol{\eta}}_2, \hat{\mathbf{q}}$ are mutually perpendicular), we obtain

$$-\frac{1}{Mv}\tilde{\psi}(0)\mathbf{b} + \left[ \frac{(n^2 - 1)}{4\pi}\mathbb{1}_2 - \boldsymbol{\alpha} \right]\mathfrak{E} = 0, \qquad (4.194)$$

where

$$\boldsymbol{\alpha} = \tilde{\mathbf{N}}\left[ \chi(\infty) - \frac{4\pi}{\varepsilon_\infty(\hat{\mathbf{q}})}\,\chi(\infty)\hat{\mathbf{Q}}\chi(\infty) \right]\mathbf{N} \qquad (4.195)$$

is a $(2 \times 2)$ matrix. Equations (4.192) and (4.194) are now in the desired form, and upon demanding nontrivial solutions for $\mathbf{b}$ and $\mathfrak{E}$ we obtain the following secular equation:

$$\left\| \begin{array}{cc} \boldsymbol{\Omega}(\mathrm{IR}) - \omega^2(\mathbf{q})\mathbb{1}_r + \dfrac{4\pi}{Mv\varepsilon_\infty(\hat{\mathbf{q}})} \mathscr{Z}(0)\hat{\mathbf{Q}}\mathscr{Z}(0) & -\tilde{\boldsymbol{\psi}}(0) \\[4mm] \dfrac{1}{Mv}\boldsymbol{\psi}(0) & \dfrac{(1-n^2)}{4\pi}\mathbb{1}_2 + \boldsymbol{\alpha} \end{array} \right\| = 0.$$

(4.196)

The above method of explicitly displaying the role of the transverse field is useful in the group-theoretic classification of polaritons to be discussed shortly.

It is to be noted that whichever form of the secular equation is adopted for computation, one always gets $(r + 2)$ branches because the problem has $(r + 2)$ degrees of freedom, two of these being associated with the transverse electric field. These branches incorporate not only the effects of the transverse field but of the longitudinal field as well.

Figure 4.10 displays the polariton spectrum of $CaWO_4$ computed using the parameters determined from the optical experiments (Table 4.3). For convenience, the spectra expected in the electrostatic approximation are also shown. Note that the $\mathbf{q} \to \infty$ limits are the frequencies $\omega_\lambda(\hat{\mathbf{q}})$ of the electrostatic approximation while the $\mathbf{q} \to 0$ limits are the longitudinal frequencies. Also observe the regular behavior of the branches at $\mathbf{q} = 0$, which contrasts with that exhibited in the electrostatic approximation.

This last feature needs further comment. To investigate the behavior of the polariton branches in the region $\mathbf{q} \approx 0$, we note that here we may take $\omega \gg cq$. (See for example Figure 4.4, where the optic frequency at $q \approx 0$ is much larger than the value $cq$). As a result, we may from (4.181) write

$$\lim_{\mathbf{q}\to 0} \mathscr{L} = -\varepsilon(\infty),$$

using which the secular determinant (4.182) reduces to

$$\left\| [\boldsymbol{\Omega}(\mathrm{IR}) - \omega^2\mathbb{1}_r] + \dfrac{4\pi}{Mv}\mathscr{Z}(0)\varepsilon^{-1}(\infty)\mathscr{L}(0) \right\| = 0,$$

which not only shows that the $\mathbf{q} \to 0$ optic frequencies are independent of $\hat{\mathbf{q}}$ but are in fact the longitudinal solutions.

It is pertinent to make here a brief reference to plasmaritons. As already remarked (Sec. 4.2.6), phonon-plasmon coupling is important when the plasmon frequency becomes comparable to phonon frequencies as happens in doped semiconductors. The coupling arises because the carrier

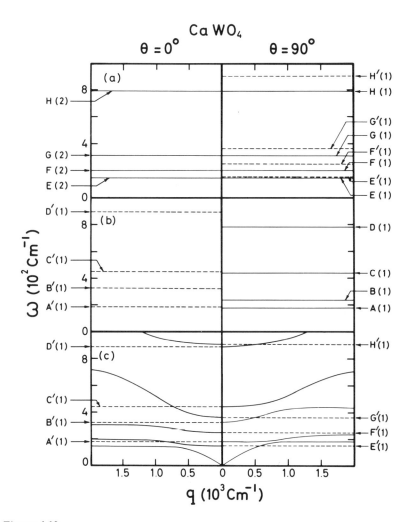

**Figure 4.10**

Dispersion curves for CaWO$_4$ in the long-wavelength limit for $\hat{q} \parallel c$ axis ($\theta = 0°$) and $\hat{q} \perp c$ axis ($\theta = 90°$). (a) and (b) summarize the electrostatic results. For convenience, the branches based on $\omega_\perp$ and $\omega_\parallel$ are shown separately, with solid lines (labeled with un-primed letters) referring to dispersion frequencies and the dashed lines (labeled with primed letters) referring to longitudinal frequencies. The numbers in parentheses indicate the degeneracies. Observe that for $\theta = 90°$ only one out of each doubly degenerate frequency $\omega_{\perp,\lambda}(0)$ is modified by the electric-field effects. (See also Figure 4.5). (c) shows the branches computed from (4.184) using measured values for the various parameters (Table 4.3). Notice that for branches affected by retardation, the $q \to \infty$ limits are the dispersion frequencies and the $q \to 0$ limits are the longitudinal frequencies. For $\theta = 90°$ there are two sets of polariton branches, one based on $\omega_{\parallel,\lambda}$ and the other on $\omega_{\perp,\lambda}$. The latter are identical to those shown for $\theta = 0°$ and are suppressed in the interest of clarity.

plasma can respond to the electric fields (both longitudinal and transverse) set up by the lattice vibrations. The resulting coupled modes are succintly described once again by Fresnel's equation (4.184) with, however, a reinterpretation of $\varepsilon(\omega)$ as

$$\varepsilon(\omega) = \varepsilon^{\text{lattice}}(\omega) + \varepsilon^{\text{electronic}}(\omega) - \mathbb{1}_3, \qquad (4.197)$$

where $\varepsilon^{\text{lattice}}(\omega)$ is the same as in (4.155) and

$$\varepsilon^{\text{electronic}}(\omega) = \mathbb{1}_3 + \frac{4\pi i}{\omega}\,\sigma(\omega), \qquad (4.198)$$

$\sigma(\omega)$ being the conductivity tensor.[31] This approach has been employed by Shah and coworkers[13] to discuss both plasmon–LO phonon coupling[32] and plasmaritons in CdS. The latter has the wurtzite structure, but since crystalline anisotropy is known to be small, $\varepsilon^{\text{lattice}}(\omega)$ was approximated by (4.14) and $\varepsilon^{\text{electronic}}(\omega)$ by

$$\varepsilon^{\text{electronic}}_{\alpha\beta}(\omega) = \delta_{\alpha\beta}\left(1 + \frac{4\pi i}{\omega}\,\sigma(\omega)\right)$$

$$= \delta_{\alpha\beta}\left(1 - \frac{\omega_p^2}{\omega^2}\,\varepsilon(\infty)\right), \qquad \text{where } \omega_p^2 = \frac{4\pi n e^2}{m^*\varepsilon(\infty)}. \qquad (4.199)$$

The expression of $\varepsilon^{\text{electronic}}(\omega)$ in terms of the collective properties of the plasma is more involved when anisotropy becomes important.

### 4.4 Group-theoretic classification of long-wavelength modes

The group-theoretic classification of long-wavelength modes in ionic crystals poses special problems, the reasons for which will become clear upon referring to Figures 4.1 and 4.2, which show the dispersion curves for ZnS. We observe that whereas in a purely short-range model there is a triple degeneracy associated with the optic modes at $\mathbf{q} = 0$, a splitting occurs in the electrostatic model. On the other hand, if group theory is applied as in Chapter 3, it predicts a triple degeneracy for the $\mathbf{q} = 0$ optic modes. It would appear that a contradiction exists between the results of group theory and those of the electrostatic model. However, the contradiction disappears if one remembers that $\mathbf{q} \to 0$ in the electrostatic approximation is to be interpreted as $q$ being much smaller than $1/a$, where $a \sim$ unit-cell dimension, but much larger than $(\omega/c)$. (This, of course, applies only to the IR-active modes.) Nevertheless the question remains as to how the $\mathbf{q} \approx 0$ results of electrostatic calculations may be labeled.

Now the tradition among spectroscopists is to apply group theory to the problem

$$\mathbf{m}\omega^2\mathbf{U} = \overline{\mathbf{M}}(0)\mathbf{U} \tag{4.200}$$

and use the results generally in all discussions of long-wavelength spectra. Obviously, in this case group theory is being used to classify modes in the absence of the macroscopic field. Since the latter in general modifies the frequencies associated with the IR-active modes, it is clearly inappropriate to use the labels based on (4.200) when macroscopic field effects are present. Indeed, no satisfactory means of labeling the long-wavelength modes of the electrostatic problem according to (4.200) exists. The reason for this is that in the electrostatic approximation it is not possible to write

$$\mathbf{M}(\mathbf{q}) = \mathbf{M}(0) + O(\mathbf{q}) + \text{higher order terms in } \mathbf{q},$$

where $O(\mathbf{q})$ represents terms of order $\mathbf{q}$. Nonetheless a hybrid scheme is occasionally used in the literature. For example $\omega_{\mathrm{TO}}$ and $\omega_{\mathrm{LO}}$ (see Figure 4.2a) are considered to belong to the three-dimensional representation $\Gamma_{15}$ of the point group $T_d$ (the point group of ZnS) which in fact is the representation to which the dispersion frequency (see Figure 4.2b) belongs.* Similarly in ZnO a hybrid labeling as illustrated in Figure 4.5 is sometimes adopted.

A more satisfactory scheme may be based on the polariton equations, since the polariton branches display regular behavior at $\mathbf{q} = 0$ (see Figure 4.10). This, however, requires an extension of the scheme considered in Chapter 3 since the vibrational field now has $(3n + 2)$ degrees of freedom per wave vector $\mathbf{q}$. We therefore start with the equations

$$\omega^2(\mathbf{q})\mathbf{m}\mathbf{U}(\mathbf{q}) = \overline{\mathbf{M}}(\mathbf{q})\mathbf{U}(\mathbf{q}) - \mathbf{Z}^\dagger(\mathbf{q})\mathbf{E}(\mathbf{q}) \tag{4.201}$$

and

$$-\frac{4\pi}{v}\mathbf{Z}(\mathbf{q})\mathbf{U}(\mathbf{q}) + \mathscr{L}\mathbf{E}(\mathbf{q}) = 0, \tag{4.202}$$

which are the generalizations of (4.128) and (4.137) in that we have now formally retained the $\mathbf{q}$ dependence of $\overline{\mathbf{M}}$ and $\mathbf{Z}$. These equations may also be viewed as extensions of (4.178) to encompass all the $(3n + 2)$ modes instead of merely $(r + 2)$. By following the same procedure as in Sec. 4.3.8, the above equations may be written in the form

$$\left[\overline{\mathbf{M}}(\mathbf{q}) - \mathbf{m}\omega^2(\mathbf{q}) + \frac{4\pi}{v\varepsilon_\infty(\hat{\mathbf{q}})}\mathbf{Z}^\dagger(\mathbf{q})\,\hat{\mathbf{Q}}\mathbf{Z}(\mathbf{q})\right]\mathbf{U}(\mathbf{q})$$
$$- \mathbf{\Psi}^\dagger(\mathbf{q})\mathfrak{E} = 0, \tag{4.203}$$

*Remember that the optic frequency computed in Figure 4.2b is the dispersion frequency since $\mathbf{E}$ has been suppressed.

$$-\frac{1}{v}\boldsymbol{\Psi}(\mathbf{q})\mathbf{U}(\mathbf{q}) + \left[\frac{(n^2 - 1)}{4\pi}\mathbb{1}_2 - \boldsymbol{\alpha}\right]\boldsymbol{\mathfrak{E}} = 0, \qquad (4.204)$$

where

$$\boldsymbol{\Psi}(\mathbf{q}) = \tilde{\mathbf{N}}\left[\mathbf{Z}(\mathbf{q}) - \frac{4\pi}{\varepsilon_\infty(\hat{\mathbf{q}})}\chi(\infty)\,\hat{\mathbf{Q}}\mathbf{Z}(\mathbf{q})\right], \qquad (4.205)$$

and all other quantities are as defined earlier. From (4.203) and (4.204) we have

$$\left\{\begin{bmatrix} \overline{\mathbf{M}}(\mathbf{q}) + \dfrac{4\pi}{v\varepsilon_\infty(\hat{\mathbf{q}})}\mathbf{Z}^\dagger(\mathbf{q})\hat{\mathbf{Q}}\mathbf{Z}(\mathbf{q}) & -\boldsymbol{\Psi}^\dagger(\mathbf{q}) \\[2ex] \dfrac{1}{v}\boldsymbol{\Psi}(\mathbf{q}) & \dfrac{1}{4\pi}\mathbb{1}_2 + \boldsymbol{\alpha} \end{bmatrix}\right.$$

$$\left. -\omega^2\begin{bmatrix} \mathbf{m} & 0 \\ 0 & 0 \end{bmatrix} - \frac{q^2c^2}{\omega^2}\begin{bmatrix} 0 & 0 \\ 0 & \mathbb{1}_2 \end{bmatrix}\right\}\begin{bmatrix} \mathbf{U} \\ \boldsymbol{\mathfrak{E}} \end{bmatrix} = 0, \qquad (4.206)$$

where all the matrices are of dimension $(3n + 2)$, and the vector

$$\begin{bmatrix} \mathbf{U} \\ \boldsymbol{\mathfrak{E}} \end{bmatrix}$$

has $(3n + 2)$ components describing the displacements and the transverse field. We thus see that the eivenvalue problem has become generalized to the form

$$\sum_i f_i(\lambda)\,\mathbf{A}_i\psi(\lambda) = 0;$$

to apply group theory to the present problem we must construct $(3n + 2)$-dimensional matrices $\mathscr{T}(\mathbf{q};\mathbf{R})$ which commute with each of the three square matrices appearing in (4.206) and which furnish a multiplier representation of $G_0(\mathbf{q})$. Once such matrices are available, then the rest of the procedure is the same as outlined in Chapter 3. It is not difficult to see that the required matrices have the form[33]

$$\mathscr{T}(\mathbf{q};\mathbf{R}) = \begin{bmatrix} \mathbf{T}(\mathbf{q};\mathbf{R}) & 0 \\ 0 & \mathscr{R} \end{bmatrix}, \qquad (4.207)$$

where

$$T_{\alpha\beta}(kk';\mathbf{q},\mathbf{R}) = \delta(k, F(k', R))R_{\alpha\beta} \qquad (4.208)$$

and $\mathscr{R}$ is a $(2 \times 2)$ matrix given by

$$\mathscr{R} = \tilde{\mathbf{N}}\mathbf{R}\mathbf{N}, \tag{4.209}$$

$\mathbf{N}$ defined as in (4.191).

To show that the matrices $\mathscr{T}(\mathbf{q};\mathbf{R})$ commute with the present dynamical matrix, we first note that Eqs. (4.203) and (4.204) are obtainable from the potential function

$$\begin{aligned}
\Phi(\mathbf{q}) &= \mathbf{u}^{\dagger}(\mathbf{q})\left[\overline{\mathbf{M}}(\mathbf{q}) + \frac{4\pi}{v\varepsilon_{\infty}(\hat{\mathbf{q}})}\mathbf{Z}^{\dagger}(\mathbf{q})\hat{\mathbf{Q}}\mathbf{Z}(\mathbf{q})\right]\mathbf{u}(\mathbf{q}) \\
&\quad - \mathbf{u}^{\dagger}(\mathbf{q})\left[\mathbf{Z}^{\dagger}(\mathbf{q}) - \frac{4\pi}{\varepsilon_{\infty}(\hat{\mathbf{q}})}\mathbf{Z}^{\dagger}(\mathbf{q})\hat{\mathbf{Q}}\chi(\infty)\right]\mathbf{e}_{\mathfrak{t}}(\mathbf{q}) \\
&\quad - \mathbf{e}_{\mathfrak{t}}^{\dagger}(\mathbf{q})\left[\mathbf{Z}(\mathbf{q}) - \frac{4\pi}{\varepsilon_{\infty}(\hat{\mathbf{q}})}\chi(\infty)\hat{\mathbf{Q}}\mathbf{Z}(\mathbf{q})\right]\mathbf{u}(\mathbf{q}) \\
&\quad - v\mathbf{e}_{\mathfrak{t}}^{\dagger}(\mathbf{q})\left[\chi(\infty) - \frac{4\pi}{\varepsilon_{\infty}(\hat{\mathbf{q}})}\chi(\infty)\hat{\mathbf{Q}}\chi(\infty)\right]\mathbf{e}_{\mathfrak{t}}(\mathbf{q}),
\end{aligned} \tag{4.210}$$

which is essentially (4.85) rewritten in terms of $\mathbf{u}(\mathbf{q})$ and $\mathbf{e}_{\mathfrak{t}}$ by replacing $\mathbf{e}$ by $\mathbf{e}_l + \mathbf{e}_{\mathfrak{t}}$, etc., as described earlier. Next we observe that by writing the displacements as (see (4.87) also)

$$\mathbf{u}\binom{l}{k} = \left\{\mathbf{U}(k|\mathbf{q})\exp\left[i\left(\mathbf{q}\cdot\mathbf{x}\binom{l}{k} - \omega t\right)\right] + \text{c.c.}\right\} \tag{4.211}$$

(see (3.14)) we may, as in (3.21), obtain

$$\mathbf{U}'(\mathbf{q}) = \mathbf{\Gamma}(\mathbf{q};\mathscr{R}_m)\mathbf{U}(\mathbf{q}).$$

where the elements of $\mathbf{\Gamma}(\mathbf{q};\mathscr{R}_m)$ are given by

$$\Gamma_{\alpha\beta}(kk'|\mathbf{q};\mathscr{R}_m) = R_{\alpha\beta}\delta(k, F(k', R))\exp\{-i\mathbf{q}\cdot[\mathbf{v}(R) + \mathbf{x}(m)]\} \tag{4.212}$$

(see Eq. (3.62). Note here $\mathbf{G} = 0$ since $\mathbf{q} \approx 0$). From (4.212) the definition of the $\mathbf{T}$ matrix follows immediately.

Likewise, starting with

$$\mathbf{e}_{\mathfrak{t}}(\mathbf{r}) = \{\mathbf{E}_{\mathfrak{t}}(\mathbf{q})\exp[i(\mathbf{q}\cdot\mathbf{r} - \omega t)] + \text{c.c.}\} \tag{4.113}$$

and remembering

$$\mathbf{e}_{\mathfrak{t},\alpha}'(\mathscr{R}_m\mathbf{r}) = \sum_{\beta} R_{\alpha\beta}e_{\mathfrak{t},\beta}(\mathbf{r}), \tag{4.214}$$

we can obtain

$$\mathbf{E}_{\mathfrak{t}}'(\mathbf{q}) = \mathbf{R}\exp\{-i\mathbf{q}\cdot[\mathbf{v}(R) + \mathbf{x}(m)]\}\mathbf{E}_{\mathfrak{t}}(\mathbf{q}). \tag{4.215}$$

Using (4.190) this gives

$$\mathfrak{E}'(\mathbf{q}) = \mathscr{R} \exp\{- i\mathbf{q}\cdot[\mathbf{v}(R) + \mathbf{x}(m)]\}\mathfrak{E}(\mathbf{q}), \tag{4.216}$$

where $\mathscr{R}$ is as defined in (4.209).

At this point we invoke, as in (3.23), the invariance

$$\Phi(\mathbf{q}) = \Phi'(\mathbf{q}),$$

where $\Phi'(\mathbf{q})$ is the transformed potential resulting from the application of an operation of $G(\mathbf{q})$ and is defined as in (4.210) but in terms of $\mathbf{u}'$ and $\mathbf{e}'_l$. This finally gives

$$\left[\overline{\mathbf{M}}(\mathbf{q}) + \frac{4\pi}{v\varepsilon_\infty(\hat{\mathbf{q}})} \mathbf{Z}^\dagger(\mathbf{q})\hat{\mathbf{Q}}\mathbf{Z}(\mathbf{q})\right]\mathbf{T}(\mathbf{q};\mathbf{R})$$

$$= \mathbf{T}(\mathbf{q};\mathbf{R})\left[\overline{\mathbf{M}}(\mathbf{q}) + \frac{4\pi}{v\varepsilon_\infty(\hat{\mathbf{q}})} \mathbf{Z}^\dagger(\mathbf{q})\hat{\mathbf{Q}}\mathbf{Z}(\mathbf{q})\right]; \tag{4.217}$$

$$\mathbf{\Psi}^\dagger(\mathbf{q})\mathscr{R} = \mathbf{T}(\mathbf{q};\mathbf{R})\mathbf{\Psi}^\dagger(\mathbf{q}); \tag{4.218}$$

$$\alpha\mathscr{R} = \mathscr{R}\alpha. \tag{4.219}$$

Furthermore,

$$\mathbf{m}\mathbf{T}(\mathbf{q};\mathbf{R}) = \mathbf{T}(\mathbf{q};\mathbf{R})\mathbf{m}. \tag{4.220}$$

With the help of the above results it is easy to prove (1) that $\mathscr{T}$ commutes with each of the matrices in Eq. (4.206) and (2) that it furnishes a multiplier representation of $G_0(\mathbf{q})$. From this point onwards the procedure of decomposition and labeling is familiar.

A difficulty arises in working with the above scheme for the case $\mathbf{q} \equiv 0$, connected with the fact that in this situation it is not possible to define transverse fields. However, because the $\mathbf{q} = 0$ phononlike modes are the longitudinal solutions of the dynamical problem (see Sec. 4.3.8), group theory for this case may be applied in the conventional fashion to the $3n$-dimensional matrix $[\overline{\mathbf{M}}(0) + (4\pi/v)\tilde{\mathbf{Z}}(0)\varepsilon^{-1}(\infty)\mathbf{Z}(0)]$, which has the same symmetry as $\overline{\mathbf{M}}(0)$. Guided by spectroscopic data, this enables one to fix the labels of the zone-center frequencies. The branch assignments for finite $\mathbf{q}$ can then be made using compatibility relations.

### 4.5. Influence of the macroscopic field on the long-wavelength external vibrations

The role of the macroscopic electric field in relation to the external modes in complex crystals merits special attention. For simplicity, we shall ignore the effects of electronic polarization and regard the dynamical matrix to be given by $\mathbf{M} = \mathbf{M}^{SR} + \mathbf{M}^C$ as in Sec. 4.2. By isolating the macroscopic field, the long-wavelength equations may be written in the fixed frame as[34]*

$$\omega_j^2(\mathbf{q}) m_\kappa U_\alpha^t\left(\kappa \left|\begin{matrix} \mathbf{q} \\ j \end{matrix}\right.\right) = \sum_{\kappa'} \sum_{\beta i} \overline{M}_{\alpha\beta}^{ti}\left(\begin{matrix} \mathbf{q} \\ \kappa\kappa' \end{matrix}\right) U_\beta^i\left(\kappa' \left|\begin{matrix} \mathbf{q} \\ j \end{matrix}\right.\right) - e\sum_{k\in\kappa} Z\left(\begin{matrix} \kappa \\ k \end{matrix}\right) E_\alpha, \quad (4.221)$$

$$\omega_j^2(\mathbf{q}) \sum_\gamma \mathscr{I}_{\alpha\gamma}(\kappa) U_\gamma^r\left(\kappa \left|\begin{matrix} \mathbf{q} \\ j \end{matrix}\right.\right) = \sum_{\kappa'} \sum_{\beta i} \overline{M}_{\alpha\beta}^{ri}\left(\begin{matrix} \mathbf{q} \\ \kappa\kappa' \end{matrix}\right) U_\beta^i\left(\kappa' \left|\begin{matrix} \mathbf{q} \\ j \end{matrix}\right.\right)$$
$$- \left[\left\{\sum_{k\in\kappa} eZ\left(\begin{matrix} \kappa \\ k \end{matrix}\right)\mathbf{x}(k)\right\} \times \mathbf{E}\right]_\alpha. \quad (4.222)$$

As usual, $\overline{\mathbf{M}}$ denotes the regular part of $\mathbf{M}$; $eZ\left(\begin{smallmatrix} \kappa \\ k \end{smallmatrix}\right)$ is the charge on the $k$th ion in the $\kappa$th molecular group.

The polarization equation is

$$\mathbf{P} = \frac{1}{v}\sum_\kappa \mathbf{p}(\kappa), \quad (4.223)$$

where $\mathbf{p}(\kappa)$ is the displacement dipole moment associated with $\kappa$ and is given by

$$\mathbf{p}(\kappa) = e\sum_{k\in\kappa} Z\left(\begin{matrix} \kappa \\ k \end{matrix}\right)\left\{\mathbf{U}^t\left(\kappa \left|\begin{matrix} \mathbf{q} \\ j \end{matrix}\right.\right) + \mathbf{U}^r\left(\kappa \left|\begin{matrix} \mathbf{q} \\ j \end{matrix}\right.\right) \times \mathbf{x}(k)\right\}. \quad (4.224)$$

The macroscopic field $\mathbf{E}$ is related to $\mathbf{P}$ as usual. It may be noted that by assumption $\mathbf{P}$ has no contribution from electronic polarization. The macroscopic field terms in Eqs. (4.221) and (4.222) represent respectively the linear force and torque on $\kappa$ produced by $\mathbf{E}$; these are of the form $\left(\sum_{k\in\kappa} eZ\left(\begin{smallmatrix} \kappa \\ k \end{smallmatrix}\right)\right)\mathbf{E}$ and $\mathbf{\Pi}(\kappa) \times \mathbf{E}$, where $\mathbf{\Pi}(\kappa)$ refers to the *intrinsic* dipole of the molecular group $\kappa$ and is given by

$$\mathbf{\Pi}(\kappa) = \sum_{k\in\kappa} eZ\left(\begin{matrix} \kappa \\ k \end{matrix}\right)\mathbf{x}(k). \quad (4.225)$$

These results are in agreement with the expectations of electrostatics. The structure of the above equations remains essentially the same when effects of electronic polarization are included, apart from the replacement of $eZ\left(\begin{smallmatrix} \kappa \\ k \end{smallmatrix}\right)$ by the tensor $\mathbf{Z}\left(\begin{smallmatrix} \kappa \\ k \end{smallmatrix}\right)$. It is interesting to note that if a long-wavelength external mode is purely librational and if it involves molecular groups that have no intrinsic dipole moment, then within the framework of the external-mode approximation one expects the concerned mode to be unaffected by the macroscopic field. Additionally, the librational mode concerned

---

*Because here we employ the matrix $\mathbf{M}(\mathbf{q})$ instead of $\mathbf{B}(\mathbf{q})$, equations (4.221) and (4.222) are slightly different from the corresponding equations of ref. 34, namely (IV.C.4) and (IV.C.6).

cannot be seen in an infrared experiment since the displacement dipole moment associated with it is zero.

These remarks may be better appreciated by considering the example of $CaWO_4$. This has a doubly degenerate mode at 202 $cm^{-1}$ which corresponds to the librations of $WO_4^{--}$ and is group-theoretically predicted to be IR active. Now the tungstate ion is tetrahedral and therefore $\Pi = 0$. Hence we do not expect to see this librational mode in an infrared experiment. It turns out, however, that the mode in question is in fact seen in infrared experiments.[28] The most likely answer to this puzzle is that there is an admixture from one of the internal vibrations of the $WO_4^{--}$ ion. This is a possibility since there is an IR-active mode in the neighborhood with a frequency of 237 $cm^{-1}$.

## 4.6 Phenomenological models which include electronic polarization

### 4.6.1 Mainly historical

Historically, the rigid-ion model was first introduced in the context of cohesive-energy calculations for alkali halides.[35] The model's great success in this respect suggested that it would be of value in describing the dynamics also, and accordingly a test was made by Kellerman[36] in 1940. Assuming the pairwise interactions of the atoms to be composed of a Coulomb potential and a short-range Born-Mayer potential as in Eq. (1.11), Kellerman computed the frequency distribution and the specific heat of NaCl and found satisfactory agreement with experiment.*

Nevertheless, it was clear even at that time that the rigid-ion model was inadequate. Whereas according to Kellerman for diatomic cubic crystals one must have

$$(\omega_{LO}^2/\omega_{TO}^2) = \varepsilon(0), \tag{4.226}$$

Lyddane, Sachs, and Teller[2] argued that the relationship is

$$(\omega_{LO}^2/\omega_{TO}^2) = \varepsilon(0)/\varepsilon(\infty), \tag{4.5}$$

which reduces to (4.226) only if the electronic polarizability is negligible. Substantiation of (4.5) by experiment was a clear pointer to the need for including the effects of electronic polarization in lattice dynamics. Indeed

---

*Lundqvist et al.[37] pointed out that this agreement is illusory in that room-temperature data were used to evaluate the constants in Kellermann's treatment, whereas low-temperature specific heats were used to compare experiment with theory. They showed that if low-temperature elastic constants are used to fix Kellerman's parameters, the agreement with experiment is much poorer.

Lyddane and Herzfeld[38] had already proposed a model allowing for such effects. Briefly, their model is as follows:

When the lattice vibrates, it leads to the creation of dipoles. The latter generate an electric field which, acting on the ions, in turn contributes to the dipole moments via electronic polarization as originally envisaged by Lorentz. For *alkali halides*, the equations of motion are*

$$m_k \omega^2(\mathbf{q}) U_\alpha(k|\mathbf{q}) = \sum_{k'\beta} M_{\alpha\beta}^{SR}\binom{\mathbf{q}}{kk'} U_\beta(k'|\mathbf{q}) - Z_k e \mathscr{E}_\alpha(k|\mathbf{q}), \qquad (4.227)$$

where $\mathscr{E}(k|\mathbf{q})$ is the amplitude of the local field acting on the ions in the $k$th sublattice. In the long-wavelength limit, this field is the same at both sublattice sites and is related to the macroscopic field by

$$\mathscr{E}(\mathbf{q}) = \mathbf{E}(\mathbf{q}) + \frac{4\pi}{3}\mathbf{P}(\mathbf{q}), \qquad (4.228)$$

where the macroscopic field and the polarization density are related as usual. In contrast to the rigid-ion model, $\mathbf{P}$ now includes the contributions due to both displacements and electronic polarization, that is,

$$\mathbf{P}(\mathbf{q}) = \frac{1}{v}\sum_k \{Z_k e \mathbf{U}(k|\mathbf{q}) + \mathbf{\mu}(k|\mathbf{q})\} \qquad (4.229)$$

where

$$\mathbf{\mu}(k|\mathbf{q}) = \alpha_k \mathscr{E}(\mathbf{q}), \qquad (4.230)$$

$\alpha_k$ denoting the electronic polarizability of the $k$th type ion (see (4.4)). Equation (4.230) embodies the *Lorentz approximation* in the sense that the electronic dipole moment is assumed to be created by the local field acting on the ion concerned, as originally envisaged by Lorentz in his study of refraction.

When calculations were made for NaCl using the electronic polarizabilities given by Pauling,[39] Lyddane and Herzfeld found that some of the frequencies were imaginary! This was quite disappointing as their model was undoubtedly a step in the right direction.†

In the meantime, evidence regarding the deficiencies of the Born model continued to accumulate. Based on this, one could identify two physical effects not present in this model but deserving of consideration. The first pertains to many-body forces, evidence for which came mainly from the

---

*The generalization to crystals of arbitrary structure is discussed later.
†Hardy[40,41] later pointed out that the Lyddane-Herzfeld model delivers real frequencies if one uses the polarizabilities deduced from experimental data by Tessman et al.[42] (see Chapter 6).

observed (though usually small) violation of the Cauchy relations (see Sec. 2.4). This led Löwdin[43] to reexamine the problem of cohesion in alkali halides from a quantum-mechanical standpoint. His analysis confirmed the presence of many-body forces and prompted Lundqvist[44-47] to include these effects systematically in lattice dynamics.

On the other hand, from a consideration of the dielectric properties and the compressibilities of NaCl-type crystals, Szigeti[48,49] and independently Odelevsky[50] deduced the following equations under the assumptions of the Lyddane-Herzfeld model:

$$\frac{\varepsilon(0) - \varepsilon(\infty)}{[\varepsilon(\infty) + 2]^2} = \frac{4\pi \mathscr{N}(Ze)^2}{9\bar{m}\omega_{TO}^2}, \tag{4.231}$$

$$\bar{m}\omega_{TO}^2\left(\frac{\varepsilon(0) + 2}{\varepsilon(\infty) + 2}\right) = \frac{6r_0}{\beta}. \tag{4.232}$$

Here $\mathscr{N}$ is the number of ion pairs per unit volume, $\bar{m}$ the reduced mass, $r_0$ the nearest-neighbor distance and $\beta$ the compressibility. Define now

$$\beta^* = 6r_0[\varepsilon(\infty) + 2]/\bar{m}\omega_{TO}^2[\varepsilon(0) + 2] \tag{4.233a}$$

and

$$(Ze)^{*2} = 9\bar{m}\omega_{TO}^2[\varepsilon(0) - \varepsilon(\infty)]/[\varepsilon(\infty) + 2]^2 4\pi \mathscr{N}. \tag{4.233b}$$

If the assumptions underlying the derivation of (4.231) and (4.232) are correct, then $\beta^*$ and $(Ze)^*$ deduced from (4.233) using experimental values for the quantities on the right-hand side must equal the observed compressibility $\beta$ and the formal ionic charge $Ze$. Table 4.4 gives $(\beta^*/\beta)$ and $(e^*/e)$ for alkali halides (in which $Z = 1$) and clearly demonstrates the inadequacy of the model.

It was argued by Szigeti that these deviations from unity, especially those in $(e^*/e)$, are mostly due to a polarization mechanism beyond that implied in (4.230). In particular he suggested that when neighboring ions approach each other, overlap forces could lead to a deformation of the charge distribution, thereby producing an additional polarization. Thus Szigeti's analysis, though confined to the long-wavelength modes, suggested that not only must electronic polarization be included in lattice dynamics but that one must in fact go beyond Lyddane and Herzfeld's attempt to do so.

Subsequent developments have been largely in the direction of incorporating Szigeti's suggestion via suitable phenomenological models. An important step was taken by Cochran and coworkers[51-54] who incorporated ideas developed earlier by Dick and Overhauser[55] in proposing a model in which each ion is visualized as consisting of a core and an outer

**Table 4.4**
Values of $(\beta^*/\beta)$ and $(e^*/e)$ derived from experimental data. (After Szigeti, refs. 48 and 49.)

| Crystal | $(\beta^*/\beta)$ | $(e^*/e)$ |
|---|---|---|
| LiF | 1.0 | 0.87 |
| NaF | 0.83 | 0.93 |
| NaCl | 0.99 | 0.74 |
| NaBr | 1.13 | 0.69 |
| NaI | 1.05 | 0.71 |
| KCl | 0.96 | 0.80 |
| KBr | 0.95 | 0.76 |
| KI | 0.99 | 0.69 |
| RbCl | 0.89 | 0.84 |
| RbBr | 0.83 | 0.82 |
| RbI | 0.66 | 0.89 |

shell of valence electrons, the latter capable of being displaced bodily with respect to the core. This is the *shell model* of lattice dynamics; its evolution is still continuing.

By adopting suggestions of Szigeti and of Born and Huang,[56] Hardy[41] gave a different description of the deformation originating from short-range forces without invoking the core-shell concept. A parallel development was the work of the Soviet school.[57-59] Starting from a quantum-mechanical consideration of the deformation of the electronic wave-functions during nuclear motions, it was concluded that dynamics is influenced by polarization mechanisms in addition to that of (4.230). Understandably, these different approaches have much in common, a point we shall elaborate later.

Presently we shall review the above-mentioned models and some of their derivatives, emphasizing the underlying physical principles and the formal structure. Their performance in explaining observed vibrational spectra is reserved for discussion in Chapter 6.

### 4.6.2 Background to the shell model—work of Dick and Overhauser

By far the most widely used among the various types of polarization models is the shell model, the origin of which can be traced to the work of Dick and Overhauser,[55] who were concerned with explaining Szigeti's observations, especially regarding $(e^*/e)$. They noted first that the interaction energy of a pair of helium atoms studied via the Heitler-London approach had a form resembling the familiar Born-Mayer term. One could generalize and argue that the repulsive energy of a pair of ions with rare-gas configuration has also the Born-Mayer form. However, when applied to ionic crystals this merely results in the Born model. To study the problem of overlap forces when the ions are polarizable, Dick and Overhauser suggested that each rare-gas configuration can be visualized as made up of an outer *spherical shell* of $n$ electrons and a core consisting of the nucleus and the remaining electrons. Further, every core is considered to be bound to its corresponding shell by a harmonic spring. The shell always retains its spherical charge but can move *bodily* with respect to the core; it is this relative motion which results in electronic polarization. When the question of overlap forces is reexamined allowing for the possibility of polarization, it is found that the interaction energy of a pair of ions has the form $Be^{-R_{ss}\alpha}$, where $R_{ss}$ is the shell-shell distance (see Figure 4.11). This is the usual Born-Mayer form except that the repulsion depends on $R_{ss}$ rather than the internuclear distance $R$.

The fact that overlap forces act via shells has important implications for the polarization mechanism.* Two such mechanisms may be identified, the first of which is illustrated in Figure 4.12. Figure 4.12a shows the disposition of a pair of ions in the unpolarized state. The application of an electric

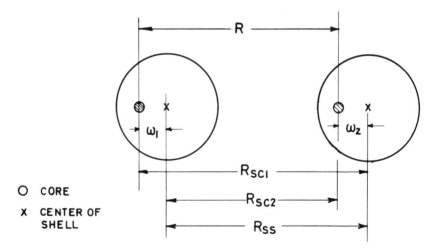

O   CORE

X   CENTER OF
    SHELL

**Figure 4.11**
Typical relative disposition of a pair of neighboring ions. The repulsive force between the ions has a Born-Mayer form but depends on the shell-shell distance $R_{ss}$ rather than the internuclear separation $R$.

---

*The connection between polarization of ions and repulsive force between them was considered independently by Yamashita and Kurosawa.[60]

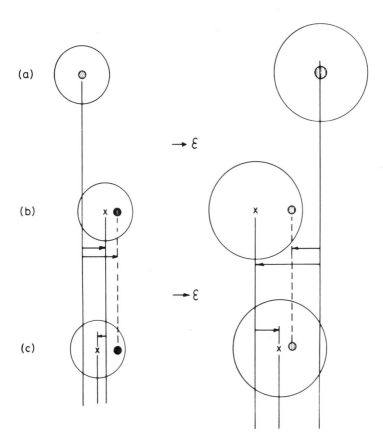

**Figure 4.12**
Disposition of adjacent ions corresponding to various situations. (a) unpolarized state; (b) in the presence of an electric field with shell-shell repulsion ignored; (c) polarized state with allowance for shell-shell repulsion. The negative ion is drawn larger.

field induces in both ions a polarization which, in the absence of overlap forces, results in the configuration depicted in Figure 4.12b. This represents the polarization of Eq. (4.230). The short-range interaction causes a further distortion leading to the situation shown in Figure 4.12c. Such a *short-range polarization* is the sort of contribution suggested by Szigeti to explain the reduction of $(e^*/e)$ to values less than unity.

Also hypothesized by Dick and Overhauser is the *exchange-charge polarization*. In the region of overlap, the Pauli exclusion principle acts to reduce the electron charge density and to redistribute this charge on the ions. The charge depletion in the overlap region may be represented as the superposition of point positive charge of suitable magnitude called the *exchange charge*, located at the center of gravity of the overlap region. This is compensated by an appropriate readjustment of the shell charges. When adjacent ions are moved with respect to one another, the exchange charge fluctuates, accompanied by a corresponding fluctuation of the shell charges. The net dipole moment per unit volume associated with this process is the exchange-charge polarization.

It is important to remember that both types of polarization arise from overlap. They merely serve to describe the overall charge distortions in terms of complementary pictures. When considering the short-range overlap energy of a pair of neighboring ions, we must remember that this energy includes the contributions from both types of polarization.

### 4.6.3 Physical picture underlying the shell model

Although Dick and Overhauser were interested primarily in providing an understanding of Szigeti's observations, their model was recognized by Cochran and coworkers to contain features relevant to a comprehensive discussion of the vibrational spectrum and the dielectric properties of insulating crystals. Of the two additional means of polarization, they retained only the short-range polarization mechanism. Further they modified the short-range interactions by allowing for core-core and core-shell forces as illustrated in Figure 4.13.

Since the shell can be displaced relative to the core, one can accommodate electronic polarization. If $Y_k e$ denotes the charge on the shell of the $\binom{l}{k}$th ion, then $\mathbf{p}\binom{l}{k} = Y_k e[\mathbf{u}^s\binom{l}{k} - \mathbf{u}^c\binom{l}{k}]$ represents the *electronic* dipole moment, where $\mathbf{u}^s$ and $\mathbf{u}^c$ denote, respectively, the displacements of the shell and the core with respect to the equilibrium position $\mathbf{x}\binom{l}{k}$. This moment has two contributions. The first arises from various long-range electrostatic forces. In favorable circumstances (for example, in cubic diatomic crystals), this contribution can in fact be expressed in the form (4.230). The second contribution is due to the short-range polarization mechanism.

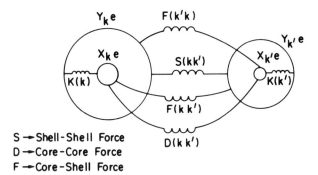

S → Shell-Shell Force
D → Core-Core Force
F → Core-Shell Force

**Figure 4.13**
Cochran's (generalized) shell model. $X_k e$ and $X_{k'} e$ denote the core charges while $Y_k e$ and $Y_{k'} e$ denote the shell charges. The springs represent the various short-range forces. In addition, one must also consider the Coulomb interaction. The forces $F(kk')$ and $F(k'k)$ are frequently taken to be equal.

It is convenient to characterize these two processes in terms of parameters $\alpha$ and $d$, respectively.* While the question of expressing these parameters in terms of lattice dynamical quantities will be taken up later, a feeling for their significance may be obtained from the following simple calculation, adapted from the work of Havinga.[61]

Consider a diatomic cubic crystal in which only one ion is polarizable (for example, NaI where one may treat Na$^+$ as a point ion). Figure 4.14 shows a typical ion pair and the short-range forces. Describing the displacements as in the figure, one may write appropriate equations of motion for the cores and shells. Here we restrict attention to the equation for the shell of the negative ion. Neglecting the mass of the shell, we have

$$0 = -kW(2) - S[U(2) + W(2) - U(1)] + Y_2 e \mathscr{E}, \qquad (4.234)$$

where $\mathscr{E}$ is the local field. The dipole moment per ion pair is

$$Ze[U(1) - U(2)] + Y_2 e W(2), \qquad Z = |Z_1| = |Z_2|,$$

which using (4.234) can be written as

$$\left(Ze + \frac{Y_2 e S}{S + k}\right)[U(1) - U(2)] + \frac{(Y_2 e)^2}{k + S}\mathscr{E}.$$

The polarization density is therefore

$$P_0 = \frac{1}{v}\left[\left(Ze + \frac{Y_2 e S}{S + k}\right)[U(1) - U(2)] + \frac{(Y_2 e)^2 \mathscr{E}}{k + S}\right].$$

*We shall later see that assigning an electronic polarizability $\alpha$ to every ion is possible only in situations where the long-range contribution to the electronic dipole moment can be expressed in the form (4.230).

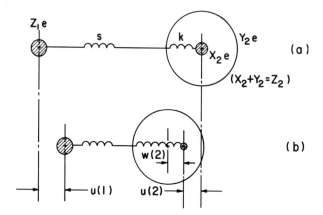

**Figure 4.14**
Schematic drawing of a pair of adjacent ions in an alkali halide in which only the negative
ion is polarizable. (a) and (b) denote the unpolarized and polarized states, respectively.
The latter leads to a simple picture for the $\mathbf{q} = 0$ vibrations.

Comparing this with (4.4) and remembering $\alpha_+ = 0$ in the present
instance, we see that

$$\alpha_- = \frac{(Y_2 e)^2}{(k + S)}. \tag{4.235}$$

An additional dipole moment $[Y_2 e \, S/(S + k)] \, \{U(1) - U(2)\}$ arises from
short-range polarization, enabling us to identify $d$ in the present model as

$$d = -\frac{Y_2 S}{(S + k)}. \tag{4.236}$$

The quantity $d$ is sometimes called the *mechanical polarizability*.

It would appear that the introduction of the shells has increased the
number of dynamical variables. However, we will see that, assuming the
shell masses to be negligible, the shell displacements are not independent
and can be eliminated.

### 4.6.4 Equations of the shell model
To deduce the equations of motion, we postulate that the harmonic potential energy is given by

$$\Phi^{(2)} = \Psi^{(2)} + \tfrac{1}{2} \sum_{lk\alpha\beta} K_{\alpha\beta}(k) \left[ u_\alpha^c\binom{l}{k} - u_\alpha^s\binom{l}{k} \right]\left[ u_\beta^c\binom{l}{k} - u_\beta^s\binom{l}{k} \right], \tag{4.237}$$

where $\Psi^{(2)}$ includes all the potential-energy change except that due to the interaction of the cores with their respective shells. The latter is represented by the second term, where $K_{\alpha\beta}(k)$ denotes the core-shell interaction of an ion in the $k$th sublattice.

Formally, we may express $\Psi^{(2)}$ as

$$
\begin{aligned}
\Psi^{(2)} = {} & \tfrac{1}{2} \sum_{lk\alpha} \sum_{l'k'\beta} \Psi_{\alpha\beta}^{cc}\begin{pmatrix} l & l' \\ k & k' \end{pmatrix} u_\alpha^c\begin{pmatrix} l \\ k \end{pmatrix} u_\beta^c\begin{pmatrix} l' \\ k' \end{pmatrix} \\
& + \tfrac{1}{2} \sum_{\substack{lk\alpha \\ (lk \neq l'k')}} \sum_{l'k'\beta} \Psi_{\alpha\beta}^{cs}\begin{pmatrix} l & l' \\ k & k' \end{pmatrix} u_\alpha^c\begin{pmatrix} l \\ k \end{pmatrix} u_\beta^s\begin{pmatrix} l' \\ k' \end{pmatrix} \\
& + \tfrac{1}{2} \sum_{\substack{lk\alpha \\ (lk \neq l'k')}} \sum_{l'k'\beta} \Psi_{\alpha\beta}^{sc}\begin{pmatrix} l & l' \\ k & k' \end{pmatrix} u_\alpha^s\begin{pmatrix} l \\ k \end{pmatrix} u_\beta^c\begin{pmatrix} l' \\ k' \end{pmatrix} \\
& + \tfrac{1}{2} \sum_{lk\alpha} \sum_{l'k'\beta} \Psi_{\alpha\beta}^{ss}\begin{pmatrix} l & l' \\ k & k' \end{pmatrix} u_\alpha^s\begin{pmatrix} l \\ k \end{pmatrix} u_\beta^s\begin{pmatrix} l' \\ k' \end{pmatrix},
\end{aligned}
\tag{4.238}
$$

where

$$
\Psi_{\alpha\beta}^{ij}\begin{pmatrix} l & l' \\ k & k' \end{pmatrix} = \frac{\partial^2 \Phi}{\partial u_\alpha^i(\substack{l \\ k}) \partial u_\beta^j(\substack{l' \\ k'})}\bigg|_0, \left[\begin{pmatrix} l \\ k \end{pmatrix} \neq \begin{pmatrix} l' \\ k' \end{pmatrix}; \; i = c, s; \; j = c, s\right], \tag{4.239}
$$

denotes the force constants of the present model. The superscripts c and s stand for core and shell, respectively. It is important to note that the displacements $\mathbf{u}^c$ and $\mathbf{u}^s$ are defined with respect to the equilibrium position. By understanding,

$$
\Psi_{\alpha\beta}^{cs}\begin{pmatrix} l & l \\ k & k \end{pmatrix} = \Psi_{\alpha\beta}^{sc}\begin{pmatrix} l & l \\ k & k \end{pmatrix} = 0 \quad \text{for every } \alpha, \beta, \text{ and } (lk). \tag{4.240}
$$

The remaining self terms $\Psi_{\alpha\beta}^{cc}(\substack{l\ l \\ k\ k})$ and $\Psi_{\alpha\beta}^{ss}(\substack{l\ l \\ k\ k})$ will be defined shortly. In writing (4.238) we remember that the various force constants have both Coulombic (long-range) and non-Coulombic (short-range) components. In the context of the former, (4.240) then merely implies that the electrostatic interaction of a core with its own shell is not counted. (Remember that the short-range part has been absorbed in $K_{\alpha\beta}(k)$.)

The force constants (4.239) have the usual properties, namely,

$$
\Psi_{\alpha\beta}^{ij}\begin{pmatrix} l & l' \\ k & k' \end{pmatrix} = \Psi_{\beta\alpha}^{ji}\begin{pmatrix} l' & l \\ k' & k \end{pmatrix}, \tag{4.241}
$$

$$
\Psi_{\alpha\beta}^{ij}\begin{pmatrix} l & l' \\ k & k' \end{pmatrix} = \Psi_{\alpha\beta}^{ij}\begin{pmatrix} l + L & l' + L \\ k & k' \end{pmatrix}, \tag{4.242}
$$

. . . .

Likewise,

$$
K_{\alpha\beta}(k) = K_{\beta\alpha}(k). \tag{4.243}
$$

From (4.241) it follows that the second and third terms on the right-hand side of (4.238) are equal.

To obtain the self terms, we formulate first the equations of motion. If we bear Eq. (4.240) in mind, the usual procedure (see (2.34)) gives

$$
m_k^c \ddot{u}_\alpha^c \binom{l}{k} = - \sum_{i=c,s} \sum_{l'k'\beta} \Psi_{\alpha\beta}^{ci} \binom{l\ \ l'}{k\ \ k'} u_\beta^i \binom{l'}{k'}
$$
$$
- \sum_\beta K_{\alpha\beta}(k) \left[ u_\beta^c \binom{l}{k} - u_\beta^s \binom{l}{k} \right], \tag{4.244}
$$

$$
m_k^s \ddot{u}_\alpha^s \binom{l}{k} = - \sum_{i=c,s} \sum_{l'k'\beta} \Psi_{\alpha\beta}^{si} \binom{l\ \ l'}{k\ \ k'} u_\beta^i \binom{l'}{k'}
$$
$$
+ \sum_\beta K_{\alpha\beta}(k) \left[ u_\beta^c \binom{l}{k} - u_\beta^s \binom{l}{k} \right]. \tag{4.245}
$$

Here $m_k^c$ ($\approx m_k$) is the mass of the core and $m_k^s$ ($\approx 0$) is the mass of the shell.

We now imagine the equilibrium crystal to be given an infinitesimal translation $\varepsilon$ in a specified Cartesian direction $\alpha$, that is,

$$
u_\beta^c \binom{l}{k} = u_\beta^s \binom{l}{k} = \varepsilon \delta_{\alpha\beta} \text{ for every } \beta \text{ and} \binom{l}{k}.
$$

Introducing this in the equations of motion and remembering that uniform translation does not produce restoring forces, we obtain

$$
\Psi_{\alpha\beta}^{cc} \binom{l\ \ l}{k\ \ k} = - \sum_{l'k'}{}' \left\{ \Psi_{\alpha\beta}^{cc} \binom{l\ \ l'}{k\ \ k'} + \Psi_{\alpha\beta}^{cs} \binom{l\ \ l'}{k\ \ k'} \right\}, \tag{4.246}
$$

$$
\Psi_{\alpha\beta}^{ss} \binom{l\ \ l}{k\ \ k} = - \sum_{l'k'}{}' \left\{ \Psi_{\alpha\beta}^{sc} \binom{l\ \ l'}{k\ \ k'} + \Psi_{\alpha\beta}^{ss} \binom{l\ \ l'}{k\ \ k'} \right\}. \tag{4.247}
$$

Noting next that

$$
\Psi_{\alpha\beta}^{cc} \binom{l\ \ l}{k\ \ k} = \Psi_{\beta\alpha}^{cc} \binom{l\ \ l}{k\ \ k}, \quad \Psi_{\alpha\beta}^{ss} \binom{l\ \ l}{k\ \ k} = \Psi_{\beta\alpha}^{ss} \binom{l\ \ l}{k\ \ k}
$$

and using (4.241), we may write the above equations in the alternate form

$$
\Psi_{\alpha\beta}^{cc} \binom{l\ \ l}{k\ \ k} = - \sum_{l'k'}{}' \left\{ \Psi_{\alpha\beta}^{cc} \binom{l'\ \ l}{k'\ \ k} + \Psi_{\alpha\beta}^{sc} \binom{l'\ \ l}{k'\ \ k} \right\}, \tag{4.248}
$$

$$
\Psi_{\alpha\beta}^{ss} \binom{l\ \ l}{k\ \ k} = - \sum_{l'k'}{}' \left\{ \Psi_{\alpha\beta}^{cs} \binom{l'\ \ l}{k'\ \ k} + \Psi_{\alpha\beta}^{ss} \binom{l'\ \ l}{k'\ \ k} \right\}. \tag{4.249}
$$

The rotational sum rules may be deduced in a similar fashion (see Sec. 2.2).

Instead of the displacements $\mathbf{u}^c$ and $\mathbf{u}^s$, it is customary and convenient to use

$$\mathbf{u}\binom{l}{k} \equiv \mathbf{u}^c\binom{l}{k} \quad \text{and} \quad \mathbf{w}\binom{l}{k} = \mathbf{u}^s\binom{l}{k} - \mathbf{u}^c\binom{l}{k} \tag{4.250}$$

to allow the electronic dipole to be compactly expressed as

$$\mathbf{p}\binom{l}{k} = Y_k e \mathbf{w}\binom{l}{k}. \tag{4.251}$$

In terms of the new displacements defined in (4.250), the harmonic potential energy has the following form:

$$
\begin{aligned}
\Phi^{(2)} = \tfrac{1}{2} \sum_{\substack{lk\alpha \ l'k'\beta \\ (lk \neq l'k')}} \Bigg[ & \{\Psi_{\alpha\beta}^{cc} + \Psi_{\alpha\beta}^{cs} + \Psi_{\alpha\beta}^{sc} + \Psi_{\alpha\beta}^{ss}\} \\
& \times \left\{ u_\alpha\binom{l}{k} u_\beta\binom{l'}{k'} - u_\alpha\binom{l}{k} u_\beta\binom{l}{k} \right\} \\
& + \{\Psi_{\alpha\beta}^{cs} + \Psi_{\alpha\beta}^{ss}\} u_\alpha\binom{l}{k} w_\beta\binom{l'}{k'} - \{\Psi_{\alpha\beta}^{sc} + \Psi_{\alpha\beta}^{ss}\} u_\alpha\binom{l}{k} w_\beta\binom{l}{k} \\
& + \{\Psi_{\alpha\beta}^{sc} + \Psi_{\alpha\beta}^{ss}\} \left\{ w_\alpha\binom{l}{k} u_\beta\binom{l'}{k'} - w_\alpha\binom{l}{k} u_\beta\binom{l}{k} \right\} \\
& + \Psi_{\alpha\beta}^{ss} w_\alpha\binom{l}{k} w_\beta\binom{l'}{k'} - \{\Psi_{\alpha\beta}^{sc} + \Psi_{\alpha\beta}^{ss}\} w_\alpha\binom{l}{k} w_\beta\binom{l}{k} \Bigg] \\
& + \tfrac{1}{2} \sum_{lk\alpha\beta} K_{\alpha\beta}(k) w_\alpha\binom{l}{k} w_\beta\binom{l}{k}. 
\end{aligned}
\tag{4.252}
$$

For notational convenience, we have suppressed the indices $\binom{l \ l'}{k \ k'}$ associated with the force constants.

Introduce now the following definitions for $\binom{l}{k} \neq \binom{l'}{k'}$:

$$\Psi_{\alpha\beta}^{AA} = \Psi_{\alpha\beta}^{cc} + \Psi_{\alpha\beta}^{cs} + \Psi_{\alpha\beta}^{sc} + \Psi_{\alpha\beta}^{ss}, \tag{4.253}$$
$$\Psi_{\alpha\beta}^{AD} = \Psi_{\alpha\beta}^{cs} + \Psi_{\alpha\beta}^{ss}, \tag{4.254}$$
$$\Psi_{\alpha\beta}^{DA} = \Psi_{\alpha\beta}^{sc} + \Psi_{\alpha\beta}^{ss}, \tag{4.255}$$
$$\Psi_{\alpha\beta}^{DD} = \Psi_{\alpha\beta}^{ss}. \tag{4.256}$$

Remembering (4.241) and the above definitions and manipulating the dummy indices, we may write (4.252) as

$$
\begin{aligned}
\Phi^{(2)} = \tfrac{1}{2} \sum_{lk\alpha \ l'k'\beta} \Bigg[ & \Psi_{\alpha\beta}^{AA}\binom{l \ l'}{k \ k'} u_\alpha\binom{l}{k} u_\beta\binom{l'}{k'} + \Psi_{\alpha\beta}^{AD}\binom{l \ l'}{k \ k'} u_\alpha\binom{l}{k} w_\beta\binom{l'}{k'} \\
& + \Psi_{\alpha\beta}^{DA}\binom{l \ l'}{k \ k'} w_\alpha\binom{l}{k} u_\beta\binom{l'}{k'} + \Psi_{\alpha\beta}^{DD}\binom{l \ l'}{k \ k'} w_\alpha\binom{l}{k} w_\beta\binom{l'}{k'} \Bigg] \\
& + \tfrac{1}{2} \sum_{lk\alpha\beta} K_{\alpha\beta}(k) w_\alpha\binom{l}{k} w_\beta\binom{l}{k},
\end{aligned}
\tag{4.257}
$$

where the self terms have the following definitions (see (4.246–4.249)):

$$\Psi^{AA}_{\alpha\beta}\begin{pmatrix} l & l \\ k & k \end{pmatrix} = -\sum_{l'k'}{}' \Psi^{AA}_{\alpha\beta}\begin{pmatrix} l & l' \\ k & k' \end{pmatrix}, \tag{4.258}$$

$$\Psi^{AD}_{\alpha\beta}\begin{pmatrix} l & l \\ k & k \end{pmatrix} = -\sum_{l'k'}{}' \Psi^{DA}_{\alpha\beta}\begin{pmatrix} l & l' \\ k & k' \end{pmatrix}, \tag{4.259}$$

$$\Psi^{DA}_{\alpha\beta}\begin{pmatrix} l & l \\ k & k \end{pmatrix} = -\sum_{l'k'}{}' \Psi^{DA}_{\alpha\beta}\begin{pmatrix} l & l' \\ k & k' \end{pmatrix}, \tag{4.260}$$

$$\Psi^{DD}_{\alpha\beta}\begin{pmatrix} l & l \\ k & k \end{pmatrix} = \cdot\sum_{l'k'}{}' \Psi^{DA}_{\alpha\beta}\begin{pmatrix} l & l' \\ k & k' \end{pmatrix}. \tag{4.261}$$

Observe that the self terms for $\Psi^{AD}$, $\Psi^{DA}$, and $\Psi^{DD}$ are equal. For some purposes, it is convenient to write (4.257) explicitly as the sum of short- and long-range parts, that is,

$$\Phi^{(2)} = \Phi^{(2)SR} + \Phi^{(2)LR}. \tag{4.262}$$

The short-range part is given by

$$\Phi^{(2)SR} = \tfrac{1}{2}\sum_{lk\alpha}\sum_{l'k'\beta}\left\{ \Psi^{AA,SR}_{\alpha\beta}\begin{pmatrix} l & l' \\ k & k' \end{pmatrix}u_\alpha\begin{pmatrix} l \\ k \end{pmatrix}u_\beta\begin{pmatrix} l' \\ k' \end{pmatrix}\right.$$
$$+ \Psi^{AD,SR}_{\alpha\beta}\begin{pmatrix} l & l' \\ k & k' \end{pmatrix}u_\alpha\begin{pmatrix} l \\ k \end{pmatrix}w_\beta\begin{pmatrix} l' \\ k' \end{pmatrix} + \Psi^{DA,SR}_{\alpha\beta}\begin{pmatrix} l & l' \\ k & k' \end{pmatrix}$$
$$\times w_\alpha\begin{pmatrix} l \\ k \end{pmatrix}u_\beta\begin{pmatrix} l' \\ k' \end{pmatrix} + \left.\Psi^{DD,SR}_{\alpha\beta}\begin{pmatrix} l & l' \\ k & k' \end{pmatrix}w_\alpha\begin{pmatrix} l \\ k \end{pmatrix}w_\beta\begin{pmatrix} l' \\ k' \end{pmatrix}\right\}$$
$$+ \tfrac{1}{2}\sum_{lk\alpha\beta} K_{\alpha\beta}(k)w_\alpha\begin{pmatrix} l \\ k \end{pmatrix}w_\beta\begin{pmatrix} l \\ k \end{pmatrix}, \tag{4.263}$$

where the self terms $\Psi^{AA,SR}_{\alpha\beta}\begin{pmatrix} l & l \\ k & k \end{pmatrix}$, etc., are to be evaluated using (4.258–4.261). To write the long-range part we first note that the Coulombic force constants of the present model are given by (for $\begin{pmatrix} l \\ k \end{pmatrix} \neq \begin{pmatrix} l' \\ k' \end{pmatrix}$):

$$\Psi^{AA,C}_{\alpha\beta}\begin{pmatrix} l & l' \\ k & k' \end{pmatrix} = -Z_k Z_{k'} e^2 \mathscr{D}, \tag{4.264}$$

$$\Psi^{AD,C}_{\alpha\beta}\begin{pmatrix} l & l' \\ k & k' \end{pmatrix} = -Z_k Y_{k'} e^2 \mathscr{D}, \tag{4.265}$$

$$\Psi^{DA,C}_{\alpha\beta}\begin{pmatrix} l & l' \\ k & k' \end{pmatrix} = -Y_k Z_{k'} e^2 \mathscr{D}, \tag{4.266}$$

$$\Psi^{DD,C}_{\alpha\beta}\begin{pmatrix} l & l' \\ k & k' \end{pmatrix} = -Y_k Y_{k'} e^2 \mathscr{D}, \tag{4.267}$$

where $\mathscr{D}$ stands for

$$\frac{\partial^2}{\partial r_\alpha \partial r_\beta}\left(\frac{1}{r}\right)\bigg|_{r=|\mathbf{x}\left(\begin{smallmatrix} l' & l \\ k' & k \end{smallmatrix}\right)|}. \tag{4.268}$$

(For notational convenience we drop temporarily the labels $_{\alpha\beta}\binom{l\ l'}{k\ k'}$ associated with $\mathscr{D}$.)

Using the above and the sum rules for expressing the self terms, we have

$$
\begin{aligned}
\Phi^{(2)\mathrm{LR}} = &-\tfrac{1}{2}\sum_{lk\alpha}\sum_{l'k'\beta}\left\{Z_k eu_\alpha\binom{l}{k}+Y_k ew_\alpha\binom{l}{k}\right\}\mathscr{D}\left\{Z_{k'}eu_\beta\binom{l'}{k'}+Y_{k'}ew_\beta\binom{l'}{k'}\right\} \\
&\quad {\scriptstyle(lk\neq l'k')} \\
&+\tfrac{1}{2}\sum_{lk\alpha}\sum_{l'k'\beta}\left[Z_k Z_{k'}e^2\mathscr{D}u_\alpha\binom{l}{k}u_\beta\binom{l}{k}+Y_k Z_{k'}e^2\mathscr{D}u_\alpha\binom{l}{k}w_\beta\binom{l}{k}\right. \\
&\quad {\scriptstyle(lk\neq l'k')} \\
&\left.+Y_k Z_{k'}e^2\mathscr{D}w_\alpha\binom{l}{k}u_\beta\binom{l}{k}+Y_k Z_{k'}e^2\mathscr{D}w_\alpha\binom{l}{k}w_\beta\binom{l}{k}\right]. \quad (4.269)
\end{aligned}
$$

For convenience we have grouped the "self contribution" into the second term. Using (4.258–4.261) to calculate separately the self terms of the short-range and Coulombic parts assumes (reasonably) that the two types of forces are additive. We observe in passing that the "self contribution" in (4.269) vanishes in diagonally cubic crystals. (See Appendix 2).

Equation (4.257) specifies the harmonic potential energy for a crystal composed of "atoms" (symbol A) and induced "dipoles" (symbol D). As will be clear shortly, the latter arise from both short- and long-range effects. It is worth noting that the dipoles can themselves interact, with short-range forces characterized by the constants $\Psi_{\alpha\beta}^{\mathrm{DD,SR}}$.

We now reformulate the equations of motions in terms of $\mathbf{u}$ and $\mathbf{w}$ via the equations

$$
m_k\ddot{u}_\alpha\binom{l}{k} = -\frac{\partial\Phi^{(2)}}{\partial u_\alpha\binom{l}{k}}, \qquad (4.270)
$$

$$
0 \approx \frac{\partial\Phi^{(2)}}{\partial w_\alpha\binom{l}{k}}. \qquad (4.271)
$$

These give

$$
m_k\ddot{u}_\alpha\binom{l}{k} = -\sum_{l'k'\beta}\left\{\Psi_{\alpha\beta}^{\mathrm{AA}}\binom{l\ l'}{k\ k'}u_\beta\binom{l'}{k'}+\Psi_{\alpha\beta}^{\mathrm{AD}}\binom{l\ l'}{k\ k'}w_\beta\binom{l'}{k'}\right\}, \qquad (4.272)
$$

$$
\begin{aligned}
0 = &-\sum_{l'k'\beta}\left\{\Psi_{\alpha\beta}^{\mathrm{DA}}\binom{l\ l'}{k\ k'}u_\beta\binom{l'}{k'}+\Psi_{\alpha\beta}^{\mathrm{DD}}\binom{l\ l'}{k\ k'}w_\beta\binom{l'}{k'}\right\} \\
&-\sum_\beta K_{\alpha\beta}(k)w_\beta\binom{l}{k}.^* \qquad (4.273)
\end{aligned}
$$

In writing (4.271) and (4.273) we have considered the shell masses negligible. The assumption of massless shells is equivalent to the adiabatic approximation since, lacking inertia, the electrons in the shells can follow the core motions faithfully. One may also view (4.273) as a minimization

---

*These equations can also be derived from (4.244) and (4.245) by suitable algebra.

condition on the deformation energy with respect to the induced dipole moments.

Equation (4.273) is sometimes referred to as the *shell equation*. From it we get

$$
\mathbf{p}\binom{l}{k} = Y_k e \mathbf{w}\binom{l}{k} = Y_k e \left[ -\mathbf{K}(k) - \mathbf{\Psi}^{DD}\binom{l\ l}{k\ k} \right]^{-1}
$$
$$
\times \left\{ \sum_{l'k'} \mathbf{\Psi}^{DA}\binom{l\ l'}{k\ k'} \mathbf{u}\binom{l'}{k'} + \sum_{l'k'}{}' \mathbf{\Psi}^{DD}\binom{l\ l'}{k\ k'} \mathbf{w}\binom{l'}{k'} \right\}, \tag{4.274}
$$

which expresses explicitly two facts: (1) that the electronic dipole moment induced at site $\binom{l}{k}$ depends on the displacements at other sites as well as the electronic moments located there, and (2) that the induced electronic moments are determined by both long- and short-range forces since $\mathbf{\Psi}^{DA}$ and $\mathbf{\Psi}^{DD}$ have components from both. Thus, in a sense, Eq. (4.273) represents a self-consistency requirement on the induced dipoles.

As usual we now seek to solve Eqs. (4.272) and (4.273) through the use of wavelike solutions:

$$
\mathbf{u}\binom{l}{k} = \mathbf{U}(k|\mathbf{q}) \exp\left\{ i\left[ \mathbf{q}\cdot\mathbf{x}\binom{l}{k} - \omega(\mathbf{q})t \right] \right\} \tag{4.275}
$$

$$
\mathbf{w}\binom{l}{k} = \mathbf{W}(k|\mathbf{q}) \exp\left\{ i\left[ \mathbf{q}\cdot\mathbf{x}\binom{l}{k} - \omega(\mathbf{q})t \right] \right\} \tag{4.276}
$$

We then obtain

$$
\omega^2(\mathbf{q})m_k U_\alpha(k|\mathbf{q}) = \sum_{k'\beta} \left\{ M_{\alpha\beta}^{AA}\binom{\mathbf{q}}{kk'} U_\beta(k'|\mathbf{q}) + M_{\alpha\beta}^{AD}\binom{\mathbf{q}}{kk'} W_\beta(k'|\mathbf{q}) \right\}, \tag{4.277}
$$

$$
0 = \sum_{k'\beta} \left\{ M_{\alpha\beta}^{DA}\binom{\mathbf{q}}{kk'} U_\beta(k'|\mathbf{q}) + \left[ M_{\alpha\beta}^{DD}\binom{\mathbf{q}}{kk'} + \delta_{kk'} K_{\alpha\beta}(k) \right] W_\beta(k'|\mathbf{q}) \right\}, \tag{4.278}
$$

where

$$
M_{\alpha\beta}^{\lambda\mu}\binom{\mathbf{q}}{kk'} = \sum_{l'} \Psi_{\alpha\beta}^{\lambda\mu}\binom{0\ l'}{k\ k'} \exp\left[ i\mathbf{q}\cdot\mathbf{x}\binom{l'\ 0}{k'\ k} \right], \quad \lambda = A, D; \mu = A, D. \tag{4.279}
$$

We next introduce the following commonly used notation:

$$
R_{\alpha\beta}\binom{\mathbf{q}}{kk'} = M_{\alpha\beta}^{AA,SR}\binom{\mathbf{q}}{kk'}
$$

$$T_{\alpha\beta}\begin{pmatrix} \mathbf{q} \\ kk' \end{pmatrix} = M_{\alpha\beta}^{\mathrm{AD,SR}}\begin{pmatrix} \mathbf{q} \\ kk' \end{pmatrix}$$

$$S_{\alpha\beta}\begin{pmatrix} \mathbf{q} \\ kk' \end{pmatrix} = \left[ M_{\alpha\beta}^{\mathrm{DD,SR}}\begin{pmatrix} \mathbf{q} \\ kk' \end{pmatrix} + \delta_{kk'} K_{\alpha\beta}(k) \right]. \tag{4.280}$$

Using the $\hat{\mathbf{C}}$ and $\mathbf{C}$ matrices defined in Appendix 2 (see (A2.23–A2.24)) we write the Coulomb matrices of the present model as follows:

$$M_{\alpha\beta}^{\mathrm{AA,C}}\begin{pmatrix} \mathbf{q} \\ kk' \end{pmatrix} = Z_k Z_{k'} e^2 \hat{C}_{\alpha\beta}\begin{pmatrix} \mathbf{q} \\ kk' \end{pmatrix}$$

$$= Z_k Z_{k'} e^2 C_{\alpha\beta}\begin{pmatrix} \mathbf{q} \\ kk' \end{pmatrix},\ k' \neq k; \tag{4.281a}$$

$$M_{\alpha\beta}^{\mathrm{AA,C}}\begin{pmatrix} \mathbf{q} \\ kk \end{pmatrix} = Z_k^2 e^2 \hat{C}_{\alpha\beta}\begin{pmatrix} \mathbf{q} \\ kk \end{pmatrix} - \sum_{k''} Z_k Z_{k''} e^2 \hat{C}_{\alpha\beta}\begin{pmatrix} 0 \\ kk'' \end{pmatrix}$$

$$= Z_k^2 e^2 C_{\alpha\beta}\begin{pmatrix} \mathbf{q} \\ kk \end{pmatrix}; \tag{4.281b}$$

$$M_{\alpha\beta}^{\mathrm{AD,C}}\begin{pmatrix} \mathbf{q} \\ kk' \end{pmatrix} = Z_k Y_{k'} e^2 C_{\alpha\beta}\begin{pmatrix} \mathbf{q} \\ kk' \end{pmatrix},\ k' \neq k; \tag{4.282a}$$

$$M_{\alpha\beta}^{\mathrm{AD,C}}\begin{pmatrix} \mathbf{q} \\ kk \end{pmatrix} = Z_k Y_k e^2 \hat{C}_{\alpha\beta}\begin{pmatrix} \mathbf{q} \\ kk \end{pmatrix} - \sum_{k''} Y_k Z_{k''} e^2 \hat{C}_{\alpha\beta}\begin{pmatrix} 0 \\ kk'' \end{pmatrix}$$

$$= Z_k Y_k e^2 C_{\alpha\beta}\begin{pmatrix} \mathbf{q} \\ kk \end{pmatrix}; \tag{4.282b}$$

$$M_{\alpha\beta}^{\mathrm{DA,C}}\begin{pmatrix} \mathbf{q} \\ kk' \end{pmatrix} = Y_k Z_{k'} e^2 C_{\alpha\beta}\begin{pmatrix} \mathbf{q} \\ kk' \end{pmatrix},\ k' \neq k; \tag{4.283a}$$

$$M_{\alpha\beta}^{\mathrm{DA,C}}\begin{pmatrix} \mathbf{q} \\ kk \end{pmatrix} = Y_k Z_k e^2 \hat{C}_{\alpha\beta}\begin{pmatrix} \mathbf{q} \\ kk \end{pmatrix} - \sum_{k''} Y_k Z_{k''} e^2 \hat{C}_{\alpha\beta}\begin{pmatrix} 0 \\ kk'' \end{pmatrix}$$

$$= Y_k Z_k e^2 C_{\alpha\beta}\begin{pmatrix} \mathbf{q} \\ kk \end{pmatrix}; \tag{4.283b}$$

$$M_{\alpha\beta}^{\mathrm{DD,C}}\begin{pmatrix} \mathbf{q} \\ kk' \end{pmatrix} = Y_k Y_{k'} e^2 C_{\alpha\beta}\begin{pmatrix} \mathbf{q} \\ kk' \end{pmatrix},\ k' \neq k; \tag{4.284a}$$

$$M_{\alpha\beta}^{\mathrm{DD,C}}\begin{pmatrix} \mathbf{q} \\ kk \end{pmatrix} = Y_k^2 e^2 \hat{C}_{\alpha\beta}\begin{pmatrix} \mathbf{q} \\ kk \end{pmatrix} - \sum_{k''} Y_k Z_{k''} e^2 \hat{C}_{\alpha\beta}\begin{pmatrix} 0 \\ kk'' \end{pmatrix}$$

$$= Y_k^2 e^2 C_{\alpha\beta}\begin{pmatrix} \mathbf{q} \\ kk \end{pmatrix} + \sum_{k''} Y_k Z_{k''} \left( \frac{Y_k}{Z_k} - 1 \right) e^2 \hat{C}_{\alpha\beta}\begin{pmatrix} 0 \\ kk'' \end{pmatrix}. \tag{4.284b}$$

The above matrices may be written as

$$\mathbf{M}^{\mathrm{AA,C}}(\mathbf{q}) = \mathbf{Z}_d \mathbf{C} \mathbf{Z}_d, \tag{4.285}$$

$$\mathbf{M}^{\mathrm{AD,C}}(\mathbf{q}) = \mathbf{Z}_d \mathbf{C} \mathbf{Y}_d, \tag{4.286}$$

$$\mathbf{M}^{\mathrm{DA,C}}(\mathbf{q}) = \mathbf{Y}_d \mathbf{C} \mathbf{Z}_d, \tag{4.287}$$

$$M_{\alpha\beta}^{DD,C}\begin{pmatrix} \mathbf{q} \\ kk' \end{pmatrix} = (\mathbf{Y}_d \mathbf{C} \mathbf{Y}_d)_{\alpha\beta}\begin{pmatrix} \mathbf{q} \\ kk' \end{pmatrix}$$

$$+ \delta_{kk'} \sum_{k''} Y_k Z_{k''} e^2 \left( \frac{Y_k}{Z_k} - 1 \right) \hat{C}_{\alpha\beta}\begin{pmatrix} 0 \\ kk'' \end{pmatrix}, \tag{4.288}$$

where $\mathbf{Y}_d$ is defined similarly to $\mathbf{Z}_d$ (see Eq. (A2.26)). Observe that in diagonally cubic crystals $\hat{C}_{\alpha\beta}\begin{pmatrix} 0 \\ kk'' \end{pmatrix} = -(4\pi/3v)\delta_{\alpha\beta}$ so that, due to the electrical neutrality of the primitive cell, the second terms of (4.284b) and (4.288) vanish, allowing $\mathbf{M}^{DD,C}(\mathbf{q})$ to be written in the same form as (4.285–4.287).

Following Cochran[62] and Traylor et al.[63] we define a matrix $\boldsymbol{\zeta}$ by

$$\zeta_{\alpha\beta}\begin{pmatrix} \mathbf{q} \\ kk' \end{pmatrix} = S_{\alpha\beta}\begin{pmatrix} \mathbf{q} \\ kk' \end{pmatrix} + \delta_{kk'} \sum_{k''} Y_k Z_{k''} e^2 \left( \frac{Y_k}{Z_k} - 1 \right) \hat{C}_{\alpha\beta}\begin{pmatrix} 0 \\ kk'' \end{pmatrix}, \tag{4.289}$$

which permits us to reexpress the equations of motion in the form often found in the literature:

$$\omega^2(\mathbf{q})m\mathbf{U} = [\mathbf{R} + \mathbf{Z}_d \mathbf{C} \mathbf{Z}_d]\mathbf{U} + [\mathbf{T} + \mathbf{Z}_d \mathbf{C} \mathbf{Y}_d]\mathbf{W} \tag{4.290}$$

$$0 = [\mathbf{T}^\dagger + \mathbf{Y}_d \mathbf{C} \mathbf{Z}_d]\mathbf{U} + [\boldsymbol{\zeta} + \mathbf{Y}_d \mathbf{C} \mathbf{Y}_d]\mathbf{W}. \tag{4.291}$$

In writing (4.291), we have used the relation

$$\mathbf{M}^{DA,SR}(\mathbf{q}) = \mathbf{T}^\dagger.$$

Using (4.291) to eliminate $\mathbf{W}$, we obtain finally

$$m\omega^2(\mathbf{q})\mathbf{U} = [\mathbf{R} + \mathbf{Z}_d \mathbf{C} \mathbf{Z}_d]\mathbf{U}$$
$$- \{[\mathbf{T} + \mathbf{Z}_d \mathbf{C} \mathbf{Y}_d][\boldsymbol{\zeta} + \mathbf{Y}_d \mathbf{C} \mathbf{Y}_d]^{-1}[\mathbf{T}^\dagger + \mathbf{Y}_d \mathbf{C} \mathbf{Z}_d]\}\mathbf{U}. \tag{4.292}$$

Two important conclusions may be drawn from this result.

1. If we make the reasonable identification that the term $[\mathbf{R} + \mathbf{Z}_d \mathbf{C} \mathbf{Z}_d]$ is the dynamical matrix appropriate to the Born model, then the term in curly brackets represents explicitly the additional contribution arising from electronic polarization.

2. The long-range contributions associated with both terms have regular and nonregular parts in the sense discussed earlier. These arise from the fact that each term in $\mathbf{C}$ contains a contribution of the form $(4\pi/v) \times (q_\alpha q_\beta/|\mathbf{q}|^2)$. Whereas the nonregular part associated with the first term has the form previously given in (4.28) (as expected), that arising from the second term has a complicated structure. Together the two nonregular terms lead to the $\mathbf{Z}(\mathbf{q})$ matrix introduced in (4.76).

An explicit expression for $\mathbf{Z}(\mathbf{q})$ in this model may be obtained as follows. We have from (4.277–4.278)

$$\mathbf{m}\omega^2\mathbf{U} = \mathbf{M}^{AA}\mathbf{U} + \mathbf{M}^{AD}\mathbf{W}, \tag{4.293}$$
$$0 = \mathbf{M}^{DA}\mathbf{U} + \mathbf{M}^{DD}\mathbf{W}. \tag{4.294}$$

Next we define a $3n$-component polarization vector $\mathbf{P}$ as

$$\mathbf{P} = \frac{1}{v}(\mathbf{Z}_d\mathbf{U} + \mathbf{Y}_d\mathbf{W}). \tag{4.295}$$

The macroscopic field is then given by

$$\mathbf{E} = -4\pi\,\hat{\mathbf{q}}\left(\hat{\mathbf{q}}\cdot\sum_k\mathbf{P}(k)\right), \tag{4.296}$$

where $\mathbf{P}(k)$ is a three-dimensional vector taken from $\mathbf{P}$:

$$\mathbf{P} = \begin{pmatrix} \mathbf{P}(1) \\ \vdots \\ \mathbf{P}(k) \\ \vdots \\ \mathbf{P}(n) \end{pmatrix}. \tag{4.297}$$

The macroscopic field is the same at every site in the primitive cell, enabling us to define a $3n$-dimensional macroscopic field vector $\mathbf{E}_{mac}$ by

$$E_{mac,\alpha}(k) = E_\alpha. \tag{4.298}$$

We then can write (4.290–4.291) as

$$\mathbf{m}\omega^2\mathbf{U} = (\mathbf{R} + \overline{\mathbf{M}}^{AA,C})\mathbf{U} + (\mathbf{T} + \overline{\mathbf{M}}^{AD,C})\mathbf{W} - \mathbf{Z}_d\mathbf{E}_{mac}, \tag{4.299}$$
$$0 = (\mathbf{T}^\dagger + \overline{\mathbf{M}}^{DA,C})\mathbf{U} + (\zeta + \overline{\mathbf{M}}^{DD,C})\mathbf{W} - \mathbf{Y}_d\mathbf{E}_{mac}. \tag{4.300}$$

The matrices $\overline{\mathbf{M}}^{AD,C}$, etc. are derived from the matrices $\mathbf{M}^{AD,C}$, etc. by removing the nonregular parts.

Using (4.300) to eliminate $\mathbf{W}$, we may write (4.299) as

$$\mathbf{m}\omega^2\mathbf{U} = [(\mathbf{R} + \overline{\mathbf{M}}^{AA,C}) - (\mathbf{T} + \overline{\mathbf{M}}^{AD,C})(\zeta + \overline{\mathbf{M}}^{DD,C})^{-1}$$
$$\times (\mathbf{T}^\dagger + \overline{\mathbf{M}}^{DA,C})]\mathbf{U} - [\mathbf{Z}_d - (\mathbf{T} + \overline{\mathbf{M}}^{AD,C})(\zeta + \overline{\mathbf{M}}^{DD,C})^{-1}\mathbf{Y}_d]$$
$$\times \mathbf{E}_{mac}, \tag{4.301}$$

in which the first square bracket on the right-hand side is entirely regular. Comparison with (4.76) then shows that in the shell model,

$$\overline{\mathbf{M}}(\mathbf{q}) = (\mathbf{R} + \overline{\mathbf{M}}^{AA,C}) - (\mathbf{T} + \overline{\mathbf{M}}^{AD,C})(\zeta + \overline{\mathbf{M}}^{DD,C})^{-1}$$
$$\times (\mathbf{T}^\dagger + \overline{\mathbf{M}}^{DA,C}). \tag{4.302}$$

To obtain an expression for the quantity $\mathbf{Z}^\dagger(\mathbf{q})$ occurring in (4.76), it is necessary to recast the second term on the right-hand side of (4.301) in the form $\mathbf{AE}$, where $\mathbf{A}$ is a $(3n \times 3)$ matrix assembled from the $(3n \times 3n)$

matrix $[\mathbf{Z}_d - (\mathbf{T} + \overline{\mathbf{M}}^{AD,C})(\zeta + \overline{\mathbf{M}}^{DD})^{-1}\mathbf{Y}_d]$ by appropriate summation in each row. When this is done,

$$[\mathbf{Z}^\dagger(\mathbf{q})]_{\alpha\beta}(k) = \sum_{k'} [\mathbf{Z}_d - (\mathbf{T} + \overline{\mathbf{M}}^{AD,C})\,(\zeta + \overline{\mathbf{M}}^{DD,C})^{-1}\mathbf{Y}_d]_{\alpha\beta}\binom{\mathbf{q}}{kk'}.$$

(4.303)

One may independently deduce $\mathbf{Z}(\mathbf{q})$ from Eq. (4.295) by an analogous procedure and verify that the result is consistent with (4.303).

These identifications in conjunction with the results derived in Sec. 4.3 permit explicit expressions for the elastic and piezoelectric constants in the shell model to be derived.

It is useful to recast some of the preceding discussion in terms of local fields, as we did earlier for the rigid-ion model. Using a treatment analogous to that given in Appendix 2, the Coulombic parts of the equations of motion (4.272) and (4.273) can be reexpressed in terms of suitably defined local fields to give:

$$m_k \ddot{u}_\alpha\binom{l}{k} = -\sum_{l'k'\beta}\left[\Psi_{\alpha\beta}^{AA,SR}\binom{l\ l'}{k\ k'}u_\beta\binom{l'}{k'} + \Psi_{\alpha\beta}^{AD,SR}\binom{l\ l'}{k\ k'}w_\beta\binom{l'}{k'}\right]$$
$$+ Z_k e \mathscr{E}_\alpha^c\binom{l}{k};$$

(4.304)

$$0 = -\sum_{l'k'\beta}\left[\Psi_{\alpha\beta}^{DA,SR}\binom{l\ l'}{k\ k'}u_\beta\binom{l'}{k'} + \left\{\Psi_{\alpha\beta}^{DD,SR}\binom{l\ l'}{k\ k'}\right.\right.$$
$$\left.\left. + \delta_{ll'}\delta_{kk'}K_{\alpha\beta}(k)\right\}w_\beta\binom{l'}{k'}\right] + Y_k e \mathscr{E}_\alpha^s\binom{l}{k}.$$

(4.305)

The local field $\mathscr{E}^c\binom{l}{k}$ acting on the core at $\binom{l}{k}$ is composed of
1. the field due to displacement dipoles $Z_{k'}e\mathbf{u}\binom{l'}{k'}$ and electronic dipoles $Y_{k'}e\mathbf{w}\binom{l'}{k'}$ associated with other sites and
2. the field due to the dipoles $-\{Z_k e\mathbf{u}\binom{l}{k} + Z_k e(Y_k/Z_k)\mathbf{w}\binom{l}{k}\}$. The latter are a manifestation of the self-term effects. Explicitly $\mathscr{E}^c\binom{l}{k}$ is given by

$$\mathscr{E}_\alpha^c\binom{l}{k} = \sum_{\substack{l'k'\beta \\ (lk \neq l'k')}} \mathscr{D}\left\{Z_{k'}eu_\beta\binom{l'}{k'} - Z_k eu_\beta\binom{l}{k}\right.$$
$$\left. + Y_{k'}ew_\beta\binom{l'}{k'} - Z_k e(Y_k/Z_k)w_\beta\binom{l}{k}\right\}.$$

(4.306)

In a similar fashion $\mathscr{E}^s\binom{l}{k}$ denotes the local field acting on the shell $\binom{l}{k}$ and, like $\mathscr{E}^c\binom{l}{k}$, is composed of two parts. The first of these is the same as above; the second, again closely connected with self-term effects, differs from the corresponding term in (4.306):

$$\mathscr{E}^s_\alpha\binom{l}{k} = \sum_{\substack{l'k'\beta \\ (lk \neq l'k')}} \mathscr{D}\left\{ Z_{k'}eu_\beta\binom{l'}{k'} - Z_{k'}eu_\beta\binom{l}{k} + Y_{k'}ew_\beta\binom{l'}{k'} - Z_{k'}ew_\beta\binom{l}{k} \right\}.$$

$$(4.307)$$

The different forms in which the effects of the self contribution appear in the last two equations is a peculiar feature of the shell model and reflect the differing status of the core displacement **u** and *relative* core-shell displacement **w**.

In diagonally cubic crystals

$$\mathscr{E}^c_\alpha\binom{l}{k} = \mathscr{E}^s_\alpha\binom{l}{k} = \sum_{\substack{l'k'\beta \\ (lk \neq l'k')}} \mathscr{D}\left\{ Z_{k'}eu_\beta\binom{l'}{k'} + Y_{k'}ew_\beta\binom{l'}{k'} \right\}, \qquad (4.308)$$

since the "self contribution" from Coulomb forces vanishes.

It is useful to note that $\Phi^{(2)\mathrm{LR}}$ itself can be expressed in terms of local fields as

$$\Phi^{(2)\mathrm{LR}} = -\tfrac{1}{2}\sum_{lk\alpha}\left[ Z_k eu_\alpha\binom{l}{k}\mathscr{E}^c_\alpha\binom{l}{k} + Y_k ew_\alpha\binom{l}{k}\mathscr{E}^s_\alpha\binom{l}{k} \right], \qquad (4.309)$$

and, using this, Eqs. (4.304) and (4.305) may be derived from the usual equations

$$m_k\ddot{u}_\alpha\binom{l}{k} = -\frac{\partial\Phi^{(2)}}{\partial u_\alpha\binom{l}{k}}, \quad 0 = \frac{\partial\Phi^{(2)}}{\partial w_\alpha\binom{l}{k}}, \qquad (4.310)$$

if one remembers that $\mathscr{E}^c$ and $\mathscr{E}^s$ themselves depend on the displacements as in (4.306–4.307).

The **q**-space versions of (4.304) and (4.305) are obtained in a straightforward fashion to be:

$$m\omega^2\mathbf{U} = \mathbf{R}\mathbf{U} + \mathbf{T}\mathbf{W} - \mathbf{Z_d}\mathscr{E}^c \qquad (4.311)$$

$$0 = \mathbf{T}^\dagger\mathbf{U} + \mathbf{S}\mathbf{W} - \mathbf{Y_d}\mathscr{E}^s, \qquad (4.312)$$

where $\mathscr{E}^c$ and $\mathscr{E}^s$ are $3n$-component column matrices with the following components:

$$\mathscr{E}^c_\alpha(k|\mathbf{q}) = -\sum_{k'\beta} C_{\alpha\beta}\binom{\mathbf{q}}{kk'}\{Z_{k'}eU_\beta(k'|\mathbf{q}) + Y_{k'}eW_\beta(k'|\mathbf{q})\} \qquad (4.313)$$

$$\mathscr{E}^s_\alpha(k|\mathbf{q}) = -\sum_{k'\beta}\left[ C_{\alpha\beta}\binom{\mathbf{q}}{kk'}\{Z_{k'}eU_\beta(k'|\mathbf{q}) + Y_{k'}eW_\beta(k'|\mathbf{q})\} \right.$$
$$\left. + \delta_{kk'}\left\{\sum_{k''}\left(\frac{Y_k}{Z_k} - 1\right)Z_{k''}e\hat{C}_{\alpha\beta}\binom{0}{kk''}\right\}W_\beta(k'|\mathbf{q}) \right]. \qquad (4.314)$$

Equations (4.311) and (4.312) have been employed by Cowley[64] to calculate the polarizability $\alpha$ of the unit cell for diagonally cubic crystals (for which $\mathscr{E}^c = \mathscr{E}^s$). The high-frequency polarizability $\alpha(\infty)$ deduced (by suppressing the lattice contribution) is related to $\varepsilon(\infty)$ by the Lorenz-Lorentz[65] formula*

$$\frac{4\pi}{3v}\alpha(\infty) = \frac{\varepsilon(\infty) - 1}{\varepsilon(\infty) + 2}. \tag{4.315}$$

The low-frequency polarizability $\alpha(0)$ is related to $\varepsilon(0)$ by the Clausius-Mossotti[65] relation†

$$\frac{4\pi}{3v}\alpha(0) = \frac{\varepsilon(0) - 1}{\varepsilon(0) + 2}. \tag{4.316}$$

Here we take note of an interesting feature of the shell equation (4.312). Let us rewrite it as

$$\mathbf{W} = (-\mathbf{S}^{-1}\mathbf{T}^{\dagger})\mathbf{U} + \mathbf{S}^{-1}\mathbf{Y}_d\mathscr{E}^s, \tag{4.317}$$

and deduce from it the amplitude of the electronic dipole moment, associated with wave vector $\mathbf{q}$, resident on the $k$th ion in the cell:

$$p_\alpha(k|\mathbf{q}) = Y_k e W_\alpha(k|\mathbf{q}) = Y_k e \sum_{k'\beta} [-\mathbf{S}^{-1}\mathbf{T}^{\dagger}]_{\alpha\beta}(kk') U_\beta(k'|\mathbf{q})$$
$$+ Y_k e \sum_{k'\beta} [\mathbf{S}^{-1}\mathbf{Y}_d]_{\alpha\beta}(kk') \mathscr{E}^s_\beta(k'|\mathbf{q}). \tag{4.318}$$

It is important to note that the long-range contribution (that is, the second term) is not quite in the form (4.230) in the sense that, for every ion of type $k$, it involves the local fields acting on types $k' \neq k$ as well. We emphasize this point because the $\mathbf{q} \to 0$ limit of (4.318) has been employed on occasion to define the electronic polarizability $\alpha_k$ (as modified by the crystalline environment) in diagonally cubic crystals. Here, in addition to $\mathscr{E}^c = \mathscr{E}^s$,

$$\mathscr{E}^s(k|\mathbf{q} \to 0) = \mathscr{E}, \text{ say, for all } k. \tag{4.319}$$

In this limit the second term on the right-hand side of (4.318) may therefore be written as

$$Y_k e \sum_{\beta k'} \{[\mathbf{S}^{-1}\mathbf{Y}_d]_{\alpha\beta}(kk')\}\mathscr{E}_\beta = \sum_\beta [\boldsymbol{\alpha}_k]_{\alpha\beta}\mathscr{E}_\beta. \tag{4.320}$$

The circumstances noted above do not obtain in general, so that for crystals of arbitrary structure it is not possible to define $\boldsymbol{\alpha}_k$ by this procedure.

---

*For a historical footnote concerning these relations see M. Born and E. Wolf, *Principles of Optics* (Pergamon, New York, 1970) 4th Ed., p. 87.
†Ibid.

The physical reason for this difficulty is quite easily seen. Examining (4.318), we find that the electronic dipole moment of the $k$th ion is influenced not only by the local field acting on the $k$th ion itself but also by that acting on the neighbors, because the dipoles induced on ions $k'$ ($\neq k$) can affect the moment of the $k$th ion itself via short-range shell-shell forces. If such forces are absent

$$S_{\alpha\beta}\left(\begin{matrix} \mathbf{q} \\ kk' \end{matrix}\right) = \delta_{kk'} K_{\alpha\beta}(k)$$

(see (4.280)), that is, $\mathbf{S}$ is diagonal in the sublattice index $k$. Therefore the matrix $\mathbf{S}^{-1}\mathbf{Y}_d$ is likewise (remember $\mathbf{Y}_d$ is also diagonal) and the second term on the right-hand side of (4.318) can be written in the form

$$\sum_\beta [\alpha_k]_{\alpha\beta} \mathscr{E}_\beta^s(k|\mathbf{q}).$$

Thus in other than diagonally cubic crystals the Lorentz approximation holds only if the shell-shell forces are absent.

### 4.6.5 Mashkevich-Tolpygo model

Many of the features of the shell model were in fact anticipated much earlier by Tolpygo.[57] Motivated by the work of Szigeti[48,49] and Odelevski,[50] Tolpygo examined quantum mechanically the electronic contribution to the vibrational energy in alkali halides. He then deduced that, subject to certain approximations, $\Phi^{(2)}$ can be expressed as a quadratic form in ionic displacements $\mathbf{u}\binom{l}{k}$ and induced dipoles $\mathbf{p}\binom{l}{k}$. Later Mashkevich and Tolpygo[58] showed that diamond-type crystals could be treated similarly.

By considering the potential energy function for cubic diatomic crystals as proposed by Mashkevich and Tolpygo, we may appreciate the close similarity of their work to that of Cochran and coworkers:

$$
\begin{aligned}
\Phi^{(2)} = \tfrac{1}{2}\sum_{lk\alpha}\Bigg[ & \Bigg\{ \sum_{l'k'\beta} \phi_{\alpha\beta}^R \binom{l\ l'}{k\ k'} u_\alpha \binom{l}{k} u_\beta \binom{l'}{k'} \\
& + \phi_{\alpha\beta}^T \binom{l\ l'}{k\ k'} u_\alpha \binom{l}{k} p_\beta \binom{l'}{k'} + \phi_{\alpha\beta}^T \binom{l\ l'}{k\ k'} p_\alpha \binom{l}{k} u_\beta \binom{l'}{k'} \\
& + \phi_{\alpha\beta}^S \binom{l\ l'}{k\ k'} p_\alpha \binom{l}{k} p_\beta \binom{l'}{k'} \Bigg\} + \alpha_k^{-1} p_\alpha^2 \binom{l}{k} \Bigg] \\
& - \tfrac{1}{2}\sum_{lk\alpha}\left[ p_\alpha\binom{l}{k} + Z_k e u_\alpha\binom{l}{k} \right] \mathscr{E}_\alpha\binom{l}{k}.
\end{aligned}
\tag{4.321}
$$

Here $\phi_{\alpha\beta}^R$, $\phi_{\alpha\beta}^T$, and $\phi_{\alpha\beta}^S$ denote short-range force constants, and $\alpha_k$ represents the electronic polarizability of the ion in the $k$th sublattice. The

Coulombic part of the interaction energy is expressed entirely by the second group of terms and arises from the electrostatic interaction between the various ionic charges, between the ions and the induced dipoles, and amongst the dipoles themselves. For convenience this part of the energy is expressed in terms of the local field $\mathscr{E}\binom{l}{k}$ acting at the site $\binom{l}{k}$. This is given by

$$\mathscr{E}_\alpha\binom{l}{k} = \sum_{\substack{l'k'\beta \\ (l'k' \neq lk)}} \mathscr{D}\left[ Z_{k'}eu_\beta\binom{l'}{k'} + p_\beta\binom{l'}{k'} \right] \tag{4.322}$$

and is seen to be of the same form as (4.308).

The equations of motion for $\mathbf{u}\binom{l}{k}$ and $\mathbf{p}\binom{l}{k}$ are given by

$$m_k \ddot{u}_\alpha\binom{l}{k} = -\frac{\partial \Phi^{(2)}}{\partial u_\alpha\binom{l}{k}}, \tag{4.323}$$

$$0 = -\frac{\partial \Phi^{(2)}}{\partial p_\alpha\binom{l}{k}}. \tag{4.324}$$

The latter embodies the adiabatic assumption and leads to a self-consistency requirement on the induced dipoles exactly as in the shell model.

To see explicitly the equivalence of the present model and the shell model, we simply note that if we adopt the shell-core picture to describe the induced dipoles, then $\mathbf{p}\binom{l}{k} = Y_k e\mathbf{w}\binom{l}{k}$. Substituting this in (4.321) we get

$$\begin{aligned}
\Phi^{(2)} = \tfrac{1}{2}\sum_{lk\alpha}\Bigg[ \sum_{l'k'\beta} \Bigg\{ &\phi^{\mathrm{R}}_{\alpha\beta}\binom{l\ l'}{k\ k'}u_\alpha\binom{l}{k}u_\beta\binom{l'}{k'} \\
&+ Y_{k'}e\phi^{\mathrm{T}}_{\alpha\beta}\binom{l\ l'}{k\ k'}u_\alpha\binom{l}{k}w_\beta\binom{l'}{k'} + Y_k e\phi^{\mathrm{T}}_{\alpha\beta}\binom{l\ l'}{k\ k'}w_\alpha\binom{l}{k}u_\beta\binom{l'}{k'} \\
&+ Y_k Y_{k'}e^2\phi^{\mathrm{S}}_{\alpha\beta}\binom{l\ l'}{k\ k'}w_\alpha\binom{l}{k}w_\beta\binom{l'}{k'} \Bigg\} + \frac{(Y_k e)^2}{\alpha_k}w_\alpha^2\binom{l}{k}\Bigg] \\
&- \tfrac{1}{2}\sum_{lk\alpha}\Bigg[ Z_k eu_\alpha\binom{l}{k} + Y_k ew_\alpha\binom{l}{k}\Bigg]\mathscr{E}_\alpha\binom{l}{k}.
\end{aligned} \tag{4.325}$$

Next we tailor (4.262) to cubic diatomic crystals using the form (4.309) for $\Phi^{(2)\mathrm{LR}}$ together with the simplifications cited in (4.308). Noting that $\mathbf{K}(k) = K(k)\mathbb{1}_3$ for cubic crystals, this gives

$$\begin{aligned}
\Phi^{(2)} = \tfrac{1}{2}\sum_{lk\alpha}\Bigg[ \sum_{l'k'}\Bigg\{ &\Psi^{\mathrm{AA,SR}}_{\alpha\beta}\binom{l\ l'}{k\ k'}u_\alpha\binom{l}{k}u_\beta\binom{l'}{k'} \\
&+ \Psi^{\mathrm{AD,SR}}_{\alpha\beta}\binom{l\ l'}{k\ k'}u_\alpha\binom{l}{k}w_\beta\binom{l'}{k'} + \Psi^{\mathrm{DA,SR}}_{\alpha\beta}\binom{l\ l'}{k\ k'}w_\alpha\binom{l}{k}u_\beta\binom{l'}{k'}
\end{aligned}$$

$$+ \ \Psi_{\alpha\beta}^{DD,SR}\begin{pmatrix} l & l' \\ k & k' \end{pmatrix} w_\alpha\begin{pmatrix} l \\ k \end{pmatrix} w_\beta\begin{pmatrix} l' \\ k' \end{pmatrix} \Bigg\} + K(k)w_\alpha^2\begin{pmatrix} l \\ k \end{pmatrix} \Bigg]$$

$$- \tfrac{1}{2}\sum_{lk\alpha} \Bigg[ Z_k e u_\alpha\begin{pmatrix} l \\ k \end{pmatrix} + Y_k e w_\alpha\begin{pmatrix} l \\ k \end{pmatrix} \Bigg] \mathscr{E}_\alpha\begin{pmatrix} l \\ k \end{pmatrix}. \tag{4.326}$$

If we make the assumption

$$\Psi_{\alpha\beta}^{AD,SR}\begin{pmatrix} l & l' \\ k & k' \end{pmatrix} = \Psi_{\alpha\beta}^{DA,SR}\begin{pmatrix} l & l' \\ k & k' \end{pmatrix}$$

(see Fig. 4.13) and compare (4.325) and (4.326), the equivalence of Tolpygo's approach with the shell model is apparent.

In practical applications, the potential function (4.321) is treated as phenomenological, the quantities $\phi_{\alpha\beta}^R$, $\phi_{\alpha\beta}^T$, $\phi_{\alpha\beta}^S$, and $\alpha_k$ being treated as parameters.

A historical footnote must be added. In his original work on alkali halides, Tolpygo[57] did not include in $\Phi^{(2)}$ the term involving $\phi_{\alpha\beta}^S(k\ k') \times p_\alpha(\tfrac{l}{k})p_\beta(\tfrac{l'}{k'})$, that is, short-range coupling between neighboring dipoles was not allowed for. A similar omission was also made in the case of the crystals of the diamond family. The importance of the omitted terms was first emphasized by Cochran[52] during the process of establishing the equivalence of the shell model with the work of the Soviet school. Tolpygo[59] subsequently amended his work, upon which the equivalence of his theory with the shell model became complete.

### 4.6.6 Deformation-dipole model

Another approach which allows for polarization effects is that due to Hardy,[41] whose model is often referred to as the *deformation-dipole model*. The idea underlying the deformation-dipole concept may be understood by referring to Figure 4.15, which shows a typical negative ion in an alkali halide, together with its nearest neighbors. Strong overlap produces overall charge rearrangement leading to concentrations of positive charges at the points marked. When the ions are relatively displaced, the overlap changes and leads to a different charge distribution. The latter can be described formally in terms of various multipoles. However, only the dipolar part is retained in the model, and the induced dipoles are assumed to reside on the ions themselves. Thus if the negative ion in Figure 4.15 is moved a distance $u_x$ in the $x$ direction, then one supposes that it results in the ion acquiring a net moment

$$-2au_x - 4bu_x. \tag{4.327}$$

The contribution $-2au_x$ arises from the changes in overlap of the pairs

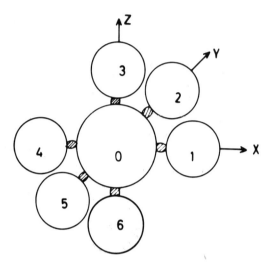

**Figure 4.15**
Deformation dipoles surrounding a negative ion in alkali halides as envisaged by Hardy.
(After Hardy, ref. 41.)

$(0, 1)$ and $(0, 4)$, while $-4bu_x$ is associated with the pairs formed by combining 0 with the remaining first neighbors. That different constants $a$ and $b$ must appear for the two sets of pairs is understandable since they involve radial and tangential forces respectively. Born and Huang[56] have shown how the above concept may be given more formal expression.

Let the dipole moment between a positive and a negative ion be denoted by $m(r)$, a function of their separation only. The sign of $m(r)$ is chosen positive if the moment is directed from negative towards positive ion. If $\mathbf{u}(0)$ denotes the displacement of the central ion and $\mathbf{u}(i)$, $(i = 1, \ldots, 6)$, that of the neighbors, then the total deformation-dipole moment is

$$\sum_{i=1}^{6} m(|\mathbf{x}(i) + \mathbf{u}(i) - \mathbf{u}(0)|) \left[ \frac{\mathbf{x}(i) + \mathbf{u}(i) - \mathbf{u}(0)}{|\mathbf{x}(i) + \mathbf{u}(i) - \mathbf{u}(0)|} \right], \tag{4.328}$$

where the $\mathbf{x}(i)$'s denote the equilibrium positions. The unit vectors in the square brackets serve to specify the directions of the individual contributions. As stated above, the total moment (4.328) is assumed to reside at the equilibrium position of the negative ion. Since the displacements are usually small, in practice one retains in (4.328) only the part linear in displacements. To obtain this we use the expansions

$$m(|\mathbf{x} + \mathbf{u}|) \approx m(|\mathbf{x}|) + \frac{m'(|\mathbf{x}|)}{|\mathbf{x}|} (\mathbf{u} \cdot \mathbf{x})$$

and

$$\frac{1}{|\mathbf{x} + \mathbf{u}|} \approx \frac{1}{|\mathbf{x}|}\left(1 - \frac{\mathbf{u} \cdot \mathbf{x}}{x^2}\right)$$

in (4.328) and retain only terms linear in the displacement. The $\alpha$th component of the deformation moment is then

$$v_\alpha = \sum_i \sum_\beta \left[-\frac{m}{x(i)}\left\{\delta_{\alpha\beta} - \frac{x_\alpha(i)x_\beta(i)}{x^2(i)}\right\} - m'\frac{x_\alpha(i)x_\beta(i)}{x^2(i)}\right](u_\beta(0) - u_\beta(i)).$$

(4.329)

For $\alpha = x$, the above expression reduces to (4.327) when we set $\mathbf{u}(0) = (u_x, 0, 0)$, $\mathbf{u}(i) = 0 (i = 1, \ldots, 6)$, $m = br_0$ ($r_0$ being the nearest-neighbor distance) and $m' = a$. The $y$ and $z$ components vanish.

For convenience let us now introduce a function $\psi(r)$ such that $\psi'(r) = m(r)$. Then the term in square brackets in (4.329) has the form

$$\left[-\frac{\psi'(r)}{r}\left\{\delta_{\alpha\beta} - \frac{r_\alpha r_\beta}{r^2}\right\} - \psi''(r)\frac{r_\alpha r_\beta}{r^2}\right].$$

(4.330)

By expressing the deformation dipole moment in terms of $\psi(r)$, one can treat the latter on the same footing as a short-range potential $V(r)$ and use suitably parameterized forms for it (see Eq. (2.225)).

In our discussion we shall not use (4.329) but rather express the distortion moment by

$$v_\alpha\binom{l}{k} = \sum_{\substack{l'k'\beta \\ (l'k' \neq lk)}} \gamma_{\alpha\beta}\binom{l \ l'}{k \ k'}\left[u_\beta\binom{l'}{k'} - u_\beta\binom{l}{k}\right].$$

(4.331)

In practice, only a small set of neighbors is included in the summation. Observe that $\mathbf{v}$ vanishes if the crystal is given a uniform displacement. The tensor $\gamma$ is assumed to have the properties

$$\gamma_{\alpha\beta}\binom{l \ l'}{k \ k'} = \gamma_{\beta\alpha}\binom{l' \ l}{k' \ k}$$

(4.332a)

$$\gamma_{\alpha\beta}\binom{l \ l'}{k \ k'} = \gamma_{\alpha\beta}\binom{l+L \ \ l'+L}{k \ \ \ \ k'}.$$

(4.332b)

Supposing that it has elements

$$\gamma_{\alpha\beta}\binom{l \ l'}{k \ k'} = -\frac{\partial^2\psi(r)}{\partial r_\alpha \partial r_\beta}\bigg|_{r=|x(\begin{smallmatrix}l' \ l \\ k' \ k\end{smallmatrix})|}$$

(4.333)

(see (2.225)), we can recover from (4.331) the Born-Huang result (4.329).

The deformation moment specified by (4.331) represents only a part of the electronic moment associated with the site $\binom{l}{k}$. An additional moment $\boldsymbol{\mu}\binom{l}{k}$ due to the local field and given by

$$\boldsymbol{\mu}\binom{l}{k} = \alpha_k \mathscr{E}\binom{l}{k} \tag{4.334}$$

(see (4.230)) is assumed to exist, so that the net electronic moment is

$$\mathbf{p}\binom{l}{k} = \boldsymbol{\mu}\binom{l}{k} + \mathbf{v}\binom{l}{k}. \tag{4.335}$$

Originally, the deformation-dipole model was formulated explicitly for the alkali halides. In view of its recent application to crystals of more complicated structures,[66] we shall present here a generalized treatment applicable to ionic crystals of arbitrary structure. Our presentation differs slightly from that of Hardy but has the advantage of being readily comparable with the earlier discussion.

We start with the expression

$$\Phi^{(2)} = \left( \left[ -\tfrac{1}{2} \sum_{\substack{lk\alpha \ l'k'\beta \\ (lk \neq l'k')}} \left\{ Z_k e u_\alpha\binom{l}{k} + p_\alpha\binom{l}{k} \right\} \mathscr{D} \left\{ Z_{k'} e u_\beta\binom{l'}{k'} + p_\beta\binom{l'}{k'} \right\} \right] \right.$$
$$\left. + \left[ \sum_{\substack{lk\alpha \ l'k'\beta \\ (lk \neq l'k')}} \left\{ \tfrac{1}{2} Z_k e u_\alpha\binom{l}{k} \mathscr{D} Z_{k'} e u_\beta\binom{l}{k} + Z_{k'} e \mathscr{D} u_\alpha\binom{l}{k} p_\beta\binom{l}{k} \right\} \right] \right)$$
$$+ \tfrac{1}{2} \sum_{lk\alpha} \alpha_k^{-1} \mu_\alpha^2\binom{l}{k} + \tfrac{1}{2} \sum_{lk\alpha \ l'k'\beta} \phi_{\alpha\beta}^{\mathrm{SR}}\binom{l \ l'}{k \ k'} u_\alpha\binom{l}{k} u_\beta\binom{l'}{k'}. \tag{4.336}$$

Here the terms in square brackets represent the electrostatic part of the energy, the form of which may be confirmed as follows.

Consider a set of charges $Z_k e$ and dipoles $\mathbf{p}\binom{l}{k}$ distributed at sites $\binom{l}{k}$. The electrostatic energy of this array of monopoles and dipoles is given by

$$\Phi^{\mathrm{LR}} = \tfrac{1}{2} \sum_{\substack{lk \ l'k' \\ (lk \neq l'k')}} \frac{Z_k Z_{k'} e^2}{\left|\mathbf{r}\binom{l \ l'}{k \ k'}\right|}$$
$$- \sum_{\substack{lk \ l'k' \\ (lk \neq l'k')}} \mathbf{p}\binom{l}{k} \cdot \frac{Z_{k'} e \mathbf{r}\binom{l \ l'}{k \ k'}}{\left|\mathbf{r}\binom{l \ l'}{k \ k'}\right|^3}$$
$$- \tfrac{1}{2} \sum_{\substack{lk \ l'k' \\ (lk \neq l'k')}} \mathbf{p}\binom{l}{k} \cdot \nabla_{lk} \left\{ \frac{\mathbf{p}\binom{l'}{k'} \cdot \mathbf{r}\binom{l' \ l}{k' \ k}}{\left|\mathbf{r}\binom{l \ l'}{k \ k'}\right|^3} \right\}, \tag{4.337}$$

where

$$\mathbf{r}\binom{l' \ l}{k' \ k} = \mathbf{x}\binom{l' \ l}{k' \ k} + \mathbf{u}\binom{l'}{k'} - \mathbf{u}\binom{l}{k}.$$

The first term on the right-hand side of (4.337) describes the monopole-monopole interactions. The second term represents the interaction between the point dipoles $\mathbf{p}\binom{l}{k}$ and the field due to the monopoles.* The last term describes the interaction of $\mathbf{p}\binom{l}{k}$ with the field generated by all the other dipoles. Using now the approximations

$$\frac{1}{|\mathbf{x} + \boldsymbol{\varepsilon}|} \approx \frac{1}{x} - \frac{\boldsymbol{\varepsilon} \cdot \mathbf{x}}{x^3} + \sum_{\alpha\beta} \frac{3\varepsilon_\alpha \varepsilon_\beta x_\alpha x_\beta - \varepsilon_\alpha \varepsilon_\beta \delta_{\alpha\beta} x^2}{2x^5}$$

and

$$\frac{1}{|\mathbf{x} + \boldsymbol{\varepsilon}|^3} \approx \frac{1}{x^3}\left[1 - \frac{3\boldsymbol{\varepsilon} \cdot \mathbf{x}}{x^2}\right],$$

where $\boldsymbol{\varepsilon}$ is a small quantity, it is straightforward to establish that the second order term $\Phi^{(2)\text{LR}}$ of (4.337) is the same as that within the square brackets in (4.336).

The self term in (4.336) has a structure different from that in (4.269), where the dipoles were introduced via the shell-core picture. This does not imply a contradiction but merely highlights the manner in which translational invariance operates in different physical pictures. Also observe that the "self energy" of the dipole has the form $\alpha_k^{-1}\boldsymbol{\mu}^2\binom{l}{k}$, rather than $(K(k)/Y_k^2 e^2)\mathbf{p}^2\binom{l}{k}$ as in the shell model (see Eq. (4.326)).

The introduction of deformation dipoles introduces many-body forces. For instance, the term

$$\frac{1}{2}\sum_{lk\alpha}\sum_{l'k'\beta}\sum_{l''k''\gamma} \mathscr{D}_{\alpha\beta}\binom{l\ \ l'}{k\ \ k'}\gamma_{\beta\delta}\binom{l'\ \ l''}{k'\ \ k''}Z_k eu_\alpha\binom{l}{k}\left\{u_\delta\binom{l''}{k''} - u_\delta\binom{l'}{k'}\right\}$$

involves three-body forces since the interaction energy of a pair $\binom{l\ l'}{k\ k'}$ depends on a third set of atoms $\binom{l''}{k''}$. Similarly, one may identify four-body interactions from terms in which the $\gamma$ tensor occurs twice.

The equations of motion are given by

$$m_k \ddot{u}_\alpha\binom{l}{k} = -\frac{\partial \Phi^{(2)}}{\partial u_\alpha\binom{l}{k}}, \tag{4.338}$$

$$0 = \frac{\partial \Phi^{(2)}}{\partial \mu_\alpha\binom{l}{k}}. \tag{4.339}$$

*This term can also be written as

$$-\sum_{\substack{lk\ \ l'k'\\(lk \neq l'k')}} Z_k e\left\{\frac{\mathbf{p}\binom{l'}{k'} \cdot \mathbf{r}\binom{l'l}{k'k}}{|\mathbf{r}\binom{l'l}{k'k}|^3}\right\}$$

and interpreted as the interaction between the monopoles $Z_k e$ and the *potential* due to the dipoles.

We shall later see that (4.339) leads to a self-consistency requirement on $\mu\binom{l}{k}$, enabling us to eliminate the latter from the equations of motion.

Using the first of the above equations, we obtain after some lengthy but straightforward algebra

$$
\begin{aligned}
m_k \ddot{u}_\alpha\binom{l}{k} = &\left[ -\sum_{l'k'\beta} \phi^{SR}_{\alpha\beta}\binom{l\ \ l'}{k\ \ k'} u_\beta\binom{l'}{k'} \right] \\
&+ \left[ \sum_{\substack{l'k'\\(\neq lk)}} \sum_\beta Z_k Z_{k'} e^2 \mathcal{D}_{\alpha\beta}\binom{l\ \ l'}{k\ \ k'} \left\{ u_\beta\binom{l'}{k'} - u_\beta\binom{l}{k} \right\} \right] \\
&+ \left[ \sum_{\substack{l_1k_1\\(\neq lk)}} \sum_\delta \sum_{\substack{l'k'\\(\neq l_1k_1)}} \sum_\beta \gamma_{\alpha\delta}\binom{l\ \ l_1}{k\ \ k_1} \mathcal{D}_{\delta\beta}\binom{l_1\ \ l'}{k_1\ \ k'} Z_{k'} e \left\{ u_\beta\binom{l'}{k'} - u_\beta\binom{l_1}{k_1} \right\} \right. \\
&\left. - \sum_{\substack{l_1k_1\\(\neq lk)}} \sum_\delta \sum_{\substack{l'k'\\(\neq lk)}} \sum_\beta \gamma_{\alpha\delta}\binom{l_1\ \ l}{k_1\ \ k} \mathcal{D}_{\delta\beta}\binom{l\ \ l'}{k\ \ k'} Z_{k'} e \left\{ u_\beta\binom{l'}{k'} - u_\beta\binom{l}{k} \right\} \right] \\
&+ \left[ \sum_{\substack{l'k'\\(\neq lk)}} \sum_\beta \sum_{\substack{l_1k_1\\(\neq l'k')}} \sum_\delta Z_k e \mathcal{D}_{\alpha\beta}\binom{l\ \ l'}{k\ \ k'} \gamma_{\beta\delta}\binom{l'\ \ l_1}{k'\ \ k_1} \left\{ u_\delta\binom{l_1}{k_1} - u_\delta\binom{l'}{k'} \right\} \right. \\
&\left. - \sum_{\substack{l'k'\\(\neq lk)}} \sum_\beta \sum_{\substack{l_1k_1\\(\neq lk)}} \sum_\delta Z_k e \mathcal{D}_{\alpha\beta}\binom{l'\ \ l}{k'\ \ k} \gamma_{\beta\delta}\binom{l\ \ l_1}{k\ \ k_1} \left\{ u_\delta\binom{l_1}{k_1} - u_\delta\binom{l}{k} \right\} \right] \\
&+ \left[ \sum_{\substack{l_1k_1\\(\neq lk)}} \sum_\delta \sum_{\substack{l'k'\\(\neq l_1k_1)}} \sum_\beta \sum_{\substack{l_2k_2\\(\neq l'k')}} \sum_\eta \gamma_{\alpha\delta}\binom{l\ \ l_1}{k\ \ k_1} \mathcal{D}_{\delta\beta}\binom{l_1\ \ l'}{k_1\ \ k'} \gamma_{\beta\eta}\binom{l'\ \ l_2}{k'\ \ k_2} \right. \\
&\times \left\{ u_\eta\binom{l_2}{k_2} - u_\eta\binom{l'}{k'} \right\} - \sum_{\substack{l_1k_1\\(\neq lk)}} \sum_\delta \sum_{\substack{l'k'\\(\neq lk)}} \sum_\beta \sum_{\substack{l_2k_2\\(\neq l'k')}} \sum_\eta \gamma_{\alpha\delta}\binom{l_1\ \ l}{k_1\ \ k} \\
&\left. \times \mathcal{D}_{\delta\beta}\binom{l\ \ l'}{k\ \ k'} \gamma_{\beta\eta}\binom{l'\ \ l_2}{k'\ \ k_2} \left\{ u_\eta\binom{l_2}{k_2} - u_\eta\binom{l'}{k'} \right\} \right] \\
&+ \left[ \sum_{\substack{l'k'\\(\neq lk)}} \sum_\beta Z_k e \mathcal{D}_{\alpha\beta}\binom{l\ \ l'}{k\ \ k'} \mu_\beta\binom{l'}{k'} \right. \\
&\left. - \sum_{\substack{l'k'\\(\neq lk)}} \sum_\beta Z_{k'} e \mathcal{D}_{\alpha\beta}\binom{l\ \ l'}{k\ \ k'} \mu_\beta\binom{l}{k} \right] \\
&+ \left[ \sum_{\substack{l_1k_1\\(\neq lk)}} \sum_\delta \sum_{\substack{l'k'\\(\neq l_1k_1)}} \sum_\beta \gamma_{\alpha\delta}\binom{l\ \ l_1}{k\ \ k_1} \mathcal{D}_{\delta\beta}\binom{l_1\ \ l'}{k_1\ \ k'} \mu_\beta\binom{l'}{k'} \right. \\
&\left. - \sum_{\substack{l_1k_1\\(\neq lk)}} \sum_\delta \sum_{\substack{l'k'\\(\neq lk)}} \sum_\beta \gamma_{\alpha\delta}\binom{l_1\ \ l}{k_1\ \ k} \mathcal{D}_{\delta\beta}\binom{l\ \ l'}{k\ \ k'} \mu_\beta\binom{l'}{k'} \right].
\end{aligned} \tag{4.340}
$$

Upon introducing wavelike solutions

$$u_\alpha\binom{l}{k} = U_\alpha(k|\mathbf{q}) \exp\left\{i\left[\mathbf{q}\cdot\mathbf{x}\binom{l}{k} - \omega t\right]\right\}$$

$$\mu_\alpha\binom{l}{k} = \mu_\alpha(k|\mathbf{q}) \exp\left\{i\left[\mathbf{q}\cdot\mathbf{x}\binom{l}{k} - \omega t\right]\right\},$$

we obtain the following equation expressed for compactness in matrix notation:

$$m\omega^2\mathbf{U} = \mathbf{R}\mathbf{U} + \mathbf{Z}_d\mathbf{C}\mathbf{Z}_d\mathbf{U} + \gamma\mathbf{C}\mathbf{Z}_d\mathbf{U}$$
$$+ \mathbf{Z}_d\mathbf{C}\gamma\mathbf{U} + \gamma\hat{\mathbf{C}}\gamma\mathbf{U} + (\mathbf{Z}_d\mathbf{C} + \gamma\hat{\mathbf{C}})\mathbf{\mu}. \tag{4.341}$$

Here $\mathbf{R}$ refers as before to the contribution from the short-range part. The matrices $\mathbf{Z}_d$, $\hat{\mathbf{C}}$, and $\mathbf{C}$ were defined earlier, while the elements of $\gamma(q)$ are given by

$$\gamma_{\alpha\beta}\binom{\mathbf{q}}{kk'} = \hat{\gamma}_{\alpha\beta}\binom{\mathbf{q}}{kk'} - \delta_{kk'}\sum_{k''}\hat{\gamma}_{\alpha\beta}\binom{0}{kk''}, \tag{4.342}$$

where

$$\hat{\gamma}_{\alpha\beta}\binom{\mathbf{q}}{kk'} = \sum_{l'}\gamma_{\alpha\beta}\binom{0\ l'}{k\ k'}\exp\left\{i\mathbf{q}\cdot\mathbf{x}\binom{l'\ 0}{k'\ k}\right\}, \quad (k \neq k'),$$

$$\hat{\gamma}_{\alpha\beta}\binom{\mathbf{q}}{kk} = \sum_{l'}'\gamma_{\alpha\beta}\binom{0\ l'}{k\ k}\exp\left\{i\mathbf{q}\cdot\mathbf{x}\binom{l'\ 0}{k\ k}\right\},$$

$$\hat{\gamma}_{\alpha\beta}\binom{0}{kk'} = \sum_{l'}\gamma_{\alpha\beta}\binom{0\ l'}{k\ k'}, \quad (k \neq k'),$$

$$\hat{\gamma}_{\alpha\beta}\binom{0}{kk} = \sum_{l'}'\gamma_{\alpha\beta}\binom{0\ l'}{k\ k}. \tag{4.343}$$

From Eq. (4.339) we get

$$\mu_\alpha\binom{l}{k} = \alpha_k\left[\sum_{\substack{l'k'\beta\\(l'k'\neq lk)}}\mathscr{D}_{\alpha\beta}\binom{l\ l'}{k\ k'}\left\{Z_{k'}eu_\beta\binom{l'}{k'}\right.\right.$$
$$\left.\left. + \left[\mu_\beta\binom{l'}{k'} + v_\beta\binom{l'}{k'}\right] - Z_k\cdot eu_\beta\binom{l}{k}\right\}\right]. \tag{4.344}$$

The multiplier of $\alpha_k$ is the local field acting on $\binom{l}{k}$, composed of:
1. the field from both displacement and electronic dipoles associated with the other sites $\binom{l'}{k'}$, and
2. the field at $\binom{l}{k}$ due to its own *displacement*.
The latter is associated with the self terms of the Coulombic force constants as discussed in Appendix 2. Equation (4.344) merely emphasizes that the relation (4.230) is implicit in the prescription (4.336) for $\Phi^{(2)}$.

The $\mathbf{q}$-space version of (4.344) is

$$\boldsymbol{\mu} = -\alpha \mathbf{C} \mathbf{Z}_d \mathbf{U} - \alpha \hat{\mathbf{C}} \boldsymbol{\mu} - \alpha \hat{\mathbf{C}} \gamma \mathbf{U}, \qquad (4.345)$$

where

$$
\boldsymbol{\alpha} = \begin{pmatrix}
\alpha_1 & & & & & \\
& \alpha_1 & & & & \\
& & \alpha_1 & & & \\
& & & \ddots & & \\
& & & & \alpha_n & \\
& & & & & \alpha_n \\
& & & & & & \alpha_n
\end{pmatrix}. \qquad (4.346)
$$

From (4.345),

$$
\begin{aligned}
\boldsymbol{\mu} &= -[\mathbb{1}_{3n} + \alpha \hat{\mathbf{C}}]^{-1} \alpha [\mathbf{C} \mathbf{Z}_d + \hat{\mathbf{C}} \gamma] \mathbf{U} \\
&= -[\boldsymbol{\alpha}^{-1} + \hat{\mathbf{C}}]^{-1} [\mathbf{C} \mathbf{Z}_d + \hat{\mathbf{C}} \gamma] \mathbf{U}.
\end{aligned} \qquad (4.347)
$$

Using (4.347) in (4.341) we finally obtain

$$
\begin{aligned}
m\omega^2 \mathbf{U} = [\mathbf{R} + \mathbf{Z}_d \mathbf{C} \mathbf{Z}_d] \mathbf{U} &+ \{ -[\mathbf{Z}_d \mathbf{C} + \gamma \hat{\mathbf{C}}][\boldsymbol{\alpha}^{-1} + \hat{\mathbf{C}}]^{-1}[\mathbf{C} \mathbf{Z}_d + \hat{\mathbf{C}} \gamma] \\
&+ [\gamma \mathbf{C} \mathbf{Z}_d + \mathbf{Z}_d \mathbf{C} \gamma + \gamma \hat{\mathbf{C}} \gamma] \} \mathbf{U}.
\end{aligned} \qquad (4.348)
$$

Observe that if we ignore the deformation dipoles by equating $\gamma$ to zero, then we recover the Born and Huang[67] generalization of the Lyddane-Herzfeld model to arbitrary crystals. If in addition we ignore electronic polarizability and set $\boldsymbol{\alpha} = 0$, then we obtain the rigid-ion result.

It is interesting to compare (4.345) with the long-range contribution to the induced dipole in Eq. (4.318). We note that, unlike the latter, (4.345) can be written as

$$\boldsymbol{\mu} = \alpha \mathscr{E}, \qquad (4.349)$$

where $\mathscr{E}$ is a column matrix with components

$$\mathscr{E}_\alpha(k|\mathbf{q}) = -\sum_{k'\beta} \left\{ Z_{k'} e C_{\alpha\beta}\binom{\mathbf{q}}{kk'} U_\beta(k'|\mathbf{q}) + \hat{C}_{\alpha\beta}\binom{\mathbf{q}}{kk'} p_\beta(k'|\mathbf{q}) \right\}, \qquad (4.350)$$

which denote the local-field amplitudes.* Based on the earlier discussion, we can now appreciate the fact that (4.349) obtains in Hardy's model because of his neglect of short-range forces between the induced dipoles.

---

*Compare (4.350) with (4.313) and (4.314). The differences arise from the different structure of the self term in the two models. For cubic diatomic crystals, however, the expressions become identical if in the shell model we replace $Y_k e \mathbf{W}(k|\mathbf{q})$ by $\mathbf{p}(k|\mathbf{q})$.

### 4.6.7 Extensions of the shell model—charge-transfer models
A number of attempts have been made to go beyond the shell model described in Sec. 4.6.3–4.6.4. The motivation for these extensions as well as their utility are best discussed in conjunction with a review of experimental results and will therefore be deferred till Chapter 6. Here we merely survey the physical content of such extended models.

Broadly speaking, we can identify two categories in the models of current interest. In the first, one retains the basic shell-model concept of spherical shells capable of rigid movement relative to their parent cores and includes one additional refinement. In the second, one allows for the possibility of shell deformation.

Most important in the former category are the charge-transfer models. Since the spirit of such models can be traced to the work of Lundqvist,[44-46] it is useful first to review his model briefly.

**Lundqvist's model.** Starting from a quantum mechanical consideration of the overlap of adjacent ions in NaCl-type crystals, Lundqvist[44-46] noted that, subject to approximations, the potential energy can be expressed as[68]

$$\Phi = \tfrac{1}{2} \sum_{lk} \sum_{\substack{l'k' \\ (lk \neq l'k')}} \frac{Z_k Z_{k'} e^2}{|\mathbf{r}(\begin{smallmatrix} l & l' \\ k & k' \end{smallmatrix})|} + \tfrac{1}{2} \sum_{lk} \sum_{l'k'} V\left(\left|\mathbf{r}\begin{pmatrix} l & l' \\ k & k' \end{pmatrix}\right|\right)$$
$$+ \sum_{lk} \sum_{l'k'} \sum_{\substack{l''k'' \\ (l''k'' \neq l'k' \neq lk)}} e f_k\left(\left|\mathbf{r}\begin{pmatrix} l & l' \\ k & k' \end{pmatrix}\right|\right) \frac{Z_{k''} e}{|\mathbf{r}(\begin{smallmatrix} l & l'' \\ k & k'' \end{smallmatrix})|}. \tag{4.351}$$

Here the first two terms denote the usual Coulomb and two-body short-range interactions familiar to us from the Born model. The function $f_k(r)$ in the third term is related to the overlap of two ions separated by a distance $r$. This term has an explicit three-body character and expresses the electrostatic interaction between the overlap-created charge depletion with ionic charges other than those associated with the ions participating in overlap. Conceptually, these charge depletions are similar to the exchange charge discussed earlier.

Based on (4.351) and assuming only nearest-neighbor overlap, the force constant for $k \neq k'$ may be expressed as

$$\phi_{\alpha\beta}\begin{pmatrix} l & l' \\ k & k' \end{pmatrix} \approx -Z_k e(1 + 6f_k(r_0)) \mathscr{D}_{\alpha\beta}(\begin{smallmatrix} l & l' \\ k & k' \end{smallmatrix}) Z_{k'} e(1 + 6f_{k'}(r_0))$$
$$- \frac{\partial^2}{\partial r_\alpha \partial r_\beta}\left[ V_1(r) + f_{k'}(r) \sum_{l_1 k_1}' \frac{Z_{k_1} e^2}{|\mathbf{x}(\begin{smallmatrix} l' & l_1 \\ k' & k_1 \end{smallmatrix})|} \right.$$

$$+ f_k(r) \sum_{l_1 k_1}{}' \frac{Z_{k_1} e^2}{|\mathbf{x}\binom{l \; l_1}{k \; k_1})|}\Big]_{r=|\mathbf{x}\binom{l \; l'}{k \; k'})|}$$

$$- \sum_{\substack{l_1 k_1 \\ (\neq lk \& l'k')}} \left\{ Z_k e^2 \left[\frac{\partial}{\partial r_\alpha}\left(\frac{1}{r}\right)\right]_{r=\mathbf{x}\binom{l_1 \; l}{k_1 \; k})} \left[\frac{\partial}{\partial r_\beta} f_{k_1}(r)\right]_{r=\mathbf{x}\binom{l' \; l_1}{k' \; k_1})} \right.$$

$$\left. + Z_{k'} e^2 \left[\frac{\partial}{\partial r_\alpha} f_{k_1}(r)\right]_{r=\mathbf{x}\binom{l \; l_1}{k \; k_1})} \left[\frac{\partial}{\partial r_\beta}\left(\frac{1}{r}\right)\right]_{r=\mathbf{x}\binom{l_1 \; l'}{k_1 \; k'})} \right\}. \tag{4.352}$$

Here $r_0$ denotes the nearest-neighbor distance, and

$$V_1\left(r\binom{l \; l'}{k \; k'}\right) = V\left(r\binom{l \; l'}{k \; k'}\right) - 2 f_k\left(r\binom{l \; l'}{k \; k'}\right) \frac{Z_{k'} e}{r\binom{l \; l'}{k \; k'}}. \tag{4.353}$$

The quantity $\sum_{l_1 k_1}{}' Z_{k_1}/|\mathbf{x}\binom{l' \; l_1}{k' \; k_1})|$ is sometimes expressed in terms of the Madelung constant $\alpha_M$ as follows:

$$\sum_{l_1 k_1}{}' \frac{Z_{k_1}}{|\mathbf{x}\binom{l' \; l_1}{k' \; k_1})|} = \frac{Z_{k'} \alpha_M}{r_0}.$$

Compared to the force constant of the Born model, namely

$$\phi_{\alpha\beta}\binom{l \; l'}{k \; k'} = -Z_k e \mathscr{D} Z_{k'} e - \frac{\partial^2 V(r)}{\partial r_\alpha \partial r_\beta}\Big|_{r=|\mathbf{x}\binom{l \; l'}{k \; k'})|},$$

we observe that three-body forces not only modify the usual Coulombic and short-range parts but also contribute explicitly to the force constant via the third term on the right-hand side of (4.352).

Lundqvist's model implies charge transfer in the sense that the three-body contribution to $\Phi$ (see (4.351)) is calculated as if a charge $e f_k(r)$ were transferred to the ion $\binom{l}{k}$, which then interacts with $Z_{k''} e$ at $\binom{l''}{k''}$ via Coulomb's law. This charge transfer is, of course, a function of the relative separation of the overlapping ions and is, therefore, a function of their respective displacements.

Lundqvist was interested primarily in the elastic constants and the long-wavelength optic frequencies and found it adequate to treat the three-body terms in (4.352) in an approximate way. The inclusion of these terms permitted him to examine deviation from the Cauchy relation (see Sec. 2.4.6).

**Verma and Singh's model.** Verma and Singh[69] have extended Lundqvist's work by evaluating in detail the three-body terms of (4.352). When this is done, the resulting dynamical equation has the following form:

$$m\omega^2 \mathbf{U} = [\mathbf{R}' + \mathbf{Z}_d(1 + 6f(r_0)) \mathbf{C} \mathbf{Z}_d(1 + 6f(r_0))$$
$$+ \mathbf{Z}_d(\mathbf{\mu}\mathbf{\lambda} + \tilde{\mathbf{\lambda}}\tilde{\mathbf{\mu}})\mathbf{Z}_d]\mathbf{U}. \tag{4.354}$$

Here $\mathbf{R}'$ is the short-range contribution to the dynamical matrix and includes all the modifications to the two-body potential $V(r)$ noted in (4.352) and (4.353). The quantity $f(r_0)$ is defined by

$$f(r_0) = f_k(r_0)Z_k/|Z_k|. \tag{4.355}$$

$\lambda$ is a $(2 \times 6)$ matrix with the following structure:*

$$\lambda = \frac{2if'(r_0)}{Z}\begin{pmatrix} 0 & 0 & 0 & \sin q_x r_0 & \sin q_y r_0 & \sin q_z r_0 \\ \sin q_x r_0 & \sin q_y r_0 & \sin q_z r_0 & 0 & 0 & 0 \end{pmatrix}, \tag{4.356}$$

where $Z = Z_+ = |Z_-|, f_+ = -f_-$, and $f' = df/dr$.
  $\mu$ is a $(6 \times 2)$ array whose elements are given by

$$\mu_\alpha\binom{q}{kk'} = -\sum_{l'}' \frac{x_\alpha\binom{l'\ 0}{k'\ k}}{|\mathbf{x}\binom{l'\ 0}{k'\ k}|^3} \exp\left\{i\mathbf{q}\cdot\mathbf{x}\binom{l'\ 0}{k'\ k}\right\}. \tag{4.357}$$

The prime on the summation above implies that if $k' = k$, then the $l' = 0$ term must be omitted.
  Noting that $[1 + 6f(r_0)]^2$ may be approximated as $1 + 12f(r_0)$, we may rewrite (4.354) as

$$m\omega^2 \mathbf{U} = [\mathbf{R}' + \mathbf{Z}_d\mathbf{C}'\mathbf{Z}_d]\mathbf{U},$$

where

$$\mathbf{C}' = [\mathbf{C}(1 + 12f(r_0)) + \mu\lambda + \tilde{\lambda}\tilde{\mu}]. \tag{4.358}$$

Thus $\mathbf{C}'$ is the effective long-range matrix that includes the effects of the explicitly three-body term as well as the modified Coulomb interactions noted in (4.352).†
  Verma and Singh have also tried to "inject" three-body effects into the shell model by replacing everywhere in Eqs. (4.290) and (4.291) the matrix $\mathbf{C}$ by $\mathbf{C}'$. This is an ad hoc prescription and does not pay detailed attention to the possible origin of charge transfer and three-body terms in a shell-model framework. The point may be appreciated by considering a similar model which we presently describe.

**Feldkamp's model.** The model of Feldkamp[70]‡ is similar in some respects to those discussed above in that it also involves the concept of charge transfer. Guided by overlap considerations, Feldkamp proposed

---

*Remember we are dealing here with the NaCl structure.
†Verma and Singh quote the result $2\mu\lambda$ (in our notation) for the three-body contribution to $\mathbf{C}'$. This appears to be due to their treating the two three-body terms in (4.352) as equal.
‡This model was introduced specifically for crystals having the zinc blende structure, but can be applied to other structures.

that during lattice vibrations, variations can occur in the shell charge expressed by

$$Y\begin{pmatrix} l \\ k \end{pmatrix}e = Y_k e - \frac{Y_v e}{r_0} \frac{\sqrt{3}}{2} \frac{Z_k}{|Z_k|} \sum_{\substack{l*k* \\ \text{over n.n.} \\ \text{of } (lk).}} \left[ \mathbf{u}^s\begin{pmatrix} l* \\ k* \end{pmatrix} - \mathbf{u}^s\begin{pmatrix} l \\ k \end{pmatrix} \right] \cdot \hat{\mathbf{x}}\begin{pmatrix} l* & l \\ k* & k \end{pmatrix}. \quad (4.359)$$

Here the second term on the right-hand side denotes explicitly the variations in the shell charge. The cube edge is $2r_0$. The parameter $Y_v$ may be considered as the magnitude of the change in the shell charge induced by a nearest neighbor's relative movement of $(2r_0/\sqrt{3})$ along the unit vector $\hat{\mathbf{x}}\begin{pmatrix} l* & l \\ k* & k \end{pmatrix}$. The physical content of (4.359) is that when the *shells* of neighboring ions are relatively displaced, an extra charge which depends on the variation in overlap appears on the *shells*. This variable charge is allowed to interact electrostatically with all the other charges in the system and leads to three- and four-body forces. It is these details concerning the origin and role of the three-body forces that Verma and Singh omit to make clear in their adaptation of Lundqvist's ideas to the shell model.

Bearing in mind (4.359), we can express the electrostatic part of the potential energy as

$$\Phi^{\text{LR}} = \frac{1}{2} \sum_{\substack{lk \ l'k' \\ (lk \neq l'k')}} \left\{ \frac{X_k X_{k'} e^2}{|\mathbf{r}^{\text{cc}}\begin{pmatrix} l & l' \\ k & k' \end{pmatrix}|} + \frac{X_k Y\begin{pmatrix} l' \\ k' \end{pmatrix} e^2}{|\mathbf{r}^{\text{cs}}\begin{pmatrix} l & l' \\ k & k' \end{pmatrix}|} \right.$$
$$\left. + \frac{Y\begin{pmatrix} l \\ k \end{pmatrix} X_{k'} e^2}{|\mathbf{r}^{\text{sc}}\begin{pmatrix} l & l' \\ k & k' \end{pmatrix}|} + \frac{Y\begin{pmatrix} l \\ k \end{pmatrix} Y\begin{pmatrix} l' \\ k' \end{pmatrix} e^2}{|\mathbf{r}^{\text{ss}}\begin{pmatrix} l & l' \\ k & k' \end{pmatrix}|} \right\}, \quad (4.360)$$

where, for example,

$$\mathbf{r}^{\text{cs}}\begin{pmatrix} l & l' \\ k & k' \end{pmatrix} = \mathbf{r}^c\begin{pmatrix} l \\ k \end{pmatrix} - \mathbf{r}^s\begin{pmatrix} l' \\ k' \end{pmatrix}.$$

Starting from (4.360), isolating the harmonic part and then examining the consequences, one finds that a simple replacement of the form $\mathbf{C} \rightarrow \mathbf{C}'$ is not possible for the shell model. One would therefore suspect that, unless special assumptions are made, a similar remark would apply also to any attempts made to adapt Lundqvist's work as such to the shell model, notwithstanding the fact that Lundqvist's approach retains only three-body terms. However we emphasize that the replacement $\mathbf{C} \rightarrow \mathbf{C}'$ is perfectly valid for Lundqvist's original model.

**Dick's exchange-charge model.** Both polarization mechanisms of Dick and Overhauser were incorporated by Marston and Dick[71] in the exchange-charge model,* in which the exchange charges are treated as follows:

1. The exchange charge is assumed to be a point charge of magnitude

$$q = -\frac{R_{ss}}{e\xi} Be^{-\alpha R_{ss}},$$

where $B$ and $\alpha$ are the usual Born-Mayer parameters, $R_{ss}$ is the nearest neighbor shell-shell distance, and $\xi$ is a dimensionless parameter.

2. These point charges are located on the line of centers of the shells and at a distance $r_+ R_{ss}/R$ from the positive ion shell center, $r_+$ being the positive-ion radius.

3. The positive (negative) ions have modified charges

$$+(-)Ze - \tfrac{1}{2} \sum_i q_i,$$

where $q_i$ is the exchange charge located on the line connecting the positive (negative) ion shell to that of its nearest neighbor $i$.

Upon incorporating the exchange-charge idea into the shell model, the potential energy becomes

$$\Phi^{(2)} = \Phi^{(2)}(\text{SM}) + \Phi^{(2)}(\text{exch. ch.}),$$

where $\Phi^{(2)}(\text{SM})$ is given by (4.257). The additional term $\Phi^{(2)}(\text{exch. ch.})$ is purely electrostatic and involves the interaction of the exchange charges among themselves and with the core and shell charges. Because of the origin of the exchange charge, care is required in excluding those interactions involving the exchange charge which are included implicitly in the short-range forces. The term $\Phi^{(2)}(\text{exch. ch.})$ is evaluated by associating the exchange charge with the shells and employing a multipole expansion. The resulting expression has a complicated structure and will therefore not be given here. The details are available in a thesis by Marston.[72]

Sinha[73] has criticized the models of both Verma and Singh and of Feldkamp as lacking (as does Hardy's model) in symmetry of cause and effect, namely that while charge transfer is assumed to arise from short-range overlap effects due to lattice motions, it is coupled to the lattice only by Coulomb interactions. The same criticism applies also to the exchange-charge model.

### 4.6.8 Deformable-shell models
The shell model discussed in Sec. 4.6.3 is a dipolar model in that charge distortions occuring during vibrations are described in terms of point

---

*This model was developed explicitly for NaCl structure crystals.

dipoles. Several attempts[74-78] have been made to go beyond the dipolar assumption by allowing for higher multipoles, particularly within the shell-model framework. Conceptually, higher-order multipoles can be accommodated by imagining the shells to undergo deformation, leading thereby to deformable-shell models. A special example of the latter is the *breathing-shell model*[75-77] in which the shell is assumed to undergo an *isotropic* deformation.

One common feature of all the deformable-shell models is that they involve, explicitly or implicitly, coordinates in addition to the usual core and shell displacements. If explicit, these coordinates can be eliminated with the help of appropriate self-consistency equations, analogous to the shell equation.

Deformable-shell models can be conveniently characterized by the following potential energy function:

$$
\Phi^{(2)} = \Phi^{(2)}(\text{SM}) + \tfrac{1}{2} \sum_{\substack{\alpha s \\ ll' \\ kk'}} \left\{ \psi_{\alpha s}^{\text{AM}} \binom{l\ l'}{k\ k'} u_\alpha \binom{l}{k} v^s \binom{l'}{k'} \right.
$$

$$
+ \psi_{s\alpha}^{\text{MA}} \binom{l\ l'}{k\ k'} v^s \binom{l}{k} u_\alpha \binom{l'}{k'} + \psi_{\alpha s}^{\text{DM}} \binom{l\ l'}{k\ k'} w_\alpha \binom{l}{k} v^s \binom{l'}{k'}
$$

$$
\left. + \psi_{s\alpha}^{\text{MD}} \binom{l\ l'}{k\ k'} v^s \binom{l}{k} w_\alpha \binom{l'}{k'} \right\} + \tfrac{1}{2} \sum_{\substack{ll' \\ kk' \\ ss'}} \psi_{ss'}^{\text{MM}} \binom{l\ l'}{k\ k'} v^s \binom{l}{k} v^{s'} \binom{l'}{k'}.
$$

$$(4.361)$$

Here $v^s\binom{l}{k}$ denotes an additional coordinate associated with the ion $\binom{l}{k}$. The total number $v$ of such independent coordinates per cell depends on the detailed model assumptions. The quantities $\psi_{\alpha s}^{\text{AM}}$, $\psi_{\alpha s}^{\text{DM}}$, and $\psi_{ss'}^{\text{MM}}$, denote appropriate force constants having the usual properties (superscript M stands for multipole). The equations of motion are derived from (4.270), (4.271), and

$$
\frac{\partial \Phi^{(2)}}{\partial v^s\binom{l}{k}} = 0.
$$

$$(4.362)$$

From (4.361) one obtains by the usual procedure

$$
m\omega^2 \mathbf{U} = [\mathbf{R} + \mathbf{Z_d CZ_d}]\mathbf{U} + [\mathbf{T} + \mathbf{Z_d CY_d}]\mathbf{W} + \mathbf{Q}^{\text{AM}}\mathbf{V}, \qquad (4.363a)
$$

$$
0 = [\mathbf{T}^\dagger + \mathbf{Y_d CZ_d}]\mathbf{U} + [\zeta + \mathbf{Y_d CY_d}]\mathbf{W} + \mathbf{Q}^{\text{DM}}\mathbf{V}, \qquad (4.363b)
$$

$$
0 = [\mathbf{Q}^{\text{AM}}]^\dagger \mathbf{U} + [\mathbf{Q}^{\text{DM}}]^\dagger \mathbf{W} + \mathbf{Q}^{\text{MM}}\mathbf{V}. \qquad (4.363c)
$$

Here $\mathbf{Q}^{\text{AM}}$ and $\mathbf{Q}^{\text{DM}}$ are $(3n \times v)$ arrays with elements

$$Q_{\alpha s}^{A/D,M}\begin{pmatrix} \mathbf{q} \\ kk' \end{pmatrix} = \sum_{l'} \psi_{\alpha s}^{A/D,M}\begin{pmatrix} 0 & l' \\ k & k' \end{pmatrix} \exp\left\{ i\mathbf{q}\cdot\mathbf{x}\begin{pmatrix} l' & 0 \\ k' & k \end{pmatrix}\right\}, \tag{4.364}$$

while $\mathbf{Q}^{MM}$ is a $(v \times v)$ matrix with elements

$$Q_{ss'}^{MM}\begin{pmatrix} \mathbf{q} \\ kk' \end{pmatrix} = \sum_{l'} \psi_{ss'}^{MM}\begin{pmatrix} 0 & l' \\ k & k' \end{pmatrix} \exp\left\{ i\mathbf{q}\cdot\mathbf{x}\begin{pmatrix} l' & 0 \\ k' & k \end{pmatrix}\right\}. \tag{4.365}$$

$\mathbf{V}$ is a $v$-component column matrix whose elements are the amplitude factors occurring in the wavelike solutions

$$v^s\begin{pmatrix} l \\ k \end{pmatrix} = V^s(k|\mathbf{q}) \exp\left\{ i\left[ \mathbf{q}\cdot\mathbf{x}\begin{pmatrix} l \\ k \end{pmatrix} - \omega t \right]\right\}. \tag{4.366}$$

Equation (4.363c) represents the self-consistency equation for the multipoles. Using this and (4.363b), one may eliminate $\mathbf{W}$ and $\mathbf{V}$ in (4.363a).

To gain a measure of insight, we now consider the relationship of the coordinates $v^s\binom{l}{k}$ to the multipole moments of the charge distribution $-e\rho(\mathbf{r})$. For this purpose we use the expansion

$$\delta(\mathbf{r} - \mathbf{r}') = \delta(\mathbf{r}) - \sum_\alpha r'_\alpha \frac{\partial}{\partial r_\alpha} \delta(\mathbf{r})$$
$$+ \tfrac{1}{2} \sum_{\alpha\beta} r'_\alpha r'_\beta \frac{\partial^2}{\partial r_\alpha \partial r_\beta} \delta(\mathbf{r})$$
$$- \tfrac{1}{6} \sum_{\alpha\beta\gamma} r'_\alpha r'_\beta r'_\gamma \frac{\partial^3}{\partial r_\alpha \partial r_\beta \partial r_\gamma} \delta(\mathbf{r}) + \cdots$$

in the relation

$$\rho(\mathbf{r}) = \int \rho(\mathbf{r}')\delta(\mathbf{r} - \mathbf{r}')\, d\mathbf{r}'.$$

This enables us to write the charge deformation $-e\Delta\rho(\mathbf{r}; lk)$ associated with the ion $\binom{l}{k}$ as

$$\Delta\rho(\mathbf{r}; lk) = -\sum_\alpha M_\alpha\binom{l}{k} \frac{\partial}{\partial r_\alpha} \delta(\mathbf{r})$$
$$+ \frac{1}{2!} \sum_{\alpha\beta} M_{\alpha\beta}\binom{l}{k} \frac{\partial^2}{\partial r_\alpha \partial r_\beta} \delta(\mathbf{r})$$
$$- \frac{1}{3!} \sum_{\alpha\beta\gamma} M_{\alpha\beta\gamma}\binom{l}{k} \frac{\partial^3}{\partial r_\alpha \partial r_\beta \partial r_\gamma} \delta(\mathbf{r}) + \cdots. \tag{4.367}$$

Here*

*The reader familiar with the idea of the adjoint space of a vector space will recognize that Eqs. (4.367) and (4.368) imply the choice 1, $r_\alpha$, $r_\alpha r_\beta$, $r_\alpha r_\beta r_\gamma$, ... for the basis functions of the vector space, and the choice $\delta(\mathbf{r})$, $\partial\delta(\mathbf{r})/\partial r_\alpha$, ... for the basis functions of the adjoint space.

$$M_\alpha\binom{l}{k} = \int r_\alpha \Delta\rho(\mathbf{r}; lk)\, d\mathbf{r},$$

$$M_{\alpha\beta}\binom{l}{k} = \int r_\alpha r_\beta \Delta\rho(\mathbf{r}; lk)\, d\mathbf{r},$$

$$M_{\alpha\beta\gamma}\binom{l}{k} = \int r_\alpha r_\beta r_\gamma \Delta\rho(\mathbf{r}; lk)\, d\mathbf{r}, \qquad (4.368)$$

with the origin for purposes of integration located at $x\binom{l}{k}$. It is further assumed that

$$\int \Delta\rho(\mathbf{r}; lk)\, d\mathbf{r} = 0,$$

which merely expresses the fact that no charge is created.

Equation (4.367) represents a multipole expansion of the charge distortion arising from ionic displacements. In the context of the earlier discussion, the quantities $M_\alpha\binom{l}{k}$ are related to the shell displacements, while $M_{\alpha\beta}\binom{l}{k}$, $M_{\alpha\beta\gamma}\binom{l}{k}$, etc. are related to $v^s\binom{l}{k}$. One could in fact retain a suitable number of multipoles and choose them as $\{v^s\binom{l}{k}\}$ themselves. It is clear that although the formal extension beyond the dipolar model is straightforward, the parameter proliferation implied in multipolar models is substantial unless extra assumptions are made. In a sense, this is what is achieved in the breathing-shell model. As usually presented, this model involves one extra coordinate $\Delta r\binom{l}{k}$ corresponding to the change in the radius of the shell. On the other hand, one could also regard it as a multipole model with the special choice† that all odd moments of $\Delta\rho$ are zero, while the even moments $M_{\lambda_1\lambda_2\cdots\lambda_{2n}}$ vanish also unless $x$, $y$, and $z$ each occur an even number of times in $(\lambda_1 \ldots \lambda_{2n})$, say $2l_1, 2l_2, 2l_3$ times, respectively (so that $2n = 2(l_1 + l_2 + l_3)$). In this case the multipole moment is given by

$$M_{\lambda_1\lambda_2\cdots\lambda_{2n}} = C_{2n} V \frac{A(l_1, l_2, l_3)}{A(n, 0, 0)}, \qquad (4.369)$$

where

$$A(a, b, c) = \frac{2\Gamma(a + \tfrac{1}{2})\Gamma(b + \tfrac{1}{2})\Gamma(c + \tfrac{1}{2})}{\Gamma(a + b + c + \tfrac{3}{2})},$$

and $C_{2n}$ is a constant having physical dimensions $[\text{length}]^{2n-1}$ and which depends on the equilibrium charge distribution. Given a charge

---

†This was (laboriously) computed by considering an isotropic deformation characterized by a parameter $V$ and then computing the moments of this distortion to first order in $V$.

distortion with such moments, it can be deduced that

$$\left[\Delta\rho + \sum_\alpha M_\alpha \frac{\partial}{\partial r_\alpha} \delta(\mathbf{r})\right] = \sum_{n=1}^{\infty} \frac{1}{2n!} C_{2n} V(\nabla^2)^n \delta(\mathbf{r}). \tag{4.370}$$

The right-hand side denotes the deformation associated with the higher-order multipoles and can be shown to be spherically symmetric. Thus the deformation considered indeed represents a spherical charge distortion. The breathing-shell model therefore allows for multipole effects (though in a restricted way) at the expense of only one additional coordinate per ion.

Using the form (4.369) for the multipole moments and noting that the coordinates in the original set $\{v^s(^l_k)\}$ can now be expressed in terms of a single coordinate $v(^l_k) = \Delta r(^l_k)$ per ion, we may write $\Phi^{(2)}$ as

$$\begin{aligned}
\Phi^{(2)} = \Phi^{(2)}(\text{SM}) + \tfrac{1}{2} \sum_{\alpha kk'll'} &\left\{ \phi_\alpha^{\text{AM}}\begin{pmatrix} l & l' \\ k & k' \end{pmatrix} u_\alpha\begin{pmatrix} l \\ k \end{pmatrix} v\begin{pmatrix} l' \\ k' \end{pmatrix} \right. \\
&+ \phi_\alpha^{\text{MA}}\begin{pmatrix} l & l' \\ k & k' \end{pmatrix} v\begin{pmatrix} l \\ k \end{pmatrix} u_\alpha\begin{pmatrix} l' \\ k' \end{pmatrix} + \phi_\alpha^{\text{DM}}\begin{pmatrix} l & l' \\ k & k' \end{pmatrix} w_\alpha\begin{pmatrix} l \\ k \end{pmatrix} v\begin{pmatrix} l' \\ k' \end{pmatrix} \\
&\left. + \phi_\alpha^{\text{MD}}\begin{pmatrix} l & l' \\ k & k' \end{pmatrix} v\begin{pmatrix} l \\ k \end{pmatrix} w_\alpha\begin{pmatrix} l' \\ k' \end{pmatrix} \right\} + \tfrac{1}{2} \sum_{ll'kk'} \phi^{\text{MM}}\begin{pmatrix} l & l' \\ k & k' \end{pmatrix} v\begin{pmatrix} l \\ k \end{pmatrix} v\begin{pmatrix} l' \\ k' \end{pmatrix}, \tag{4.371}
\end{aligned}$$

the force constants appearing above representing appropriate regroupings of the earlier ones. Various versions of the breathing-shell model have been developed for NaCl structure crystals.[75-77]

Another version of the deformable-shell model has been proposed by Basu and Sengupta[78] who treat the deformation without explicitly introducing additional coordinates into the equations of motion. The basic motivation underlying their work may be appreciated as follows:

Consider a set of three ions $A$, $B$, $C$ arranged in a row. In the usual shell-model picture, the symmetric displacement of $A$ and $C$ along the row is not expected to distort the shell of $B$. Intuitively, however, one expects that "under the influence of overlap forces from both sides, the electronic shell (of $B$) will be deformed and its charge distribution slightly altered." Basu and Sengupta therefore include an additional deformation energy of the form

$$\Phi^{(2)}(\text{def}) = \tfrac{1}{2} \sum_{lk} a(k) b^2\begin{pmatrix} l \\ k \end{pmatrix}, \tag{4.372}$$

where $a(k)$ denotes the deformation energy per unit deformation for ions on the $k$th sublattice and $b(^l_k)$ is the deformation coordinate given by

$$b\begin{pmatrix} l \\ k \end{pmatrix} = \sum_{\substack{l^*k^* \\ (\text{over n.n. of } lk)}} \beta(k) \left.\frac{d^2}{dr^2} V_{kk^*}(r)\right|_{r=r_0} \mathbf{u}^s\begin{pmatrix} l^* & l \\ k^* & k \end{pmatrix} \cdot \mathbf{x}\begin{pmatrix} l^* & l \\ k^* & k \end{pmatrix}, \tag{4.373}$$

$\beta(k)$ being a parameter and $r_0$ the nearest-neighbor distance.

Actually, an explicit shell-core picture is not necessary to conceive of a deformation energy of the form (4.372–4.373). One could still write it with, however, the replacement

$$\mathbf{u}^s\begin{pmatrix} l^* & l \\ k^* & k \end{pmatrix} \rightarrow \mathbf{u}\begin{pmatrix} l^* & l \\ k^* & k \end{pmatrix}.$$

Expression (4.372) simply represents a specific type of overlap effect usually ignored.

It is straightforward to incorporate the consequences of (4.372) into the shell model. Basu and Sengupta have done this for NaCl-type crystals, assuming that only the negative ions are polarizable. This leads to an extra energy of the form

$$\frac{1}{2}\sum_{\alpha\beta ll'} \phi_{\alpha\beta}^{\mathrm{def}\,++}\begin{pmatrix} l & l' \\ + & + \end{pmatrix} u_\alpha\begin{pmatrix} l \\ + \end{pmatrix} u_\beta\begin{pmatrix} l' \\ + \end{pmatrix}$$

$$+ \frac{1}{2}\sum_{\alpha\beta ll'} \phi_{\alpha\beta}^{\mathrm{def}\,--}\begin{pmatrix} l & l' \\ - & - \end{pmatrix} u_\alpha^s\begin{pmatrix} l \\ - \end{pmatrix} u_\beta^s\begin{pmatrix} l' \\ - \end{pmatrix}, \qquad (4.374)$$

where the deformation constants are easily expressed in terms of quantities given earlier. The modification to $\mathbf{D}(\mathbf{q})$ resulting from (4.374) is then easily incorporated.

Figure 4.16 summarizes the interrelationship of the various models.

### 4.6.9 Polarization models for homopolar crystals

One important respect in which homopolar crystals differ from ionic crystals is that the atoms do not carry any formal ionic charge. This, however, does not imply that electric field effects are not important in such crystals. Indeed they often are and arise primarily due to electronic polarization which, in some instances (for example, Se, Te), can even endow the atoms with nonvanishing apparent charges leading to IR activity, as discussed earlier. Historically, the need for considering electronic polarization effects in relation to the dynamics of homopolar crystals was brought home forcefully when experimental dispersion curves for Ge first became available.[79] A conventional analysis suggested the presence of fairly long-range forces which could be accommodated only through the concept of interaction between various multipoles induced during the vibrations. This quite naturally prompted the use of the shell model for such crystals.[52,80] The physical basis for such an application might seem questionable since the concept of cores and shells was originally invented specifically in connection with atoms/ions having the rare-gas configuration and not for covalently bonded solids. On the other hand, one can give

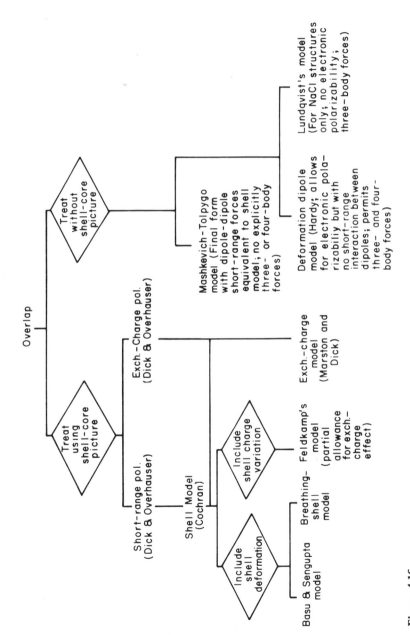

**Figure 4.16**
Genealogy of various phenomenological models.

it a partial justification by the argument that the dynamical equations in terms of shells and cores constitute merely a convenient representation, being a transform of the more physical equations involving atoms and induced dipoles. This argument rests mainly on the work of Mashkevich and Tolpygo[58] who, as noted earlier, showed that a potential function as in (4.321) is also valid for diamond-type crystals, with the ionic charge set equal to zero. In other words, one may use the mechanics of the shell model without seriously implying the existence of shells and cores.

Remembering this, the basic equations are easily written down from (4.290) and (4.291) after noting that $Z_k = 0$ and $X_k = -Y_k$. This gives

$$m\omega^2 \mathbf{U} = \mathbf{R}\mathbf{U} + \mathbf{T}\mathbf{W}, \tag{4.375}$$

$$0 = \mathbf{T}^\dagger\mathbf{U} + [\mathbf{S} + \mathbf{Y}_d\hat{\mathbf{C}}\mathbf{Y}_d]\mathbf{W}. \tag{4.376}$$

Observe that $\mathbf{S}$ and $\hat{\mathbf{C}}$ replace $\zeta$ and $\mathbf{C}$ in (4.376).

Eliminating $\mathbf{W}$, we have

$$m\omega^2 \mathbf{U} = \mathbf{R}\mathbf{U} - \mathbf{T}[\mathbf{S} + \mathbf{Y}_d\hat{\mathbf{C}}\mathbf{Y}_d]^{-1}\mathbf{T}^\dagger\mathbf{U}. \tag{4.377}$$

Although $\mathbf{Z}_d$ vanishes, $\mathbf{Z}(\mathbf{q})$ need not do so in general (for example, in Te) and is given by

$$[\mathbf{Z}^\dagger(\mathbf{q})]_{\alpha\beta}(k) = \sum_{k'} [-\mathbf{T}(\mathbf{S} + \mathbf{M}^{DD,C})^{-1}\mathbf{Y}_d]_{\alpha\beta}\binom{\mathbf{q}}{kk'}. \tag{4.378}$$

It is interesting to note in passing that while the shell-core picture could also be legitimately applied to the rare-gas solids, it never has been because the polarization effects in such crystals are too weak to warrant explicit introduction of the shell-model concept. The effects of induced dipoles are sufficiently well accommodated in the Lennard-Jones potential which, as noted in Sec. 2.9, provides a fairly good description of the observed dispersion curves.

### 4.6.10 Relationship of the shell model to microscopic theories of lattice dynamics

Before concluding this chapter, we wish to reconsider the shell model from a standpoint that emphasizes its relationship to microscopic theories of lattice dynamics (see Chapter 7). In the latter, the harmonic part of the potential energy may be written in the form (see (7.59))

$$\Phi^{(2)} = \Phi_{nn}^{(2)} + \tfrac{1}{2}\int \Delta\rho(\mathbf{r})V_{ne}^{(1)}(\mathbf{r})\,d\mathbf{r} + \int \rho^{(0)}(\mathbf{r})V_{ne}^{(2)}(\mathbf{r})\,d\mathbf{r}. \tag{4.379}$$

Here $\rho^{(0)}$ is the unperturbed electron density in the crystal and $\Delta\rho$ the first-order change consequent to nuclear displacements; $V_{ne}^{(1)}$ and $V_{ne}^{(2)}$ are, respectively, the first- and second-order changes in the potential energy experienced by an electron at the point $\mathbf{r}$ due to all the nuclei. The first term on the right-hand side above denotes the potential energy contribution arising from the direct Coulomb interaction between the nuclei, while the others represent the electronic contribution; the last term is closely related to the fulfillment of translation invariance. In writing $\Phi^{(2)}$ in this form we are visualizing the crystal as composed of bare nuclei and electrons, so that the electron densities refer to those arising from *all* the electrons, that is, both core and valence electrons. From a practical viewpoint, it is more convenient (as in the shell model) to regard the crystal as consisting of ion cores plus valence electrons, especially as the electron density changes associated with polarization are largely confined to the latter. In this case one may make suitable approximations and write $\Phi^{(2)}$ as the sum of three terms analogous to those of (4.379):

$$\Phi^{(2)} = \Phi_{ii}^{(2)} + \tfrac{1}{2} \int \Delta\rho(\mathbf{r}) V_{ie}^{(1)}(\mathbf{r}) \, d\mathbf{r} + \int \rho^0(\mathbf{r}) V_{ie}^{(2)}(\mathbf{r}) \, d\mathbf{r}, \qquad (4.380)$$

where $\rho$ now refers to the valence electron density and $V_{ie}$ is the total (effective) ion-electron interaction. The second term on the right-hand side thus denotes the polarization part of the potential energy.

Supplementing (4.380) is a self-consistency requirement on $\Delta\rho$ of the form

$$\Delta\rho(\mathbf{r}) = \int \chi_{sc}(\mathbf{r}, \mathbf{r}') V(\mathbf{r}', \mathbf{r}'') \Delta\rho(\mathbf{r}'') \, d\mathbf{r}' \, d\mathbf{r}'' - e \int \chi_{sc}(\mathbf{r}, \mathbf{r}') V_{ext}(\mathbf{r}') \, d\mathbf{r}',$$

$$(4.381)$$

where $\chi_{sc}(\mathbf{r}, \mathbf{r}')$ is the nonlocal (screened) dielectric susceptibility of the system; $V_{ext}$ formally denotes a weak external potential acting on the system, but in the present context $-eV_{ext} = V_{ie}^{(1)}$. The quantity $V(\mathbf{r}', \mathbf{r}'')$ represents the potential energy of interaction between the valence electrons at $\mathbf{r}'$ and $\mathbf{r}''$ and is essentially Coulombic, except for small separations $|\mathbf{r}' - \mathbf{r}''|$ where exchange effects cause deviations.

Returning to (4.380), we remark that $\Delta\rho$ and $V_{ie}^{(1)}$ are of first order in displacements and that $V_{ie}^{(2)}$ is of second order. By solving for $\Delta\rho$ from (4.381) and suitably expressing all the other quantities, we may recover from (4.380) an expression for $\Phi^{(2)}$ which is explicitly of second order in displacements, at which point the contact with conventional Born-von Kármán formalism becomes evident. In using the microscopic theory, however, one obtains the dynamical matrix directly by evaluating $\Delta\rho$ and $\chi_{sc}$ in terms of electronic wave functions.

We wish now to consider phenomenological expressions for $\chi_{sc}$ and $\Delta\rho$ in the context of the shell model and to examine subsequently the significance of the self-consistency equation. The latter can be manipulated to the form

$$\Delta\rho(\mathbf{r}) = -e \int \chi_{sc}(\mathbf{r}, \mathbf{r}') V_t(\mathbf{r}') \, d\mathbf{r}', \qquad (4.382)$$

where $V_t$ is the total self-consistent potential acting inside the medium and is the sum of the external potential and that induced by the density fluctuation of the electrons themselves. Although according to (4.382) $\chi_{sc}$ really relates $\Delta\rho$ to $V_t$, one may calculate $\chi_{sc}$ operationally using (4.381) by first imagining the interaction $V(\mathbf{r}', \mathbf{r}'')$ to be "switched off" and then computing the density response $\Delta\rho_{NI}$ induced by a weak external potential with which only the electrons can interact. In other words, $\chi_{sc}$ is obtainable via

$$\Delta\rho_{NI}(\mathbf{r}) = -e \int \chi_{sc}(\mathbf{r}, \mathbf{r}') V_{ext}(\mathbf{r}') \, d\mathbf{r}', \qquad (4.383)$$

where the subscript NI indicates that the density fluctuation is calculated as if no electron-electron interaction were present. The point may be better appreciated by noting the structural similarity of (4.381) to the Lorentz equation $\mathbf{p}_j = \alpha_j \mathscr{E}$ for the dipole moment induced in an ion $j$ of polarizability $\alpha_j$ by a local field $\mathscr{E}$. Considering for simplicity a diagonally cubic crystal where $\mathscr{E} = \mathbf{E} + (4\pi/3)\mathbf{P}$ (see Eq. (A2.53)), we may write[81]

$$\mathbf{p}_j = \alpha_j \left( \frac{4\pi\mathbf{P}}{3} + \mathbf{E} \right) = \alpha_j \left( \frac{4\pi}{3} \sum_i N_i \alpha_i \mathscr{E} + \mathbf{E} \right), \qquad (4.384)$$

where $N_i$ is the number of atoms of type $i$ per unit volume. The two terms on the right-hand side of (4.381) are conceptually similar to those of (4.384), with the first representing the "local-field correction."* Furthermore, $\alpha$ can be regarded formally as the dipole moment induced per unit external field in the absence of local-field correction. This is also the spirit of the operational definition (4.383) for $\chi_{sc}$.

Let us now employ (4.383) to deduce the form of $\chi_{sc}$ within the framework of the shell model. For this purpose, we first imagine that all interactions between valence electrons have been suppressed and then subject the crystal to a weak external potential $V_{ext}(\mathbf{r})$. This produces at the various sites dipole moments given by

$$p_\alpha \binom{l}{k} = -\sum_\beta \alpha_{\alpha\beta}(k) \left[ \frac{\partial}{\partial r_\beta} V_{ext}(\mathbf{r}) \right]_{\mathbf{r} = \mathbf{x}\binom{l}{k}}, \qquad (4.385)$$

---

*It must be remembered, however, that whereas (4.384) considers only the correction due to polarization caused by the local field, (4.381) includes implicitly the effects of short-range polarization also, on account of the possible departure of $V(\mathbf{r}, \mathbf{r}')$ from a Coulomb potential.

where $\boldsymbol{\alpha}\binom{l}{k}$ is the polarizability of the ion at $\binom{l}{k}$. In the shell-model picture and in the absence of valence electron interactions,

$$\boldsymbol{\alpha}\binom{l}{k} = \boldsymbol{\alpha}\binom{0}{k} = \boldsymbol{\alpha}(k) = (Y_k e)^2 \mathbf{K}^{-1}(k). \tag{4.386}$$

Equation (4.385) ignores the effects at $\binom{l}{k}$ produced by the dipoles induced at the various other sites. Assuming that the charge distortion associated with each ion is confined to its own proximity cell,* the electron density associated with the above set of dipoles is given by (see (4.367))

$$\Delta\rho_{\mathrm{NI}}(\mathbf{r}) = -\frac{1}{e}\sum_{lk\alpha\beta}\left[\frac{\partial}{\partial r_\alpha}\delta\left(\mathbf{r}-\mathbf{x}\binom{l}{k}\right)\right]\alpha_{\alpha\beta}(k)\left[\frac{\partial}{\partial r''_\beta}V_{\mathrm{ext}}(\mathbf{r}'')\right]_{\mathbf{r}''=\mathbf{x}\binom{l}{k}}$$

$$= \frac{1}{e}\int\sum_{lk\alpha\beta}\left[\frac{\partial}{\partial r_\alpha}\delta\left(\mathbf{r}-\mathbf{x}\binom{l}{k}\right)\right]\alpha_{\alpha\beta}(k)\frac{\partial}{\partial r'_\beta}\delta\left(\mathbf{r}'-\mathbf{x}\binom{l}{k}\right)V_{\mathrm{ext}}(\mathbf{r}')\,d\mathbf{r}'.$$

Upon comparison with (4.383), we find that $\chi_{\mathrm{sc}}(\mathbf{r},\mathbf{r}')$ for the shell model is given by

$$\chi_{\mathrm{sc}}(\mathbf{r},\mathbf{r}') = -\frac{1}{e^2}\sum_{lk\alpha\beta}\frac{\partial}{\partial r_\alpha}\delta\left(\mathbf{r}-\mathbf{x}\binom{l}{k}\right)\alpha_{\alpha\beta}(k)\frac{\partial}{\partial r'_\beta}\delta\left(\mathbf{r}'-\mathbf{x}\binom{l}{k}\right)$$

$$= -\sum_{lk\alpha\beta}\frac{\partial}{\partial r_\alpha}\delta\left(\mathbf{r}-\mathbf{x}\binom{l}{k}\right)Y_k^2[\mathbf{K}^{-1}(k)]_{\alpha\beta}\frac{\partial}{\partial r'_\beta}\delta\left(\mathbf{r}'-\mathbf{x}\binom{l}{k}\right). \tag{4.387}$$

We next turn our attention to the form of $\Delta\rho$. Remembering that valence charge deformation is described in the shell model by permitting the shells to have displacements different from those of the ion cores, we may write

$$\Delta\rho(\mathbf{r}) = \sum_{lk}Y_k\left\{\delta\left(\mathbf{r}-\mathbf{x}\binom{l}{k}-\mathbf{u}^s\binom{l}{k}\right)-\delta\left(\mathbf{r}-\mathbf{x}\binom{l}{k}\right)\right\}$$

$$= -\sum_{lk\alpha}Y_k u_\alpha^s\binom{l}{k}\frac{\partial}{\partial r_\alpha}\delta\left(\mathbf{r}-\mathbf{x}\binom{l}{k}\right). \tag{4.388}$$

Expressions (4.387) and (4.388) may now be used in (4.380) and (4.381) to recover all the results obtained earlier (Sec. 4.6.4). In particular, it is of interest to reexamine the self-consistency equation. Noting first that

$$V_{\mathrm{ext}}(\mathbf{r}) = -\frac{1}{e}V_{\mathrm{ie}}^{(1)}(\mathbf{r}) = \frac{1}{e}\sum_{lk\alpha}\frac{\partial U^{\mathrm{P}}}{\partial r_\alpha}\left(\mathbf{r}-\mathbf{x}\binom{l}{k}\right)\Big|_0 u_\alpha^c\binom{l}{k},$$

---

*A *proximity cell* is that region about a lattice site which is closer to that site than to any other. If the sites lie on a Bravais lattice, the proximity cell is a Wigner-Seitz cell. See P. T. Landsberg in *Solid State Theory*, edited by P. T. Landsberg (Wiley-Interscience, New York and London, 1969), p. 65.

where $U^P(\mathbf{r} - \mathbf{x}\binom{l}{k})$ is the (effective) potential energy of an electron at $\mathbf{r}$ due to an ion at $\mathbf{x}\binom{l}{k}$, we find after substituting (4.388) in (4.381) that

$$-\sum_{lk\alpha} Y_k u_\alpha^s\binom{l}{k} \frac{\partial}{\partial r_\alpha} \delta\left(\mathbf{r} - \mathbf{x}\binom{l}{k}\right)$$

$$= \int \chi_{sc}(\mathbf{r}, \mathbf{r}') V(\mathbf{r}', \mathbf{r}'') \left\{ -\sum_{l'k'\beta} Y_{k'} u_\beta^s\binom{l'}{k'} \frac{\partial}{\partial r_\beta''} \delta\left(\mathbf{r}'' - \mathbf{x}\binom{l'}{k'}\right) \right\} d\mathbf{r}' \, d\mathbf{r}''$$

$$+ \int \chi_{sc}(\mathbf{r}, \mathbf{r}') \left\{ -\sum_{l'k'\gamma} \frac{\partial}{\partial r_\gamma'} U^P\left(\mathbf{r}' - \mathbf{x}\binom{l'}{k'}\right) u_\gamma^c\binom{l'}{k'} \right\} d\mathbf{r}'. \quad (4.389)$$

Together with the result deduced in (4.387) for $\chi_{sc}$, this gives the shell-model version of the self-consistency requirement on the *induced electron density*. On the other hand, the shell model specifies a self-consistency requirement on the shell displacements. To obtain the latter condition, we first substitute for $\chi_{sc}(\mathbf{r}, \mathbf{r}')$ from (4.387), multiply (4.389) throughout by $r_\mu$, and then integrate over the proximity cell around $\binom{L'}{K'}$. Upon using the result*

$$\int_{\mathbf{r} \in \text{cell about }(L'K')} r_\mu \frac{\partial}{\partial r_\alpha} \delta\left(\mathbf{r} - \mathbf{x}\binom{l}{k}\right) d\mathbf{r} = 0 \quad \text{if } \binom{l}{k} \neq \binom{L'}{K'}$$

$$= -\delta_{\alpha\mu} \quad \text{if } \binom{l}{k} = \binom{L'}{K'},$$

we find

$$Y_{K'} u_\mu^s\binom{L'}{K'} = \int \sum_\beta Y_{K'}^2 [\mathbf{K}^{-1}(K')]_{\mu\beta} \frac{\partial}{\partial r_\beta'} \delta\left(\mathbf{r}' - \mathbf{x}\binom{L'}{K'}\right) V(\mathbf{r}', \mathbf{r}'')$$

$$\times \left\{ -\sum_{l'k'\gamma} Y_{k'} u_\gamma^s\binom{l'}{k'} \frac{\partial}{\partial r_\gamma''} \delta\left(\mathbf{r}'' - \mathbf{x}\binom{l'}{k'}\right) \right\} d\mathbf{r}' \, d\mathbf{r}''$$

$$+ \int \sum_\beta Y_{K'}^2 [\mathbf{K}^{-1}(K')]_{\mu\beta} \frac{\partial}{\partial r_\beta'} \delta\left(\mathbf{r}' - \mathbf{x}\binom{L'}{K'}\right)$$

$$\times \left\{ -\sum_{l'k'\gamma} \frac{\partial}{\partial r_\gamma'} U^P\left(\mathbf{r}' - \mathbf{x}\binom{l'}{k'}\right) u_\gamma^c\binom{l'}{k'} \right\} d\mathbf{r}'. \quad (4.390)$$

*This result is a special case of the property

$$\int f(\mathbf{r}) \frac{\partial}{\partial r_\alpha} \delta(\mathbf{r} - \mathbf{x}) d\mathbf{r} = -\frac{\partial f(\mathbf{r})}{\partial r_\alpha}\bigg|_{\mathbf{r} = \mathbf{x}}$$

which is an extension to three dimensions of the definition of derivatives of the $\delta$-function, namely,

$$\int_{-\infty}^{\infty} f(x) \frac{d^m}{dx^m} \delta(x) dx = (-1)^m \frac{d^m}{dx^m} f(x)\bigg|_{x=0}.$$

See A. Messiah, *Quantum Mechanics* (North-Holland, Amsterdam, 1961), Vol. I, App. A.

After integrations and rearrangement we get

$$\sum_{\mu} K_{\lambda\mu}(K')u^s_{\mu}\binom{L'}{K'} = -\sum_{l'k'\gamma} Y_{K'}Y_{k'} \frac{\partial}{\partial r'_{\lambda}} \frac{\partial}{\partial r''_{\gamma}} V(\mathbf{r}', \mathbf{r}'')\Big|_{\substack{\mathbf{r}'=\mathbf{x}\binom{L'}{K'}\\ \mathbf{r}''=\mathbf{x}\binom{l'}{k'}}} u^s_{\gamma}\binom{l'}{k'}$$

$$+ \sum_{l'k'\gamma} Y_{K'} \frac{\partial}{\partial r'_{\lambda}} \frac{\partial}{\partial r'_{\gamma}} U^P\left(\mathbf{r}' - \mathbf{x}\binom{l'}{k'}\right)\Big|_{\mathbf{r}'=\mathbf{x}\binom{L'}{K'}} u^c_{\gamma}\binom{l'}{k'}.$$

$$(4.391)$$

If we now split off from the second summation on the right-hand side of (4.391) the term with $(l'k') = (L'K')$ and recognize that in the shell model the coupling of an electron shell with its own core is described by the matrix $\mathbf{K}$ (see (4.237)), we see that

$$Y_{K'} \frac{\partial}{\partial r'_{\lambda}} \frac{\partial}{\partial r'_{\gamma}} U^P\left(\mathbf{r}' - \mathbf{x}\binom{L'}{K'}\right)\Big|_{\mathbf{r}'\to\mathbf{x}\binom{L'}{K'}} \quad \text{corresponds to } K_{\lambda\gamma}(K').$$

Accordingly, (4.391) can be rearranged to yield

$$\sum_{\mu} K_{\lambda\mu}(K')\left[u^s_{\mu}\binom{L'}{K'} - u^c_{\mu}\binom{L'}{K'}\right]$$

$$= -\sum_{l'k'\gamma} Y_{K'}Y_{k'} \frac{\partial}{\partial r'_{\lambda}} \frac{\partial}{\partial r''_{\gamma}} V(\mathbf{r}', \mathbf{r}'')\Big|_{\substack{\mathbf{r}'=\mathbf{x}\binom{L'}{K'}\\ \mathbf{r}''=\mathbf{x}\binom{l'}{k'}}} u^s_{\gamma}\binom{l'}{k'}$$

$$+ \sum_{\substack{l'k'\gamma\\(l'k'\neq L'K')}} Y_{K'} \frac{\partial}{\partial r'_{\lambda}} \frac{\partial}{\partial r'_{\gamma}} U^P\left(\mathbf{r}' - \mathbf{x}\binom{l'}{k'}\right)\Big|_{\mathbf{r}'=\mathbf{x}\binom{L'}{K'}} u^c_{\gamma}\binom{l'}{k'}. \quad (4.392)$$

Comparing this last equation with (4.245), we can see readily that the shell equation is really a disguised form of the self-consistency equation (4.381) of microscopic theory. What is even more significant is that the shell model replaces the *integral* equation for $\Delta\rho$ by an *algebraic* equation for the shell displacements. As we shall see in Chapter 7, such a transformation is a priori not possible in microscopic theory unless, guided by phenomenological considerations, one assumes suitable models for $\chi_{sc}(\mathbf{r}, \mathbf{r}')$. The utility of (4.387) and in particular its Fourier transform $\chi_{sc}(\mathbf{Q}, \mathbf{Q}')$ will be apparent then.

Finally, let us mention that one can relate $\Phi^{(2)}$ as given in (4.380) to the phenomenological expression (4.237). Details are given elsewhere.[82]

## References

1. See Appendix 2; also BH, p. 249.

2. R. H. Lyddane, R. G. Sachs, and E. Teller, Phys. Rev. **59**, 673 (1941).

3. BH, p. 397.

4. BH, p. 100.

5. BH, p. 339.

6. BH, p. 89; K. Huang, Proc. Roy. Soc. (London) **A208**, 352 (1951).

7. BH, p. 330.

8. C. A. Arguello, D. L. Rousseau, and S. P. S. Porto, Phys. Rev. **181**, 1351 (1969).

9. J. P. Mathieu and L. Couture-Mathieu, J. Phys. Radium **13**, 271 (1952).

10. A. Mooradian and A. L. McWhorter, Phys. Rev. Letters **19**, 849 (1967).

11. Kittel, p. 233.

12. W. Cochran, R. A. Cowley, G. Dolling, and M. M. Elcombe, Proc. Roy. Soc. (London) **A293**, 433 (1966).

13. J. Shah, T. C. Damen, J. F. Scott, and R. C. C. Leite, Phys. Rev. **B3**, 4238 (1971).

14. R. M. Pick, Advan. Phys. **19**, 269 (1970).

15. W. Cochran and R. A. Cowley, J. Phys. Chem. Solids **23**, 447 (1962).

16. J. F. Nye, *Physical Properties of Crystals* (Oxford University Press, London, 1964).

17. W. P. Mason, *Crystal Physics of Interaction Processes* (Academic Press, New York, 1966); C. S. Smith, Solid State Phys. **6**, 175 (1958).

18. BH, p. 262.

19. R. Zallen, Phys. Rev. **173**, 824 (1968).

20. Charles Cullen, *Matrices and Linear Transformations* (Addison-Wesley, Reading, Mass., 1966).

21. BH, p. 264.

22. W. Cochran, Advan. Phys. **10**, 401 (1961).

23. R. P. Feynman, R. B. Leighton, and M. Sands, *The Feynman Lectures on Physics* (Addison-Wesley, Reading, Mass., 1964) Vol. I, Chapter 31.

24. V. M. Agranovich and V. L. Ginzberg, *Spatial Dispersion in Crystal Optics and the Theory of Excitons* (Wiley-Interscience, New York, 1966).

25. BH, p. 336.

26. BH, p. 337.

27. J. P. Russell and R. Loudon, Proc. Phys. Soc. (London) **85**, 1029 (1965); S. P. S. Porto and J. F. Scott, Phys. Rev. **157**, 716 (1967).

28. A. S. Barker, Phys. Rev. **135**, A742 (1964).

29. R. Zallen, M. L. Slade, and A. T. Ward, Phys. Rev. **B3**, 4257 (1971).

30. BH, p. 334.

31. D. Pines and P. Nozieres, *The Theory of Quantum Liquids* (Benjamin, New York, 1966), p. 253.

32. J. F. Scott, T. C. Damen, R. C. C. Leite, and J. Shah, Phys. Rev. **B1**, 4330 (1970).

33. V. C. Sahni and G. Venkataraman, in Nusimovici, p. 121.

34. G. Venkataraman and V. C. Sahni, Rev. Mod. Phys. **42**, 409 (1970).

35. Kittel, p. 87.

36. E. W. Kellermann, Phil. Trans. Roy. Soc. London **A238**, 513 (1940).

37. S. O. Lundqvist, V. Lundstrom, E. Tenerz, and I. Waller, Ark. Fysik **15**, 193 (1959).

38. R. H. Lyddane and K. F. Herzfeld, Phys. Rev. **54**, 846 (1938).

39. L. Pauling, Proc. Roy. Soc. (London) **A114**, 181 (1927).

40. J. R. Hardy, Phil. Mag. **4**, 1278 (1959).

41. J. R. Hardy, Phil. Mag. **7**, 315 (1962).

42. J. R. Tessman, A. H. Kahn, and W. Shockley, Phys. Rev. **92**, 890 (1953).

43. P. O. Löwdin, *A Theoretical Investigation into Some Properties of Ionic Crystals* (Almqvist and Wiksells, Uppsala, 1948); see also Advan. Phys. **5**, 1 (1956).

44. S. O. Lundqvist, Ark. Fysik **6**, 25 (1952).

45. S. O. Lundqvist, Ark. Fysik **9**, 435 (1955).

46. S. O. Lundqvist, Ark. Fysik **12**, 263 (1957).

47. S. O. Lundqvist, Ark. Fysik **19**, 113 (1961).

48. B. Szigeti, Trans. Faraday Soc. **45**, 155 (1949).

49. B. Szigeti, Proc. Roy. Soc. (London) **A204**, 51 (1950); see also BH, p. 109 ff.

50. V. I. Odelevski, Izv. Akad. Nauk. Sev. Fiz. SSSR **14**, 232 (1950).

51. W. Cochran, Phys. Rev. Letters **2**, 495 (1959).

52. W. Cochran, Proc. Roy. Soc. (London) **A253**, 260 (1959).

53. A. D. B. Woods, W. Cochran, and B. N. Brockhouse, Phys. Rev. **119**, 980 (1960).

54. R. A. Cowley, W. Cochran, B. N. Brockhouse, and A. D. B. Woods, Phys. Rev. **131**, 1030 (1963).

55. B. G. Dick and A. W. Overhauser, Phys. Rev. **112**, 90 (1958).

56. BH, p. 113 ff.

57. K. B. Tolpygo, Zh. Eksp. Teor. Fiz. **20**, 497 (1950).

58. V. S. Mashkevich and K. B. Tolpygo, Zh. Eksp. Teor. Fiz. **32**, 520 (1957). [Sov. Phys.—JETP **5**, 435 (1957)].

59. K. B. Tolpygo, Fiz. Tverd. Tela **3**, 943 (1961). [Sov. Phys.—Solid State **3**, 685 (1961)].

60. J. Yamashita and T. Kurosawa, J. Phys. Soc. Japan **10**, 610 (1955).

61. E. E. Havinga, Phys. Rev. **119**, 1193 (1960).

62. See reference 22 above, especially the remarks following Eq. (2.10).

63. J. G. Traylor, H. G. Smith, R. M. Nicklow, and M. K. Wilkinson, Phys. Rev. **B3**, 3457 (1971).

64. R. A. Cowley, Proc. Roy. Soc. (London) **A268**, 121 (1962).

65. Kittel, p. 382.

66. M. A. Nusimovici, M. Balkanski, and J. L. Birman, Phys. Rev. **B1**, 595 (1970).

67. BH, p. 272.

68. See Eq. (7) in reference 45 above.

69. M. P. Verma and R. K. Singh, Phys. Stat. Sol. **33**, 769 (1969); **36**, 335 (1969); **38**, 851 (1970); Phys. Rev. **B2**, 4288 (1970).

70. L. A. Feldkamp, J. Phys. Chem. Solids **33**, 711 (1972).

71. R. L. Marston and B. G. Dick, Solid State Comm. **5**, 731 (1967); see also B. G. Dick, Phys. Rev. **129**, 1583 (1963).

72. R. L. Marston, Ph.D. Thesis, University of Utah (1967).

73. S. K. Sinha, Critical Reviews in Solid State Sciences **3**, 273 (1973).

74. R. A. Cowley, Proc. Roy. Soc. (London) **A268**, 109 (1962).

75. U. Schröder, Solid State Comm. **4**, 347 (1966); V. Nusslein and U. Schröder, Phys. Stat. Sol. **21**, 309 (1967).

76. M. J. L. Sangster, G. Peckham, and D. H. Saunderson, J. Phys. C **3**, 1026 (1970).

77. J. S. Melvin, J. D. Pirie, and T. Smith, Phys. Rev. **175**, 1082 (1968).

78. A. N. Basu and S. Sengupta, Phys. Stat. Sol. **29**, 367 (1968).

79. B. N. Brockhouse and P. K. Iyengar, Phys. Rev. **111**, 747 (1958).

80. G. Dolling, in Chalk River proceedings, Vol. II, p. 37.

81. Kittel, p. 382.

82. V. C. Sahni and G. Venkataraman (to be published).

# 5 Experimental Study of Lattice Vibrations

Although our main preoccupation in this volume is with lattice vibrations per se, it is of interest and value to know how they are studied experimentally. Of the several techniques developed for that purpose, we shall confine our attention to the more direct ones, although historically it was an indirect method (measurement of specific heat) which stimulated interest in lattice dynamics. We shall discuss slow-neutron scattering, infrared absorption, light scattering, and x-ray diffuse scattering, the presently important sources of experimental information. A few other promising methods will be mentioned briefly toward the end.

## 5.1 Slow-neutron scattering

### 5.1.1 Qualitative description

The neutron is essentially a nuclear particle and as such interacts with the nucleus via nuclear forces. This interaction leads to scattering which is isotropic (for low-energy neutrons),* spin dependent, and which can be characterized by a *scattering amplitude a*. Though the latter is not accessible from basic theory, it has been determined experimentally for many elements and for most may be considered real.

To appreciate how neutron scattering can be used to study phonons, consider first the interaction between a plane wave $e^{ikz}$ of neutrons and an *isolated bound* nucleus. As a result of the neutron–nucleus interaction, a scattered wave of the form $(a/r)e^{ikr}$ is produced. The differential cross section $(d\sigma/d\Omega)$ for scattering into unit solid angle in this case simply equals $a^2$; in other words, the scattering is isotropic, as stated earlier. On the other hand, if there are several nuclei then there will be several scattered waves and these may interfere. In the case of a single crystal (assuming for the moment that the nuclei are frozen in a periodic lattice), the interference effects are rather strong and scattering takes place only under very restricted conditions, the well-known Bragg conditions. In reality, of course, the atoms in a crystal are in a state of motion, and we know from Chapter 2 that the displacements can be represented as a superposition of traveling waves. These periodic modulations produced by the traveling waves give rise to their own characteristic reflections, similar to the Bragg reflections with the difference that the reflected waves are also "Doppler shifted." This occurrence of "modified Bragg reflections" due to the vibrational modulations of the lattice is similar to the ghost lines produced by a periodic ruling defect in an optical grating. By observing these satellite

---

*This isotropy stems from the fact that the range of the neutron-nucleus force ($\sim 10^{-13}$ cm) is very small compared to typical neutron wavelengths ($\sim 10^{-8}$ cm). Only $s$-wave scattering is important in these circumstances.

reflections and further measuring the "Doppler shifts" associated with them, the wave vector $\mathbf{q}$ and the frequency $\omega_j(\mathbf{q})$ of the phonon can be determined simultaneously. By repeating the experiment under varied conditions the dispersion relation $\omega = \omega_j(\mathbf{q})$ may be mapped.

At this stage it is necessary to recognize that the spin dependence of the scattering amplitude and the possible presence of several isotopes for each nuclear species causes the scattering amplitude even in a monatomic system to fluctuate randomly from site to site among its various possible values. Under these circumstances, one can split the scattering from each nucleus into two parts, called *coherent* and *incoherent*.[1] The corresponding scattering amplitudes, denoted by $a_{coh}$ and $a_{inc}$, respectively, are defined as

$$a_{coh} = \langle a \rangle, \qquad a_{inc} = \{\langle a^2 \rangle - \langle a \rangle^2\}^{1/2},$$

where $\langle \cdots \rangle$ denotes an appropriate average over the spin states and the isotopic distribution. To appreciate the significance of splitting the scattering in the above fashion, consider the differential cross section for an assembly of $N$ fixed nuclei having the same atomic number. We have

$$\frac{d\sigma}{d\Omega} = \left| \sum_{i=1}^{N} a_i e^{-i\mathbf{Q} \cdot \mathbf{r}_i} \right|^2 = \sum_i a_i^2 + \sum_{i(\neq j)} a_i a_j e^{-i\mathbf{Q} \cdot (\mathbf{r}_i - \mathbf{r}_j)}, \qquad (5.1)$$

where $\mathbf{r}_i$ is the position of the $i$th nucleus, and $\mathbf{Q} = \mathbf{k}' - \mathbf{k}$, $\mathbf{k}$ and $\mathbf{k}'$ being the incident and scattered neutron wave vectors, respectively. (Note in this case, $|\mathbf{k}| = |\mathbf{k}'|$.) Upon formally performing the average indicated by $\langle \cdots \rangle$, (5.1) becomes

$$\begin{aligned}\frac{d\sigma}{d\Omega} &= N\langle a^2 \rangle + \sum_{i(\neq j)} \langle a \rangle^2 e^{-i\mathbf{Q} \cdot (\mathbf{r}_i - \mathbf{r}_j)} \\ &= N\{\langle a^2 \rangle - \langle a \rangle^2\} + \left| \sum_i \langle a \rangle e^{-i\mathbf{Q} \cdot \mathbf{r}_i} \right|^2 \\ &= Na_{inc}^2 + \left| \sum_i a_{coh} e^{-i\mathbf{Q} \cdot \mathbf{r}_i} \right|^2,\end{aligned} \qquad (5.2)$$

where it is assumed that the scattering amplitudes of two *distinct* sites $i, j$ are not correlated. From (5.2) we observe that the total coherent scattering from the assembly is equal to the squared modulus of the sum of the coherently scattered wave amplitudes from the various nuclei. It therefore involves the interference of waves scattered by different nuclei. This is also true when the nuclei are free to move, so that coherent scattering essentially leads to information about the *cooperative motions of different nuclei*. The total incoherent intensity, on the other hand, is essentially the sum of the *intensities* contributed by the various nuclei separately and can therefore at best lead to a knowledge of *single-particle motions*. Both types of scattering have been exploited in the study of vibration spectra, although the former more than the latter.

### 5.1.2 Coherent scattering

We now examine coherent scattering in greater detail.[1-4] Now the partial differential coherent scattering cross section for neutrons of initial energy $E$ and momentum $\hbar\mathbf{k}$ to be scattered with energy $E'$ and momentum $\hbar\mathbf{k}'$ by a single crystal may be expressed formally as

$$\left(\frac{d^2\sigma}{d\Omega dE'}\right)_{\text{coh}} = \left(\frac{d^2\sigma}{d\Omega dE'}\right)^{(0)}_{\text{coh}} + \left(\frac{d^2\sigma}{d\Omega dE'}\right)^{(1)}_{\text{coh}} + \cdots, \tag{5.3}$$

where the different terms on the right-hand side denote, respectively, the contributions associated with the zero-phonon process, the one-phonon process, etc. Explicit calculation shows that

$$\left(\frac{d^2\sigma}{d\Omega dE'}\right)^{(0)}_{\text{coh}} = \frac{1}{\hbar}\frac{k'}{k}N^2 \, |F^{(0)}(\mathbf{Q})|^2 \Delta(\mathbf{Q})\delta(\omega), \tag{5.4}$$

where

$$\hbar\omega = \frac{\hbar^2}{2m_{\text{n}}}(k'^2 - k^2), \tag{5.5a}$$

$$\mathbf{Q} = \mathbf{k}' - \mathbf{k}, \tag{5.5b}$$

and $\Delta(\mathbf{Q})$ is defind in Eq. (A.1.3). Further $m_{\text{n}}$ is the neutron mass, and

$$F^{(0)}(\mathbf{Q}) = \sum_{\kappa=1}^{n} a^{\kappa}_{\text{coh}}e^{-W_{\kappa}(\mathbf{Q})}e^{-i\mathbf{Q}\cdot\mathbf{x}(\kappa)}. \tag{5.6}$$

Here $a^{\kappa}_{\text{coh}}$ is the coherent scattering amplitude of the nuclear species on the $\kappa$th sublattice, and

$$W_{\kappa}(\mathbf{Q}) = \tfrac{1}{2}\left\langle\left[\mathbf{Q}\cdot\mathbf{u}\begin{pmatrix} l \\ \kappa \end{pmatrix}\right]^2\right\rangle_T. \tag{5.7}$$

The quantity $e^{-W_{\kappa}(\mathbf{Q})}$ is called the Debye-Waller factor. Equation (5.4) describes Bragg scattering by neutrons and forms the basis for neutron crystallography.

Here we are interested in the first-order term of (5.3) which can be expressed as

$$\left(\frac{d^2\sigma}{d\Omega dE'}\right)^{(1)}_{\text{coh}} = \frac{1}{\hbar}\frac{k'}{k}N\sum_{\mathbf{q}j}|F^{(1)}(\mathbf{Q},\mathbf{q},j)|^2\Delta(\mathbf{Q}-\mathbf{q})$$

$$\times\left(\left[1+\left\langle n\begin{pmatrix}\mathbf{q}\\j\end{pmatrix}\right\rangle_T\right]\delta(\omega+\omega_j(\mathbf{q}))\right.$$

$$\left.+\left\langle n\begin{pmatrix}\mathbf{q}\\j\end{pmatrix}\right\rangle_T\delta(\omega-\omega_j(\mathbf{q}))\right) \tag{5.8}$$

with

$$F^{(1)} = \sum_{\kappa} \left(\frac{\hbar}{2m_\kappa \omega_j(\mathbf{q})}\right)^{1/2} a_{\text{coh}}^\kappa e^{-W_\kappa(\mathbf{Q})} e^{-i\mathbf{Q}\cdot\mathbf{x}(\kappa)} \left[\mathbf{Q} \cdot \mathbf{e}\left(\kappa \left|\begin{matrix}\mathbf{q}\\j\end{matrix}\right.\right)\right]. \qquad (5.9)$$

The overall form of (5.8) is similar to that for Bragg scattering, (5.4). We observe that the scattering is again singular and is governed by the equations

$$\mathbf{q} + \mathbf{G} = \mathbf{Q}, \qquad\qquad\qquad\qquad\qquad (5.10a)$$

$$\omega = \pm\omega_j(\mathbf{q}), \qquad\qquad\qquad\qquad\qquad (5.10b)$$

which are analogous to the conditions $\mathbf{Q} = \mathbf{G}$, $\omega = 0$ governing Bragg scattering.* Furthermore, just as the Bragg intensity is dictated by $|F^{(0)}|^2$, the intensity of scattering in the present case is governed by $|F^{(1)}|^2$. Equations (5.10a) and (5.10b) are frequently termed the conservation laws for quasi momentum and energy, respectively. Observe that the first-order process can lead either to phonon creation ($\omega = -\omega_j(\mathbf{q})$) or to phonon annihilation ($\omega = \omega_j(\mathbf{q})$). The temperature factors associated with the two processes naturally are different.

The experimental study of phonon dispersion relations by neutron scattering is based on Eqs. (5.10). The experiments are invariably carried out using neutron beams from reactors[2,6–8] which provide a copious supply of neutrons of energy $\sim 50$ meV and corresponding wave vectors $\sim 10^8$ cm$^{-1}$. The ranges available are ideally suited for phonon investigations, since they overlap with the ranges of typical phonon energies and wave vectors. However, reactor neutron beams are polychromatic and for convenience of experimentation first must be rendered monochromatic. The experiment then consists of scattering such a monochromatic beam off the specimen, analyzing the energy spectrum of the neutrons scattered at a particular angle, and then applying Eq. (5.10) to extract the phonon frequency and wave vector.

There are several ways of observing the one-phonon coherent scattering of neutrons. The one most often used is the so-called *constant-$\mathbf{Q}$ technique*.[2] In this, one first selects for study a particular phonon $(\mathbf{q}, j)$ in the irreducible prism. Next one decides which point $\mathbf{Q} = \mathbf{G} + \mathbf{q}$ in reciprocal space is optimum for the observation. The choice of this point is guided by instrumental considerations as well as by the factor $F^{(1)}$, often estimated from simple models for the dynamics. The experimental variables $\mathbf{k}$, $\mathbf{k}'$ and the crystal orientation are now adjusted so as to hold $\mathbf{Q}$ constant while

*Bragg's law $\lambda = 2d \sin \theta$ can easily be shown equivalent to the equations $\mathbf{Q} = \mathbf{G}$, $\omega = 0$.[5]

the scattered intensity is measured at a series of $\omega$ values. (See also Figure 5.1.) When the value of $\omega$ matches $\omega_j(\mathbf{q})$, then both the conservation equations (5.10) are satisfied and scattering occurs with an intensity proportional to $|F^{(1)}|^2$, resulting in a peak in the scattered neutron intensity. Ideally, the peak would be a $\delta$-function, but in practice it is broadened by resolution and in some cases by other effects, such as anharmonicity. A background due to multiple scattering and multiphonon effects may also be present. In spite of these perturbations, scattered neutron distributions often display clear one-phonon peaks from which the frequencies of the concerned phonons may be deduced. By repeating the experiment under appropriate conditions, complete sets of dispersion curves may be built up. It may be noted that the conservation equations (5.10) merely specify the wave vector and the frequency of the phonon. Information about the polarization of the phonon is contained in the structure factor $F^{(1)}$. Figure 5.1 shows a typical scattered neutron peak produced by one-phonon scattering in ZnS.[9]

The great merit of the neutron-scattering technique is that it permits the observation of phonons over the entire BZ. Further, the neutron's relatively small probability of being absorbed permits it to penetrate into the bulk, making scattering measurements insensitive to surface conditions. Another advantage is that the method may be applied equally well to both

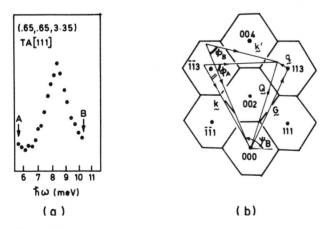

(a)               (b)

**Figure 5.1**
(a) shows the neutron scattering resonance corresponding to the TA phonon in ZnS along the [111] direction, with $(q/q_{max}) = 0.7$. Plotted here is the scattered intensity as a function of $\omega$ with $\mathbf{Q}$ held fixed. (b) shows the location in reciprocal space corresponding to the experiment. The vector diagrams correspond to the disposition of the various vectors at the low- and high-frequency ends of the scan. (After Feldkamp, ref. 9.)

metals and insulators. Against this is the necessity of using large crystals (typically several $cm^3$) because of the relatively low neutron fluxes currently available. Further, it is presently impossible, for purely experimental reasons, to investigate with high $\mathbf{q}$-space resolution the long-wavelength region ($q \sim 10^4$ $cm^{-1}$) which, as noted in the previous chapter, we know to be interesting in crystals exhibiting infrared activity. In spite of these shortcomings, slow-neutron coherent scattering has proved to be the most valuable technique for exploring vibration spectra.

The higher-phonon contributions to coherent scattering are important only at elevated temperatures. They have mainly a nuisance value (!) and will not be considered here.

### 5.1.3 Incoherent scattering

As already mentioned, incoherent scattering has also been exploited to some extent to obtain information about the vibration spectra of solids. In general, incoherent scattering experiments are not as informative for lattice vibration studies as those exploiting coherent scattering. The most fruitful application of the incoherent scattering technique consists of the study of narrow bands in the frequency spectra (such as bands due to librations) of hydrogenous compounds.[10]

To understand the principle of the method, let us consider the following phonon expansion for the partial differential cross section for incoherent scattering, analogous to (5.3):

$$\left(\frac{d^2\sigma}{d\Omega dE'}\right)_{inc} = \left(\frac{d^2\sigma}{d\Omega dE'}\right)^{(0)}_{inc} + \left(\frac{d^2\sigma}{d\Omega dE'}\right)^{(1)}_{inc} + \cdots. \tag{5.11}$$

Here

$$\left(\frac{d^2\sigma}{d\Omega dE'}\right)^{(0)}_{inc} = \frac{1}{\hbar}\frac{k'}{k}N\left(\sum_\kappa [a^\kappa_{inc}]^2 e^{-2W_\kappa(\mathbf{Q})}\right)\delta(\omega), \tag{5.12}$$

$$\left(\frac{d^2\sigma}{d\Omega dE'}\right)^{(1)}_{inc} = \frac{1}{\hbar}\frac{k'}{k}N\sum_\kappa [a^\kappa_{inc}]^2 e^{-2W_\kappa(\mathbf{Q})}\frac{3n\hbar}{2m_\kappa|\omega|}$$
$$\times \left\{\mathbf{Q}\cdot\mathbf{G}^\kappa(\omega)\cdot\mathbf{Q}\frac{1}{e^x - 1} + \mathbf{Q}\cdot\mathbf{G}^\kappa_-(\omega)\cdot\mathbf{Q}\frac{e^x}{e^x - 1}\right\},$$

$$x = \frac{\hbar|\omega|}{k_B T}. \tag{5.13}$$

The elements of the tensors[11] $\mathbf{G}^\kappa(\omega)$ and $\mathbf{G}^\kappa_-(\omega)$ are defined by (remember $\omega_j(\mathbf{q})$ is positive by convention)

$$G^{\kappa}_{\alpha\beta}(\omega) = \frac{1}{3nN} \sum_{\mathbf{q}j} e^*_\alpha\left(\kappa \left| \begin{matrix} \mathbf{q} \\ j \end{matrix} \right. \right) e_\beta\left(\kappa \left| \begin{matrix} \mathbf{q} \\ j \end{matrix} \right. \right) \delta\{\omega - \omega_j(\mathbf{q})\};$$  (5.14a)

$$G^{\kappa}_{-\alpha\beta}(\omega) = \frac{1}{3nN} \sum_{\mathbf{q}j} e^*_\alpha\left(\kappa \left| \begin{matrix} \mathbf{q} \\ j \end{matrix} \right. \right) e_\beta\left(\kappa \left| \begin{matrix} \mathbf{q} \\ j \end{matrix} \right. \right) \delta\{\omega + \omega_j(\mathbf{q})\},$$  (5.14b)

and have the properties

$$G^{\kappa}_{\alpha\beta}(\omega) = G^{\kappa}_{-\alpha\beta}(-\omega),$$  (5.15)

$$\sum_\kappa \sum_\alpha G^{\kappa}_{\alpha\alpha}(\omega) = \frac{1}{3nN} \sum_{\mathbf{q}j} \delta\{\omega - \omega_j(\mathbf{q})\} = g(\omega).$$  (5.16)

The zero-phonon term (5.12) represents elastic scattering and is not important as far as crystal dynamics is concerned, apart from the Debye-Waller factor which, in some instances, can be studied to advantage.[12]

The one-phonon incoherent scattering is essentially proportional to the tensor $\mathbf{G}^\kappa(\omega)$ which, as seen from (5.16), is related to $g(\omega)$. To this extent, features in $g(\omega)$ may be expected to appear in incoherent scattering, and it is this fact that experimentalists seek to exploit.

Figure 5.2 shows a typical example of incoherent scattering spectra.[13] Plotted in this figure are distributions of neutrons scattered through $90°$ from $Ba(ClO_3)_2 \cdot H_2O$. The scattering here is predominantly from protons and is incoherent.† The peaks in the figure are related to peaks in $g(\omega)$ associated with the librations of the water molecule. The primitive cell of barium chlorate monohydrate contains two formula units and therefore two water molecules. Fortunately, not only are the planes of the water molecules parallel, but so also are the two H–H vectors. Considering the librational degrees of freedom, one would expect six librational branches. It is believed, however, that the intermolecular forces are weak so that the six branches effectively appear as three. For the same reason it is expected that these branches will be rather flat. Hence one expects three well-defined peaks in $g(\omega)$ due to librations of the water molecule about its principal axes and, correspondingly, three peaks in the scattered neutron spectrum.

Figure 5.2a shows the spectrum observed in a polycrystalline sample, while 5.2b and 5.2c show the spectra obtained with a single crystal in different orientations. Because of the factor $\mathbf{Q} \cdot \mathbf{G}^\kappa(\omega) \cdot \mathbf{Q}$, different modes are excited under different orientations, and it was possible by analyzing the data to assign the peak at $477 \text{ cm}^{-1}$ uniquely to the rocking mode $R$.

†This arises from the spin dependence of the neutron-proton scattering.

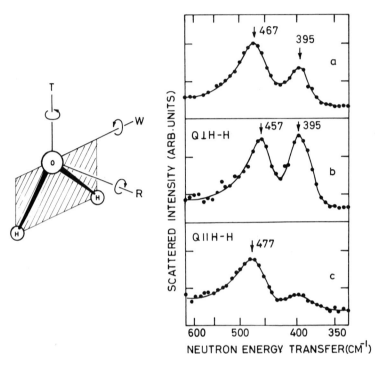

**Figure 5.2**
Librational peaks in Ba(ClO$_3$)$_2$ ·H$_2$O observed at 120 K and 90° scattering angle. (a) shows the pattern observed with a polycrystalline sample; (b) was obtained with a single crystal with **Q** ⊥ H-H vector and also the plane of the molecule. This shows the twisting (T) and waving (W) modes. The rocking (R) mode is observed with **Q** ∥ H-H and is shown in (c). (After Thaper et al., ref. 13.)

## 5.2 Infrared absorption

### 5.2.1 Basic ideas

Although the neutron has proved to be a very powerful probe for the study of vibration spectra, it does not have a monopoly in this respect. Indeed, the use of photons belonging to various parts of the electromagnetic spectrum for the same purpose predates considerably the use of neutrons, extending in some cases as far back as the beginning of this century.

Now the frequency of infrared radiation is of the same order as the vibrational frequencies of crystal lattices. Consequently when such radiation is directed on a crystal, it can couple with the oscillating dipole moments in the crystal and transfer energy. By studying the absorption process, therefore, one can hope to learn something about the vibration spectrum. In practice such experiments are possible only in insulating crystals because the conductivity of metals suppresses the vibrational effects.*

Infrared experiments fall basically into two categories, absorption and reflection. In an absorption experiment, the transmission of a *thin* slab of the specimen is investigated as a function of frequency, and vibrational information is extracted from the observed transmission resonances. In a reflection experiment one can reconstruct parts of the vibration spectrum from the reflected intensity when a monochromatic beam is incident normally on a *thick* (effectively semi-infinite) slab of the crystal.

Crucial to an understanding of infrared experiments is an appreciation of the behavior of electromagnetic waves inside a dielectric medium.† The equations governing these waves are

$$\mathbf{E}(\mathbf{q}) = \frac{4\pi}{n^2 - 1}\{\mathbf{P}(\mathbf{q}) - n^2\hat{\mathbf{q}}[\hat{\mathbf{q}} \cdot \mathbf{P}(\mathbf{q})]\} \tag{5.17}$$

and

$$\mathbf{D} = \varepsilon\mathbf{E}, \tag{5.18}$$

already familiar to us (see (4.78b) and (4.167)). From these we obtain Fresnel's equation (see (4.185))

$$\|\varepsilon(\omega) - n^2[\mathbb{1}_3 - \hat{\mathbf{Q}}]\| = 0 \tag{5.19}$$

---

*Another way of stating this is that pure vibrational spectra can be observed in absorption only when the incident frequency is smaller than the energy gap between filled and unfilled electronic energy states. In metals there is no such gap whence photon absorption is always accompanied by electronic transitions.

†For simplicity we shall assume that the crystal has no optical activity.

which is often given in books on optics in the form*

$$\sum_\alpha \frac{\hat{q}_\alpha^2}{\left(\dfrac{1}{n^2} - \dfrac{1}{\varepsilon_\alpha}\right)} = 0, \tag{5.20}$$

where $\varepsilon_\alpha$ ($\alpha = x, y, z$) denotes a component of $\boldsymbol{\varepsilon}$ in the principal-axis representation.

Solving either (5.19) or (5.20) for a given $\hat{\mathbf{q}}$ leads to two values $n^{(1)}(\hat{\mathbf{q}}, \omega)$ and $n^{(2)}(\hat{\mathbf{q}}, \omega)$, respectively, for the refractive index.† It is interesting that although we are ignoring terms of order $|\mathbf{q}|$ (and above) in both the dielectric function and the refractive index, the direction $\hat{\mathbf{q}}$ is important for the latter.

A useful geometrical construction that readily permits the determination of the two values of $n(\hat{\mathbf{q}}, \omega)$ is the *indicatrix* ellipsoid[14]

$$\frac{x^2}{n_x^2} + \frac{y^2}{n_y^2} + \frac{z^2}{n_z^2} = 1, \tag{5.21}$$

where $n_\alpha^2 = \varepsilon_\alpha$ and the axes are chosen to coincide with the principal axes of $\boldsymbol{\varepsilon}$. The required values of the refractive index are given by the semimajor and semiminor axes of the ellipse that results when a section of the indicatrix is taken through the origin and normal to $\hat{\mathbf{q}}$. Further, the **D** vectors associated with the waves characterized by the two refractive indices are directed along the corresponding axes of the ellipse.

The above résumé of basic crystal optics disregards absorption. However, the latter is readily allowed for by formally permitting $n$ to be complex, that is, by replacing $n$ by $\bar{n}$ where

$$\bar{n}(\hat{\mathbf{q}}, \omega) = n(\hat{\mathbf{q}}, \omega) + i\mathcal{K}(\hat{\mathbf{q}}, \omega) \tag{5.22}$$

($n$ and $\mathcal{K}$ real). Correspondingly, the real dielectric function considered thus far must be replaced by

$$\varepsilon(\omega) = \varepsilon'(\omega) + i\varepsilon''(\omega).‡ \tag{5.23}$$

The real and imaginary parts of $\bar{n}$ are jointly referred to as *optical constants*, while $\mathcal{K}$ is termed the *extinction coefficient*. It is relevant to note that allowing the refractive index to be complex causes an alteration of the usual phase factor

---

*For a derivation of this form see BH, p. 408.
†By convention $n$ is defined as positive so that only the positive square root of $n^2$ is considered.
‡In general, the principal axes of the real and imaginary parts need not coincide. Fortunately, however, they do so for all except monoclinic and triclinic classes.

$$\exp\{i(\mathbf{q} \cdot \mathbf{r} - \omega t)\} = \exp\left\{i\left(\frac{\omega n}{c}\hat{\mathbf{q}} \cdot \mathbf{r} - \omega t\right)\right\}$$

to

$$\exp\left\{i\left(\frac{\omega \bar{n}}{c}\hat{\mathbf{q}} \cdot \mathbf{r} - \omega t\right)\right\} = \exp\left\{i\left(\frac{\omega n}{c}\hat{\mathbf{q}} \cdot \mathbf{r} - \omega t\right)\right\}\exp\left(-\frac{\omega \mathcal{K}}{c}\hat{\mathbf{q}} \cdot \mathbf{r}\right),$$

which implies an attenuation of the wave amplitude inside the medium, and hence absorption.

Based on the above we can argue that if a plane electromagnetic wave is incident normally on a slab cut from a single crystal, then upon entering the medium it splits into two linearly polarized waves. The directions of polarization of these waves and also their associated refractive indices can be determined by taking a section of the indicatrix normal to $\hat{\mathbf{q}}$. In particular, if the incident beam is so polarized as to have its electric vector parallel to one of the two permitted directions for $\mathbf{D}$, then only the corresponding wave will be propagated inside the slab.

Let us now consider the normal incidence of such a linearly polarized beam on and its subsequent passage through a slab. Born and Huang[15] have shown that if the thickness $d$ of the slab is small compared to the wavelength $\lambda$, then the transmission $T$ through the slab is given by

$$T = \frac{\text{intensity of emerging beam}}{\text{intensity of incident beam}}$$

$$= \frac{1}{1 + (2\omega d/c)n(\hat{\mathbf{q}}, \omega)\mathcal{K}(\hat{\mathbf{q}}, \omega)}. \qquad (5.24)$$

For cubic crystals

$$\bar{n}^2 = \varepsilon = \varepsilon' + i\varepsilon'', \qquad (5.25)$$

so that (5.24) simplifies to

$$T = \frac{1}{1 + (\omega d/c)\varepsilon''(\omega)}. \qquad (5.26)$$

Remembering the connection between refraction and the dielectric properties, it becomes clear from (5.24) that optical absorption in insulating crystals is crucially dependent on $\varepsilon(\omega)$ and through the latter on the vibrational properties, for frequencies in the infrared. It must also be evident that the complex nature of $\varepsilon$, which was ignored in Chapter 4, is vital to a discussion of optical absorption.

The extension of (4.155) to the complex form is easily achieved by a method due to Born and Huang. Noting (for example from (5.26)) that a complex dielectric dispersion formula implies absorption and in turn energy dissipation, we introduce an ad hoc damping term into the equations of motion (4.153) as follows:

$$\omega^2 \mathbf{b} = [\mathbf{\Omega}(\text{IR}) - i\omega\boldsymbol{\gamma}]\mathbf{b} - \mathscr{Z}(0)\mathbf{E}.$$

The elements of the matrix $[\mathbf{\Omega}(\text{IR}) - i\omega\boldsymbol{\gamma}]$ are given by $\omega_\lambda^2(0) - i\omega\gamma_\lambda$, where $\gamma_\lambda$ is a positive quantity and denotes the damping parameter for the $\lambda$th IR-active mode. Use of the above equation together with (4.154) then gives in place of (4.155):

$$\varepsilon_{\alpha\beta}(\omega) = \varepsilon_{\alpha\beta}(\infty) + \frac{4\pi}{Mv} \sum_{\lambda=1}^{r} \frac{\mathscr{Z}_{\alpha\lambda}(0)\mathscr{Z}_{\beta\lambda}(0)}{\omega_\lambda^2(0) - \omega^2 - i\omega\gamma_\lambda}. \tag{5.27a}$$

It is easily seen that $\varepsilon_{\alpha\beta}(\omega)$ is analytic* in the upper half of the complex $\omega$ plane, whence it is possible to derive the following results[16] (often referred to as the Kramers-Kronig (K–K) relations):†

$$\varepsilon'_{\alpha\beta}(\omega) - \varepsilon_{\alpha\beta}(\infty) = \frac{2}{\pi}\mathscr{P} \int_0^\infty \frac{\varepsilon''_{\alpha\beta}(\omega')\omega' \, d\omega'}{\omega'^2 - \omega^2} \tag{5.28a}$$

$$\varepsilon''_{\alpha\beta}(\omega) = -\frac{2\omega}{\pi}\mathscr{P} \int_0^\infty \frac{\varepsilon'_{\alpha\beta}(\omega') - \varepsilon_{\alpha\beta}(\infty)}{\omega'^2 - \omega^2} d\omega'. \tag{5.28b}$$

In these relations

$$\mathscr{P} \int_0^\infty \frac{f(\omega') \, d\omega'}{\omega'^2 - \omega^2}$$

means

$$\lim_{\delta \to 0} \left( \int_0^{\omega - \delta} \frac{f(\omega') \, d\omega'}{\omega'^2 - \omega^2} + \int_{\omega + \delta}^\infty \frac{f(\omega') \, d\omega'}{\omega'^2 - \omega^2} \right).$$

It must be emphasized that the validity of the K–K relations rests only on the analytical properties of $\boldsymbol{\varepsilon}(\omega)$ and in fact applies to any causal response function (including a generalization of (5.27a) to include multiphonon contributions).

Reverting to (5.27a), let us rationalize the denominator of the second term on the right-hand side and let $\gamma_\lambda \to 0$. Using the results

$$\lim_{\alpha \to 0} \frac{1}{\pi} \frac{\alpha}{\alpha^2 + x^2} = \delta(x)$$

---

*It is interesting that if $\gamma_\lambda$ is negative, then one has a dynamic instability as the displacements have the time dependence $\exp\{-i\omega_\lambda(0)t - \gamma_\lambda t\}$. In this case $\varepsilon_{\alpha\beta}(\omega)$ is no longer analytic in the upper half-plane and the K–K relations (5.28) do not apply. This is of interest in the context of certain phase transitions.

†Here $\varepsilon(\infty)$ has the same meaning as in the previous chapter and is assumed to be real.

and

$$\delta[f(x)] = \sum_n \frac{1}{|(df/dx)_{x_n}|} \delta(x - x_n), \qquad f(x_n) = 0,$$

we obtain the "harmonic" result

$$\varepsilon_{\alpha\beta}(\omega) = \varepsilon_{\alpha\beta}(\infty) + \frac{4\pi}{Mv} \sum_{\lambda=1}^{r} \frac{\mathscr{Z}_{\alpha\lambda}(0)\mathscr{Z}_{\beta\lambda}(0)}{\omega_\lambda^2(0) - \omega^2}$$
$$- \frac{i\pi}{2} \sum_{\lambda=1}^{r} \frac{4\pi}{Mv} \frac{\mathscr{Z}_{\alpha\lambda}(0)\mathscr{Z}_{\beta\lambda}(0)}{\omega_\lambda(0)} \{\delta[\omega_\lambda(0) + \omega] - \delta[\omega_\lambda(0) - \omega]\}.$$

(5.27b)

From our present standpoint, (5.27b) is too restrictive as it describes only the one-phonon contribution to the dielectric function. This is a consequence of the polarization **P** depending only *linearly* on the displacements (see Eq. (4.154)). On the other hand, it is now well established from experiments that there can also be multiphonon contributions to the absorption. To permit a discussion of such effects, it is necessary to consider a more general form of $\varepsilon(\omega)$ derived from a quantum treatment. This gives[17]*

$$\varepsilon(\omega) = 1 + 4\pi\chi(\omega),$$

with

$$\chi_{\alpha\beta}(\omega) - \chi_{\alpha\beta}(\infty) = \frac{1}{Nvh} \sum_i \sum_f p_i(T) \left\{ \frac{\langle i|M_\beta|f\rangle\langle f|M_\alpha|i\rangle}{(\omega_f - \omega_i) + \omega} \right.$$
$$\left. - i\pi\langle i|M_\beta|f\rangle\langle f|M_\alpha|i\rangle\delta[(\omega_f - \omega_i) + \omega] \right\}$$
$$+ \frac{1}{Nvh} \sum_{i'} \sum_{f'} p_{i'}(T) \left\{ \frac{\langle i'|M_\alpha|f'\rangle\langle f'|M_\beta|i'\rangle}{(\omega_{f'} - \omega_{i'}) - \omega} \right.$$
$$\left. + i\pi\langle i'|M_\alpha|f'\rangle\langle f'|M_\beta|i'\rangle\delta[(\omega_{f'} - \omega_{i'}) - \omega] \right\}.$$

(5.29)

Here $|i\rangle$ and $|f\rangle$ denote the initial and final vibrational states of the crystal corresponding to energies $\hbar\omega_i$ and $\hbar\omega_f$, respectively; $p_i(T)$ is the population factor for the state $|i\rangle$ (similar meanings apply to the primed quantities); **M** is the dipole moment operator for the crystal and, under the adiabatic assumption, is a function of the nuclear positions alone.[18] This

---

*We have made slight notational changes as compared to the expression cited in ref. 17. Also we have explicitly indicated thermal averaging.

permits us to make the following Taylor expansion (see (2.19)):

$$M_\alpha\left(\left\{\mathbf{r}\binom{l}{k}\right\}\right) = M_\alpha^{(0)} + M_\alpha^{(1)} + M_\alpha^{(2)} + \cdots, \tag{5.30}$$

where

$$M_\alpha^{(0)} = M_\alpha\left(\left\{\mathbf{x}\binom{l}{k}\right\}\right), \tag{5.31a}$$

$$M_\alpha^{(1)} = \sum_{lk\beta} \left.\frac{\partial M_\alpha}{\partial u_\beta\binom{l}{k}}\right|_0 u_\beta\binom{l}{k} = \sum_{lk\beta} M_{\alpha,\beta}\binom{l}{k} u_\beta\binom{l}{k}, \tag{5.31b}$$

$$M_\alpha^{(2)} = \frac{1}{2}\sum_{lk\beta}\sum_{l'k'\gamma} \left.\frac{\partial^2 M_\alpha}{\partial u_\beta\binom{l}{k}\partial u_\gamma\binom{l'}{k'}}\right|_0 u_\beta\binom{l}{k}u_\gamma\binom{l'}{k'}$$

$$= \frac{1}{2}\sum_{lk\beta}\sum_{l'k'\gamma} M_{\alpha,\beta\gamma}\binom{l\ l'}{k\ k'} u_\beta\binom{l}{k}u_\gamma\binom{l'}{k'}, \tag{5.31c}$$

etc. Next we express (5.30) in $\mathbf{q}$ space using the transformation

$$u_\alpha\binom{l}{k} = \frac{1}{\sqrt{Nm_k}}\sum_{\mathbf{q}j} e_\alpha\left(k\left|\begin{matrix}\mathbf{q}\\j\end{matrix}\right.\right)Q\binom{\mathbf{q}}{j}e^{i\mathbf{q}\cdot\mathbf{x}(l)}$$

(see (2.60)). This gives[19]

$$M_\alpha = M_\alpha^{(0)} + \sqrt{N}\sum_{\mathbf{q}j} M_\alpha\binom{\mathbf{q}}{j}Q\binom{\mathbf{q}}{j}\Delta(\mathbf{q})$$

$$+ \frac{1}{2}\sum_{\mathbf{q}j}\sum_{\mathbf{q}'j'} M_\alpha\binom{\mathbf{q}\ \mathbf{q}'}{j\ j'}Q\binom{\mathbf{q}}{j}Q\binom{\mathbf{q}'}{j'}\Delta(\mathbf{q}+\mathbf{q}') + \cdots, \tag{5.32}$$

where

$$M_\alpha\binom{\mathbf{q}}{j} = \sum_{k\beta}\frac{1}{\sqrt{m_k}}M_{\alpha,\beta}\binom{0}{k}e_\beta\left(k\left|\begin{matrix}\mathbf{q}\\j\end{matrix}\right.\right), \tag{5.33a}$$

$$M_\alpha\binom{\mathbf{q}\ \mathbf{q}'}{j\ j'} = \sum_{\substack{l'kk'\\ \beta\gamma}}\frac{1}{\sqrt{m_km_{k'}}}M_{\alpha,\beta\gamma}\binom{0\ l'}{k\ k'}e_\beta\left(k\left|\begin{matrix}\mathbf{q}\\j\end{matrix}\right.\right)e_\gamma\left(k'\left|\begin{matrix}\mathbf{q}'\\j'\end{matrix}\right.\right)e^{i\mathbf{q}'\cdot\mathbf{x}(l')}, \tag{5.33b}$$

etc. Bearing in mind the $\mathbf{q}$-space restriction, we find from (5.32) that the *first-order* contribution to the dipole moment arising from the *displacements* is $\mathbf{M}\binom{0}{j}$. On the other hand, from (4.148) we know that the amplitude of the displacive moment due to the long-wavelength mode $\binom{0}{j}$ is given by $\mathscr{L}_j$. Recalling the differences in the eigenvectors entering the definitions of the two quantities, we get

$$\mathscr{L}_j = \sqrt{M}\,\mathbf{M}\binom{0}{j}. \tag{5.34}$$

In particular, for the rigid-ion model[20]

$$M_\alpha = M_\alpha^{(0)} + \sum_{lk} Z_k e u_\alpha \binom{l}{k},$$

(5.35)

implying

$$M_{\alpha,\beta}\binom{l}{k} = Z_k e \delta_{\alpha\beta}, \qquad M_{\alpha,\beta\gamma}\binom{l\ l'}{k\ k'} = 0;$$

(5.36)

the other higher-order coefficients vanish similarly. Thus the quadratic and the higher-power terms of the displacement in (5.30) arise specifically due to electronic polarization. However, they cannot be calculated using the phenomenological models discussed in the previous chapter, since those models can yield only the linear contribution. The nonlinear terms characterize *electrical anharmonicity*[21] and require a more general treatment of the polarization problem.

Returning to (5.29), one must now employ the expansion (5.32) for $\mathbf{M}$ and then use the results of Sec. 2.6 to evaluate the matrix elements. This will deliver $\chi(\omega)$ in a suitable form for use in Fresnel's equation and an eventual calculation of absorption. By way of indicating the final structure of $\chi$, we give below the result for $\chi''(\omega)$, $(\omega > 0)$:

$$
\begin{aligned}
\chi''_{\alpha\beta}(\omega) = {} & \frac{\pi}{v} \sum_{\mathbf{q}j} M_\alpha\binom{\mathbf{q}}{j} M_\beta\binom{\mathbf{q}}{j} \frac{1}{2\omega_j(\mathbf{q})} \Delta(\mathbf{q}) \delta[\omega - \omega_j(\mathbf{q})] \\
& + \frac{\pi\hbar}{2Nv} \sum_{\mathbf{q}j} \sum_{\mathbf{q}'j'} M_\alpha\binom{\mathbf{q}\ \mathbf{q}'}{j\ j'} M_\beta\binom{\mathbf{q}\ \mathbf{q}'}{j'\ j} \frac{1}{4\omega_j(\mathbf{q})\omega_{j'}(\mathbf{q}')} \Delta(\mathbf{q} + \mathbf{q}') \\
& \times \left\{ \left(1 + \left\langle n\binom{\mathbf{q}}{j}\right\rangle_T + \left\langle n\binom{\mathbf{q}'}{j'}\right\rangle_T \right) \delta\{\omega - [\omega_j(\mathbf{q}) + \omega_{j'}(\mathbf{q}')]\} \right. \\
& \left. + 2\left(\left\langle n\binom{\mathbf{q}'}{j'}\right\rangle_T - \left\langle n\binom{\mathbf{q}}{j}\right\rangle_T \right) \delta\{\omega - [\omega_j(\mathbf{q}) - \omega_{j'}(\mathbf{q}')]\} \right\} \\
& + \cdots.
\end{aligned}
$$

(5.37)

The structural similarity of (5.37) to the phonon expansion formulas (5.3) and (5.11) can hardly escape notice. Again, the different terms of (5.37) lead to processes involving various numbers of phonons.* Observe, incidentally, that the one-phonon contribution to $\chi''$ obtained above is consistent with that in (5.27b) when use is made of (5.34).

### 5.2.2 One-phonon process
First-order absorption occurs due to the creation of a phonon by a photon. Owing to the factor $\Delta(\mathbf{q})$, only the long-wavelength phonons can lead to

*In the case of molecules, where a similar expansion is possible, an additional contribution to $\chi''$ arises from $M^{(0)}$ and describes the absorption associated with rotational transitions.

absorption in the first order. Further, recalling (5.34) and the discussion of Sec. 4.3, it is clear that first-order absorption can be produced only by those modes $\binom{0}{j}$ for which $\mathscr{L}_j$ is nonvanishing. The frequencies at which the absorption resonances appear are dictated by the maxima in the product $n(\hat{\mathbf{q}}, \omega)\mathscr{K}(\hat{\mathbf{q}}, \omega)$. After somewhat lengthy algebra it can be shown that, for one-phonon processes and provided the damping is small, these maxima appear at the limiting frequencies $\omega_\lambda(\hat{\mathbf{q}})$ (see Sec. 4.3) for modes $(\hat{\mathbf{q}}, \lambda)$ which have a nonvanishing component of dipole moment along the electric field of the incident radiation. Later we shall discuss an example which illustrates this feature. For the present we may write the conservation rules for the one-phonon process as

$$\mathbf{q} \approx 0$$
$$\omega_f - \omega_i = \omega = \omega_\lambda(\hat{\mathbf{q}}). \tag{5.38}$$

Observe that the one-phonon absorption intensity is temperature independent.

Figure 5.3 shows the results of a typical absorption experiment performed on GaAs.[22] The data illustrate the need for very thin samples and the difficulty of making accurate transmission measurements. A notable feature is that the resonance line is not a $\delta$-function but is broadened by anharmonicity.

Generally, first-order infrared processes are studied in reflection rather than transmission. In a reflection experiment, a monochromatic beam of

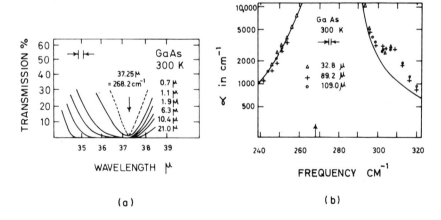

(a)                                    (b)

**Figure 5.3**
(a) Transmission for thin films of GaAs at 300 K. (b) Absorption coefficient $\alpha(\omega)$ deduced from transmission experiments. Owing to the difficulty of making accurate measurements around the resonance, only the data for the wings are shown. The solid line is based on the oscillator formula discussed in the text. (After Iwasa et al., ref. 22.)

frequency $\omega$ is directed normally on a thick slab of the crystal and the reflectivity $R(\hat{\mathbf{q}}, \omega)$ is observed. From optics[23]

$$R(\hat{\mathbf{q}}, \omega) = \left| \frac{\bar{n}(\hat{\mathbf{q}}, \omega) - 1}{\bar{n}(\hat{\mathbf{q}}, \omega) + 1} \right|^2. \tag{5.39}$$

Since $\bar{n}$ is related to $\varepsilon$, it follows that a study of $R(\hat{\mathbf{q}}, \omega)$ can lead to knowledge of the long-wavelength vibration frequencies.

Figure 5.4 shows the results of a typical reflection experiment.[22] The characteristic feature is the presence of a flat-topped band extending from $\omega_{TO}$ to $\omega_{LO}$. Physically this is easy to understand by referring to Figure 5.5 wherein polariton curves are sketched alongside those for the optical constants and the dielectric function. From that figure it is clear that the crystal cannot sustain waves with electromagnetic field components if the frequency $\omega$ lies between $\omega_{TO}$ and $\omega_{LO}$. Hence if electromagnetic waves with frequencies in this interval are introduced into the crystal from outside, they are reflected.

Two methods of analyzing reflection data have been used. The first utilizes oscillator formulas such as (5.27) but modified phenomenologically to allow for damping. Expressions for cubic and uniaxial crystals, to which reflection experiments have so far largely been confined, are:

*cubic*

$$\varepsilon(\omega) = \varepsilon(\infty) + \sum_{\lambda} \frac{S_{\lambda}\omega_{\lambda}^2(0)}{\omega_{\lambda}^2(0) - \omega^2 - i\gamma_{\lambda}\omega}, \tag{5.40}$$

*uniaxial*

$$\varepsilon_{\parallel}(\omega) = \varepsilon_{\parallel}(\infty) + \sum_{\lambda} \frac{S_{\parallel,\lambda}\omega_{\parallel,\lambda}^2(0)}{\omega_{\parallel,\lambda}^2(0) - \omega^2 - i\gamma_{\parallel,\lambda}\omega} \tag{5.41}$$

$$\varepsilon_{\perp}(\omega) = \varepsilon_{\perp}(\infty) + \sum_{\lambda} \frac{S_{\perp,\lambda}\omega_{\perp,\lambda}^2(0)}{\omega_{\perp,\lambda}^2(0) - \omega^2 - i\gamma_{\perp,\lambda}\omega}. \tag{5.42}$$

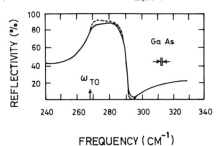

**Figure 5.4**
Reflection spectrum observed from GaAs (solid line). The dotted line is based on the oscillator formula. (After Iwasa et al., ref. 22.)

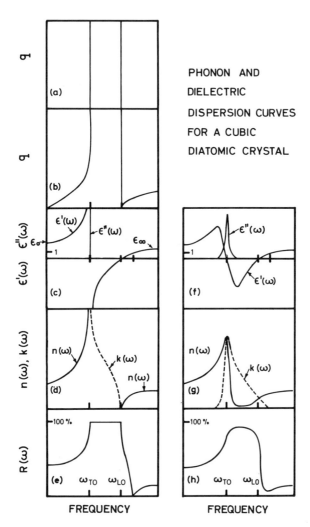

**Figure 5.5**
Schematic plots of various quantities associated with reflection of light from a cubic diatomic crystal. (a) Optic modes in the electrostatic approximation (long-wavelength part). (b) Polariton dispersion curves. (c) Real and imaginary parts of the complex dielectric function in the harmonic approximation. (d) Optical constants in the harmonic case. (e) Reflective spectrum based on (c) and (d). The plots (f), (g), and (h) are curves similar to those in (c), (d), and (e), respectively, but with smaller damping.

Here $S$ denotes the oscillator strength and $\gamma$ the damping factor. In the analysis the dispersion frequencies, oscillator strengths, and damping factors are regarded as parameters and the reflection spectrum is then fit using Fresnel's equation (5.19), the reflectivity equation (5.39), and the oscillator formulas. The longitudinal frequencies are subsequently obtained as the zeros of $\varepsilon'_{\alpha}(\omega)$ (see (4.165)). This method was used to analyze the absorption data of Figure 5.3 and the reflection data of Figure 5.5. In this case (5.40) reduces to

$$\varepsilon(\omega) = \varepsilon(\infty) + \frac{[\varepsilon(0) - \varepsilon(\infty)]\omega_{TO}^2}{\omega_{TO}^2 - \omega^2 - i\gamma\omega}, \tag{5.43}$$

and suitable values for the parameters $\varepsilon(0)$, $\varepsilon(\infty)$, $\omega_{TO}$, $\omega_{LO}$, and $\gamma$ satisfactorily fit the observed spectra.

The second method utilizes the K–K relations. One first writes the reflection *amplitude* $\mathscr{R}(\hat{\mathbf{q}}, \omega)$ as $r(\hat{\mathbf{q}}, \omega)\exp[i\theta(\hat{\mathbf{q}}, \omega)]$ so that

$$R(\hat{\mathbf{q}}, \omega) = \mathscr{R}(\hat{\mathbf{q}}, \omega)\mathscr{R}^*(\hat{\mathbf{q}}, \omega) = r^2(\hat{\mathbf{q}}, \omega).$$

$r(\hat{\mathbf{q}}, \omega)$ can thus be computed from the measured spectrum and, if known over the entire frequency range, can lead to $\theta(\hat{\mathbf{q}}, \omega)$ via the K–K relation for the complex function $\ln\mathscr{R}(\hat{\mathbf{q}}, \omega) = \ln r(\hat{\mathbf{q}}, \omega) + i\theta(\hat{\mathbf{q}}, \omega)$:

$$\theta(\hat{\mathbf{q}}, \omega) = -\frac{2\omega}{\pi}\mathscr{P}\int_0^{\infty}\frac{\ln r(\hat{\mathbf{q}}, \omega')\,d\omega'}{\omega'^2 - \omega^2}. \tag{5.44}$$

Once the quantities $r(\hat{\mathbf{q}}, \omega)$ and $\theta(\hat{\mathbf{q}}, \omega)$ are known, the optical constants may be deduced from

$$\mathscr{R}(\hat{\mathbf{q}}, \omega) = \frac{\bar{n}(\hat{\mathbf{q}}, \omega) - 1}{\bar{n}(\hat{\mathbf{q}}, \omega) + 1}. \tag{5.45}$$

If the beam is incident so that $\hat{\mathbf{q}}$ is parallel to one of the principal axes of $\boldsymbol{\varepsilon}(\omega)$ (the usual situation), then a knowledge of $\bar{n}(\hat{\mathbf{q}}, \omega)$ immediately leads to $\varepsilon_{\alpha}(\omega)$. By arranging incidence successively along the various principal-axis directions, $\boldsymbol{\varepsilon}(\omega)$ and hence the dispersion and longitudinal frequencies may be completely determined.

Figure 5.6 shows a typical example of reflection spectra from uniaxial crystals.[24] In such cases one usually carries out two measurements. In each the incident beam is parallel to the $y$ axis (or $x$ axis); the polarizations, however, are different. These are adjusted such that the electric vector is either parallel to the $z$ axis (extraordinary wave) or perpendicular to the $z$ axis (ordinary wave). The dielectric functions relevant to these experiments are (5.41) and (5.42), respectively, and between the two all

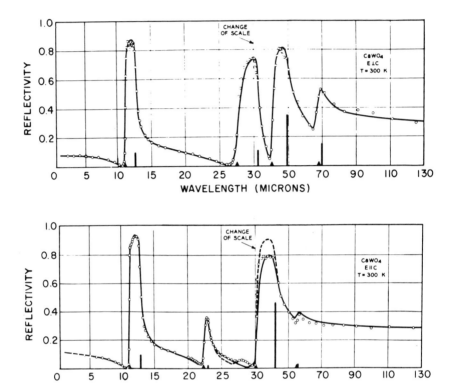

**Figure 5.6**
Reflectivity spectra for $CaWO_4$. The curves show the best fit obtained with dispersion
formulas. The dark vertical bars give the infrared mode strengths and frequencies while
the triangles indicate the frequencies of the longitudinal modes. The dashed curve is an
alternate fit. (After Barker, ref. 24.)

the dispersion and longitudinal frequencies may be determined (compare Figure 5.6 with Table 4.3).

Occasionally measurements in uniaxial crystals are made with the incident beam making varying angles $\theta$ with the $z$ axis. One example is given in Figure 5.7, which shows some of the reflection bands in $NaNO_3$.[25] The data were analyzed by computing $\bar{n}_\theta(\omega)$ via Eq. (5.19), taking

$$\varepsilon_x(\omega) = [\bar{n}_x(\omega)]^2 = n^2 + \frac{\rho}{\omega_0^2 - \omega^2 - i\gamma\omega},$$
$$\varepsilon_y(\omega) = [\bar{n}_y(\omega)]^2 = \varepsilon_x(\omega),$$
$$\varepsilon_z(\omega) = [\bar{n}_z(\omega)]^2 = n^2. \tag{5.46}$$

These formulas represent a simplification of (5.41) and (5.42) in that they allow for a contribution from only *one* oscillator.

Figure 5.8a,b shows plots of the real and imaginary parts of $[\bar{n}_\theta(\omega)]^2$ for $\omega_0 = 1352.8$ cm$^{-1}$, $n = 1.5$, $\rho = 0.605 \times 10^{-4}$ cm$^{-2}$ and for $\gamma = 0$ and 6.9 cm$^{-1}$, the values given by Ketelaar et al.[24] These curves bear close similarity to those for the real and the imaginary parts of typical oscillator functions. However, we observe that $[\bar{n}_\theta(\omega)]^2$ can be equated to the dielectric function only for such values of $\theta$ as correspond to the principal-axis directions. The curves of Figure 5.8b are closely related to the profile of the absorption line for various values of $\theta$, the angular variation of the absorption frequency being given by

$$\omega^2(\text{abs}) = \omega_0^2\cos^2\theta + (\omega_{0,l})^2\sin^2\theta.$$

This is precisely the variation expected for the limiting frequency $\omega_\lambda(\theta)$

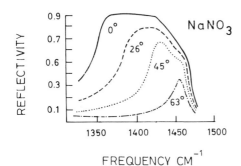

**Figure 5.7**
Reflection spectra for $NaNO_3$ in the neighborhood of one of the planar frequencies with incident light propagating as an extraordinary wave. The angle denotes that between the normal to the crystal face and the $c$ axis. (After Ketelaar et al., ref. 25.)

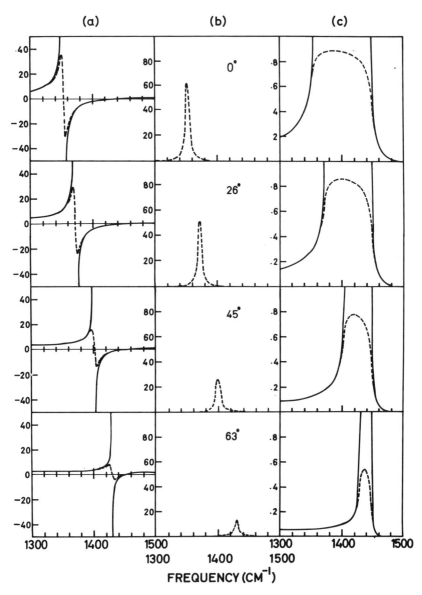

**Figure 5.8**
(a) shows the real part of $\bar{n}_\theta^2$ for $NaNO_3$ for zero and finite damping. (b) shows the imaginary part for finite damping. (For zero damping one obtains $\delta$-functions). The curves of (a) and (b) closely resemble those usually given for the real and imaginary parts of the dielectric function. However $\bar{n}_\theta^2$ can be so identified only for $\theta = 0°$ and $90°$. The curves of (b) are closely related to the profile of the absorption line. (c) shows the reflectivities calculated using the data of (a) and (b). The latter set of curves resemble the experimental data given in Figure 5.7.

computed as discussed in Sec. 4.3. The reflectivity spectra based on the above model for the optical constants are shown in Figure 5.8c, and these bear similarity to the measured spectra (see Figure 5.7).

### 5.2.3 Second-order absorption

Second-order infrared absorption involves two phonons, and, in contrast to the first-order process, the wave vectors of the individual phonons need not be zero although the sum of their wave vectors must vanish. It is this relaxation of the restriction on $\mathbf{q}$ which makes the second-order process attractive.

Several variations of the two-phonon process are possible. The photon can be absorbed and two phonons $\binom{\mathbf{q}}{j}$ and $\binom{-\mathbf{q}}{j'}$ emitted; such a process is called the *sum mode*. In particular, if the two phonons belong to the same branch, that is, $j = j'$ (or the same set of degenerate branches), then the combination is termed an *overtone*. One can also have a *difference mode* in which a phonon $\binom{\mathbf{q}}{j}$ is created and a phonon $\binom{-\mathbf{q}}{j'}$ is annihilated. Two-phonon absorption is temperature-dependent as is clear from Eq. (5.37); in particular, the difference process is important only when the temperature is high enough or the frequency $\omega_{j'}(\mathbf{q})$ low enough to provide for a substantial population of the phonon $\binom{-\mathbf{q}}{j'}$.

The conservation equations for the various processes are

$$\mathbf{q} + \mathbf{q}' = 0 \tag{5.47a}$$

$$\begin{array}{ll} \omega = \omega_j(\mathbf{q}) + \omega_{j'}(\mathbf{q}') & \text{sum mode} \\ \omega = \omega_j(\mathbf{q}) - \omega_{j'}(\mathbf{q}') & \text{diff. mode} \end{array} \Bigg\} \quad (\omega_j > \omega_{j'}). \tag{5.47b}$$

Experimental investigation of second-order processes is invariably done via absorption rather than reflection, mainly because attenuation problems are not so severe as in the first order. Furthermore, the rather small contribution of second-order processes to reflectivity makes reflection experiments unsuitable.

At the present time, second-order absorption has been investigated principally in the diamond and zinc blende lattices,[26] and to some extent in the wurtzite structure crystals.[27] They have also been observed as sidebands of the main absorption peak in alkali halides.[28]

As already explained, two-phonon absorption arising due to the operator $\mathbf{M}\binom{\mathbf{q}}{j}\binom{-\mathbf{q}}{j'}$ is related to nonlinear processes in electronic polarization, that is, electrical anharmonicity. Indeed the observation of such processes in diamond and MgO was the first available clue that vibrations could induce electronic polarization.[29] On the other hand, in crystals exhibiting infrared activity, two-phonon absorption could also arise due to the usual mechanical anharmonicity (associated with cubic and higher-order terms in the expansion of $\Phi$) by means of the incident photon first creating an IR-active phonon ($\mathbf{q} \approx 0$) which subsequently decays into two phonons $\binom{\mathbf{q}_1}{j_1}$ and $\binom{-\mathbf{q}_1}{j_2}$ via anharmonic interaction.

The relative importance of these two processes in crystals where they are both possible has been the subject of debate. Burstein,[30] for example, is of the view that in zinc blende crystals the scope for mechanical anharmonicity is small on account of their "open" structures. He suggests, on the other hand, that mechanical anharmonicity is the dominant contributing factor in alkali halides, electrical anharmonicity being negligible since the ions have the rare-gas configuration. However, Szigeti[31] disputes the latter viewpoint and argues that electrical anharmonicity is not negligible in alkali halides.

Figure 5.9 shows a typical example of second-order absorption spectra.[26] The early attempts at analyzing such data amounted to identifying peaks in the spectra as a combination of a small set of frequencies, typically four. Underlying such an approach was the idea that $g(\omega)$ could be characterized in terms of four or so bands of frequencies, and that peaks in the two-phonon spectra represented various combinations of these "lattice bands." This approach was often used especially when the data was of such a quality that only the peaks could be identified clearly, all other fine structure being blurred. However, the method is inadequate in the sense that one can often extract different sets of frequencies, all of which explain the data equally well.

We gain a better understanding of the origin of the structure in two-phonon spectra if we momentarily regard $\mathbf{M}\binom{\mathbf{q}}{j}\,{}^{-\mathbf{q}}_{j'}$ in (5.37) as $\mathbf{q}$-independent.* Fixing our attention on a particular $\omega$ and a particular mode, say the sum mode, we see that the intensity of absorption is controlled by all the phonon pairs $\binom{\mathbf{q}}{j}\,{}^{-\mathbf{q}}_{j'}$, the sum of whose frequencies matches $\omega$. Further, approximating the temperature factor as unity (valid at low temperatures), we notice that the sum over $\mathbf{q}$ in (5.37) simplifies to

$$\sum_{\mathbf{q}} \delta\{\omega - [\omega_j(\mathbf{q}) + \omega_{j'}(\mathbf{q})]\}.$$

Introduce now a quantity $g^+_{jj'}(\omega)$ defined by

$$
\begin{aligned}
g^+_{jj'}(\omega) &= \frac{1}{(3n)^2 N} \sum_{\mathbf{q}} \delta\{\omega - [\omega_j(\mathbf{q}) + \omega_{j'}(\mathbf{q})]\} \\
&= \frac{v}{(3n)^2 (2\pi)^3} \int_S \frac{dS}{|\mathrm{grad}\,\omega|}\Bigg|_{\omega = \omega_j(\mathbf{q}) + \omega_{j'}(\mathbf{q})},
\end{aligned}
\tag{5.48}
$$

*We follow here the usual practice of regarding absorption as being dependent primarily on $\chi''$. This ignores the detailed interrelationship between the refractive and the dielectric properties (see (5.23)), but is valid for cubic crystals (see (5.25)). The approximation being made is perhaps not too bad for second-order processes.

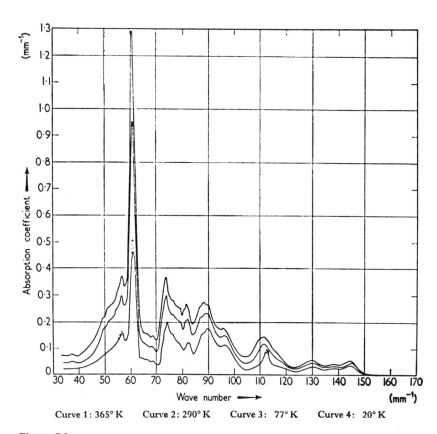

Curve 1: 365° K        Curve 2: 290° K        Curve 3:  77° K        Curve 4:  20° K

**Figure 5.9**
Absorption coefficient for silicon. (After Johnson, ref. 26.)

where $S$ is the constant-frequency surface corresponding to the combination frequency. The quantity $g_{jj'}^+(\omega)$ is similar to $g_j(\omega)$ introduced in Sec. 2.3 except that here we are dealing with a combination of two branches rather than a single branch. For this reason, $\Sigma_{jj'}g_{jj'}^+(\omega)$ is often called the *combined* or *joint frequency distribution*. An analogous quantity can be defined for the difference mode. Like $g_j(\omega)$, $g_{jj'}^+(\omega)$ also has singularities, the most important of which arise when $|\text{grad } \omega|_{\omega = \omega_j(\mathbf{q}) + \omega_{j'}(\mathbf{q})} = 0$. These singularities may be classified as in Sec. 2.3.3 to provide an idea of the combined frequency distribution in the neighborhood of these critical (pair) frequencies. The *structure* in the absorption spectra arises from these singularities and, since $|\text{grad } \omega_j(\mathbf{q}) + \text{grad } \omega_{j'}(\mathbf{q})|$ vanishes whenever both grad $\omega_j(\mathbf{q})$ and grad $\omega_{j'}(\mathbf{q})$ vanish, an analysis of the data will yield the critical frequencies. It is this aspect that makes second-order spectra interesting in spite of their continuous nature. It must be remembered, however, that the absorption spectrum is *not* strictly proportional to the joint frequency spectrum. If we introduce a function $\mathcal{D}_{\alpha\beta}^+(\omega)$ defined by

$$\mathcal{D}_{\alpha\beta}^+(\omega) = \sum_{\mathbf{q}jj'} M_\alpha \begin{pmatrix} \mathbf{q} & -\mathbf{q} \\ j & j' \end{pmatrix} M_\beta^* \begin{pmatrix} \mathbf{q} & -\mathbf{q} \\ j & j' \end{pmatrix} \delta\{\omega - [\omega_j(\mathbf{q}) + \omega_{j'}(\mathbf{q})]\}$$
$$\times \left[ 1 + \left\langle n\begin{pmatrix} \mathbf{q} \\ j \end{pmatrix}\right\rangle_T + \left\langle n\begin{pmatrix} \mathbf{q} \\ j' \end{pmatrix}\right\rangle_T \right] \frac{1}{\omega_j(\mathbf{q})\omega_{j'}(\mathbf{q})}, \qquad (5.49)$$

then the absorption spectrum for the sum modes is proportional to $\boldsymbol{\eta} \cdot \mathcal{D}^+(\omega) \cdot \boldsymbol{\eta}$, where $\boldsymbol{\eta}$ is a unit vector in the direction of the electric field of the incident radiation. In other words, the spectrum is really proportional to the *dipole-moment-weighted* combined frequency spectrum. (Compare with the amplitude-weighted frequency distribution defined in (5.14).) Even so, $\mathcal{D}^+(\omega)$ exhibits essentially the same singularities as $\Sigma_{jj'}g_{jj'}^+(\omega)$. However, some of these may be absent owing to selection rules, that is, $M\begin{pmatrix} \mathbf{q} & -\mathbf{q} \\ j & j' \end{pmatrix}$ may vanish for a particular combination of critical phonons.

A full critical-point analysis of two-phonon spectra proceeds roughly as follows: One constructs a reasonably representative model for the dynamics. Based on this, one can locate the critical frequencies of the combination spectra and the associated structure as outlined in Secs. 2.3 and 2.9. Equipped thus with a rough knowledge of the critical frequencies, the shapes of the joint frequency spectrum in the neighborhood of these frequencies and the absences expected due to selection rules, one may analyze the data systematically to extract precise numbers for the critical frequencies.

A critical-point analysis of infrared data on Si was performed by Johnson and Loudon.[32] An idea of the state of the art is provided by

**Table 5.1**
Critical-point energies for Si(in cm$^{-1}$).

|         | From IR data[a] | From model[b] | From neutron data[c] |
| ------- | --------------- | ------------- | -------------------- |
| Γ       | 522             | 498           | 518                  |
| *L* TO  | 491             | 494           | 489                  |
| LO      | 426             | 449           | 420                  |
| LA      | 374             | 294           | 378                  |
| TA      | 114             | 106           | 114                  |
| *X* TO  | 463             | 494           | 463                  |
| L       | 412             | 378           | 411                  |
| TA      | 149             | 133           | 150                  |

Note: The accuracies of IR data are typically $\sim 2$ cm$^{-1}$. For neutron data the accuracies for TA frequencies are $\sim 2$ cm$^{-1}$, while those for the others are in the range 7–11 cm$^{-1}$.
[a]Reference 32.
[b]F. A. Johnson and W. Cochran, Proc. Int. Conf. on Semiconductors, Exeter, 1962, p. 498. This model provided the starting point of the analysis.
[c]G. Dolling, in Chalk River proceedings, Vol. 2, p. 37.

Table 5.1, which summarizes the critical frequencies used as starting guides and those extracted from the infrared data. Also shown for comparison are the critical frequencies as determined from neutron-scattering experiments. The latter set generally has larger errors than that extracted from optical data.

The example cited suffices to show that the analysis of two-phonon data can, in principle, yield the critical frequencies. A proper analysis, however, requires a good input lattice dynamical model and a knowledge of the selection rules. In spite of this, unambiguous assignments may not always be possible. The availability of other information, for example, Raman data or neutron data, is of course helpful.

## 5.3 Raman Scattering

### 5.3.1 Basic ideas and phonon-expansion formula
Vibration spectra can also be studied by photon scattering experiments. In this and the following section we shall restrict attention to such investigations involving photons in the optical region. Later we shall briefly consider x-ray scattering.

When monochromatic light is scattered by a crystal, it frequently appears with a different frequency, the change in frequency having been brought about by the exchange of energy between the incident photon and the various dynamical processes in the medium. Our concern here is restricted to exchange of energy with phonons. If a single acoustic phonon is responsible for the scattering, then the process is termed *Brillouin scattering*. On the other hand, if optic phonons are involved, then the phenomenon is referred to as *Raman scattering*. All multiphonon processes are also collectively referred to as Raman scattering.

Phenomenologically, one can view Raman (and Brillouin) scattering as a two-step process. One supposes that the electric field $\mathbf{E} = \mathbf{E}_0 e^{i(\mathbf{\kappa \cdot r} - \omega t)}$ of the incident light induces a time-varying dipole moment associated with the entire crystal given by

$$\mathscr{P} = \alpha \mathbf{E},$$

where $\alpha$ is a second-rank tensor and is referred to as the electronic *polarizability* of the medium. This polarizability itself fluctuates at frequencies corresponding to the various natural frequencies of the system (and their combinations) so that $\mathscr{P}$ has frequency components $\omega' = \omega \pm \omega_v$, where $\omega_v$ denotes one of the characteristic frequencies associated with $\alpha$. The scattered light is now viewed as the radiation emitted by the induced moment. Scattered frequencies having $\omega' < \omega$ are called *Stokes* frequencies, while those for which $\omega' > \omega$ are called *anti-Stokes* frequencies.

In many respects, Raman scattering is analogous to neutron scattering. The photon initially has a momentum $\hbar\mathbf{\kappa}$ and energy $\hbar\omega$ which become altered after scattering to $\hbar\mathbf{\kappa}'$ and $\hbar\omega'$ respectively. The photon thus exchanges with the crystal both energy and momentum, supplied or accepted, as the case may be, by the vibrational quanta. The major qualitative difference as compared to neutron scattering is that since both $|\mathbf{\kappa}|$ and $|\mathbf{\kappa}'|$ are $\sim 10^5$ cm$^{-1}$, the wave-vector transfer $\mathbf{Q} = \mathbf{\kappa}' - \mathbf{\kappa}$ is small and to a first approximation may be regarded as zero on the scale of BZ dimensions. It is important to remember that $\mathbf{\kappa}$ and $\mathbf{\kappa}'$ refer to wave vectors *inside* the crystal and thus incorporate the modifications produced by the refracting properties of the medium.

In a quantum-mechanical treatment of Raman scattering, $\alpha$ is regarded as an operator. It is a function of nuclear coordinates[33] and can therefore be expanded formally as (see Eq. (5.30))

$$\alpha_{\alpha\beta}\left(\left\{\mathbf{r}\binom{l}{k}\right\}\right) = \alpha_{\alpha\beta}\left(\left\{\mathbf{x}\binom{l}{k}\right\}\right) + \sum_{lk\gamma}\alpha_{\alpha\beta,\gamma}\binom{l}{k}u_\gamma\binom{l}{k}$$

$$+ \frac{1}{2}\sum_{lk\gamma}\sum_{l'k'\delta}\alpha_{\alpha\beta,\gamma\delta}\binom{l\ l'}{k\ k'}u_\gamma\binom{l}{k}u_\delta\binom{l'}{k'} + \cdots \qquad (5.50)$$

Using this, it is possible to express the differential scattering cross section per unit cell due to phonons in terms of an expansion as follows:

$$\left(\frac{d^2\sigma}{d\Omega d\omega'}\right) = \left(\frac{d^2\sigma}{d\Omega d\omega'}\right)^{(1)} + \left(\frac{d^2\sigma}{d\Omega d\omega'}\right)^{(2)} + \cdots, \tag{5.51}$$

where

$$\left(\frac{d^2\sigma}{d\Omega d\omega'}\right)^{(1)} = \frac{\omega'^4}{c^4}\sum_{\mathbf{q}j}\left|\boldsymbol{\eta}^s \cdot \boldsymbol{\alpha}\!\left(\begin{matrix}\mathbf{q}\\j\end{matrix}\right)\cdot\boldsymbol{\eta}^i\right|^2 \frac{\hbar}{2\omega_j(\mathbf{q})}\Delta(\mathbf{q})$$
$$\times\left[\left(1 + \left\langle n\!\left(\begin{matrix}\mathbf{q}\\j\end{matrix}\right)\right\rangle_T\right)\delta\{\omega - \omega' - \omega_j(\mathbf{q})\}\right.$$
$$\left.+ \left\langle n\!\left(\begin{matrix}\mathbf{q}\\j\end{matrix}\right)\right\rangle_T \delta\{\omega - \omega' + \omega_j(\mathbf{q})\}\right]; \tag{5.52}$$

$$\left(\frac{d^2\sigma}{d\Omega d\omega'}\right)^{(2)} = \frac{1}{2N}\frac{\omega'^4}{c^4}\sum_{\mathbf{q}j}\sum_{\mathbf{q}'j'}\left|\boldsymbol{\eta}^s \cdot \boldsymbol{\alpha}\!\left(\begin{matrix}\mathbf{q}&\mathbf{q}'\\j&j'\end{matrix}\right)\cdot\boldsymbol{\eta}^i\right|^2 \frac{\hbar^2}{4\omega_j(\mathbf{q})\omega_{j'}(\mathbf{q}')}$$
$$\times\Delta(\mathbf{q} + \mathbf{q}')\left[\left(1 + \left\langle n\!\left(\begin{matrix}\mathbf{q}\\j\end{matrix}\right)\right\rangle_T\right)\left(1 + \left\langle n\!\left(\begin{matrix}\mathbf{q}'\\j'\end{matrix}\right)\right\rangle_T\right)\right.$$
$$\times\delta\{\omega - \omega' - [\omega_j(\mathbf{q}) + \omega_{j'}(\mathbf{q}')]\}$$
$$+\left(1 + \left\langle n\!\left(\begin{matrix}\mathbf{q}\\j\end{matrix}\right)\right\rangle_T\right)\left\langle n\!\left(\begin{matrix}\mathbf{q}'\\j'\end{matrix}\right)\right\rangle_T$$
$$\times\delta\{\omega - \omega' + [\omega_{j'}(\mathbf{q}') - \omega_j(\mathbf{q})]\}$$
$$+\left(1 + \left\langle n\!\left(\begin{matrix}\mathbf{q}'\\j'\end{matrix}\right)\right\rangle_T\right)\left\langle n\!\left(\begin{matrix}\mathbf{q}\\j\end{matrix}\right)\right\rangle_T$$
$$\times\delta\{\omega - \omega' + [\omega_j(\mathbf{q}) - \omega_{j'}(\mathbf{q}')]\}$$
$$+\left\langle n\!\left(\begin{matrix}\mathbf{q}\\j\end{matrix}\right)\right\rangle_T\left\langle n\!\left(\begin{matrix}\mathbf{q}'\\j'\end{matrix}\right)\right\rangle_T$$
$$\left.\times\delta\{\omega - \omega' + [\omega_j(\mathbf{q}) + \omega_{j'}(\mathbf{q}')]\}\right]. \tag{5.53}$$

In these expressions, $\boldsymbol{\eta}^i$ and $\boldsymbol{\eta}^s$ denote unit vectors in the directions of polarization of the incident and the scattered light beams, and $(\omega'/\omega)$ is assumed $\approx 1$. Further,[34]

$$\alpha_{\alpha\beta}\!\left(\begin{matrix}\mathbf{q}\\j\end{matrix}\right) = \sum_{\gamma k}(m_k)^{-1/2}\alpha_{\alpha\beta,\gamma}\!\left(\begin{matrix}0\\k\end{matrix}\right)e_\gamma\!\left(k\Big|\begin{matrix}\mathbf{q}\\j\end{matrix}\right) \tag{5.54}$$

and

$$\alpha_{\alpha\beta}\!\left(\begin{matrix}\mathbf{q}&\mathbf{q}'\\j&j'\end{matrix}\right) = \sum_{\substack{l'kk'\\\gamma\delta}}(m_k m_{k'})^{-1/2}\alpha_{\alpha\beta,\gamma\delta}\!\left(\begin{matrix}0&l'\\k&k'\end{matrix}\right)$$
$$\times e_\gamma\!\left(k\Big|\begin{matrix}\mathbf{q}\\j\end{matrix}\right)e_\delta\!\left(k'\Big|\begin{matrix}\mathbf{q}'\\j'\end{matrix}\right)\exp[i\mathbf{q}' \cdot \mathbf{x}(l')]. \tag{5.55}$$

All other quantities have their usual meanings. It must be pointed out that Eqs. (5.51–5.55) are based on the assumption $\mathbf{Q} = \mathbf{\kappa}' - \mathbf{\kappa} \approx 0$ which, of course, is not strictly true. The consequences of relaxing this assumption are pointed out in Sec. 5.3.2. The different terms in (5.51) correspond to light scattering involving various numbers of phonons and bear obvious similarities to the phonon expansion formulas for neutron scattering and IR absorption.

### 5.3.2 First-order Raman scattering

First-order Raman scattering is a one-phonon process governed by the conservation equations (assuming $\mathbf{Q} \equiv 0$)

$$\mathbf{Q} = \mathbf{q} = 0, \tag{5.56a}$$

$$\omega - \omega' = \omega_j(0) \qquad \text{Stokes}, \tag{5.56b}$$

$$\omega - \omega' = -\omega_j(0) \qquad \text{anti-Stokes}. \tag{5.56c}$$

The temperature dependences of the intensity of the Stokes and anti-Stokes lines are similar to those of one-phonon neutron coherent scattering.

Phonons $\binom{0}{j}$ which can be seen in Raman scattering experiments are said to be *Raman active*, which means that at least one element of the polarizability tensor $\boldsymbol{\alpha}\binom{0}{j}$ is nonzero. The structure of $\boldsymbol{\alpha}\binom{0}{j}$ can be determined from group theory and has been tabulated by Loudon[35] for all the crystallographic point groups. A useful rule of thumb in connection with Raman activity is the following: In crystals that have a center of inversion, that is, have an element of the type $\{\mathbf{I}|\mathbf{v}(I)\}$ in the space group, the $\mathbf{q} = 0$ vibrations can be classified as having even parity or odd parity, that is, we can construct eigenvectors $\mathbf{e}\binom{0}{sa\lambda}$ such that for a given $(s, a)$

$$\mathbf{T}(\mathbf{q} = 0; \mathbf{I})\mathbf{e}\binom{0}{sa\lambda} = \pm\mathbf{e}\binom{0}{sa\lambda} \text{ for all } \lambda.$$

The Raman-active modes all have even parity and the IR-active modes have odd parity.* This is the so-called *mutual exclusion rule* well known in molecular spectroscopy. For instance in $CaWO_4$, which has inversion symmetry, the even-parity modes carry the labels $A_g$, $B_g$, and $E_g$ and are all Raman active. The odd-parity modes are of the types $A_u$, $B_u$, and $E_u$ of which, as previously noted (see Sec. 4.3.7), the $A_u$ and $E_u$ types are IR active. In crystals without a center of inversion (zinc blende, for example),

---

*The converse is not true. Some crystals have even parity modes which are not Raman active. Similarly there are instances where some of the odd-parity modes are not IR active.

such mutual exclusion does not apply and a given mode may be both IR and Raman active.

Raman scattering experiments are usually arranged so that the scattering angle is $90°$. In such experiments, the wave-vector transfer $Q$ is $\sim 10^5$ $cm^{-1}$. This, while close to the zone center on the scale of BZ dimensions, is nevertheless large enough to avoid the polariton region. We shall consider first a few examples of Raman scattering by modes that are not IR active. Figure 5.10 shows a few spectra[36] of this type, pertaining to $CaWO_4$, obtained by shining plane-polarized light from a laser onto the crystal and viewing the scattered light at right angles with a grating spectrometer. The polarization $\eta^i$ of the incident light was controlled by a quarter-wave plate, and the desired polarization $\eta^s$ of the scattered light was secured using a polarizer. The various curves in the figure correspond to different orientations of the polarizations and to different beam directions relative to the crystal axes. These were adjusted on the basis of the symmetry of the polarizability tensor. For example, for the $A_g$ modes $\alpha(^0_j)$ has the form

$$\begin{pmatrix} a & & \\ & a & \\ & & b \end{pmatrix}, \tag{5.57}$$

where $a$ and $b$ are constants. To observe the $A_g$ modes, the incident beam was directed along the $x$ axis with $\eta^i$ along $z$, and the scattered beam was observed along $y$ with $\eta^s$ again along $z$. The configuration is abbreviated as $X(ZZ)Y$ in the figure. It is easily seen that $\eta^i \cdot \alpha \cdot \eta^s$ with $\alpha$ as in (5.57) is nonvanishing. Similar considerations show that the $B_g$ and $E_g$ modes do not contribute in this configuration. The other traces in Figure 5.10 were obtained similarly.

In Raman scattering by IR-active modes the finite value of $\mathbf{Q}$ and hence of $\mathbf{q}$ becomes significant, since the frequencies are direction dependent. Remembering that $\mathbf{Q} = \mathbf{q}$† is large enough to clear the polariton region, we may say that $90°$-scattering experiments measure essentially $\omega_\lambda(\hat{\mathbf{q}})$, that is, the limiting frequencies along $\hat{\mathbf{q}}$ determined in the electrostatic approximation. The angular variation shown in Figure 4.6 was obtained from such an experiment.

The intensity formula in such cases also requires modification because $\alpha(^0_j)$ as considered previously pertains really to the dispersion frequencies $\omega_\lambda(0)$. Since the macroscopic electric field produces mode mixing (see Sec. 4.3), the polarizability tensor for IR-active modes must correspondingly be modified. The details are available in the article by Loudon.[35]

†Since $\mathbf{Q}$ is small, the wave-vector-conservation equation is $\mathbf{Q} = \mathbf{q}$ and not $\mathbf{Q} = \mathbf{q} + \mathbf{G}$ as in neutron scattering.

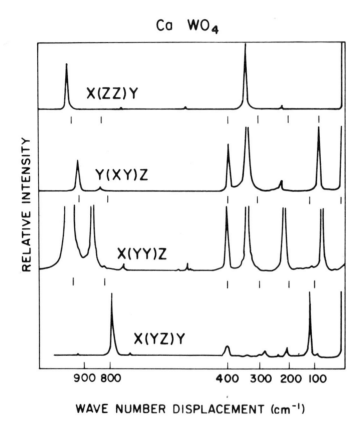

**Figure 5.10**

First-order Raman spectrum of CaWO₄. The notation $X(ZZ)Y$ is explained in the text. (After Porto and Scott, ref. 36.)

The recent advent of the laser with its high directivity has enabled the exploration of Raman spectra at near-forward angles ($\sim 4°$ and less). In such experiments, $\mathbf{Q}$ can be made small enough to lie in the polariton region, so that the dispersion curves for polaritons may be mapped in the same way as are phonons in neutron scattering. An example of results so obtained appears in Figure 5.11.[37] Forward Raman scattering has also been employed to study plasmaritons.[38]

In the past, Raman scattering experiments were restricted to insulating crystals since they are usually transparent. For the scattering geometry normally employed, transparency was a necessary requirement. Recently, however, first-order Raman scattering has been observed from a few metals and alloys by backscattering techniques.[39]

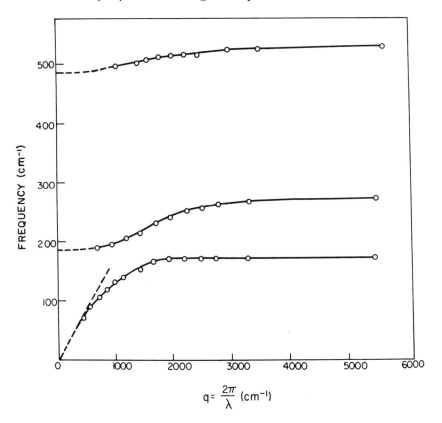

**Figure 5.11**
Polariton spectrum in $BaTiO_3$. (After Burstein et al., ref. 37.)

### 5.3.3 Second-order Raman scattering

Second-order Raman scattering arises from the interaction of a photon
with two phonons and gives rise to a continuous spectrum containing the
critical features associated with the combined frequency distribution. This
is clearly seen in the recently obtained second-order spectrum of diamond[40] shown in Figure 5.12. As in second-order absorption, the observed spectrum is not strictly proportional to the joint frequency spectrum. If from (5.53) we write for the sum modes, Stokes processes

$$\left(\frac{d^2\sigma}{d\Omega d\omega'}\right)^{(2)} = \frac{\hbar^2\omega'^4}{8c^4 N} \sum_{\alpha\beta\gamma\delta} \eta_\alpha^s \eta_\beta^s \eta_\gamma^i \eta_\delta^i \mathscr{P}_{\alpha\gamma,\beta\delta}^+(\omega - \omega'), \qquad (5.58)$$

where

$$\mathscr{P}_{\alpha\gamma,\beta\delta}^+(\omega - \omega') = \sum_{\mathbf{q}jj'} \alpha_{\alpha\gamma}\begin{pmatrix} \mathbf{q} & -\mathbf{q} \\ j & j' \end{pmatrix} \alpha_{\beta\delta}^*\begin{pmatrix} \mathbf{q} & -\mathbf{q} \\ j & j' \end{pmatrix}$$

$$\times \left[1 + \left\langle n\begin{pmatrix}\mathbf{q}\\j\end{pmatrix}\right\rangle_T\right]\left[1 + \left\langle n\begin{pmatrix}\mathbf{q}\\j'\end{pmatrix}\right\rangle_T\right]$$

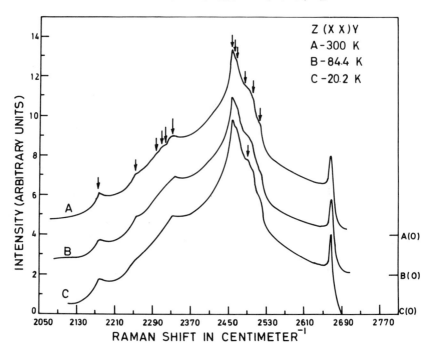

**Figure 5.12**
Second-order Raman spectrum of diamond. (After Solin and Ramdas, ref. 40.)

$$\times \; \delta\{\omega - \omega' - [\omega_j(\mathbf{q}) + \omega_{j'}(\mathbf{q})]\} \; \frac{1}{\omega_j(\mathbf{q})\omega_{j'}(\mathbf{q})}, \quad (5.59)$$

then we see that the second-order spectrum is proportional to the *polarizability-weighted* joint frequency spectrum.

Thus, like infrared spectra, second-order Raman spectra are a potential source of critical frequencies. For example, the data of Figure 5.12 have been analyzed to extract the critical frequencies shown in Table 5.2.[40,41] Once again we notice that the optical experiments yield higher accuracy than does neutron scattering. However, such an unambiguous assignment is possible only in favorable circumstances.

**Table 5.2**
Critical frequencies for diamond (in $cm^{-1}$).

|  | From Raman data[a] | From neutron data[b] |
|---|---|---|
| $\Gamma$ | $1332 \pm 0.5$ | |
| $X$ TO | 1069 | $1072 \pm 26$ |
| L | 1185 | $1184 \pm 21$ |
| TA | 807 | $807 \pm 32$ |
| $L$ TO | 1206 | $1210 \pm 37$ |
| LO | 1252 | $1242 \pm 37$ |
| TA | 536 | $552 \pm 16$ |
| LA | 1006 | $1035 \pm 32$ |
| $W$ TO | 999 | $993 \pm 53$ |
| L | 1179 | $1168 \pm 53$ |
| TA | 908 | $918 \pm 11$ |
| $\Sigma^c$ O* | 1230 | $1231 \pm 32$ |
| O* | 1109 | $1120 \pm 21$ |
| O | 1045 | $1046 \pm 21$ |
| A | 988 | $982 \pm 11$ |
| A* | 980 | $993 \pm 16$ |
| A | | $748 \pm 16$ |

[a]Reference 40. Error in these numbers is $\pm 5 \; cm^{-1}$ except as noted.
[b]Reference 41.
[c]The CP occurs at $(q/q_{max}) \sim 0.7$. Only the starred phonons show critical behavior in one-phonon dispersion curves.

## 5.4 Brillouin scattering

As previously mentioned, first-order light scattering can also occur via the creation or annihilation of an acoustic phonon as was first predicted by Brillouin.[42] He pointed out that thermally excited acoustic waves cause dielectric fluctuations in the medium, having wavelengths and frequencies characteristic of acoustic waves. When light is incident on a crystal, it may be "Bragg reflected" by these fluctuations, besides undergoing a Doppler shift as in the case of neutron scattering. Since the frequency of the acoustic wave varies with $\mathbf{q}$, the wave-vector transfer $\mathbf{Q}$, though small, cannot be ignored as was done in the case of Raman scattering. Thus the conservation equations pertaining to Brillouin scattering are

$$\mathbf{\kappa}' - \mathbf{\kappa} = \mathbf{Q} = \mathbf{q} \tag{5.60a}$$

$$\omega' - \omega = \pm\omega_j(\mathbf{q}) = \pm|\mathbf{q}|c_j(\hat{\mathbf{q}}), \qquad j = 1, 2, 3, \tag{5.60b}$$

where $c_j(\hat{\mathbf{q}})$ is the sound velocity for the $j$th branch for the direction $\hat{\mathbf{q}}$. From these equations it is easily established that, since $|\mathbf{\kappa}' - \mathbf{\kappa}| \sim 10^5$ cm$^{-1}$, the frequency shifts will be $\sim 0.1$–5 cm$^{-1}$ for typical sound velocities. Hence the change in the length of the wave vector due to scattering is very small and so for optically isotropic crystals we may write

$$\Delta\omega = \omega' - \omega = \pm2|\mathbf{\kappa}|c_j(\hat{\mathbf{q}})\sin\frac{\theta}{2} = \pm2\left(\frac{n\omega}{c}\right)c_j(\hat{\mathbf{q}})\sin\frac{\theta}{2}, \tag{5.61}$$

where $n$ is the refractive index and $\theta$ is the scattering angle. For anisotropic crystals Eq. (5.61) becomes[43]

$$\Delta\omega = \pm\{\kappa^2 + \kappa'^2 - 2\kappa\kappa'\cos\theta\}^{1/2}c_j(\hat{\mathbf{q}})$$

$$= \pm\frac{\omega}{c}\{n_i^2 + n_s^2 - 2n_in_s\cos\theta\}^{1/2}c_j(\hat{\mathbf{q}}), \tag{5.62}$$

where $n_i$ and $n_s$ are the refractive indices corresponding to the incident and scattered wave vectors.

Equations (5.60) and (5.61) predict that the Brillouin spectrum of a cubic material has three doublets, that is, three Stokes lines and three anti-Stokes lines. For a given branch, the Stokes and the anti-Stokes peaks should have almost equal intensity since, for long-wavelength acoustic modes, $\hbar\omega_j(\mathbf{q}) \ll k_BT$. On the other hand, for anisotropic crystals one can expect a maximum of twelve Brillouin doublets owing to the double-valued nature of the refractive index (see Eq. (5.20)). By employing polarized incident radiation and polarization analysis of the scattered radiation, one can select desired subsets.

As in the case of first-order Raman scattering, the intensity of Brillouin scattering is dictated by the polarizability tensor $\alpha\binom{q}{j}$, $(\mathbf{q},j)$ now pertaining to the acoustic modes. Since the polarization fluctuations are produced by an elastic straining of the crystal, $\alpha\binom{q}{j}$ and therefore the Brillouin intensity can be related to the photoelastic constants of the medium.[44]

Figure 5.13 shows a typical Brillouin spectrum.[45] Such measurements are usually employed to determine the sound velocity and hence the elastic constants. The latter may also be determined from ultrasonic experiments by studying the propagation of externally generated sound waves of frequencies $\sim 10^7$ Hz. Consequently, it is of interest to compare these values with those obtained from Brillouin scattering experiments, since the latter sample acoustic waves with frequencies $\sim 10^9$ Hz. Any variations in the elastic constants imply variations in the sound velocities and hence dispersion of acoustic waves. Such dispersion is not contained in the harmonic theory but may be caused by anharmonicity. In this context, the comparison of the Brillouin data ($q \sim 10^5$ cm$^{-1}$) with "low $q$" neutron data ($q \sim 10^7$ cm$^{-1}$) is also of interest.[46,47]

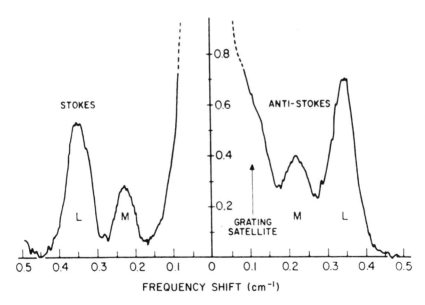

**Figure 5.13**
Brillouin spectrum of RbCl. $\kappa$, $\kappa'$, and $\mathbf{q}$ are all in the ($1\bar{1}0$) plane, with $\mathbf{q}$ making an angle of 40° to the [001] direction. Instead of the three expected doublets, only two are seen due to intensity considerations. (After Benedek and Fritsch, ref. 45.)

## 5.5 X-ray scattering

We shall now consider the study of vibration spectra by x-ray scattering. Though this technique is difficult to employ, it has been used successfully, especially prior to the advent of neutron scattering. Unlike light scattering experiments which in first order yield information only about zone-center phonons, x-ray scattering, like neutron scattering, can sample phonons throughout the BZ, since both the incident and scattered wave vectors are $\sim 10^8$ cm$^{-1}$. Indeed, the principles of x-ray and neutron scattering are very similar.[48,49]

Consider first the interaction of x rays with an isolated atom; as with light, the scattering is mainly by the electrons. However, on account of x rays' comparatively higher frequencies, the polarization effects are small and to a first approximation may be ignored. The scattering by an atom can therefore be characterized as due to a rigid charge distribution.* For the usual situation of unpolarized incident radiation and unanalyzed scattered radiation, the scattering amplitude $a(\mathbf{Q})$ is given by[50]

$$a(\mathbf{Q}) = \frac{e^2}{mc^2}\left(\frac{1 + \cos^2\theta}{2}\right)^{1/2} f(\mathbf{Q}),  \tag{5.63}$$

where $\theta$ is the scattering angle and

$$f(\mathbf{Q}) = \int \rho(\mathbf{r})e^{-i\mathbf{Q}\cdot\mathbf{r}}\, d\mathbf{r}  \tag{5.64}$$

is the Fourier transform of the electron density $\rho(\mathbf{r})$.

By replacing $a_{coh}^\kappa$ with $a^\kappa(\mathbf{Q})$, the x-ray scattering by the crystal can be expressed in a phonon expansion just as in (5.3), with the one-phonon part given as before by (5.8). It might appear from this that to measure dispersion curves one need merely repeat for x rays what is done in the case of neutrons. However, this is impractical because the fractional energy change suffered by x rays on scattering by phonons ($\sim 10^{-5}$) would require spectral resolution beyond that currently available in the x-ray region.† Instead one measures at each scattering angle the scattered intensity contributed by phonons of all branches, subject to the usual conservation rules (5.10). By careful analysis of the data it may then be possible to extract the dispersion relations.

Figure 5.14 shows a typical example of such intensity data.[51] The

---

*This part of the scattering, known as *Thomson scattering*, is unimportant at optical frequencies where electronic polarization processes dominate. Typically polarization effects decrease with increasing frequencies. Compare, for example, $\varepsilon(0)$ and $\varepsilon(\infty)$.

†Contrast this with the situation in neutron scattering where fractional energy changes are often $\sim 0.1$–$1.0$.

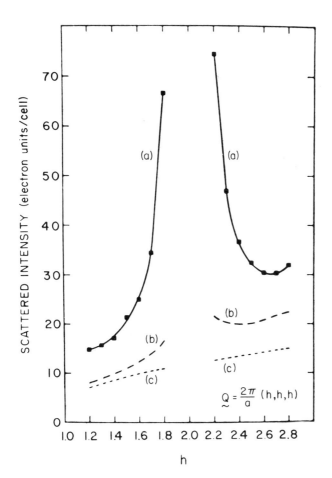

**Figure 5.14**
X-ray intensity along the [111] direction in NaCl. If there were no thermal vibrations, then there would only be a sharp Bragg peak at (222) and no intensity elsewhere. However, due to thermal agitation there is a diffuse tail to the Bragg peak. The curve (a) is the total measured intensity; (b) and (c) denote respectively the Compton plus multiphonon and Compton contributions. The one-phonon intensity is given by the difference $(a - b)$. (After Buyers and Smith, ref. 51.)

results are for NaCl in the [111] direction and were obtained around the reciprocal lattice point (222). The measured absolute intensity must be corrected for Compton scattering and the multiphonon contribution. These are evaluated theoretically, the latter from a suitable dynamical model. Because of the $\mathbf{Q} \cdot \mathbf{e}$ factor in $F^{(1)}$ (see Eq. (5.9)), the intensity along [111] is contributed only by the longitudinal modes—both acoustic and optic. By comparing the intensities at points such as $(222 + \mathbf{q})$ and $(333 - \mathbf{q})$, it was possible to compute the frequencies $\omega_{LA}(\mathbf{q})$ and $\omega_{LO}(\mathbf{q})$. The dispersion relations deduced in this fashion are shown in Figure 5.15.

The lack of a spectroscopic analysis of the scattered radiation considerably increases the difficulty of extracting the dispersion relations and restricts the technique to crystals of simple structures. It is not surprising therefore that the neutron scattering method has overshadowed the x-ray scattering technique. One crystal in which x rays have a clear advantage over neutrons is vanadium. The scattering of neutrons from vanadium is almost entirely incoherent and hence only the frequency distribution can be measured by neutrons.[52] The determination of the dispersion relations has been done by x rays, the only other technique available. Considerable interest is attached to this measurement[53] in view of the influence of electron-phonon interactions on the dispersion curves. This aspect will be considered further in the next chapter.

## 5.6 Some new techniques

The techniques reviewed in the foregoing sections constitute the most important sources of experimental phonon spectra. Thanks to the ingenuity of experimenters, however, several new methods have been developed, the more promising among which we now discuss briefly.

### 5.6.1 Impurity absorption
Considerable interest, both theoretical and experimental, has been shown recently in the vibration of impurity atoms in crystals.[54] A fringe benefit of such experiments has been information about the dynamics of the host lattice itself. One of the consequences of the introduction of defects or chemical impurities is the disruption of the crystal's translational periodicity which in turn relaxes the wave-vector-conservation selection rule in photon-phonon interactions. In particular, one-phonon absorption occurs not only due to $\mathbf{q} \approx 0$ phonons but due to phonons throughout the BZ so that, in fact, the absorption band reflects the density of states, leading to the possibility of direct extraction of critical frequencies. An example is provided by the work of Angress et al.,[55] who studied the infrared ab-

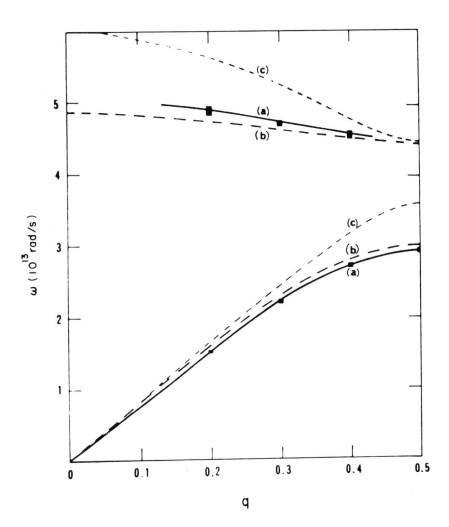

**Figure 5.15**
Longitudinal branches along [111] for NaCl. (a) measured, (b) predicted by the shell
model, and (c) predicted by the rigid-ion model (After Buyers and Smith, ref. 51.)

sorption spectrum of silicon doped with boron and phosphorous. Unfortunately, the introduction of these impurities results in the production of free carriers which themselves contribute to the absorption, thereby masking the vibrational effects. In their work, Angress et al. minimized the effects of free carriers by irradiation with a small dose of fast electrons.

Figure 5.16 shows the absorption spectrum of Si doped with B and P in a concentration of $\sim 5 \times 10^{19}$ cm$^{-3}$. Of the considerable visible structure, four features are definitely assigned to critical frequencies on the basis of their frequencies and shapes. Table 5.3 compares the critical frequencies of Si deduced by this technique with values obtained from other sources. The agreement is encouraging and augurs well for this method, especially as the critical frequencies are obtained directly, rather than from their combinations.

### 5.6.2 Emittance studies

It is well known that the wavelength distribution of the power per unit area emitted by an ideal blackbody at temperature $T$ is given[56] by the Planck formula

$$W_\lambda(T) = \frac{2\pi c^2 h}{\lambda^5} \frac{1}{\exp(hc/k_{\mathrm{B}}T\lambda) - 1}. \tag{5.65}$$

In the case of a real body, the emission is modified over that given by (5.65) to the extent of the spectral emittance function of the body. In other words, the body emits radiation in the amount

$$W_\lambda'(T) = E_\lambda W_\lambda(T), \tag{5.66}$$

where $E_\lambda$ is the *spectral emittance*.

Now an ideal blackbody absorbs completely all radiation falling on it. By contrast, a nonideal body absorbs only a fraction of the incident energy, the absorption relative to an ideal blackbody being given by the *spectral absorptance $A_\lambda$*. By Kirchhoff's laws,[56]

$$E_\lambda = A_\lambda. \tag{5.67}$$

Consider now a quasi-transparent sample having plane parallel sides and specular surfaces upon which radiation is incident normally. The incident radiation can either be reflected, transmitted or absorbed. In particular, the absorptance is given by[56]

$$A_\lambda = E_\lambda = \frac{(1 - |\mathscr{R}(\lambda)|^2)(1 - \exp[-\alpha(\lambda)t])}{1 - |\mathscr{R}(\lambda)|^2 \exp[-\alpha(\lambda)t]}, \tag{5.68}$$

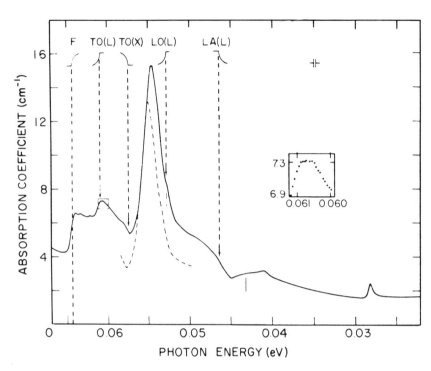

**Figure 5.16**
Absorption spectrum of silicon doped with B and P. Full curve, 290 K; dashed curve,
80 K. Critical-point energies and shapes of some singularities are shown. (After Angress
et al., ref. 55.)

**Table 5.3**
Critical frequencies in Si (in $cm^{-1}$).

|        | From impurity absorption[a] | From two-phonon data[b] | From neutron scattering[b] |
|--------|------------------------------|--------------------------|-----------------------------|
| $L$ TO | 491                          | 491                      | 489                         |
| LO     | 426                          | 426                      | 420                         |
| LA     | 377                          | 374                      | 378                         |
| $X$ TO | 462                          | 463                      | 463                         |

[a]Reference 55.
[b]See Table 5.1.

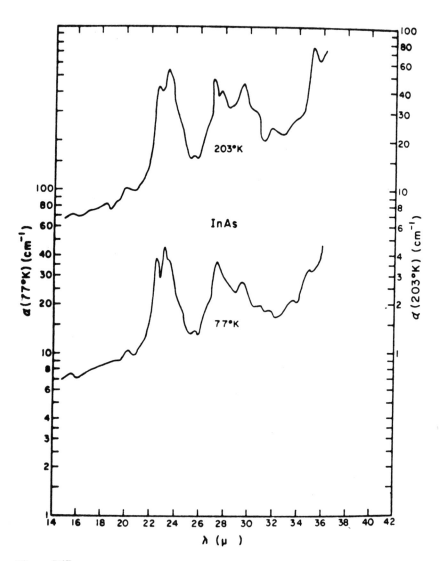

**Figure 5.17**
Absorption coefficient of InAs in the range 14–38 microns deduced from emittance studies. (After Steirwalt and Potter, ref. 57.)

where $\mathscr{R}(\lambda) = [\bar{n}(\lambda) - 1]/[\bar{n}(\lambda) + 1]$ is the reflection amplitude, $\alpha(\lambda)$ the absorption coefficient, $t$ the thickness, and $\bar{n}$ the refractive index. Equation (5.68) shows that $\alpha(\lambda)$ can be extracted by measuring not only the absorption in the sample, as discussed earlier, but also by studying the spectral emittance of the sample.

Steirwalt and Potter[57] have exploited the method to measure $E_\lambda$ and thereby $\alpha(\lambda)$ for a number of III–V semiconducting compounds. Their experiments consist of comparing as a function of wavelength the radiation emitted by the sample with that from a blackbody at the same temperature. This gives $E_\lambda$ (see Eq. (5.66)) from which $\alpha(\lambda)$ is determined via (5.68). Figure 5.17 shows results for InAs.[57] The spectral region covered in the figure spans primarily the two-phonon part of the absorption spectrum, and the structure seen is related to that in the joint frequency spectrum. By an analysis of this structure, Steirwalt and Potter extracted the critical frequencies shown in Table 5.4.

### 5.6.3 Dispersion curves from the study of polytypes

Several substances have the ability to crystallize in a number of modifications in all of which two dimensions of the unit cell are the same while the third, though variable, is an integral multiple of a common unit. This phenomenon is referred to as *polytypism*.[58] The various polytypic modifications can all be regarded as built up of layers of structures stacked parallel to each other at constant intervals along the variable dimension.

Among the substances in which polytypism is found are ZnS, PbS, and SiC. The latter provides the most outstanding example of polytypism, occurring in over 40 modifications. In each of these the hexagonal unit cell has $a = b = 3.078\text{Å}$, while $c$ is a multiple of $c_0 = 2.518\text{Å}$, ranging from

**Table 5.4**
Critical frequencies in InAs (in $cm^{-1}$) (from ref. 57).

| | | |
|---|---|---|
| $\Gamma$ LO | 246 | |
| | TO | 221 |
| $L$ LO | 194 | |
| | TO | 216 |
| | LA | 148 |
| | TA | 73 |
| $X$ LO | 164 | |
| | TO | 212 |
| | LA | 145 |
| | TA | 171 |

$c \sim 5\text{Å}$ (two-layered hexagonal polytype) to $c \sim 1500\text{Å}$ (594-layered rhombohedral polytype).

The existence of polytypes in SiC has been ingeneously exploited by Feldman et al.,[59] who by using optical techniques alone were able to map completely the dispersion curves for SiC along the $c$ direction. The basic concept on which their method is based is that of the large zone, best explained by considering a specific example.

Polytype $6H$† is a modification with a primitive hexagonal lattice and six layers along the $c$ axis as the repeat unit. The boundary of the BZ along the $c$ direction is thus at $(\pi/6c_0)$. The primitive cell contains 12 atoms and consequently the dispersion relations involve 36 branches. However, if these are remapped into a larger zone with boundary at $(\pi/c_0)$, then only 6 branches are obtained. Owing to the degeneracies that exist for $\mathbf{q}\|c$ axis, these 6 branches effectively appear as 4.

Figure 5.18 illustrates schematically the dispersion curves in the two schemes. The merit of the large-zone representation is that its dimension

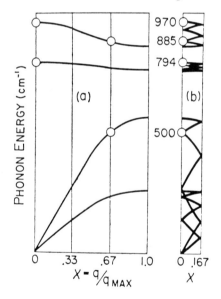

**Figure 5.18**
Schematic drawing of the dispersion curves for $6H$ SiC in the axial direction. (a) shows the large-zone representation. Two branches are doubly degenerate while the other two are nondegenerate. (b) shows the same dispersion curves but folded into the BZ. Open circles indicate four optical modes measured in IR experiments. The abscissa is $x = (q/q_{max})$ where $q_{max} = 6\pi/c$. (After Feldman et al., ref. 59.)

†The symbol $6H$ is the "Ramsdell notation"[58] employed in polytype classification.

is the same for all polytypes, equal to $L(\pi/Lc_0) = (\pi/c_0)$, where $L$ is the number of layers in the stacking sequence. Since $a$ and $c_0$ are the same for all polytypes, it is reasonable to expect on physical grounds that all poly-types exhibit nearly the same phonon spectrum in the large-zone repre-sentation. Phonons measured in first-order optical experiments, while re-ferring to $\mathbf{q} \approx 0$ in the BZ, may correspond either to $\mathbf{q} \neq 0$ or $\mathbf{q} \approx 0$ when translated into the large zone. The phonons in the former category appear as weak events while those in the latter are quite intense. By systematically examining several polytypes with Raman scattering, Feldman et al.[59] mapped the dispersion curves for SiC along the $c$ direction as shown in Figure 5.19, which illustrates the observations for various polytypes in the large-zone representation.

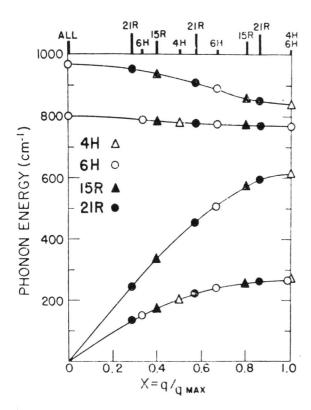

**Figure 5.19**
Dispersion curves for SiC obtained from Raman scattering experiments on various poly-types. For each polytype, the Raman accessible values of $x = (q/q_{max})$ are marked at the top. (After Feldman et al., ref. 59.)

### 5.6.4 Tunneling spectroscopy

Another technique that gives information about phonon spectra as a useful by-product is tunneling spectroscopy.[60] Tunneling is a quantum-mechanical phenomenon associated with the penetration of particles (electrons in this case) through a potential barrier without acquiring enough energy to surmount it. Tunneling experiments leading to vibrational information have been carried out in two systems, semiconductors and superconductors. In the case of semiconductors, one studies the tunneling of electrons across the band gap from the valence to the conduction band. Such a process, considered a long time ago by Zener in connection with dielectric breakdown, is not very probable. However, by producing a PN junction within the body of a homopolar semiconductor, one forms a *tunnel diode* in which the probability for such an event is greatly enhanced, especially if an external potential is applied. The experiment consists essentially of studying the current-voltage ($I$-$V$) characteristics of the diode. While the tunneling current is controlled primarily by the density of electronic states in the valence and conduction bands, it is also influenced by the phonon spectrum via electron-phonon coupling. The latter manifests itself in the $I$-$V$ characteristics as small kinks, better seen in plots of the second derivative ($d^2I/dV^2$). A typical plot for Si,[61] Figure 5.20, exhibits peaks and kinks attributable to specific phonons ($\mathbf{q}, j$) or combination of phonons which assist the tunneling. These phonons can be assigned with the help of band structure information. The results obtained are in reasonable agreement with neutron scattering data.

Tunneling in superconductors is also explored with diodes, typically consisting of a metal film–oxide film–metal film sandwich. It was discovered by Giaever[62] that such a diode behaves rather like the Esaki tunnel diode discussed earlier, except that here the tunneling of electrons occurs across the superconducting gap. A fine structure due to the electron-phonon interaction is seen in the diode characteristics at very low temperature, especially in the so-called strong-coupling superconductors. By a detailed analysis[63] of the data, it is possible to extract a quantity $F(\omega) = A^2(\omega)g(\omega)$ (where $A^2(\omega)$ is called the electron-phonon coupling parameter) which displays singularities associated with critical points.

Figure 5.21 shows a comparison of $F(\omega)$ for white tin deduced from tunneling experiments[64] with the dispersion curves determined by neutron scattering.[65,66] Generally speaking, close correspondence can be established between the critical features of $F(\omega)$ and the critical frequencies deduced from the dispersion curves. Such a relationship is, however, lacking in some cases; for example, the tunneling peak at 8 meV has no obvious origin in the dispersion curves and is, therefore, clearly due to a

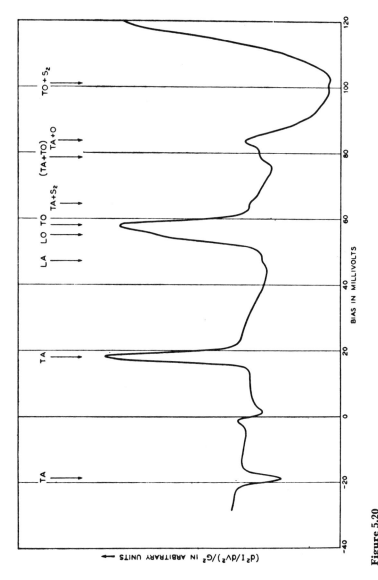

**Figure 5.20**
Second derivative of the tunneling current in Si measured at helium temperature. (After Chynoweth et al., ref. 61.)

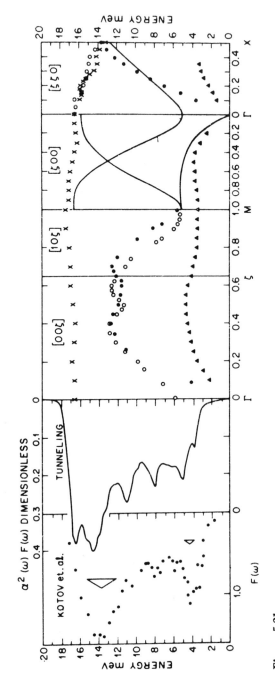

**Figure 5.21**
Comparison of $F(\omega)$ for white tin obtained from tunneling experiments with the dispersion curves and frequency distribution determined by neutron scattering. (After Rowell and Dynes, ref. 64.)

branch in an off-symmetry direction. Note, however, the good correspondence between the features observed in $F(\omega)$ and those seen in the
frequency spectrum directly measured by neutron spectroscopy.[67] In
this connection it should be noted that a frequency distribution computed
using a force-constant model based on values deduced by fitting to dispersion curves for symmetry directions alone (the usual situation) may not
be compatible with the dispersion curves and critical points for off-
symmetry directions. This difficulty was acutely felt in lead, for example,
where not only did the highly structured character of the measured dispersion curves make difficult the task of achieving good fits for the symmetry-
direction data,[68] but even a model that gave a passably acceptable fit
for these data was later found to generate a $g(\omega)$ which compared
poorly[69] with the critical features of the measured $F(\omega)$. Whereas the
model well predicted frequencies associated with critical points occurring
in symmetry directions, as expected by virtue of the initial fit, it fared
poorly with regard to off-symmetry-direction critical frequencies. Thus
tunneling experiments are useful supplements to the dispersion curve
measurements insofar as the latter traditionally are confined to symmetry
directions.

### 5.6.5 Vibronic studies

Vibronic transition is the name given to the electronic transition of an ion
accompanied by a vibrational transition in the system to which the ion is
coupled. The study of such electronic-cum-vibrational transitions is well
known in molecular spectroscopy.[70] Recently, analogous studies have
also been made in crystalline solids, leading to information about crystal
dynamics. Typically one introduces into the crystal under investigation an
optically active ion which serves as the probe, and then investigates the
electronic transitions of the concerned ion either in emission or in absorption. Under favorable conditions, such transitions are accompanied by
well-delineated phonon sidebands, the structure in which leads to information about the critical frequencies of the host lattice. In this respect
vibronic studies are an extension of the impurity infrared absorption experiments discussed earlier.

   As a typical example, we consider the experiments on KBr by Timusk
and Buchanan.[71] In these studies $Sm^{++}$ was introduced into KBr and
served as the probe. The rare-earth ions enter substitutionally at the positive ion sites and exhibit several sharp lines in their fluorescence spectra
extending from 6890Å to near infrared which correspond to transitions
within the $4f^6$ shell. On the low-energy sides of the sharp lines are weak
continuous spectra arising from the interaction of the rare-earth ion with

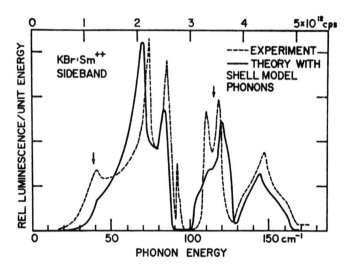

**Figure 5.22**
Experimentally observed and theoretically calculated spectrum of the sideband associated with the 6890 Å line of Sm$^{++}$ in KBr. The arrows indicate local modes. The peak at 92 cm$^{-1}$ is believed to be due to an impurity. (After Timusk and Buchanan, ref. 71.)

the phonons of the host lattice. Figure 5.22 shows the sideband of the $^5D_0 \to {}^7F_0$ line at 6890Å which, we observe, exhibits considerable structure. In point of fact, the 6890Å line is strictly forbidden as it corresponds to a transition between two (electronic) states of even parity. However, the presence of electric fields lowers the cubic symmetry, making the transition possible. The fields include the static one arising from the positive-ion vacancy, necessary for electrical neutrality,* and the time-varying fields associated with the vibrations of the host lattice. It is the latter which are mainly responsible for the sideband seen in Figure 5.22. The figure also shows the emission spectrum calculated under the assumption that the dynamics can be described in terms of the shell model. All the major features in the observed spectrum are reasonably well reproduced in the calculations, and can be correlated with critical features in the frequency spectrum of the host lattice. It is worth pointing out that, since the vibronic spectrum explicitly involves the electric fields associated with the vibrations, it affords a more stringent test of the dipolar models than a study of phonon dispersion curves alone.[72] This is because the solutions for the dipole moments themselves are eliminated in the calculation of the final

*Remember the samarium ion is doubly charged while the potassium ion is only singly charged.

dynamical matrix (see (4.292)), and the possibility exists that a dipole model which fits phonon frequencies may not necessarily be physically valid.

Vibronic spectra for several other crystals have also been reported, especially for those of the fluorite structure.[73]

## References

1. W. Marshall and S. Lovesey, *Theory of Thermal Neutron Scattering* (Oxford University Press, London, 1971).

2. B. N. Brockhouse, in Bak, p. 221.

3. W. M. Lomer and G. G. Low, in *Thermal Neutron Scattering*, edited by P. A. Egelstaff (Academic Press, New York, 1965).

4. I. I. Gurevich and L. V. Tarasov, *Low Energy Neutron Physics* (North Holland, Amsterdam, 1968).

5. Kittel, Chapter 2.

6. B. N. Brockhouse, S. Hautecler, and H. Stiller, in *The Interaction of Radiation with Solids*, edited by R. Strumane et al. (North Holland, Amsterdam, 1964), p. 580.

7. B. N. Brockhouse, in Stevenson, p. 110.

8. G. Dolling and A. D. B. Woods, in *Thermal Neutron Scattering*, edited by P. A. Egelstaff (Academic Press, New York, 1965).

9. L. A. Feldkamp, Ph. D. Thesis, University of Michigan (1969).

10. H. Boutin and S. Yip, *Molecular Spectroscopy with Neutrons* (MIT Press, Cambridge, Mass., 1969).

11. G. C. Summerfield, J. M. Carpenter, and N. A. Lurie, Lecture notes on Thermal Neutron Scattering, University of Michigan (1968).

12. G. Venkataraman, K. Usha Deniz, P. K. Iyengar, A. P. Roy, and P. R. Vijayaraghavan, J. Phys. Chem. Solids **27**, 1103 (1966).

13. C. L. Thaper, B. A. Dasannacharya, A. Sequeira, and P. K. Iyengar, Solid State Comm. **8**, 497 (1970).

14. J. F. Nye, *Physical Properties of Crystals* (Oxford University Press, London), p. 235.

15. BH, p. 124.

16. L. D. Landau and E. M. Lifshitz, *Statistical Physics* (Addison-Wesley, Reading, Mass., 1969), p. 389.

17. BH, p. 364.

18. BH, p. 204.

19. BH, p. 305.

20. BH, p. 212.

21. G. Herzberg, *Infrared and Raman Spectra of Polyatomic Molecules* (Van Nostrand, New York, 1945).

22. S. Iwasa, I. Balslev, and E. Burstein, in *Physics of Semiconductors*, edited by M. Hulin (Dunod, Paris, 1964), p. 1077.

23. M. Born and E. Wolf, *Principles of Optics* (Pergamon, New York, 1965) p. 670.

24. A. S. Barker Jr., Phys. Rev. **135**, A742 (1964).

25. J. A. A. Ketelaar, C. Haas, and J. Fahrenfort, Physica **20**, 1259 (1954).

26. F. A. Johnson, in *Progress in Semiconductors*, edited by A. F. Gibson and R. E. Burgess (Temple Press, London, 1965), p. 181, Vol. 9.

27. M. Balkanski, in Wallis, p. 345.

28. C. Smart, G. R. Wilkinson, A. M. Karo, and J. R. Hardy, in Wallis, p. 387.

29. M. Lax and E. Burstein, Phys. Rev. **97**, 39 (1955).

30. E. Burstein, in Wallis, p. 315.

31. B. Szigeti, in Wallis, p. 405.

32. F. A. Johnson and R. Loudon, Proc. Roy. Soc. (London) **A281**, 274 (1964).

33. BH, p. 219.

34. BH, p. 305.

35. R. Loudon, Advan. Phys. **13**, 423 (1964).

36. S. P. S. Porto and J. F. Scott, Phys. Rev. **157**, 716 (1967).

37. E. Burstein, S. Ushioda, A. Pinczuk, and J. F. Scott, in *Light Scattering Spectra of Solids*, edited by G. B. Wright (Springer, New York, 1969), p. 43.

38. J. Shah, T. C. Damen, J. F. Scott, and R. C. C. Leite, Phys. Rev. **B3**, 4238 (1971).

39. J. H. Parker Jr., D. W. Feldman, and M. Ashkin, in *Light Scattering Spectra of Solids* edited by G. B. Wright (Springer, New York, 1969), p. 389.

40. S. A. Solin and A. K. Ramdas, Phys. Rev. **B1**, 1687 (1970).

41. J. L. Warren, J. L. Yarnell, G. Dolling, and R. A. Cowley, Phys. Rev. **158**, 805 (1966).

42. L. Brillouin, Compt. Rend. **158**, 1331 (1914); Ann. Phys. (Paris) **17**, 88 (1922).

43. V. Chandrasekharan, Proc. Indian Acad. Sci. **32A**, 379 (1950); **33A**, 183 (1951).

44. BH, p. 375.

45. G. B. Benedek and K. Fritsch, Phys. Rev. **149**, 647 (1966).

46. R. A. Cowley, Proc. Phys. Soc. (London) **90**, 1127 (1967).

47. E. C. Svensson and W. J. L. Buyers, Phys. Rev. **165**, 1063 (1968).

48. W. Cochran, Rept. Progr. Phys. **26**, 1 (1963).

49. W. Cochran, in Stevenson, p. 153; see also T. Smith in the same volume, p. 161.

50. R. W. James, *Optical Principles of the Diffraction of X-rays* (Bell, London, 1950), pp. 31, 109.

51. W. J. L. Buyers and T. Smith, Phys. Rev. **150**, 758 (1966).

52. W. Gläser, F. Carvallo, and G. Ehret, in Bombay proceedings, Vol. I, p. 99.

53. R. Colella and B. W. Batterman, Phys. Rev. **B1**, 3913 (1970).

54. R. J. Elliott, in Stevenson, p. 377.

55. J. F. Angress, A. R. Goodwin, and S. D. Smith, Proc. Roy. Soc. (London) **A287**, 64 (1965).

56. D. L. Steirwalt and R. F. Potter, in *Semiconductors and Semimetals*, edited by R. K. Willardson and A. C. Beer (Academic Press, New York, 1967), Vol. 3, p. 71.

57. D. L. Steirwalt and R. F. Potter, Phys. Rev. **137**, A1007 (1965).

58. A. R. Verma and P. Krishna, *Polymorphism and Polytypism in Crystals* (Wiley, New York, 1966).

59. D. W. Feldman, J. H. Parker Jr., W. J. Choyke, and Lyle Patrick, Phys. Rev. **170**, 698 (1968); **173**, 787 (1968).

60. L. Esaki, in *Electronic Structures in Solids*, edited by E. D. Haidemenakis (Plenum, New York, 1969), p. 1.

61. A. G. Chynoweth, R. A. Logan, and D. E. Thomas, Phys. Rev. **125**, 877 (1962).

62. I. Giaever, Phys. Rev. Letters **5**, 147 (1960).

63. W. L. McMillan and J. M. Rowell, in *Superconductivity*, edited by R. D. Parks (Dekker, New York, 1969), Chap. 11.

64. J. M. Rowell and R. C. Dynes, in Nusimovici, p. 150.

65. J. M. Rowe, Phys. Rev. **163**, 547 (1967).

66. D. L. Price, Proc. Roy. Soc. (London) **A300**, 25 (1967).

67. B. A. Kotov, N. M. Okuneva, A. R. Regel, and A. L. Shakh Budagov, Fiz. Tver. Tela **9**, 1227 (1967) [Sov. Phys.—Solid State **9**, 255 (1968)]; B. A. Kotov, N. M. Okuneva, and É. L. Plachenova, Fiz. Tver. Tela **10**, 513 (1968) [Sov. Phys.—Solid State **10**, 402 (1968)].

68. B. N. Brockhouse, T. Arase, G. Caglioti, K. R. Rao, and A. D. B. Woods, Phys. Rev. **128**, 1099 (1962).

69. R. C. Dynes, J. P. Carbotte, and E. J. Woll Jr., Solid State Comm. **6**, 101 (1968).

70. G. Herzberg, *Electronic Spectra and Electronic Structure of Polyatomic Molecules* (Van Nostrand, New York, 1966).

71. T. Timusk and M. Buchanan, Phys. Rev. **164**, 345 (1967).

72. S. K. Sinha, Critical Reviews in Solid State Sciences **3**, 273 (1973).

73. D. H. Kühner, H. V. Lauer, and W. E. Bron, Phys. Rev. **B5**, 4112 (1972); see also other papers due to this group cited in this reference.

# 6 Survey of Experimental Results

A large body of experimental results on lattice dynamics has come into being in the past fifteen or so years. Particularly noteworthy is the availability of dispersion curves for a wide class of crystals, undoubtedly the driving force behind the rapid evolution of lattice dynamical models during recent years.

Since we have already described the methodology of lattice dynamical calculations and the principal models, we are in a position to review the experimental data in conjunction with the models and to examine what light has been shed on the nature of interatomic forces in solids.

In treating the experimental data we shall concern ourselves primarily with phonon dispersion curves. Usually these are available only for symmetry directions, though, in a few cases, some off-symmetry data is also available. By contrast, direct experimental information on eigenvectors is almost nonexistent.

Before launching into the survey, a few general observations:

1. Model calculations are made for a variety of reasons including (a) providing a rough prediction of the dispersion curves prior to or while performing an experiment, (b) providing an *interpolation scheme* to aid in computing various bulk properties such as thermodynamic quantities, (c) predicting the critical frequencies and shapes likely to be observed in second-order optical spectra, and (d) analyzing dispersion curve data for physical content, usually to relate it to other measured quantities. Hence in assessing a model it is important to remember the context in which it was devised. Our concern here is restricted to models in the last category.

2. A popular method of analyzing experimental data is to treat the force constants as adjustable parameters and deduce their values by forcing the calculated curves to fit those measured. In favorable cases one endeavors to interpret the values so obtained, a practice quite familiar in molecular spectroscopy. It is well to remember that such force constants are in general not unique. To appreciate why, let us first note that starting from

$$D_{\alpha\beta}\begin{pmatrix} \mathbf{q} \\ kk' \end{pmatrix} = \frac{1}{\sqrt{m_k m_{k'}}} \sum_{l'} \phi_{\alpha\beta}\begin{pmatrix} 0 & l' \\ k & k' \end{pmatrix} e^{i\mathbf{q}\cdot\mathbf{x}(l')} \tag{6.1}$$

we obtain by inversion (see Eq. (A1.6)),

$$\phi_{\alpha\beta}\begin{pmatrix} 0 & l' \\ k & k' \end{pmatrix} = \frac{\sqrt{m_k m_{k'}}}{N} \sum_{\substack{\mathbf{q} \\ \text{over BZ}}} D_{\alpha\beta}\begin{pmatrix} \mathbf{q} \\ kk' \end{pmatrix} e^{-i\mathbf{q}\cdot\mathbf{x}(l')}, \tag{6.2}$$

which shows that to deduce the force constants we really must know the dynamical matrix throughout the BZ. Since

$$\mathbf{D}(\mathbf{q}) = \mathbf{e}(\mathbf{q})\Omega(\mathbf{q})\mathbf{e}^{\dagger}(\mathbf{q}) \tag{6.3}$$

(see (2.55)), we must know both the frequencies *and* the eigenvectors throughout the BZ! If the parameterized force constants are of finite range, Eq. (6.1) can be terminated after summing over a few neighbors; correspondingly the inversion (6.2) would require eigenvalue and eigenvector information only over a coarse mesh in the BZ. Even this demand can hardly ever be met for lack of eigenvector information. However, for models in which the parameterized forces are of *highly restricted range*, the number of parameters to be determined may be relatively small and it may then be possible to deduce their values from eigenfrequencies for symmetry directions alone.[1] In general, the elements of the dynamical matrix cannot be determined uniquely from experiments, which in turn leads to nonuniqueness in the values of the derived force constants. This problem has been illustrated by Leigh et al.[2] and Cochran.[3]*

3. Sometimes a force-constant fit to experimental data is achieved only at the expense of a large number of parameters. In such instances, the actual values probably have little physical significance and are best regarded as a convenient means of expressing the dynamical matrix, enabling the calculation of normal mode frequencies for any arbitrary $q$. The dynamical matrix based on such a parameter set is sometimes referred to as an *interpolation formula*. Often this is constructed from an analysis of dispersion curves for symmetry directions alone. This, however, does not guarantee that the formula can make a satisfactory prediction for off-symmetry directions, so that it is advisable to include some off-symmetry data while devising an interpolation formula.

We now review the experimental data for materials of simple structures. For convenience, we shall consider the crystals according to their bonding.

## 6.1 Results for rare-gas solids

It is generally believed that the interatomic potential in rare-gas systems can be represented by the *same* (LJ) potential in all the states.‡ Since the potential parameters are believed to be fairly well known from an analysis of a wide class of experiments, one would expect to be able to predict the dispersion curves with substantial accuracy. Experiments on these crystals were eagerly awaited for many years but became feasible only recently when the technique of growing large single crystals was mastered.

---

*Another kind of nonuniqueness often encountered has its origin in the computational procedures employed. Force-constant fitting is usually done using a nonlinear least-squares method; depending on the starting values for the adjustable parameters, one may encounter several minima in parameter space.

‡See, however, Gillis et al., Phys. Rev. **165**, 951 (1968), who suggest that the potential parameters may change on going from the gas phase to the crystal.

Dispersion curves for He,[4-7], Ne[8], A[9,10], and Kr[11] have now been measured using neutron scattering. In all cases except helium, a straightforward calculation using the Born-von Kármán formalism and an appropriate potential function predicts the measured dispersion curves fairly satisfactorily (see Sec. 2.9).

In the measurements on krypton, anharmonic effects were found to be important[12] since the temperature (79 K) was close to melting. Högberg and Bohlin[12] therefore computed the modified frequencies $\omega_j(\mathbf{q})$ + $\Delta\omega_j(\mathbf{q})$ (see (2.184)), taking into account the third- and fourth-order contribution to the potential energy, and found better agreement with experiment than is furnished by a quasi-harmonic calculation.

For helium the Born–von Kármán approach totally fails by predicting imaginary frequencies.[13] This failure is primarily a consequence of the large zero-point motions which, as noted in Chapter 2, cannot in this case be ignored. Helium thus demands an entirely new approach. We postpone a discussion of the special theories developed for helium until the next chapter. Here we note that these seek to allow for

1. the large-amplitude motions, including those associated with zero-point oscillations, and

2. the fact that neighboring atoms correlate their motions to avoid bumping into each other.

These effects are handled quantum mechanically and the resulting theories have proved reasonably successful in explaining experimental data. A typical comparison between theory and experiment[4] is shown in Figure 6.1.

The quantum treatment developed for helium has been applied to other rare-gas solids also. (Such calculations usually ignore the effects due to the correlated motions of neighbors, since they are important only in helium.) It appears, however, that such theories are not indispensible for crystals other than helium, especially at temperatures far removed from melting. The point is well brought out by Figure 6.2. Displayed here are the measured curves for Ne (at 4.7 K corresponding to a lattice spacing of 4.462 Å),[8] those computed as in Sec. 2.9 using an appropriate LJ potential, and those computed using the lattice dynamical theory as developed for a quantum solid.[14] Considering that, apart from helium, the quantum effects are largest in neon,[15] the near equality of the various sets of curves of Figure 6.2 implies that it is not essential to regard the heavier rare-gas crystals as quantum solids for discussing the dynamics, provided the temperature is low enough.*

---

*It must, however, be mentioned that data for neon taken at 4.7 K and with a lattice spacing of 4.402 Å show some evidence of quantum effects.

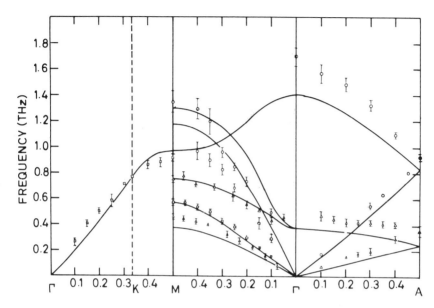

**Figure 6.1**
Dispersion curves of hcp He⁴ at a molar volume of 16.0 cm³ at 4.2 K. The solid lines are based on theoretical calculations. (After Reese et al., ref. 4.)

Though hydrogen is not a member of the rare-gas family, it perhaps deserves a few remarks. Solid hydrogen is a molecular crystal having a hexagonal structure. Dispersion curves have been measured by Nielson and Carneiro[16] using neutron spectrometry for a sample in the para-hydrogen state. In this state the incoherent scattering, which otherwise obscures the observation of coherent scattering of neutrons by phonons, is substantially diminished. The results are in reasonable agreement with a quantum-solid calculation by Klein and Koehler[17] using the self-consistent-field method (Sec. 7.2). Comparison with a similar calculation by Biem and Mertens[18] using the random-phase-approximation method (Sec. 7.2) reveals greater disagreement. The latter authors have also attempted to incorporate the effects of rotations in their model. It may be recalled that, in solid hydrogen, the molecular rotations are almost like those in a gas so that the formalism of Sec. 2.7 is inapplicable.

## 6.2 Results for metals and alloys

### 6.2.1 Early viewpoint regarding the dynamics of metals
Metals offer an interesting example of how the availability of accurate data together with the discovery of unexpected features profoundly altered the

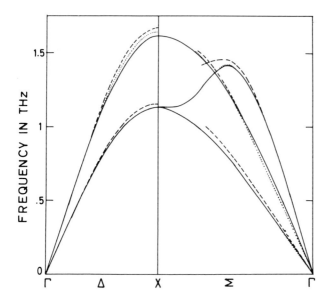

**Figure 6.2**
Dispersion curves for neon at 4.7 K corresponding to a lattice spacing of 4.462 Å. The solid lines are the experimental results, while the dotted lines are obtained by a conventional calculation and the dashed lines by treating neon as a quantum solid. (After Leake et al., ref. 8.)

course of modeling. The early attempts at calculating the dynamics were motivated by specific-heat and elastic-constant data. It was recognized even then that the presence of conduction electrons implies the forces of interaction between two atoms in the crystal to be of relatively short range, longer perhaps than the van der Waals forces existing in rare-gas solids but definitely shorter than the electrostatic forces in ionic crystals. This led early workers[19,20] to employ simple force-constant-type calculations as considered in Sec. 2.9. The force constants were treated as adjustable parameters and the interactions were restricted to third neighbors at most. Variants for the force field, such as central forces, axially symmetric forces, and tensor forces (see Sec. 2.8) were tried, and the parameters were usually determined by fitting to observed elastic constants. These early models were sufficient for the intended purposes; in particular they served to emphasize the inadequacy of the Debye frequency spectrum. The drastic assumption regarding the range of forces was, however, subjected to a detailed test only in the last decade when extensive experimental results became available.

## 6.2.2 Results for bcc metals

The important bcc metals are

Li, Na, K, Rb, Cs,
V, Nb, Ta,
Cr, Mo, W,
Fe,
Pt.

The alkali metals form an important group because their conduction electrons behave as if nearly free, facilitating detailed theoretical analysis. Dispersion data for almost the entire family is available,[21-31,50] and it is found that they are equally well described by central, axially symmetric, and tensor-force models. While a short-range model extending up to third neighbors is adequate for a moderately good fit, there is little doubt that forces out to at least fifth neighbors play a significant role. An interesting inference is that the first-neighbor forces are repulsive. Now the force-constant matrix for the nearest-neighbor pair $((0, 0, 0,), a(\tfrac{1}{2}, \tfrac{1}{2}, \tfrac{1}{2}))$ can be written as

$$\begin{pmatrix} \alpha_1 & \beta_1 & \beta_1 \\ \beta_1 & \alpha_1 & \beta_1 \\ \beta_1 & \beta_1 & \alpha_1 \end{pmatrix}.$$

Further, if $\alpha_1$ and $\beta_1$ are derivable from a pair potential $V(r)$, then

$$\left. \frac{dV(r)}{dr} \right|_{r=\text{n.n. distance}} = -\frac{\sqrt{3}}{2} a(\alpha_1 - \beta_1) \tag{6.4}$$

(see Eq. (2.225)). Experimentally it is found that $V'(r)$ is negative for $r$ corresponding to the nearest-neighbor distance, which implies that if $V(r)$ has the form sketched in Figure 1.1, then the nearest neighbor is located slightly inward from the potential minimum.

A brief comment on certain aspects of the equilibrium condition vis-à-vis axially symmetric forces is pertinent here. In Na, for example, it is found that the observed dispersion curves can be fit using axially symmetric force constants.[22] However, as noted in Sec. 2.8, by definition the two-body potential $V(r)$ from which the observed axially symmetric force constants could conceivably be derived does not entirely account for the crystal potential $\Phi$. Since three-body forces, etc., do not manifestly appear in the dynamics, evidently $\Phi$ must have the form

$$\Phi = \tfrac{1}{2} \sum_{\substack{ll' \\ (l \neq l')}} V(|\mathbf{r}(l) - \mathbf{r}(l')|) + E(v),$$

where $E(v)$ is a volume-dependent term (see Sec. 1.2). Thus in the axially symmetric model the forces responsible for the lattice vibration spectra do not by themselves hold the crystal in equilibrium. If these forces alone are considered, then an artifact like external pressure must be assumed to be applied to the crystal to hold it in equilibrium.[32]*

The dynamics of the transition metals are interesting in view of the possible role of the $d$ electrons in influencing the dynamics. Data is presently available for all the transition metals listed above, that for V being obtained using x rays. It is noteworthy that, in contrast to the alkalis, $V'(r)$ is positive at the nearest neighbor so that the first-neighbor forces are attractive. Another striking feature of the dispersion curves is the presence of structure not present in those of the alkali metals. Typical anomalies may be seen in Figure 6.3[34] and are believed to arise from the influence of the conduction electrons.

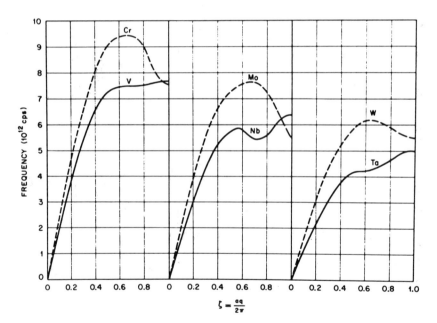

**Figure 6.3**
Dispersion curves for the longitudinal branch along [100] in various transition elements. (After Smith, ref. 34.)

*Cochran[33] considered a nearest-neighbor axially symmetric model for Ge and showed that if the crystal were finite the surface atoms would fly apart leading to a disruption of the crystal. This can be countered by the application of a suitable external pressure. However, caution is required in this procedure in the case of crystals of symmetry lower than cubic,[33] as it might turn out that the external stresses required for equilibrium are such as to lower the crystal symmetry.

### 6.2.3 Results for fcc metals

Some of the important fcc metals are

Ca, Sr,
Rh, Ir,
Ni, Pd,
Cu, Ag, Au,
Al,
Pb,
Th.

Of these, Al and Pb behave like nearly free-electron metals. In all the rest, the $d$ electrons appear to cause a departure from a free-electron characteristic.

Aluminum has been studied repeatedly by several groups,[35-40] leading to very precise data, an example of which is shown in Figure 6.4. Analysis suggests that forces out to the eighth neighbor are important.[41] The results for Pb are shown in Figure 6.5.[42] Comparison with Al immedi-

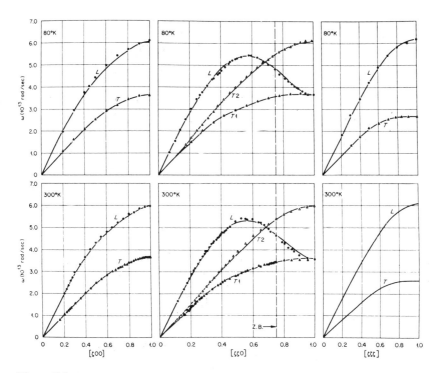

**Figure 6.4**
Dispersion curves for Al. (After Gilat and Nicklow, ref. 41.)

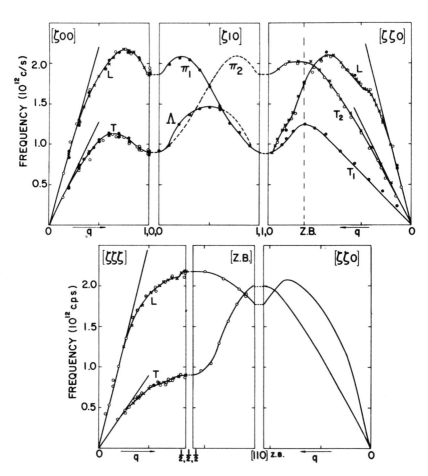

**Figure 6.5**
Dispersion curves for Pb. (After Brockhouse et al., ref. 42.)

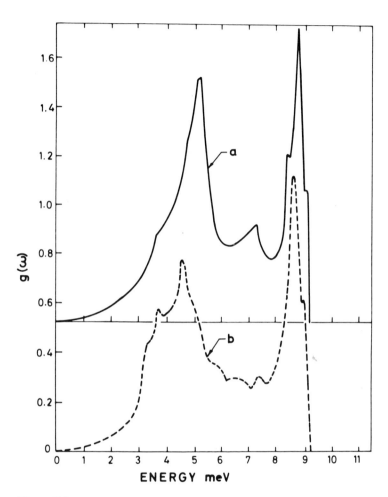

**Figure 6.6**
Frequency distribution for Pb: (a) calculations of Gilat (ref. 43) based on an interpolation
formula; (b) measurements of Stedman et al. (ref. 44).

ately reveals the presence of several anomalies (to be commented upon later). Force-constant analysis of the data for Pb has generally proved a failure. Models of several kinds with forces going out as far as the tenth neighbor have been tried without success. Indeed there is reason to believe that forces could extend to twenty neighbors.[42] The case of Pb also illustrates the point made earlier about the inadequacy of interpolation formulas based only on symmetry-direction data. Figure 6.6a shows the frequency spectrum for Pb as calculated by Gilat[43] from dispersion curve data[42] while Fig. 6.6b displays the spectrum obtained by Stedman et al.[44] by sampling experimentally the phonon frequencies throughout the BZ. While Gilat's spectrum predicts the critical features associated with the symmetry directions reasonably well, other features are not so well predicted.

Compared to lead, the noble metals are well behaved. In Cu, for example, the first-neighbor interactions dominate,[45-47] and in fact the dispersion curves can be fit to an accuracy of a few percent using only nearest-neighbor interactions. Nevertheless, forces extending to the sixth neighbor are required to obtain a complete fit.[46] A similar situation exists in Ag,[48] Th,[49] Ni,[51] and Pd,[52] the dynamics for the latter two being dominated by strong, axially symmetric forces between nearest neighbors.

An interesting innovation is due to Hanke[53] who has used the shell model to discuss the dynamics of Cu, Ag, Au, Ta, Pd, and Ni. The essential idea is to use the shell-core concept to simulate polarization effects in the noble and transition metals. While the use of such a dipolar model for metals might at first seem inappropriate, Hanke has argued that the applicability of the model can be made plausible by microscopic considerations (see Sec. 7.1.4).

### 6.2.4 Results for hcp metals
Some of the important hcp metals are

Be, Mg,
Sc, Y, La,
Ti, Zr, Hf,
Tc, Re,
Co,
Zn, Cd,
Tb, Ho, Er.

The ideal hcp structure has a $(c/a)$ ratio of 1.633. Most metals show a departure from the ideal, rather large in some cases (for example, Zn: 1.855). On the other hand, Mg and Co have close to ideal structures.

Dispersion curves for most of the metals listed above have now been measured.[54-69] Typical examples are presented in Figure 6.7; the curves for other metals look generally similar. Note the separation of the optic branches at $q = 0$. In the upper-frequency mode, the various hexagonal planes vibrate against each other, that is, the atomic motions are perpendicular to the hexagonal planes. In the notation of Chapter 4, these frequencies may be designated $\omega_{\parallel}$ and $\omega_{\perp}$, respectively, their frequency difference being a manifestation of crystalline anisotropy.

Data analysis of these metals most often involves the so-called modified axially symmetric model[70] which amounts to the usual model with concessions to crystalline anisotropy. However, Roy et al.[71] have recently shown from a consideration of the ordering of frequencies at the zone boundary in the [1120] direction that, in many hcp metals, the force field has a tensor-type admixture. Irrespective of the type, the range of the force field extends typically to about the sixth neighbors.

An interesting surprise has been the results for technetium,[65] shown in Figure 6.8 alongside those for comparable metals. The degeneracy of $\omega_{\parallel}$ and $\omega_{\perp}$ in Tc is believed to result from the peculiar role played by the con-

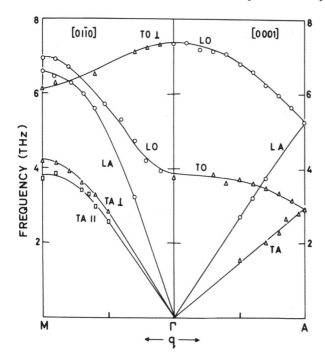

**Figure 6.7**
Dispersion curves for Mg. (After Iyengar et al., ref. 55.)

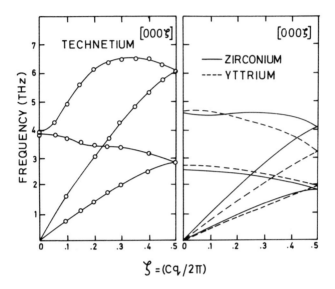

**Figure 6.8**
Dispersion curves along [0001] for Zr, Y, and Tc. (After Smith et al., ref. 65.)

duction electrons in that metal, though no detailed explanation has yet been proposed.

In general, experimental results for metals have demonstrated that while the conduction electrons do screen the ion-ion interactions so as to produce a finite range, the range is definitely not as short as originally envisaged. Further, the conduction electrons leave their imprint rather significantly in many cases, and such effects cannot easily be accommodated by simple force-constant models. Rather, they strongly suggest the need for models (such as the shell model) which more explicitly treat the role played by the conduction electrons. Steps in this direction have already been taken (see Chapter 7).  ·

### 6.2.5 Reconstruction of the interatomic potential

Assuming that the dynamics of metals can be derived from an interatomic pair potential, one might ask whether it is possible to reconstruct the potential from the values of the force constants deduced from experiment. From Eq. (2.225) it should be evident that it is not, since force constants merely sample the derivatives $V'(r)$ and $V''(r)$ at values of $r$ corresponding to the positions of the various neighbors. This is emphasized in Figure 6.9[72] which shows two potentials[73] for sodium deduced from the same dispersion curve data.[22]

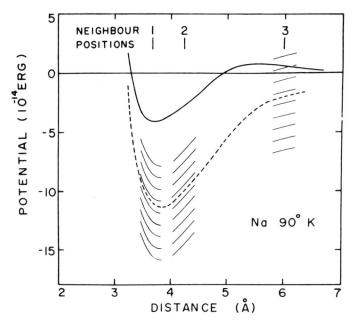

**Figure 6.9**
Interatomic potential in sodium. The short segments are pieces consistent with the observed force constants. The dashed line is a possible candidate for $V(r)$ which is consistent with the force constants and yields the correct binding energy. The solid line is the potential deduced by Cochran (refs. 32 and 73) from the same experimental data but by a different analysis. (After Brockhouse et al., ref. 72.)

A similar nonuniqueness has been noted for K by Cowley et al.[23] who observed that their data could be inverted to construct widely differing potentials ranging from oscillatory to those with barely any oscillations.

### 6.2.6 Homology

The normal-mode frequencies of elements in the same column of the periodic table are sometimes related to each other by simple scale factors. This *homology* is rather striking in the case of alkali metals. For example, Brockhouse et al.[72] note that a set of 82 normal modes common to Na and K yields

$$\left\langle \frac{\omega(\text{Na}, 90 \text{ K})}{\omega(\text{K}, 9 \text{ K})} \right\rangle = 1.635 \pm 0.005,$$

with a standard deviation of 0.040 for an individual ratio from the mean. If exact homology were to hold, then the frequency ratio for atoms of types 1 and 2 would have the value

$$\left\langle \frac{\omega_1}{\omega_2} \right\rangle = \left[ \frac{M_2 a_2^2}{M_1 a_1^2} \right]^{1/2} ,$$

where $M_1$, $M_2$ are atomic masses, and $a_1$, $a_2$ are lattice constants. For sodium and potassium this ratio is 1.61, remarkably close to that deduced from experiment. Similarly, the following results[24] are obtained for some other combinations of alkali metals.

$$\left\langle \frac{\omega(K)}{\omega(Rb)} \right\rangle : 1.67 \text{ (from expt.), } 1.59 \text{ (calc.)}$$

$$\left\langle \frac{\omega(Na)}{\omega(Rb)} \right\rangle : 2.73 \text{ (from expt.), } 2.56 \text{ (calc.).}$$

Brockhouse et al.[72] have also drawn attention to the connection of the homology rule to the Lindemann formula,[74] a slightly modified form of which reads

$$\left\langle \frac{\omega_1}{\omega_2} \right\rangle = \left[ \frac{M_2 a_2^{\,2} T_{m_1}}{M_1 a_1^{\,2} T_{m_2}} \right]^{1/2} ,$$

where $T_{m_i}$ is the melting temperature. For Na–K the Lindemann formula gives a frequency ratio 1.67, again in good agreement with experiment.

### 6.2.7 Role of conduction electrons

Earlier we remarked that many dispersion curves display features attributed to conduction electrons. That the latter can influence lattice vibrations has been recognized for many years. As early as 1936, Fuchs[75] considered their role in connection with the calculation of the elastic constants of the alkali metals and of Cu, with particular reference to the failure of the Cauchy relations. Starting from considerations previously employed by Wigner and Seitz[76] in calculating the cohesive energies of the alkalis, Fuchs argued that one could divide the contributions to the potential energy of the crystal into two parts, one associated with pairwise interactions between ions and the other associated with volume-dependent effects ascribable to the conduction electrons (see Sec. 2.4). This led him to deduce the relations (2.154) which show that volume-dependent forces lead to a violation of Cauchy relations. Fuchs further demonstrated that when a crystal is subjected to a uniform compression, forces due to both the pair potential and the volume-dependent potential are called into play in determining the bulk modulus, given by

$$\kappa = \frac{(C_{11}^{\mathrm{p}} + C_{11}^{\mathrm{v}} + 2C_{12}^{\mathrm{p}} + 2C_{12}^{\mathrm{v}})}{3}$$

(in the notation of Sec. 2.4).

De Launay[77] used the ideas of Fuchs to incorporate phenomenologically the modifications to the dispersion curves of simple cubic metals which arise from electron gas effects. He supposed that the conduction electron gas responds in phase to the longitudinal component of the lattice vibrational waves but is left unaffected by the transverse (shear) components. Since the waves traversing a crystal are, in general, neither purely longitudinal nor purely transverse, they are first separated into their longitudinal and transverse components; the electron gas is now assumed to affect only the longitudinal components. The net result is to modify the otherwise central force constants such that the deviation from the Cauchy relation is equal to bulk modulus of the electron gas.

Later Bhatia[78] offered a slightly different scheme for including volume-dependent effects. He argued that the force acting on an ion during its motion could be written as $(\mathbf{F}_1 + \mathbf{F}_2)$, where $\mathbf{F}_1$ is due to the pair potentials and is calculated as usual in terms of force constants. The part arising from volume-dependent forces is assumed to be given by

$$\mathbf{F}_2 = -e \operatorname{grad} \phi,$$

where $\phi$ is the electrostatic potential, evaluated with the Thomas-Fermi description for screening effects. The effects on the elastic constants, dispersion curves and $g(\omega)$ were examined. Some improvements to the Bhatia model were later made by Sharma and Joshi.[79]

Real understanding of the effects of conduction electrons began when the interplay of theory and experiment offered a detailed picture. The first step was taken by Toya[80] in 1957 when he calculated the dispersion curves for several alkalis, treating them as an array of positive ions immersed in a sea of conduction electrons. The modifications to the electronic wave functions consequent to ionic motion and their subsequent influence on the normal-mode frequencies were handled by Hartree-Fock theory. We shall give greater consideration to such theories in the next chapter.

Toya's work coincided with the period when rapid strides were being made in the understanding of many-body effects in electron gas systems. These considerations were widely applied to conduction electrons in metals, especially nearly free-electron (NFE) metals like the alkalis.

For instance, Langer and Vosko[81] showed that a charged impurity introduced into an electron gas produces a local disturbance, and the induced charge density shows oscillatory behavior, leading in turn to the asymptotic form (1.9) for the potential presented by the impurity in the medium. Analogous considerations have been employed by Cochran[32] and Harrison[82] who argue that the interatomic potential in a NFE metal can be written as

$$V(r) = \frac{Z^2 e^2}{r} - V^{\text{el}}(r)$$

$$= \frac{Z^2 e^2}{r} - \frac{1}{(2\pi)^3} \int V^{\text{el}}(Q) e^{i\mathbf{Q}\cdot\mathbf{r}} \, d\mathbf{Q}$$

$$= \frac{Z^2 e^2}{r} - \frac{1}{(2\pi)^3} \int \frac{U^2(Q)}{(4\pi e^2/Q^2 v^2)} \, (1 - \varepsilon^{-1}(Q)) e^{i\mathbf{Q}\cdot\mathbf{r}} \, d\mathbf{Q} \tag{6.5}$$

(see Eq. (1.8)). Here the second term on the right-hand side gives the modification due to the conduction electrons and involves the Fourier transform of the ion-electron pseudopotential (see Sec. 7.1) and the dielectric function $\varepsilon(Q)$. More will be said concerning the latter in Chapter 7. Presently we note merely that, neglecting exchange effects,

$$\varepsilon(Q) = 1 + \frac{4\pi\rho^{(0)} e^2}{Q^2} \frac{3}{2E_F} \left( \frac{1}{2} + \frac{4k_F^2 - Q^2}{8k_F Q} \ln \left| \frac{Q + 2k_F}{Q - 2k_F} \right| \right), \tag{6.6}$$

where $k_F$ is the Fermi wave vector, $E_F$ the Fermi energy, and $\rho^{(0)}$ the electron density. Note that the derivative of $\varepsilon(Q)$ has a logarithmic singularity at $Q = 2k_F$; the latter is responsible for the asymptotic oscillations in $V(r)$ noted in (1.9).

Koenig[83] has suggested that an oscillatory $V(r)$ must leave its imprint on the force constants. Now in the case of simple cubic metals, motions corresponding to waves along the principal symmetry directions involve rigid displacement of atomic planes which are normal to $\mathbf{q}$. The normal-mode problem consequently is formally equivalent to a linear chain and may be analyzed in terms of interplanar force constants.[84] Figure 6.10 displays the interplanar force constants derived from experiments on Na.[22] Also shown with a solid line is the variation expected on the basis of a potential oscillating asymptotically as in (1.9). It appears that the force constants indeed carry evidence of the singularity in $\varepsilon(Q)$.

### 6.2.8 Kohn anomalies
It is illuminating to examine conduction-electron effects in $\mathbf{q}$ space. Basically this requires an analysis of the type due to Toya. Alternately one can assume ad hoc that the pair potential is given by (6.5) and use it to calculate the dynamics. Either way it emerges, as first pointed out by

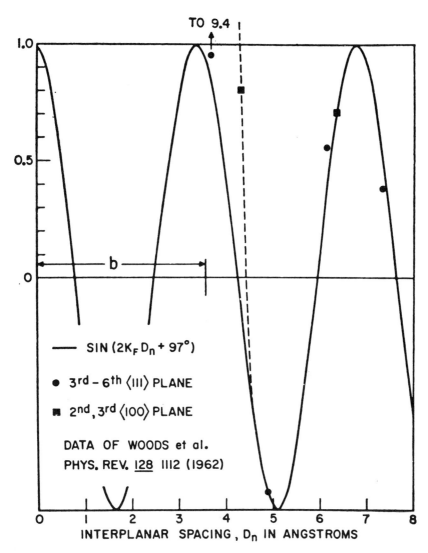

**Figure 6.10**
Interplanar force constants in sodium. The data points are based on the experimental results of Woods et al. (ref. 22). The solid line is a computed curve based on an oscillatory potential. (After Koenig, ref. 83.)

Kohn,[85] that whenever the condition

$$\mathbf{q} + \mathbf{G} = 2\mathbf{k}_F \tag{6.7}$$

is satisfied (in the extended-zone scheme), an anomaly in the form of
an infinite slope must occur in the dispersion relation $\omega = \omega_j(\mathbf{q})$. Based
on Eq. (6.7), Figure 6.11 shows how such *Kohn anomalies* can be located in
a typical case. Shortly after being predicted, Kohn anomalies were
detected in Pb[86,87] (see Fig. 6.5), and since then have been reported in
several other metals.

The technical reasons for the anomaly may be understood as follows:
Assume that $V(r)$ has the form given in (6.5). The dynamical matrix for a
Bravais lattice like Pb can then be written as

$$D_{\alpha\beta}(\mathbf{q}) = D_{\alpha\beta}^C(\mathbf{q}) + D_{\alpha\beta}^{el}(\mathbf{q}), \tag{6.8}$$

where

$$D_{\alpha\beta}^C(\mathbf{q}) = \frac{4\pi Z^2 e^2}{m_{ion}v}\left[\sum_{\mathbf{G}}\frac{(\mathbf{q}+\mathbf{G})_\alpha(\mathbf{q}+\mathbf{G})_\beta}{|\mathbf{q}+\mathbf{G}|^2} - \sum_{\mathbf{G}(\neq 0)}\frac{G_\alpha G_\beta}{|\mathbf{G}|^2}\right] \tag{6.9}$$

denotes the contribution associated with the direct Coulomb interaction

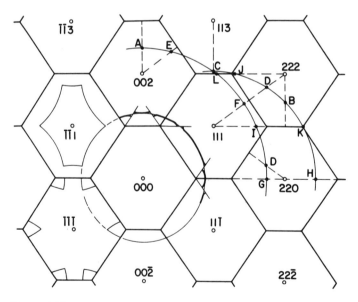

**Figure 6.11**
Kohn construction for Pb. Shown here are sections of the Fermi surface in both the
reduced and extended zone schemes in the free-electron approximation. Some of the possible anomalies are shown with the help of an arc of radius $2k_F$. (After Brockhouse et al.,
ref. 86.)

between the ions. (See Eq. (A2.17) with $\eta = \infty$ and specialized to a Bravais lattice.) The contribution associated with electronic screening is given by

$$
D^{el}_{\alpha\beta}(\mathbf{q}) = -\frac{1}{m_{ion}v}\left[\sum_{\mathbf{G}}(\mathbf{q} + \mathbf{G})_\alpha(\mathbf{q} + \mathbf{G})_\beta V^{el}(|\mathbf{q} + \mathbf{G}|)\right.
$$
$$
\left. - \sum_{\mathbf{G}(\neq 0)} G_\alpha G_\beta V^{el}(|\mathbf{G}|)\right],
\tag{6.10}
$$

(see also (2.87)). From the above equations* one has

$$
\omega_j^2(\mathbf{q}) = [\omega_j^C(\mathbf{q})]^2 - [\omega_j^{el}(\mathbf{q})]^2,
\tag{6.11}
$$

where

$$
[\omega_j^C(\mathbf{q})]^2 = \mathbf{e}^\dagger\!\begin{pmatrix}\mathbf{q}\\j\end{pmatrix}\mathbf{D}^c(\mathbf{q})\mathbf{e}\begin{pmatrix}\mathbf{q}\\j\end{pmatrix}, \qquad -[\omega_j^{el}(\mathbf{q})]^2 = \mathbf{e}^\dagger\!\begin{pmatrix}\mathbf{q}\\j\end{pmatrix}\mathbf{D}^{el}(\mathbf{q})\mathbf{e}\begin{pmatrix}\mathbf{q}\\j\end{pmatrix}.
$$

We observe from (6.11) that screening has a tendency to reduce the frequencies. Furthermore, because

$$
\frac{d\varepsilon(Q)}{dQ}\bigg|_{Q=2k_F} = -\infty,
$$

grad $\omega_j(\mathbf{q})$ has a corresponding singularity. These arguments may be generalized to an anisotropic Fermi surface, leading to (6.7) for the location of the anomaly.

The physical reason for the logarithmic singularity in $\varepsilon(Q)$ may be appreciated by considering the dielectric susceptibility of an electron gas. Recall first the result[88]

$$
\alpha = \sum_j{}'\frac{2e^2|x_{ij}|^2}{E_i - E_j}
\tag{6.12}
$$

for atomic polarizability, where $E_i$, $E_j$ refer to the energies of the levels $i, j$ and $x_{ij}$ is the matrix element of the coordinate operator $x$. Here $\alpha$ is a measure of the deformability of the electron cloud of the atom and depends on the "ease" with which virtual transitions $i \rightarrow j$ can occur, which in turn depends on the energy denominator associated with the virtual transition.

Analogously, for an electron gas one may write[89]

$$
\varepsilon(Q) = 1 + 4\pi\alpha(Q)
\tag{6.13}
$$

---

*The $\mathbf{G}=0$ contributions associated with the second terms on the right-hand sides of (6.9) and (6.10) are both ill-defined. However they cancel each other and for this reason have been omitted above.

with

$$\alpha(Q) = \frac{2e^2}{\Omega Q^2} \sum_{k < k_F} \frac{1}{E(\mathbf{Q} + \mathbf{k}) - E(\mathbf{k})}, \tag{6.14}$$

where $E(\mathbf{Q}) = \hbar^2 Q^2 / 2m$ and $\Omega$ is the normalization volume. Again $\alpha$ is determined mainly by the energy denominator associated with virtual transitions. For $Q < 2k_F$ a large number of transitions from one part of the Fermi surface to another are possible. Having small or vanishing denominators, these make sizable contributions to $\alpha$. For $Q > 2k_F$ such elastic scattering restricted to the Fermi surface is no longer possible, and the polarizability is reduced; in other words, the electron gas appears to stiffen. The logarithmic singularity in $\varepsilon(Q)$ is thus connected with the abrupt change in the number of possible virtual transitions.

In principle, Kohn anomalies should be observable in all metals. Their strength, however, varies from metal to metal, and in practice anomalies have been seen only in a few cases. Certain broad criteria have emerged for their observation:

1. Metals with large electron-phonon interactions are good candidates. Considering that such interactions are strong in the superconductors, especially those with a high transition temperature $T_c$, it is not surprising that strong anomalies are seen in Pb ($T_c = 7.2$ K) and Nb ($T_c = 9.2$ K).

2. Conditions are favorable in metals in which

$$[\omega_j^{obs}(\mathbf{q})]^2 \ll [\omega_j^C(\mathbf{q})]^2. \tag{6.15}$$

This is readily understood by referring to (6.11), from which we note that for the above condition to obtain we must have

$$\omega_j^C(\mathbf{q}) \approx \omega_j^{el}(\mathbf{q}),$$

that is, screening must be very strong. If

$$\omega_p^2 = \frac{4\pi Z^2 e^2}{v m_{ion}},$$

the square of the plasma frequency of the ionic system, is used as a measure of $[\omega_j^C(\mathbf{q})]^2$, and $\bar{\omega}^{obs}$ denotes a sort of average of observed frequencies, then (6.15) implies

$$\omega_p \gg \bar{\omega}^{obs}.$$

For example $(\omega_p / \bar{\omega}^{obs})$ is $\sim 9$ for Pb while $\sim 2$ for Na, consistent with the presence of anomalies in Pb and their absence in Na.

Wakabayashi et al.[62] have pointed out that anomalies can occur in a form more complicated than envisaged by Kohn, particularly so if the Fermi surface consists of two pieces of relatively large area separated by approximately the same **q**. In this case, $\alpha(\mathbf{Q})$ can build up into a large peak for some values of **Q** without an actual singularity, and these peaks may leave their marks on the dispersion curves.

Hanke and Bilz[90] have argued that anomalies could also arise from the off-diagonal elements of the screening function. The essential idea is that screening is described by a matrix $\varepsilon(\mathbf{Q}, \mathbf{Q}')$ (**Q** and **Q**' label the rows and columns) rather than by $\varepsilon(Q)$, and that anomalies could arise due to terms with $\mathbf{Q} \neq \mathbf{Q}'$. These anomalies also are related to band structure but in a complicated way. It has been suggested that the anomalies seen in group V and group VI elements (see Fig. 6.3) are of this origin.

### 6.2.9 Results for alloys

The dynamics of several alloys—ordered, disordered, and dilute—have also been studied. Among the ordered alloys, mention may be made of $\beta$-brass (CuZn)[91] and Fe$_3$Al.[92] $\beta$-brass has the CsCl structure; however, since the masses of Cu and Zn are very close to each other, one would expect the dynamics of $\beta$-brass to be rather like those of a bcc metal. As Figure 6.12 shows, this is indeed the case. The small differences seen in

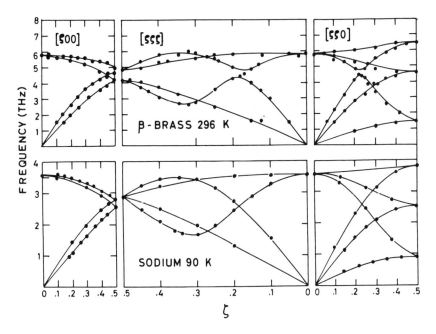

**Figure 6.12**
Dispersion curves for $\beta$-brass (ref. 91). Also shown for comparison are the results for sodium (ref. 22), plotted treating the cubic unit cell as basic.

that figure between the spectra of brass and of Na are due mainly to the slight asymmetry between Cu and Zn. In passing, it is interesting to observe that, in contrast to those in ionic crystals of the CsCl structure, the optic modes at $\mathbf{q} = 0$ in CuZn are degenerate, a result of the quenching of the macroscopic field by the conduction electrons.

One would suspect at first sight that phononlike excitations in disordered alloys could not exist owing to gross disturbance of the periodic structure. However, most of the alloys studied thus far involve neighboring elements in the periodic table, so that at least in terms of mass differences the disorder is no different from that caused in pure metals by admixture of isotopes. Ignoring mass and chemical differences one expects these alloys to exhibit well-defined vibrational excitations appropriate to an average or "virtual" crystal of the same structure, made up of suitably defined "average atoms." Clearly this is an idealization, and the true

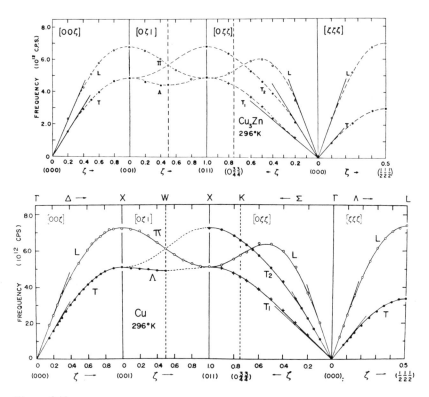

**Figure 6.13**
Dispersion curves for $\alpha$-brass. Also shown for comparison are the results for Cu. (After refs. 93 and 46.)

crystal is the virtual crystal with superposed random effects, the latter leading to a damping of the vibrational waves.

Results obtained for several alloys have supported this picture. Figure 6.13, for example, displays the similarity of the dispersion curves for α-brass[93] ($Cu_3Zn$) and for Cu.[46] In fact, a Born–von Kármán fit to α-brass showed that, as in Cu, the first-neighbor forces dominate.[93]

The validity of the virtual crystal concept has opened up the possibility of studying alloys as a function of conduction electron concentration. As an example, we cite the experiments[72,94] of Brockhouse and coworkers on binary and ternary alloys of Pb, Tl, and Bi. In all cases well-defined dispersion curves appropriate to the average structure (fcc) were obtained. As seen in Figure 6.14, the measured curves show a strong dependence on electron concentration. That the dispersion curves are primarily dependent on the electron concentration is further emphasized by the results (Figure 6.15) for $Pb_{0.4}Tl_{0.6}$ and $Bi_{0.2}Tl_{0.8}$, both of which have 3.4 conduction electrons per average atom.

Several Kohn anomalies were detected in these alloys. Their positions were interpreted in terms of a rigid-band model, according to which the

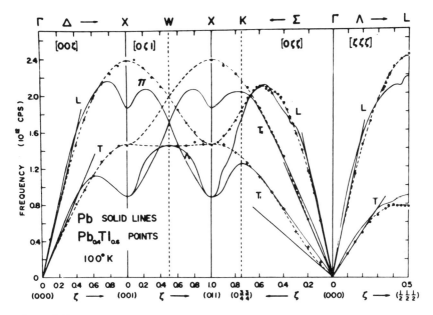

**Figure 6.14**
Dispersion curves for $Pb_{0.4}Tl_{0.6}$. Also shown for comparison are the results for Pb. (After Brockhouse et al., ref. 72.)

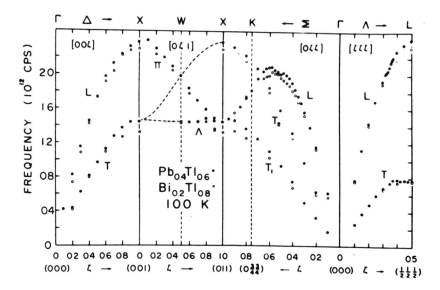

**Figure 6.15**
Dispersion curves for two binary alloys having the same electron concentration. (After Brockhouse et al., ref. 72.)

band structure of the entire series of alloys is the same as that of Pb, the only difference being in the position of the Fermi level, which is concentration dependent. The observed anomalies in the Pb–Tl alloys seem to corroborate this view.

Force-constant analysis for the Pb-Tl-Bi alloys yielded good fits with 8-neighbor models. Forces of longer range, however, appear to be present also. In general, the interaction with distant neighbors seems to become smaller as the electron concentration in the alloys is reduced.

Work along similar lines with similar results and conclusions have been reported for Mo-Nb alloys.[95]

Recently Kamatikahara and Brockhouse[96] have carried out measurements on the disordered alloy $Ni_{0.55}Pd_{0.45}$, an alloy which falls in a category altogether different from those considered above. Here the components are chemically similar but have substantially different mass. Nevertheless, well-defined dispersion curves appropriate to the mean crystal were observed.

Results have also been reported for a few dilute alloys such as Au in Cu[97] and Al in Cu.[98] Of interest in these investigations were specific effects associated with impurities, a topic beyond our scope.

## 6.3 Results for crystals having the NaCl structure

### 6.3.1 General features of dispersion curves for alkali halides

As the most purely ionic crystals in existence, the alkali halides represent the group of materials whose dynamics and related properties have commanded the greatest attention.[99–110]* Figures 6.16 and 6.17 display representative data from which the following features may be noted.

1. The small $\mathbf{q}$ behavior of the acoustic branches differs little from crystal to crystal, the reason being that the velocities of the various elastic waves propagating in a given direction generally have the same ordering irrespective of material† (LiF being a conspicuous exception).

2. In RbBr, two branches belonging to IMR $\Sigma_1$ are seen to approach strikingly closely along [110]. Interestingly, group theory provides no help in deciding whether to represent such data as shown as the "repulsion" of branches or as crossing of branches. Either choice would meet the compatibility requirement at $X$. However, a detailed consideration of the possibility of an accidental degeneracy at this point (see refs. 19 and 20 cited in Chapter 3) suggests that it is unlikely, so that the branches are represented as repelling each other.

3. As the polarizability of the negative ion increases, the LA [100] branch shows an increasing tendency to dip near the zone boundary. (See Figure 6.18 for polarizabilities.)

4. In those materials whose anion and cation masses are reasonably close (NaF, for example), the longitudinal acoustic and optic branches approach very closely. On the other hand, if the masses are quite different (for example, RbF) there is a clear separation and a corresponding gap in the frequency spectrum.

### 6.3.2 Electronic polarizability of ions in alkali halides

Since electronic polarizability plays an important part in determining the vibration spectra of alkali halides, it is pertinent to discuss first the polarizability of the individual ions in such crystals. Now from the Lorenz-Lorentz formula

$$\frac{4\pi\alpha}{3v} = \frac{\varepsilon(\infty) - 1}{\varepsilon(\infty) + 2}$$

---

*Alkali halides of the CsCl structure have also been studied, but to a lesser extent. See, for example, A. A. Z. Ahmad, H. G. Smith, N. Wakabayashi, and M. K. Wilkinson, Phys. Rev. **B6**, 3956 (1972).

†See Table 4.1 for a summary of velocities of elastic waves in the principal symmetry directions. This table is applicable to NaCl-type crystals with the additional proviso $C_{44}^{E} = C_{44}^{D} = C_{44}$.

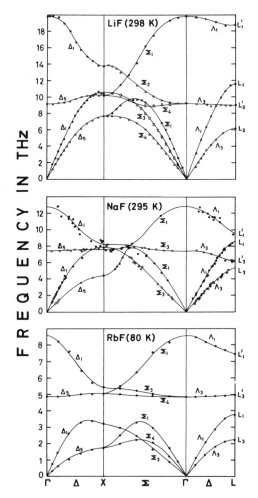

**Figure 6.16**
Dispersion curves for some alkali fluorides. (After refs. 99, 100, and 107.)

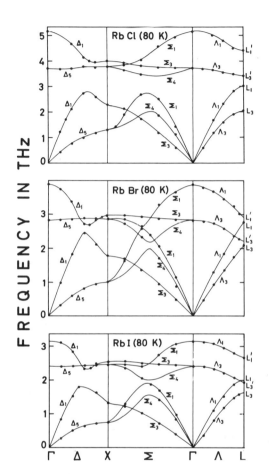

**Figure 6.17**
Dispersion curves for some rubidium halides. (After refs. 108–110.)

one may deduce for each salt a crystal polarizability $\alpha$ using the measured high-frequency dielectric constant. If $\alpha_+$ and $\alpha_-$ denote the polarizabilities of the individual ions, then to the extent that additivity holds one must have $\alpha = \alpha_+ + \alpha_-$. In fact, if this were true for all the alkali halides, then it would be possible to assign the various ionic species unique values, from which the crystal polarizability for any alkali halide could be predicted. Tessman, Kahn, and Shockley[111] (TKS) attempted to deduce such a set from an analysis of experimental data.

The TKS results are displayed in Figure 6.18 in the form of a Veiel diagram.[112] Assuming $\alpha(\text{LiF}) = \alpha(\text{Li}) + \alpha(\text{F})$, one first marks off LiF, assigning for $\alpha(\text{Li})$ the value 0.029 suggested by Pauling[113] and deriving $\alpha(\text{F})$ from the above equation using the experimental value of $\alpha(\text{LiF})$. The various lithium salts are then plotted along a vertical line using the experimental results for $\alpha(\text{Li}X) - \alpha(\text{LiF})$, $(X = \text{Cl}, \text{Br}, \text{I})$. The alkali fluorides are similarly marked off along a horizontal line starting from LiF. To complete the figure, points for the other salts are located by drawing arcs of appropriate lengths from appropriate points in the vertical and horizontal edges drawn earlier. For instance, KBr is located at the intersection of two arcs of radii $(\alpha(\text{KBr}) - \alpha(\text{LiBr}))$ and $(\alpha(\text{KBr}) - \alpha(\text{KF}))$ drawn from the points corresponding to LiBr and KF. If the additivity law were valid, the points would lie on a perfectly rectangular array; in fact they do not. What TKS have attempted is to fit (by least squares) a rectangular grid, shown by solid lines in Figure 6.18. The polarizabilities of the individual species corresponding to this fitted array are also indicated. It may be seen that the negative ion polarizability generally exceeds that of the positive ion and that the negative-ion polarizability increases toward the heavier halogens.

### 6.3.3 Models for NaI—a canonical example

We shall now review modeling experience for alkali halides using NaI as an example, our choice being partly influenced by the high polarizability of this salt.

Figure 6.19 shows the results of Woods et al.[103] for this crystal. Also shown by a dashed line is a calculation based on the rigid-ion model (see Sec. 4.2). This calculation involves repulsive forces of the Born-Mayer form between nearest neighbors and Coulomb interactions between all ion pairs (see Eq. (1.11)). The short-range force-constant matrix between the ion at the origin (taken as the positive ion) and its nearest neighbor at $r_0(1, 0, 0)$ is given by

$$\frac{e^2}{2v}\begin{bmatrix} A & & \\ & B & \\ & & B \end{bmatrix},$$
(6.16)

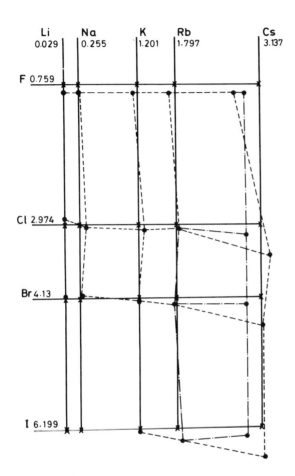

**Figure 6.18**

Electronic polarizabilities of alkali and halogen ions presented in the form of a Veiel diagram. The dots represent points derived from the total polarizability $\alpha$ (deduced from $\varepsilon(\infty)$) as discussed in the text. With respect to the origin (not shown), the coordinates of the dot corresponding to LiF are (0.029, 0.880), where 0.029 is Pauling's value for the polarizability of $Li^{+}$. In the case of the cesium halides, results for both the NaCl (dashed lines) and CsCl (dot-dash lines) structures are shown. The rectangular grid defined by the crosses corresponds to the fit made by TKS. The polarizabilities of the individual ions corresponding to this fitted grid are also shown.

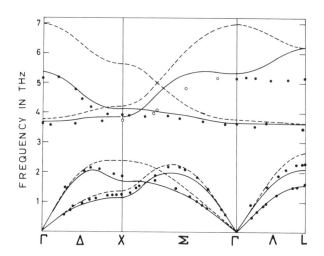

**Figure 6.19**
Dispersion curves for NaI. The circles are the results of experiment (ref. 103); the dashed lines and the solid lines are calculated curves using respectively the rigid-ion model and the simple shell model (ref. 103).

where

$$\frac{e^2 A}{2v} = \left. \frac{d^2 V^{SR}_{+-}(r)}{dr^2}\right|_{r=r_0}, \qquad \frac{e^2 B}{2v} = \frac{1}{r_0}\left. \frac{dV^{SR}_{+-}(r)}{dr}\right|_{r=r_0} \tag{6.17}$$

Here $2r_0$ is the lattice constant so that $v = 2r_0^3$. The factor $e^2/2v$ is introduced for numerical convenience. The ionic charge $Z$ is given its formal value of unity, and $B$ is determined from the equilibrium condition[114] to be (see also (2.228))

$$B = -\tfrac{2}{3}\alpha_M Z^2 = -1.165Z^2, \tag{6.18}$$

where $\alpha_M$ is the Madelung constant (see p. 260). The value of $A$ is determined from elasticity data, using the following expressions (see Eq. (4.126); in the NaCl structure, the quantities $(\alpha\lambda, \beta\gamma)$ all vanish):

$$C_{11} = C_{\alpha\alpha,\alpha\alpha} = [\alpha\alpha, \alpha\alpha],$$
$$C_{12} = C_{\alpha\alpha,\beta\beta} = 2[\alpha\beta, \alpha\beta] - [\alpha\alpha, \beta\beta],$$
$$C_{44} = C_{\alpha\beta,\alpha\beta} = [\alpha\alpha, \beta\beta],$$
$$[\alpha\alpha, \alpha\alpha] = \frac{e^2}{vr_0}\left[\frac{A}{2} - 2.56Z^2\right],$$
$$[\alpha\alpha, \beta\beta] = \frac{e^2}{vr_0}\left[\frac{B}{2} + 1.28Z^2\right],$$
$$[\alpha\beta, \alpha\beta] = \frac{e^2}{vr_0}[0.696Z^2]. \tag{6.19}$$

We observe from Figure 6.19 that while the rigid-ion model is able to re-produce the gross features (including the LO-TO splitting at the zone center), it is quantitatively inadequate.

Considerable improvement is obtained if polarization effects are included even in a restricted way. This can be seen by referring to the solid lines in Figure 6.19, which represent calculations based on a simple version of the shell model (SSM).[103] It is assumed here that

1. the short-range forces act entirely through shells;
2. they are restricted to nearest neighbors and are derivable from pair potentials, and
3. only the negative ion is polarizable.

Assumption 1 implies (in the notation of Sec. 4.6)

$$\Psi_{\alpha\beta}^{sc,SR} = \Psi_{\alpha\beta}^{cs,SR} = \Psi_{\alpha\beta}^{cc,SR} = 0,$$

so that

$$\Psi_{\alpha\beta}^{AA,SR} = \Psi_{\alpha\beta}^{DD,SR} = \Psi_{\alpha\beta}^{AD,SR} = \Psi_{\alpha\beta}^{DA,SR}. \tag{6.20}$$

The structure of these force-constant tensors for the representative nearest-neighbor pair $((000), r_0(100))$ is thus that of (6.16). From (6.20) it follows that

$$\mathbf{R} = \mathbf{M}^{DD,SR}(\mathbf{q}) = \mathbf{T}.$$

In the numerical calculations, the negative ion parameters $k_-$ and $Y_-$ are replaced by new parameters $\alpha_-$ and $d_-$ defined by

$$\alpha_- - \frac{Y_-^2 e^2}{k_- + R_0}, \tag{6.21a}$$

$$d_- = -\frac{Y_- R_0}{k_- + R_0}, \tag{6.21b}$$

where

$$R_0 = R_{\alpha\alpha}\left(\begin{matrix}\mathbf{q} = 0\\ + +\end{matrix}\right) = R_{\alpha\alpha}\left(\begin{matrix}\mathbf{q} = 0\\ - -\end{matrix}\right) = \frac{e^2}{v}(A + 2B). \tag{6.21c}$$

The elastic constants in this model are still given by (6.19), while the long-wavelength optic frequencies are given by

$$\bar{m}\omega_{TO}^2 = \left(R_0 - \frac{e^2 d_-^2}{\alpha_-}\right) - \frac{4\pi e^2(Z - d_-)^2[\varepsilon(\infty) + 2]}{9v}, \tag{6.22}$$

$$\bar{m}\omega_{LO}^2 = \left(R_0 - \frac{e^2 d_-^2}{\alpha_-}\right) + \frac{8\pi e^2(Z - d_-)^2[\varepsilon(\infty) + 2]}{9v\varepsilon(\infty)}. \tag{6.23}$$

The value of $B$ is determined via the equilibrium condition (6.18) with $Z$

taken as unity. The model thus has three parameters, $R_0$, $\alpha_-$, and $d_-$. Of these, $R_0$ is determined from (6.21c); $\alpha_-$ is obtained from the Lorenz-Lorentz formula

$$\frac{4\pi\alpha_-}{3v} = \frac{\varepsilon(\infty) - 1}{\varepsilon(\infty) + 2} \tag{6.24}$$

from the measured value of $\varepsilon(\infty)$; $d_-$ is deduced from $\varepsilon(0)$ using the Clausius-Mossotti formula

$$\frac{4\pi(\alpha_- + \alpha_{\mathrm{I}})}{3v} = \frac{\varepsilon(0) - 1}{\varepsilon(0) + 2}, \tag{6.25}$$

where $\alpha_{\mathrm{I}}$ is given by*

$$\alpha_{\mathrm{I}} = (Z - d_-)^2 e^2 \bigg/ \left( R_0 - \frac{e^2 d_-^2}{\alpha_-} \right). \tag{6.26}$$

The improvement of the simple shell model over the rigid-ion model is quite evident from Figure 6.19.† Note, however, that the simple shell model predicts the same result for the LO mode at $L$ as the rigid-ion model. Further comments on this will be made later.

In an effort to improve the overall fit, Cowley et al.[115]
1. allowed for next-nearest-neighbor interactions,
2. removed the restriction that nearest-neighbor forces are derivable from a pair potential,
3. assumed that the positive ion is also polarizable, and
4. allowed $Z$ to vary.
This elaboration naturally increased the number of adjustable parameters, facilitating the attainment of a good fit. Unfortunately, in the process some parameters acquired unphysical values. For example, the ratio $(\alpha_+/\alpha_-)$ is considerably larger than might be expected from the TKS values and, more seriously, $d_+$ is positive, which implies a positive charge on the shell of the positive ion! Cowley et al.[115] pointed out that this last feature may be due to the model attempting to simulate the charge distribution associated with the mode $LO(L)$. Figure 6.20 shows the movement of ions in this mode. Observe that since this mode corresponds to an IMR which occurs once, only one species of ions moves, namely the light positive ions (see Sec. 3.2). Consider a typical negative ion, say $B$. As explained (in Sec. 4.6) in the context of the model of Basu and Sengupta,

---

*This result can be obtained from (6.22), (6.23), (6.24), and the LST relation (4.5).
†The agreement with experiment appeared even better when the results were originally published since the optic branches $\Sigma_1$ and $\Lambda_1$ had not yet been mapped.

EQUILIBRIUM   CONFIGURATION

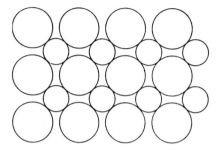

DISTORTED   CONFIGURATION

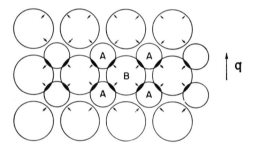

**Figure 6.20**
Two-dimensional analogue of the motion of atoms in the LO($L$) mode in NaCl-type crystals. The lower portion of the figure shows the atomic displacements for the optic mode LO($L$). Here the $B$ type ions are stationary while the $A$ type move causing overlap as shown by the shaded regions. The arrows show the direction of motion of electrons of the negative ions. (After Cowley et al., ref. 115.)

in the shell model the symmetric movement of the neighbors $A$ does not induce any polarization of $B$, which explains why the SSM and the rigid-ion model give the same LO($L$) frequency. Yet, in reality, polarization could occur due to some mechanism ignored by the shell model. If this is the case, the associated dipoles would be distributed as in the lower part of the figure. The frequency of the LO($L$) mode would clearly depend on these dipoles which cannot be accommodated by the SSM. The large discrepancy noted in Figure 6.19 at $L$ tends to support this hypothesis. Allowing the positive ions to be polarizable certainly improves the fit but with consequences noted earlier. The flexibility inherent in the parameter proliferation introduced by Cowley et al.[115] has apparently enabled the description of features lying beyond the shell model (see Fig. 6.20) but through the assignment of unphysical values to the adjustable parameters. Indeed this conclusion is reinforced by the success achieved by the breathing-shell model, to be discussed shortly.

Figure 6.21 shows the results of Karo and Hardy[116] for the deformation-dipole model. The quantities to be specified are the ionic charges, the electronic polarizabilities, the short-range force constants and the parameters pertaining to the distortion dipoles.

In the simplest version, $Z$ is taken as unity and the $\alpha_k$ are given the TKS values. Short-range forces are restricted to nearest neighbors and are assumed to be central so that the force-constant matrix for the pair $((000), r_0(100))$ has the form of (6.16). The values of $A$ and $B$ are derived from the Born-Mayer potential $V(r) = \lambda \exp(-r/\rho)$ using the values of $\lambda$ and $\rho$ quoted by Born and Huang.[117]† Distortion dipoles are attributed to the negative ions and are assumed to arise solely from nearest-neighbor overlap. They are calculated assuming (see Eq. (4.333))

$$\psi'_{++}(r) = \psi'_{+-}(r) = \psi'_{--}(r) = 0$$
$$\psi'_{-+}(r) \equiv \psi'_{-}(r) = m_-(r) = d_- R_- [\lambda e^{-r/\rho}], \tag{6.27}$$

where $R_-$ is the Zachariasen radius[118] and $d_-$ is a disposable parameter. In the actual calculations, parameters $\gamma_2$ and $\gamma'_2$ defined by

$$\gamma_2 = \frac{m_-(r_0)}{r_0}, \quad \gamma'_2 = \left[\frac{dm_-(r)}{dr}\right]_0 \tag{6.28}$$

are used to characterize the distortion dipoles. Clearly only one of these parameters can be disposable, as may be confirmed from (6.27) which gives

$$(\gamma'_2/\gamma_2) = -(r_0/\rho).$$

†To avoid confusion with force-constant parameters, we express the Born-Mayer potential here using the notation of Born and Huang.[117]

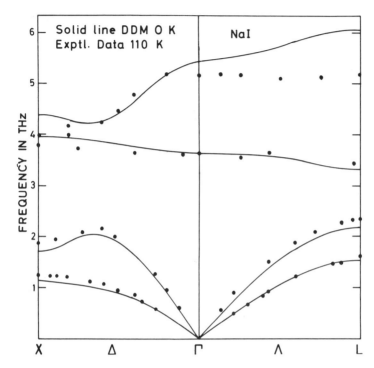

**Figure 6.21**
Comparison of the experimentally measured dispersion curves for NaI (ref. 103) with
those computed using the deformation-dipole model (ref. 116).

$\gamma_2$ is taken as the adjustable parameter and its value is fixed using the
Szigeti relation, which for this model reads

$$\frac{e^*}{e} = 1 - 2\left(\frac{r_0}{\rho} - 2\right)\frac{\gamma_2}{e}.$$

Thus all the parameters can be specified from long-wavelength data.

The overall performance of this model is of the same quality for NaI as
that of the simple shell model. It may be noted that the deformation-
dipole model predicts the LO($L$) frequency differently from the rigid-ion
model (in contrast to the SSM), since here the positive ion is allowed
electronic polarizability.

The breathing-shell model was first introduced by Schröder[119] in an
effort to alleviate the difficulty of the LO($\Lambda$) branch. Later Melvin et
al.[120] explored the mechanics of the model in greater detail and helped
clarify certain assumptions made implicitly by Schröder.

Melvin et al. assume (in common with model II of Cowley et al.[115])

that nearest-neighbor forces are noncentral so that the matrix in (6.16) now becomes

$$\frac{e^2}{2v}\begin{pmatrix} A & & \\ & B + B'' & \\ & & B + B'' \end{pmatrix}, \tag{6.29}$$

where $B''$ represents the contribution from noncentral forces. They further suppose that second-neighbor forces are important only between anions and are given by (2.225) with

$$\frac{1}{r}\frac{dV_{--}^{\text{SR}}(r)}{dr}\bigg|_{r=\sqrt{2}r_0} = \frac{e^2}{2v}B'(-),$$

$$\frac{d^2V_{--}^{\text{SR}}(r)}{dr^2}\bigg|_{r=\sqrt{2}r_0} = \frac{e^2}{2v}A'(-).$$

All short-range forces are assumed to act via shells. To incorporate the breathing effects, it is necessary (in the notation of Sec. 4.6) to specify the following constants:

$$\phi_\alpha^{\text{AM}}\begin{pmatrix} 0 & l' \\ k & k' \end{pmatrix}, \quad \phi_\alpha^{\text{DM}}\begin{pmatrix} 0 & l' \\ k & k' \end{pmatrix}, \quad \phi_\alpha^{\text{MA}}\begin{pmatrix} 0 & l' \\ k & k' \end{pmatrix},$$

$$\phi_\alpha^{\text{MD}}\begin{pmatrix} 0 & l' \\ k & k' \end{pmatrix}, \quad \phi^{\text{MM}}\begin{pmatrix} 0 & l' \\ k & k' \end{pmatrix}. \tag{6.30}$$

It is supposed that the forces leading to shell deformation are also transmitted via shells, leading to

$$\phi_\alpha^{\text{AM}}\begin{pmatrix} 0 & l' \\ k & k' \end{pmatrix} = \phi_\alpha^{\text{DM}}\begin{pmatrix} 0 & l' \\ k & k' \end{pmatrix}, \quad \phi_\alpha^{\text{MA}}\begin{pmatrix} 0 & l' \\ k & k' \end{pmatrix} = \phi_\alpha^{\text{MD}}\begin{pmatrix} 0 & l' \\ k & k' \end{pmatrix}.$$

In the calculations it is further assumed that only the negative-ion shell can deform and that such deformation is produced only by nearest-neighbor forces. Hence it is sufficient to specify†

$$\phi_\alpha^{\text{AM}}\begin{pmatrix} 0 & l' \\ + & - \end{pmatrix}, \quad \alpha = x, y, z,$$

$$\begin{pmatrix} l' \\ - \end{pmatrix} \text{ a nearest neighbor of } \begin{pmatrix} 0 \\ + \end{pmatrix}; \tag{6.31a}$$

$$\phi_\alpha^{\text{MA}}\begin{pmatrix} 0 & l' \\ - & + \end{pmatrix}, \quad \alpha = x, y, z,$$

$$\begin{pmatrix} l' \\ + \end{pmatrix} \text{ a nearest neighbor of } \begin{pmatrix} 0 \\ - \end{pmatrix}; \tag{6.31b}$$

†In the literature the subscripts 1 and 2 are often used to denote the positive and negative ions respectively. We prefer the more explicit $+$ and $-$.

and

$$\phi^{MM}\begin{pmatrix} 0 & 0 \\ - & - \end{pmatrix}. \tag{6.31c}$$

On the premise that a radial deformation is most likely produced by radial forces, the displacement of a positive ion is considered to produce a radial deformation of a neighboring negative ion shell only if the positive ion displacement has a component along the line joining the concerned neighbors. Hence, for the standard pair $((000), r_0(100))$,

$$\phi_x^{AM} = \varepsilon_+ \frac{e^2 A}{2v}, \quad \phi_y^{AM} = 0 = \phi_z^{AM},$$

$\varepsilon_+$ denoting a multiplier. Force constants for other pairs may be deduced by transformation (2.26).

Melvin et al. show that $\phi^{MM}\begin{pmatrix} 0 & 0 \\ - & - \end{pmatrix}$, which represents the resistance offered by a shell to its own deformation, can be expressed in the form

$$\phi^{MM}\begin{pmatrix} 0 & 0 \\ - & - \end{pmatrix} = \varepsilon_{--} \frac{e^2}{v} \tag{6.32}$$

with

$$\varepsilon_{--} = 3A + k^r(-), \tag{6.33}$$

where $k^r(-)$ is a constant. Further reasoning suggests that if the above restoring constant arises only due to the spring coupling a shell to its core, then

$$k^r(-) = 3k_-, \tag{6.34}$$

$k_-$ being the usual core-shell constant.

If $k^r(-)$ is specified, by (6.34) or otherwise, then the model has just the parameter $\varepsilon_+$ in addition to those of the shell model. To eliminate even this, Melvin et al. assume that nearest-neighbor interactions arise via a potential

$$V_{+-} = \psi(|\mathbf{r}(+) - \mathbf{r}(-)| - R(-)), \tag{6.35}$$

where $\mathbf{r}(+)$ and $\mathbf{r}(-)$ denote the positions of the shell centers and $R(-)$ the radius of the negative ion shell, that is, the potential is determined by those parts of the negative ion closest to the positive ion (see Figure 6.22). It then follows that

$$\phi_x^{AM}((000), r_0(100)) = \psi'' = \frac{e^2 A}{2v},$$

implying $\varepsilon_+ = 1$. In other words, a negative ion exerts the same force on

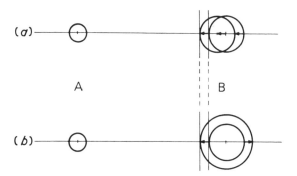

**Figure 6.22**
(a) shows the displacement of the shell of a negative ion $B$ relative to a neighboring positive ion $A$. In (b), the shell is not displaced but allowed an isotropic expansion. A suitable choice of interaction potential can be made such that the force in either case depends only on the shortest distance between ion $A$ and the shell of ion $B$. (After Sangster et al., ref. 124.)

a neighboring positive ion by a translation as by an isotropic expansion of appropriate magnitude, though the force exerted on other neighbors would be different in the two cases.

The model of Schröder is very similar to this with $\varepsilon_+ = 1$ and $k''(-) = k_-$. The results obtained by Schröder are displayed in Figure 6.23a. On the other hand, Melvin et al. have made computations assuming $\varepsilon_+ = 1$ and $k''(-) = 3k_-$ (dashed line in Figure 6.23b), which obviously yields poor agreement with experiment. However, with $k''(-) = \lambda k_-$ and $\lambda = 0.81$ determined by fitting to the LA($L$) phonon, the agreement is improved substantially (solid line in Figure 6.23b). As speculated by Melvin et al., this presumably reflects the omission of intrashell effects in the derivation of (6.34).

### 6.3.4 General modeling experience for alkali halides
The experience regarding modeling for NaI is representative of that pertaining to all the alkali halides. The simple shell model is always superior to the rigid-ion model in predicting the dispersion curves, though residual discrepancies remain. Particularly for the LO($\Lambda$) branch, the discrepancies tend to be more pronounced the greater the polarizability of the negative ion. This may be understood on the basis that ions of small polarizability such as F$^-$ have small ionic radii which should minimize the importance of quadrupole interactions, which increase with ion size. In those shell model calculations that give a good fit (including the LO($\Lambda$)

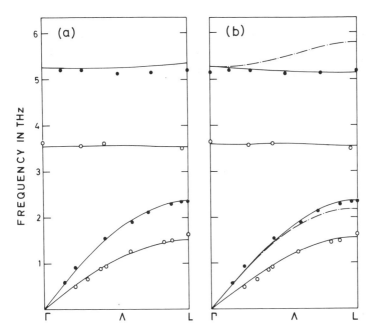

**Figure 6.23**
Comparison of the measured dispersion curves for NaI along [111] with breathing-shell model calculations. (a) shows the comparison with respect to Schröder's version (ref. 119); (b) shows a similar comparison with respect to the calculations of Melvin et al. (ref. 120). The dashed and the solid lines in b correspond to different choices for $k^r(-)$.

branch), some of the parameters have unphysical values,* especially for materials whose negative-ion polarizabilities are large.

Several calculations based on the deformation-dipole model have also appeared. In general, when short-range interactions are restricted to nearest neighbors the agreement with experimental data is similar to that of the simple shell model, discrepancies being significant for the $LO(\Lambda)$ branches of high-polarizability materials.

Allowance for shell deformation generally seems to make up for the shortcomings of the shell model. However, breathing models have a tendency to produce more of a dip in the $LO(\Lambda)$ branch near $\mathbf{q} = (2\pi/a) \times (0.8, 0, 0)$ than exists in the data.

*In this connection, see also the remarks of Sinha[121] who argues that there is an indeterminancy in the shell-model parameters due to the elimination of the dipole moments from the final lattice equations of motion.

It may also be noted that the adjustment of the charge parameter (that is, that pertaining to the ions in the static crystal) in many of the elaborate versions of shell-model calculations has no particular justification. In any case, the dispersion relations can seemingly be well described in the context of deformable-shell models, without reducing the charge from its formal value.

### 6.3.5 Concerning the Cauchy discrepancy in alkali halides

Though nearly fulfilled, the Cauchy relation $C_{12} = C_{44}$ is violated to some extent in all alkali halides.* On the basis of earlier discussion (see Sec. 2.4.6), we can conclude that, in all such crystals, noncentral forces are present, which in turn implies many-body forces between various atoms. This means that the effective force constant $\phi_{\alpha\beta}^{\text{eff}}(\begin{smallmatrix} l & l' \\ k & k' \end{smallmatrix})$ associated with a given pair $(\begin{smallmatrix} l & l' \\ k & k' \end{smallmatrix})$ depends on the disposition of other atoms $(\begin{smallmatrix} l'' \\ k'' \end{smallmatrix})$. It is relevant to note that if the shell-model equations (4.272) and (4.273) are manipulated to the form

$$m_k \ddot{u}_\alpha \begin{pmatrix} l \\ k \end{pmatrix} = -\sum_{l'k'\beta} \phi_{\alpha\beta}^{\text{eff}} \begin{pmatrix} l & l' \\ k & k' \end{pmatrix} u_\beta \begin{pmatrix} l' \\ k' \end{pmatrix}, \tag{6.36}$$

then the effective force constant so defined (after the elimination of shell displacements) in general has a many-body character. However (as the reader may verify), this many-body character does not appear in the simple shell model.

In the literature, many-body forces are occasionally mentioned in the context of the shell model. Usually these refer to the explicitly many-body character of the shell-model force constants themselves,† and not that of the effective constants appearing in (6.36).

We see thus that one may achieve a Cauchy discrepancy in a shell model calculation without involving explicitly many-body forces. For instance, if in the SSM one frees $B$ from satisfying (6.18), then it immediately leads to $C_{12} \neq C_{14}$.‡

The models of Hardy and of Lundqvist (see Sec. 4.6) always lead to Cauchy discrepancy since explicit many-body forces are built into those

---

*In referring to the Cauchy discrepancy, one often considers room-temperature data. This ignores possible temperature dependence of the elastic constants. However, in many instances where data is available the constants deduced by extrapolating to 0 K show a Cauchy discrepancy of the same sign as that observed at higher temperatures. It is this fact which often leads people to ignore temperature effects while discussing departure from Cauchy relations.

†For example, one has such constants in the Basu-Sengupta model and in the Feldkamp model; see Sec. 4.6.

‡Compare the expressions for $C_{12}$ and $C_{44}$ given in (6.19) (which apply to the SSM also), with and without $B = -1.165Z^2$.

models. It must however be recognized that different models may produce Cauchy discrepancies of either sign. This is important in constructing a model that explains both neutron and elastic data.*

### 6.3.6 Results for MgO, CaO and AgCl

Some crystals (like the alkaline-earth oxides and the silver halides) which possess the NaCl structure are substantially different from the prototype ionic crystals in exhibiting large departures from the Cauchy relations (see Table 2.1). Among this class of crystals, dispersion curves have been measured for MgO[124] CaO,[125,126] and AgCl.[127] Many-body forces are important in all of these; while in MgO and CaO these conceivably arise from covalency effects, in AgCl they probably are due to the aniso-tropy in the $d$-electron distribution produced by crystal-field splitting of the $d$ levels.[128]

In general, the shapes of the dispersion curves are similar to those of the alkali halides except for the LO($\Delta$) branch in AgCl, where the zone-boundary frequency is higher than that at the zone center. The data have been analyzed using the shell model and some of its derivatives, but with many more parameters than employed for alkali halides.

It is worth noting that the Cauchy discrepancy in AgCl $(C_{12} \approx 5C_{44})$ is of a different nature from that in MgO $(C_{12} \approx 0.6C_{44})$. Thus the breathing-shell model, which generally leads to $C_{12} < C_{44}$, is suitable for MgO[124] but inappropriate for AgCl. However, by adding "quadrupolar deformations" of the shell in addition to the "breathing deformation," Fischer et al.[129] have successfully fit the AgCl spectrum.†

### 6.3.7 Effect of free carriers

A number of crystals having the NaCl structure display dispersion curves quite dissimilar to those of the alkali halides as a result of the presence of free carriers‡ which tend to screen the macroscopic fields. One example of this may be seen in Figure 6.24a which displays the LO mode in PbTe

---

*It is necessary to remark once again that a straightforward connection between the slope of the acoustic branches at $q = 0$ and the elastic constants exists only in a harmonic theory appropriate to 0 K. In practice one tends to gloss over this fact. When fitting a model to neutron results, one frequently includes elastic-constant data as well. On the presump-tion[122] that neutron experiments for small $q$ reflect the *isothermal* elastic constants, one should in fitting use the latter, which may be deduced from adiabatic constants given by ultrasonic experiments using well-known relations.[123]

†A modified shell model of this type has also been proposed by D. Kühner, Z. Physik **230**, 108 (1970).

‡In the examples considered here, carriers arise either due to impurities or departure from stoichiometry. In either case there is a disturbance to the structure of the host lattice which, however, is ignored.

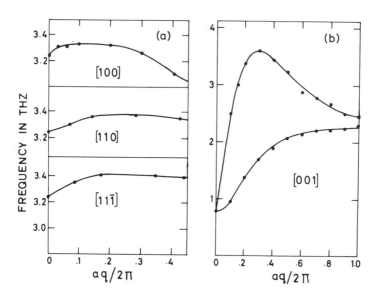

**Figure 6.24**
LO modes in PbTe and SnTe. (a) shows the LO branch in PbTe for different directions.
Observe the small dip around $\mathbf{q} = 0$. (b) shows the LO and TO branches in SnTe. Here
the dip is very pronounced and in fact at $\mathbf{q} = 0$ the LO mode is degenerate with the
TO mode. (After refs. 130 and 133.)

($\sim 1\text{--}2 \times 10^{19}$ el/cm$^3$) along various directions;[130,131] in every case
there is a dip at the zone center. In SnTe ($\sim 5 \times 10^{20}$ holes/cm$^3$), the dip
is more pronounced (see Fig. 6.24b), and the screening is sufficient to
quench the macroscopic field completely.[132,133] The same happens in
the transition-metal carbides[134-136] where also the LO-TO splitting
vanishes, as may be seen in Figure 6.25. In addition, pronounced structure
exists in the dispersion curves of those transition-metal carbides which
exhibit a fairly high superconducting transition temperature. For example,
in Figure 6.26 we note that while TaC with $T_c \sim 10$ K exhibits such
structure, HfC, which is not superconducting, does not. This is reminiscent
of the situation in several metals and alloys discussed above.

The obvious starting point for modeling is, of course, the shell model.
For PbTe, a reasonable fit was achieved only with a complex version of
the shell model involving 13 adjustable parameters.[131] However, even
with this, the fit around $\mathbf{q} \approx 0$ for the LO branches was poor, since the
screening effects of the carriers were not built into the model. The principal message of the fitting exercise was that the shell model tends to do a
poor job as one moves from classic ionic crystals to those of greater
polarizability.

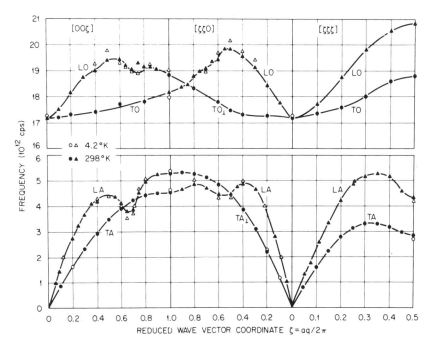

**Figure 6.25**
Dispersion curves for TaC. (After Smith and Gläser, ref. 134.)

Carriers affect lattice vibrations in two ways. First, they tend to screen the fields of the ions and, second, their collective modes can couple with the vibrational modes of the host lattice. The latter effect, however, is important only if the carrier concentration is low enough to make the plasmon frequency comparable to typical vibrational frequencies. In SnTe, the carrier concentration is high and the (static) screening effects are dominant; these have been built into the shell model by Cowley et al.[133] These authors note that since the shells and cores interact with each other in a dielectric medium provided by the carriers, one must replace every Coulomb matrix $C$ in Eqs. (4.290) and (4.291) by a screened matrix $C' = (C + E)$. The form of $E$ is analogous to that used in the microscopic theory of metals (see Sec. 7.1) and involves a static screening function $\varepsilon(Q)$ similar to that mentioned earlier.* The model constructed in this fashion has 14 disposable parameters and fitted the SnTe data reasonably well when a formula similar to (7.40) was used. The use of a Thomas-Fermi screening function for $\varepsilon(Q)$ resulted in a poorer fit.

*Cowley et al. pointed out the need for care in the replacement $C \rightarrow C'$ because of the requirements of translational invariance. It turns out that a different prescription is required for the replacement ($C \rightarrow C'$) in the term $Y_d C Y_d$, even for the NaCl structure.

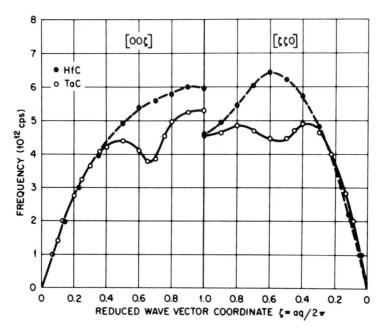

**Figure 6.26**
Longitudinal acoustic modes in TaC and HfC. (After Smith and Gläser, ref. 134.)

Weber et al.[137] have proposed for the transition-metal carbides a
shell model similarly modified for screening, with an elaboration leading
to a **q**-dependent polarizability for the metal ions, which exhibits resonance behavior in the vicinity of the anomalies. This extended shell model
has been successful in reproducing the structure observed[134–136] in TaC
and NbC and the anomalous behavior of the optic branches in HfC.

Models of the above type involving static screening are inadequate
when the coupling of plasmons with the vibrational modes becomes important. Usually the effects of such coupling are confined to the long-
wavelength region and can be allowed for phenomenologically by treating
$\varepsilon(\omega)$ in Fresnel's equation as

$$\varepsilon(\omega) = \varepsilon^{\text{lattice}}(\omega) + \varepsilon^{\text{electronic}}(\mathbf{q}, \omega) - \mathbb{1}_3, \tag{6.37}$$

which is a generalization of (4.197) to include the **q** dependence of
$\varepsilon^{\text{electronic}}$. In cubic crystals, a simplification is possible and the solutions
for the longitudinal and transverse modes may be obtained separately. The
former are found from

$$\varepsilon^{\text{lattice}}(\omega) + \varepsilon_{\|}(q, \omega) - 1 = 0, \tag{6.38}$$

where

$$\varepsilon_{\|}(q, \omega) = \hat{\mathbf{q}}^{tr}\boldsymbol{\varepsilon}^{electronic}(q, \omega)\hat{\mathbf{q}} \qquad (6.39)$$

denotes the longitudinal component of $\boldsymbol{\varepsilon}^{electronic}$. In the free-electron approximation for the carriers, $\varepsilon_{\|}$ is well represented by the Lindhard formula,[138] a generalization of the static screening function (7.40) to include dynamic screening.

The transverse solutions are obtained from

$$\frac{q^2c^2}{\omega^2} = \{\varepsilon^{lattice}(\omega) + \varepsilon_{\perp}(q, \omega) - 1\}, \qquad (6.40)$$

where $\varepsilon_{\perp}(q, \omega)$ is the wave-vector- and frequency-dependent transverse dielectric function for the carriers. However, bearing in mind that the modifications due to the carriers are confined to values of $q \sim \omega/c \sim 10^3$ cm$^{-1}$, one may make the $q \to 0$ limit and use the result

$$\lim_{q\to 0} \varepsilon_{\perp}(q, \omega) = \lim_{q\to 0} \varepsilon_{\|}(q, \omega) = 1 - \frac{\omega_p^2}{\omega^2}, \qquad (6.41)$$

the latter being the familiar Drude formula. Use of (6.41) in (6.40) then leads to the dispersion curves for plasmaritons. It is important to note that $\varepsilon_{\|}$ and $\varepsilon_{\perp}$ are in general not equal, although they are in the $q \to 0$ limit (for an isotropic system).*

An alternate way of dealing with the plasmon-phonon coupling would be to modify $\mathscr{L}$ in (4.182) to

$$\mathscr{L} = \frac{c^2q^2}{\omega^2}(\mathbb{1}_3 - \hat{\mathbf{Q}}) - [\boldsymbol{\varepsilon}(\infty) + \boldsymbol{\varepsilon}^{electronic}(\mathbf{q}, \omega) - \mathbb{1}_3] \qquad (6.42)$$

(see (4.181)). Applied to cubic diatomic crystals, for example, this enables the plasmon-LO phonon coupled modes to be obtained as the solutions of

$$\left\{(\omega_{TO}^2 - \omega^2) + \frac{4\pi}{Mv}\frac{(\mathscr{X}_{TO})^2}{\varepsilon(\infty) + \varepsilon_{\|}(q, \omega) - 1}\right\} = 0, \qquad (6.43)$$

where $\mathscr{X}_{TO}$ is defined as in (4.148). Cochran et al. have used essentially this approach in explaining their results on PbTe.† The structure observed in the LA [100] branch of Nb has been explained similarly by Ganguly and Wood,[140] using a variant of (6.38).

---

*In this context observe that the $q$ dependence of $\boldsymbol{\varepsilon}^{electronic}$ was ignored in Sec. 4.3.8. It is this fact that permitted Shah et al.[139] to use the same $\varepsilon(\omega)$ to discuss both plasmon-LO coupling and plasmaritons. Note, however, that the $q$ dependence of plasmon-LO coupling cannot be calculated in this approximation.
†The schematic curves of Figure 4.8 are in fact based on such a calculation.

## 6.4 Results for diamond-type crystals

Among the homopolar insulating crystals, those of the diamond family are the most important. Interest in the dynamics of diamond-type crystals far predates the availability of dispersion curves. As early as 1914, Born[141] showed that the diamond lattice is unstable against shear if only nearest-neighbor central forces are present, so that noncentral forces must be included. Interest increased further when Raman[142] claimed that the frequency spectra of crystals were discrete and not continuous as in the Born–von Kármán formalism, using the measured second-order Raman spectrum as supporting evidence.* In an attempt to refute Raman, Smith[143] made the first detailed calculation of the normal-mode spectrum and thence the second-order Raman spectrum of diamond, assuming a force field extending to the second neighbors.† Subsequently, Hsieh[144] reported calculations of frequency spectra for Ge and Si, using a two-neighbor model as employed by Smith. Hsieh noted that the measured elastic constants, particularly for Ge, satisfied very closely the identity

$$4C_{11}(C_{11} - C_{44}) = (C_{11} + C_{12})^2 \tag{6.44}$$

deduced by Born[141] for the diamond lattice under the assumption of only nearest-neighbor forces, leading him to suppose that the normal modes in Ge could conceivably be well described under such a restriction.‡

The first measured dispersion curves for Ge[145,146] became available in 1958, and further results for Ge[147,148], Si[148–150] diamond[151,152] and $\alpha - Sn$[153] have since appeared. Paralleling these measurements have been several optical investigations. In general the existence of measured dispersion curves has been required in order to understand properly pre-existing optical data.

As is to be expected, homology in the diamond family has attracted much attention.[148,149,151,154,155] It is found[153] that while the dispersion curves for Ge and Si scale reasonably well (relative to each other), those for $\alpha - Sn$ do so only partially and those for diamond very poorly. The latter is due to the strong covalency of diamond, while the breakdown in

---

*For a historical review of the Born-Raman controversy, see G. Venkataraman, Current Science (India) **41**, 349 (1972).

†It is interesting to note that Smith erred in deducing the structure of the force-constant matrix for the second neighbors through an incorrect application of Eq. (2.32).

‡It must be emphasized that (6.44) does not represent a sufficient condition for the vanishing of interactions beyond the first neighbor.

$\alpha-$ Sn is presumably due to the valence electrons being considerably delocalized. (In fact, $\alpha-$ Sn is technically a zero-gap semiconductor.)

Dispersion curve calculations for the diamond family fall into three types, namely those employing (1) short-range force models, (2) dipolar models, and (3) the bond-charge model. These are reviewed below.

### 6.4.1 Applicability of short-range force models

Experiments have established conclusively that a short-range force field with drastically curtailed range as assumed by Hsieh is totally inadequate. This is evident, for example, from the experimental data of Brockhouse and Iyengar[145] for Ge, a striking feature of which is the relative flatness of the TA branches, indicative of the presence of forces involving distant neighbors. The point may be appreciated readily by analogy to the one-dimensional chain. With nearest-neighbor interactions alone, one has the familiar result[156]

$$\omega^2(q) = \frac{2k}{M}(1 - \cos qa). \tag{6.45}$$

If interactions with $n$ neighbors are permitted, one has[156]

$$\omega^2(q) = \frac{2}{M}\sum_n k_n(1 - \cos nqa), \tag{6.46}$$

which results in a flatter dispersion curve than given by (6.45). The fulfillment of the Born identity (6.44) in Ge is thus an accident and served to give a misleading impression of the range of interatomic forces. This was subsequently confirmed by Herman[157] and by Pope[1] who found that tensor-force interactions to fifth neighbors (involving 16 parameters) were required to fit the experimental data. It is interesting to note that, by contrast, Aggarwal[158] found a two-neighbor tensor-force model reasonably adequate for describing the dispersion curves of diamond along [100] and [111].

The strong covalent bonding in diamond suggests that, at least for this crystal, it might be more efficacious to employ valence force constants (Sec. 2.8) rather than tensor force constants. This has been confirmed by McMurry et al.[159] who were able to obtain a better fit to the measured curves than was obtained with a shell-model calculation[151] (to be described later) which involved many more parameters. In addition, the valence constants used are close to those inferred for similar bonds occurring in carbon-containing molecules. Another interesting outcome of this work was that, when translated into the Born–von Kármán force-constant language (see Eq. (2.237)), the results indicated a fifth-neighbor interaction but no interactions with third or fourth neighbors.* This paradox can be resolved by inspecting a model of the diamond structure built up by covalent links (for example, Chapter 1, Figure 29 in Kittel). One then sees bonding links with fifth neighbors but none with the third or the fourth. Similar applications[160,161] of the valence force model to Si revealed that the fit is poorer unless more parameters are employed. Further, the ratio of bond-stretching and bond-bending constants was $\sim 50$ in contrast to the value of $\sim 10$ for diamond, indicating greater resistance to bond bending in the latter case and hence more covalency.

### 6.4.2 Shell-model analysis

Almost immediately following the experiments of Brockhouse and Iyengar on Ge, Lax[162] suggested that the origin of the long-range forces rests perhaps in the multipoles resulting from charge distortions caused by lattice vibrations. There was already evidence from the occurrence of second-order infrared absorption in diamond (see Sec. 5.2.3) that charge distortion does accompany lattice vibrations. It was natural therefore to apply the shell model; this was first done in 1959 by Cochran,[163,164] using the dynamical equation given in (4.377). A relatively simple 5-parameter model involving nearest-neighbor interactions was devised, and tailored to satisfy the Born identity (6.44). In contrast to the SSM employed for alkali halides, core-core and core-shell interactions were permitted. The resulting fit for Ge shown in Figure 6.27 seems to be rather good, probably better than it really is because of scatter in the data. This is confirmed by Dolling's work on Si[150] in which 8 parameters were required to attain a reasonable fit, even though the curves for Si are essentially scaled-up versions of those for Ge.

Dolling and Cowley[165] have reevaluated the data for Si, Ge, and diamond and have made 11-parameter shell-model fits for each. Some of the parameter values seem to lack physical significance, but the models have proven useful as interpolation formulas for the calculation of various thermodynamic and optical properties. Workers of the Soviet school[166-169] have performed several calculations using the Mash-kevich-Tolpygo theory. Most of these were made before the availability of measured dispersion curves or were based on incomplete data. The most recent calculations of Kucher and Nechiporuk[169] for diamond are based on the data for [100] and [111]. The fit in these directions is fair, but comparison with later determined [110] data is much less impressive.

---

*A somewhat similar conclusion was reached by Pope[1] in his tensor-force model analysis of Ge.

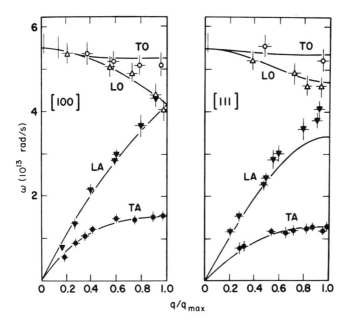

**Figure 6.27**
Comparison of the measured dispersion curves (ref. 145) for Ge with those computed using
the shell model. (After Cochran, ref. 164.)

### 6.4.3 Bond-charge model

Yet another approach to the dynamics of the diamond lattice involves the
use of the bond-charge model.[170–172] The basic premise is that a
covalent bond results in the localization of negative charge in the region
of the bond with compensating positive regions at the atomic sites, and
that the electrostatic interaction of the charged entities represents a
major part of the effect of covalent bonding on the lattice vibrations. In the
model, the negatively charged regions are approximated as points located
at the midpoints of the lines of centers between the atomic sites and are con-
strained to remain midway even during the vibrations.*

The force constants of the model consist of Coulombic and short-range
parts. The latter is assumed to arise from a two-body potential $V(r)$ acting
between nearest neighbors. The Coulombic part of the force constant associated with typical atoms $a$ and $b$ may be written

---

*The model as outlined here is specifically for crystals of the diamond structure. It could
also be applied with suitable modifications to the zinc blende, wurtzite, and similar
structures.

$$
\begin{aligned}
\phi_{\alpha\beta}^{C}(ab) = -\Bigg\{ &\frac{\partial^2 V^{ii}}{\partial r_\alpha(a)\partial r_\beta(b)}\Bigg|_0 \\
&+ \sum_{c(b)}\sum_\gamma \frac{\partial^2 V^{ib}}{\partial r_\alpha(a)\partial w_\gamma[c(b)]}\Bigg|_0 \frac{\partial w_\gamma[c(b)]}{\partial r_\beta(b)} \\
&+ \sum_{c(a)}\sum_\gamma \frac{\partial^2 V^{bi}}{\partial r_\beta(b)\partial w_\gamma[c(a)]}\Bigg|_0 \frac{\partial w_\gamma[c(a)]}{\partial r_\alpha(a)} \\
&+ \sum_{c(a)}\sum_\gamma\sum_{c'(b)}\sum_\delta \frac{\partial^2 V^{bb}}{\partial w_\gamma[c(a)]\partial w_\delta[c'(b)]}\Bigg|_0 \frac{\partial w_\gamma[c(a)]}{\partial r_\alpha(a)}\frac{\partial w_\delta[c'(b)]}{\partial r_\beta(b)}\Bigg\}.
\end{aligned}
$$

$$(6.47)$$

Here $\mathbf{r}(a)$, $\mathbf{r}(b)$ are the position vectors of the two atoms, $\mathbf{w}[c(a)]$ the position of the bond charge on the $c$th bond of atom $a$, etc., $V^{ii}$, $V^{ib}$ ($= V^{bi}$), and $V^{bb}$ are the Coulomb potentials between ionic charges, between ions and bond charges, and between bond charges, respectively. Because of the constraint placed on the bond charges, one has

$$
\frac{\partial w_\gamma[c(a)]}{\partial r_\alpha(a)} = \tfrac{1}{2}\delta_{\alpha\gamma}.
$$

The first derivative $V'(R)$ is determined via the equilibrium condition, leaving two parameters, $V''(R)$ and the ionic charge $Ze$. (This implies that the bond charge is $(-Ze/2)$, since there are two bonds per atom and the *total* charge on an atom must vanish.) By suitably adjusting these parameters, Martin computed the dispersion curves for Si and diamond. He noted that the shape of the TA [100] branch in Si (which, like that of Ge, is much flatter than that of diamond) can be interpreted as a manifestation of silicon's relatively smaller bond charge.

### 6.5  Results for crystals with zinc blende structure

Although a number of crystals occur with the zinc blende structure, dispersion curve data for this class is relatively meager.[153,173–179] From the more abundant optical data, many attempts have been made to infer zone-boundary frequencies.

Several shell model and deformation-dipole model calculations have been reported for crystals whose dispersion curves have been measured. Among these may be mentioned 14-parameter shell model calculations for GaAs by Dolling and Waugh[173] and for GaP by Yarnell et al.[174] In some cases,[175,180] the short-range forces have been parameterized in terms of valence constants, in analogy with the treatment of McMurry et al.[159] for diamond. In view of the partially covalent bonding, this scheme appears conceptually satisfying and results in the need for fewer parameters.

In general, there is no model close to being as successful, in the sense of predicting dispersion curves with a small number of parameters, as is the shell model in the case of alkali halides. The reason usually advanced for the less satisfactory nature of the dipolar models for the zinc blende structure is the substantial difference in the valence charge distribution, often leading to strong delocalization. Undoubtedly this reasoning has merit. In addition, there are reasons to believe that the inadequacy of the dipolar models stems partly from their ignoring certain features crucial to the proper description of internal strain and piezoelectric phenomena. The latter, absent in NaCl structure crystals (see Sec. 4.3.2), are usually described in terms of internal strains in a lattice of point ions. This, however, leads to a value of $e_{14}$ (the piezoelectric constant) which is roughly an order of magnitude larger than typical values unless very small charge values are used. It has been suggested[181,182]* that electronic polarization and charge transfer between neighboring ions could be important in explaining piezoelectricity, and it is in this context that the model of Feldkamp (Sec. 4.6.7) was introduced. The success of this model has not yet been demonstrated.

Recently, dispersion curve data have also been obtained for CuCl[178] and CuI.[179] The point of interest here is the possible role of $d$ electrons in the interaction between neighboring Cu ions. It has been found that such interactions are less important in CuI than in CuCl, a result intuitively to be expected since the ionic radius of $I^-$ is greater than that of $Cl^-$, leading to a larger separation for Cu ions in CuI as compared to that in CuCl.

## 6.6 Effective charge

In the study of the dynamics of insulating crystals, one frequently finds reference to various types of charges: apparent charge, effective charge, etc. The need for these various definitions arises because it is not possible to assign uniquely to every atom a charge distribution which moves rigidly with the atom during vibrations. Indeed, if this were possible, it would imply the validity of the rigid-ion model. In reality, nuclear motions always create electronic polarizations which lead to the "dressing" of the charges attributable to the atom in the equilibrium configuration. Since the modification thus produced is dependent on the circumstances con-

---

*Martin (Phys. Rev. **B5**, 1607 (1972)) has asserted that induced quadrupolar distortions of the charge density are responsible for the small piezoelectric constant in zinc-blende-structure materials. This premise awaits a lattice dynamical test.

sidered, one may attribute various charges to the same ion depending on the context! Several definitions have been introduced specifically for cubic diatomic crystals, especially in the context of analyzing for systematics over a group of isomorphic crystals. Before discussing these, we restate some of the terminology we have employed for ionic crystals of arbitrary structure.

Earlier we introduced the ionic charge $Z_k e$ appropriate to atoms in a *static* crystal. While the assignment of a value for $Z_k$ is relatively simple for the classic ionic crystals, the problem becomes progressively more difficult as the valence electrons are less and less localized. It was originally hoped that vibration spectra would shed some light on the static charge. We now know this is not possible since the static charge is always modified by polarization produced during vibrations.

In the context of dynamics, the charge parameter is more conveniently expressed through the components of the matrix $\mathbf{Z}^{\dagger}(\mathbf{q})$ introduced in (4.76). A shell-model expression for the latter was deduced in (4.303), displaying explicitly the modifications produced by polarization effects within the framework of the model. Following Cochran and Cowley,[183] we refer to $\mathbf{Z}^{\dagger}(\mathbf{q})$ as the wave-vector-dependent *apparent charge matrix*. It is composed of the *apparent charge tensors* $\mathbf{Z}(k|\mathbf{q})$ attributable to the various sublattices in a motion of wave vector $\mathbf{q}$. The $\mathbf{q} = 0$ limit of $\mathbf{Z}(k|\mathbf{q})$ and hence of $\mathbf{Z}^{\dagger}(\mathbf{q})$ is a well-defined quantity. In the literature, the quantity $\mathbf{Z}(k|\mathbf{q} = 0)$ is sometimes referred to as the effective charge tensor, a nomenclature that we shall avoid.

In discussing the long-wavelength optic modes in diagonally cubic crystals, Cowley[184] defines an *effective charge matrix*. To introduce this, let us specialize Eqs. (4.311) and (4.312) to diagonally cubic crystals in the limit $\mathbf{q} \to 0$:

$$m\omega^2 \mathbf{U}^{(0)} = \mathbf{R}^{(0)}\mathbf{U}^{(0)} + \mathbf{T}^{(0)}\mathbf{W}^{(0)} - \mathbf{Z}_d \mathscr{E},$$
$$0 = \tilde{\mathbf{T}}^{(0)}\mathbf{U}^{(0)} + \mathbf{S}^{(0)}\mathbf{W}^{(0)} - \mathbf{Y}_d \mathscr{E}.$$

(Remember $\mathscr{E}^c = \mathscr{E}^s$; see (4.308).) The last two equations can be combined to give

$$m\omega^2 \mathbf{U}^{(0)} = [\mathbf{R}^{(0)} - \mathbf{T}^{(0)}\mathbf{S}^{(0)-1}\tilde{\mathbf{T}}^{(0)}]\mathbf{U}^{(0)} - \overline{\mathbf{X}}\mathscr{E},$$

where

$$\overline{\mathbf{X}} = \mathbf{Z}_d - \mathbf{T}^{(0)}\mathbf{S}^{(0)-1}\mathbf{Y}_d \tag{6.48}$$

denotes the effective charge matrix. As in Eq. (4.30), the use of the dipolar approximation permits description of the electrostatic interactions entirely in terms of $\overline{\mathbf{X}}$ and the local fields acting on the induced dipoles.

Cowley also defines a square matrix $\mathbf{X}$ (termed by him the apparent charge matrix) which amounts to a zero-order version of

$$[\mathbf{Z}_d - (\mathbf{T} + \overline{\mathbf{M}}^{AD,C})(\mathbf{S} + \overline{\mathbf{M}}^{DD,C})^{-1}\mathbf{Y}_d] \tag{6.49}$$

(see Eq. (4.301)). In contrast to $\overline{\mathbf{X}}$, $\mathbf{X}$ is a matrix that facilitates the description of the macroscopic part of the electrical interactions. The relationship between Cowley's $\mathbf{X}$ and our result for the apparent charge matrix in the shell model is obtainable from (4.303). Observe that both $\mathbf{X}$ and $\overline{\mathbf{X}}$ incorporate the effects of short-range polarization.

In the special case of cubic diatomic crystals, one finds in the literature several types of charges, for example, the transverse (Born) effective charge, longitudinal (Callen) effective charge, and the Szigeti effective charge.

The Born charge is the one ascribable to an ion in a TO mode. Remembering that $\mathbf{E} = 0$ in the TO mode, we have from (4.9)

$$\mathbf{P}_t = b_{12}\mathbf{W}_t. \tag{6.50}$$

Moreover, from the dielectric equation (4.14),

$$b_{12}^2 = \frac{[\varepsilon(0) - \varepsilon(\infty)]}{4\pi}\omega_{TO}^2. \tag{6.51}$$

Hence, (6.50) becomes (for $b_{12} > 0$)

$$\mathbf{P}_t = \frac{1}{v}\omega_{TO}\left(\frac{[\varepsilon(0) - \varepsilon(\infty)]\overline{m}v}{4\pi}\right)^{1/2}[\mathbf{U}_t(1) - \mathbf{U}_t(2)].$$

Define now

$$e_T^{*2} = \omega_{TO}^2\left(\frac{[\varepsilon(0) - \varepsilon(\infty)]\overline{m}v}{4\pi}\right); \tag{6.52}$$

then

$$\mathbf{P}_t = \frac{1}{v}e_T^*[\mathbf{U}_t(1) - \mathbf{U}_t(2)]. \tag{6.53}$$

Comparing (6.50) and (6.53), we obtain

$$b_{12} = \frac{1}{\sqrt{\overline{m}v}}e_T^*, \tag{6.54}$$

so that from (4.8) we find that $e_T^*$ pertains to the charge interacting with the macroscopic field, enabling the long-wavelength limit of the apparent charge matrix for cubic diatomic crystals to be written as

$$\mathbf{Z}(0) = \begin{bmatrix} e_T^* & 0 & 0 & -e_T^* & 0 & 0 \\ 0 & e_T^* & 0 & 0 & -e_T^* & 0 \\ 0 & 0 & e_T^* & 0 & 0 & -e_T^* \end{bmatrix}.$$

The matrix $\mathbf{X}$ may be similarly deduced.

The Callen effective charge $e_L^*$ is defined similarly to $e_T^*$ from a consideration of the polarization density $\mathbf{P}_l$ produced in a LO mode. This gives,

$$\mathbf{P}_l = \frac{1}{v} e_L^* [\mathbf{U}_l(1) - \mathbf{U}_l(2)]$$

with

$$e_L^* = e_T^*/\varepsilon(\infty). \tag{6.55}$$

Another derived quantity frequently used is the Szigeti effective charge $e_S^*$, required as seen in (4.233b) to have the identity

$$1 = \frac{4\pi e_S^{*2}}{9\bar{m}v\omega_{TO}^2} \frac{[\varepsilon(\infty) + 2]^2}{\varepsilon(0) - \varepsilon(\infty)}. \tag{6.56}$$

The Szigeti charge is related to $\overline{\mathbf{X}}$ defined in (6.48) by

$$e_S^* = \sum_{k'=+,-} \bar{X}_{\alpha\alpha}(+k').$$

Comparison of (6.56) with (6.52) shows that

$$e_T^* = \frac{\varepsilon(\infty) + 2}{3} e_S^*. \tag{6.57a}$$

Observe that as the valence electrons are more localized, $e_T^*$ approaches $e_S^*$ since $\varepsilon(\infty) \to 1$. Using (6.55), we also have

$$e_L^* = \frac{\varepsilon(\infty) + 2}{3\varepsilon(\infty)} e_S^*. \tag{6.57b}$$

In order to study systematics, Lucovsky et al.[185] decompose $e_T^*$ as

$$e_T^* = e_l^* + e_{nl}^* \tag{6.58}$$

on the presumption that part of the valence electron charge density is localized and nonpolarizable (except by short-range forces) while the rest of it is extended. Using a suitable model in conjunction with experimental data, values for $e_l^*$ have been deduced for several crystals. It is found that the values of the ratio $(e_l^*/Z)$ plotted against the values of Phillips' ionicity[186] for zinc blende and wurtzite structure compounds fall plausibly near a straight line passing through the origin. However, such a comparison for crystals of the NaCl and CsCl structures is less clear and is taken to indicate that in these cases other factors, such as the relative ionic size, must be taken into account. There is as yet no microscopic justification for the correlations found, and in any case the division (6.58) is arbitrary.

Other attempts[187-189] have also been made to discover systematics in the effective charge and related quantities as a function of various parameters. These have been recently reviewed by Sinha.[121]

## 6.7 Ferroelectricity and lattice dynamics

Over a decade ago, Anderson[190] and Cochran[191] suggested independently that lattice vibrations play a central role in certain ferroelectric crystals. Since then there has been considerable experimental and theoretical enquiry into the detailed manner in which lattice vibrations influence ferroelectric behavior. This subject merits a complete review in itself, and we can hardly do justice to it here. We restrict ourselves, therefore, to indicating schematically how anomalous behavior of certain vibrational modes could give rise to lattice instabilities, triggering a transition to a ferroelectric phase.

For simplicity, we shall follow Cochran[191] and consider a cubic diatomic crystal* like NaCl. Using the Lyddane-Herzfeld model, Born and Huang[192] showed that the long-wavelength modes of such a crystal can be described by the equation

$$\bar{m}\ddot{\mathbf{W}} = \left\{ -k + \frac{(4\pi/3v)(Ze)^2}{1 - (4\pi\alpha/3v)} \right\} \mathbf{W} + \sqrt{\frac{\bar{m}}{v}} \frac{Ze}{1 - (4\pi\alpha/3v)} \mathbf{E}, \qquad (6.59)$$

where $\alpha = \alpha_+ + \alpha_-$ and $k$ is a phenomenological coupling parameter which can be expressed in terms of short-range force constants. Comparing with Eq. (4.8), we see that in this model,

$$b_{11} = \frac{1}{\bar{m}} \left\{ -k + \frac{(4\pi/3v)(Ze)^2}{1 - (4\pi\alpha/3v)} \right\}, \qquad b_{12} = \frac{1}{\sqrt{\bar{m}v}} \frac{Ze}{1 - (4\pi\alpha/3v)}. \qquad (6.60)$$

Recalling (4.10), we then have

$$\omega_{\text{TO}}^2 = \frac{1}{\bar{m}} \left\{ k - \frac{(4\pi/3v)(Ze)^2}{1 - (4\pi\alpha/3v)} \right\} \qquad (6.61)$$

---

*GeTe[132] is possibly an example of a crystal of the NaCl structure exhibiting a ferroelectric phase transition. SnTe is a "near ferroelectric" in the sense that it exhibits a tendency for mode instability characteristic of those signaling the onset of a ferroelectric phase transition.

so that, as expected, the TO mode frequency has mechanical and electrical contributions. Observe that the local-field correction $\{(4\pi/3v)(Ze)^2/[1 - (4\pi\alpha/3v)]\}$ tends to reduce the squared frequency of the TO mode from the value $(k/\bar{m})$ it would have in the sole presence of short-range forces.

Interestingly, the usual decrease of $v$ with decreasing temperature means that the frequency of the TO mode could conceivably reduce to zero at some temperature. To examine the problem further,[193] consider the equation of motion for the zero-frequency TO mode in the harmonic approximation:

$$\bar{m}\ddot{W} = 0. \tag{6.62}$$

The solution is

$$\mathbf{W} = \mathbf{a}t + \mathbf{b} \tag{6.63}$$

($\mathbf{a}$ and $\mathbf{b}$ are defined by initial conditions), which implies that the separation between the ions in the primitive cell increases indefinitely with time! However, at this point we must modify the analysis to include anharmonic effects, which become important for large $\mathbf{W}$. We therefore rewrite (6.59) as (taking $\mathbf{W}$ for simplicity to be along one of the Cartesian axes)

$$\bar{m}\ddot{W} = (-k_{\text{eff}}W + \beta W^3 - \gamma W^5) + \sqrt{\frac{\bar{m}}{v}}\,e_T^* E, \tag{6.64}$$

where the Born effective charge is given by

$$e_T^* = \frac{Ze}{1 - (4\pi\alpha/3v)}$$

while

$$k_{\text{eff}} = \left\{ k - \frac{(4\pi/3v)(Ze)^2}{1 - (4\pi\alpha/3v)} \right\}. \tag{6.65}$$

The new restoring force $(k_{\text{eff}}W - \beta W^3 + \gamma W^5)$ may be integrated with respect to $W$ to give the anharmonic potential governing the dynamics. Using representative values for the various parameters, Barker[193] has sketched the form of the potential at various temperatures, as shown in Figure 6.28a. With decreasing temperature the potential at first softens and becomes increasingly anharmonic. Correspondingly the vibrational frequency decreases. As the crystal is cooled further, the potential develops humps on the sides leading to subsidiary minima. At 120°C, a first-order transition to a new phase occurs. The ions no longer have $W = 0$ in the static crystal but have finite separations dictated by the positions of the

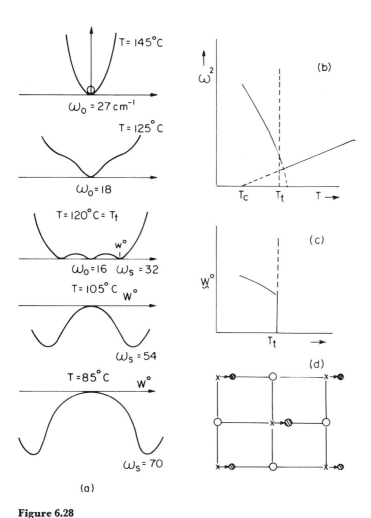

**Figure 6.28**
(a) shows a schematic drawing of the potential appropriate to the oscillator of Eq. (6.64) corresponding to different temperatures. (b) shows the temperature dependence of the transverse optic frequency. (c) sketches the temperature dependence of the distortion parameter, while (d) shows the distorted configuration appropriate to the low-temperature phase. (After Barker, ref. 193.)

emergent new minima. As a result the crystal now has a spontaneous polarization; further, the new vibrations occur in the displaced wells.

Figure 6.28b shows a schematic plot of $\omega_{TO}^2$ as the crystal is cooled through the transition. Figure 6.28c shows the behavior of $W^{(0)}$ with temperature and Figure 6.28d the distorted configuration. By considering a different form of the anharmonic restoring force, one similarly can construct an example of an instability which leads to a second-order transition.

The preceding discussion is oversimplified (especially in using a potential function rather than the free energy) but serves to illustrate how the anomalous behavior of a $\mathbf{q} = 0$ mode is related to the onset of ferroelectricity. Such a *soft mode* was first detected by Cowley[194] in $SrTiO_3$.

Modes other than at the zone center can go "soft." For example, Cochran[195] has pointed out that, in certain circumstances, a CsCl-type crystal becomes unstable against a transverse mode for which $\mathbf{q} = (2\pi/a)$ $\times (\tfrac{1}{2}, 0, 0)$. In this mode one of the sublattices is stationary, as shown in Figure 6.29a. An instability in this mode leads to a crystal distortion with new equilibrium positions as indicated in Figure 6.29b, resulting in a doubling of the unit cell and an antiferroelectric phase. Such a soft mode appears to have been detected in $SrTiO_3$.[196] A zone-boundary soft mode leading to the ferroelectric phase is also possible and is reported to have been observed in $Tb_2(MoO_4)_5$.[197]

Reviewing the experimental results as a whole, perhaps the most significant conclusion to emerge is that electronic polarization influences the vibration spectra in practically all types of solids. In covalent crystals like Ge, the relatively flat TA branches offer some indication of the pre-

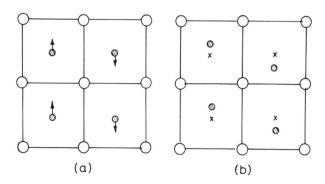

(a)                              (b)

**Figure 6.29**
(a) shows the displacement of atoms in a transverse mode of a CsCl lattice having $\mathbf{q} = (2\pi/a)(\tfrac{1}{2}, 0, 0)$. (b) shows the distorted configuration resulting from the phase transition. (After Cochran, ref. 195.)

sence of polarization effects. In the case of ionic crystals, the effects are seen very directly when experimental results are compared with rigid-ion calculations. In metals, electronic polarization not only leads to a screening of the ion-ion potential but also produces observable anomalies in the dispersion curves. All of this suggests the desirability of a unified framework for describing lattice vibrations, allowing explicitly for polarization effects but providing for differences in the polarization mechanisms associated with different types of solids. This topic is pursued further in the following chapter.

## References

1. N. K. Pope, in Wallis, p. 147.

2. R. S. Leigh, B. Szigeti, and V. K. Tewary, Proc. Roy. Soc. (London) **A320**, 505 (1971).

3. W. Cochran, Acta Cryst. **A27**, 556 (1971).

4. hcp He$^4$: T. O. Brun, S. K. Sinha, C. A. Swenson, and C. R. Tilford, in Copenhagen proceedings, Vol. I, p. 339; R. A. Reese, S. K. Sinha, T. O. Brun, and C. R. Tilford, Phys. Rev. **A3**, 1688 (1971).

5. hcp He$^4$: F. P. Lipshultz, V. J. Minkiewicz, T. A. Kitchens, G. Shirane, and R. Nathans, Phys. Rev. **174**, 267 (1968).

6. bcc He$^4$: E. B. Osgood, V. J. Minkiewicz, T. A. Kitchens, and G. Shirane, Phys. Rev. **A5**, 1537 (1972); J. Skalyo Jr., V. J. Minkiewicz, G. Shirane, and W. B. Daniels, Phys. Rev. **B6**, 4766 (1972).

7. fcc He$^4$: J. G. Traylor, C. Stassis, R. A. Reese, and S. K. Sinha, in Grenoble proceedings, p. 134.

8. Ne: J. A. Leake, W. B. Daniels, J. Skalyo, B. C. Frazer, and G. Shirane, Phys. Rev. **181**, 1251 (1969).

9. A: H. Egger, M. Gsänger, E. Lüscher and B. Dorner, Phys. Letters **28A**, 433 (1968).

10. A: D. N. Batchelder, M. F. Collins, B. C. G. Haywood, and G. R. Sidey, J. Phys. C **3**, 249 (1970).

11. Kr: W. B. Daniels, G. Shirane, B. C. Frazer, H. Umebayashi, and J. A. Leake, Phys. Rev. Letters **18**, 548 (1967).

12. T. Högberg and L. Bohlin, Solid State Comm. **5**, 951 (1967).

13. F. W. de Wette and B. R. A. Nijboer, Phys. Letters **18**, 19 (1965).

14. N. S. Gillis, N. R. Werthamer, and T. R. Koehler, Phys. Rev. **165**, 951 (1968).

15. R. A. Guyer, Solid State Phys. **23**, 413 (1969).

16. H$_2$: M. Nielsen and K. Carneiro, in Grenoble proceedings, p. 111; M. Nielsen, Phys. Rev. **B7**, 1626 (1973).

17. M. L. Klein and T. R. Koehler, J. Phys. C **3**, 102 (1970).

18. W. Biem and F. C. Mertens, in Nusimovici, p. 263.

19. P. C. Fine, Phys. Rev. **56**, 355 (1939).

20. M. Blackman, Proc. Roy. Soc. (London) **A148**, 365 (1935); **A149**, 117 (1935); **A159**, 416 (1937).

21. Li: H. G. Smith, G. Dolling, R. M. Nicklow, P. R. Vijayaraghavan, and M. K. Wilkinson, in Copenhagen proceedings, Vol. I, p. 149.

22. Na: A. D. B. Woods, B. N. Brockhouse, R. H. March, A. T. Stewart, and R. Bowers, Phys. Rev. **128**, 1112 (1962).

23. K: R. A. Cowley, A. D. B. Woods, and G. Dolling, Phys. Rev. **150**, 487 (1966).

24. Rb: J. R. D. Copley, B. N. Brockhouse, and S. H. Chen, in Copenhagen proceedings, Vol. I, p. 209; J.R.D. Copley and B. N. Brockhouse, Can. J. Phys. **51**, 657 (1973).

25. V: R. Colella and B. W. Batterman, Phys. Rev. **B1**, 3913 (1970).

26. Cr: W. M. Shaw and L. D. Muhlestein, Phys. Rev. **B4**, 969 (1971); see also H. B. Moller and A. R. Mackintosh, in Bombay proceedings, Vol. I, p. 95.

27. Fe: B. N. Brockhouse, H. E. Abou-Helal and E. D. Hallman, Solid State Comm. **5**, 211 (1967); see also V. J. Minkiewicz, G. Shirane and R. Nathans, Phys. Rev. **162**, 528 (1967).

28. Nb: Y. Nakagawa and A. D. B. Woods, in Wallis, p. 39.

29. Mo: A. D. B. Woods and S. H. Chen, Solid State Comm. **2**, 233 (1964).

30. Ta: A. D. B. Woods, Phys. Rev. **136**, A781 (1964).

31. W: S. H. Chen and B. N. Brockhouse, Solid State Comm. **2**, 73 (1964).

32. W. Cochran, in Bombay proceedings, Vol. I, p. 3.

33. W. Cochran, Critical Reviews in Solid State Sciences **2**, 1 (1971).

34. H. G. Smith, in *Superconductivity in d- and f-Band Metals*, edited by D. H. Douglass (AIP Conference Proc. **4**, New York, 1972), p. 321.

35. Al: B. N. Brockhouse and A. T. Stewart, Rev. Mod. Phys. **30**, 236 (1958).

36. Al: C. B. Walker, Phys. Rev. **103**, 547 (1956).

37. Al: R. S. Carter, H. Palevsky, and D. J. Hughes, Phys. Rev. **106**, 1168 (1957).

38. Al: K. E. Larsson, V. Dahlborg, and S. Holmyrd, Ark. Fysik **17**, 369 (1960).

39. Al: J. L. Yarnell, J. L. Warren, and S. H. Koenig, in Wallis, p. 57.

40. Al: R. Stedman and G. Nilsson, Phys. Rev. **145**, 492 (1966).

41. G. Gilat and R. M. Nicklow, Phys. Rev. **143**, 487 (1966).

42. Pb: B. N. Brockhouse, T. Arase, G. Caglioti, K. R. Rao, and A. D. B. Woods, Phys. Rev. **128**, 1099 (1962).

43. G. Gilat, Solid State Comm. **3**, 101 (1965).

44. Pb: R. Stedman, L. Almqvist, G. Nilsson, and G. Raunio, Phys. Rev. **162**, 545 (1967).

45. Cu: S. K. Sinha, Phys. Rev. **143**, 422 (1966).

46. Cu: E. C. Svensson, B. N. Brockhouse, and J. M. Rowe, Phys. Rev. **155**, 619 (1967).

47. Cu: R. M. Nicklow, G. Gilat, H. G. Smith, L. J. Raubenheimer, and M. K. Wilkinson, Phys. Rev. **164**, 922 (1967).

48. Ag: W. A. Kamitakahara and B. N. Brockhouse, Phys. Letters **29A**, 639 (1969).

49. Th: R. A. Reese, S. K. Sinha, and D. T. Peterson, Phys. Rev. **B8**, 1332 (1973).

50. Pt: R. Ohrlich and W. Drexel, in Copenhagen proceedings, Vol. I. p. 203; D. H. Dutton, B. N. Brockhouse, and A. P. Miiller, Can. J. Phys. **50**, 2915 (1972).

51. Ni: R. J. Birgeneau, J. Cordes, G. Dolling, and A. D. B. Woods, Phys. Rev. **136**, A1359 (1964).

52. Pd: A. P. Miiller and B. N. Brockhouse, Phys. Rev. Letters **20**, 798 (1968).

53. W. Hanke, in Nusimovici, p. 294.

54. Mg: M. F. Collins, Proc. Phys. Soc. (London) **80**, 362 (1962).

55. Mg: P. K. Iyengar, G. Venkataraman, P. R. Vijayaraghavan, and A. P. Roy, in Bombay proceedings, Vol. I, p. 153.

56. Mg: G. L. Squires, Proc. Phys. Soc. (London) **88**, 919 (1966); see also R. Pynn and G. L. Squires, in Copenhagen proceedings, Vol. I, p. 215.

57. Be: R. E. Schmunk, R. M. Brugger, P. D. Randolph, and K. A. Strong, Phys. Rev. **128**, 562 (1962); see also R. E. Schmunk, Phys. Rev. **149**, 450 (1966).

58. Be: C. L. Thaper, K. R. Rao, B. A. Dasannacharya, A. P. Roy, and P. K. Iyengar, in Nusimovici, p. 140; B. A. Dasannacharya, P. K. Iyengar, R. V. Nandedkar, K. R. Rao, A. P. Roy, and C. L. Thaper, Pramāna **2**, 179 (1974).

59. Zn: G. Borgonovi, G. Caglioti and J. J. Antal, Phys. Rev. **132**, 683 (1963).

60. Zn: D. L. McDonald, M. M. Elcombe, and A. W. Pryor, J. Phys. C **2**, 1857 (1969).

61. Y: S. K. Sinha, T. O. Brun, L. D. Muhlestein, and J. Sakurai, Phys. Rev. **B1**, 2430 (1970).

62. Sc: N. Wakabayashi, S. K. Sinha, and F. H. Spedding, Phys. Rev. **B4**, 2398 (1971).

63. Zr: H. F. Bezdek, R. E. Schmunk and L. Finegold, Phys. Stat. Sol. **42**, 275 (1970).

64. Ti: N. Wakabayashi (to be published).

65. Tc: H. G. Smith, N. Wakabayashi, R. M. Nicklow, and S. Mihailovich (to be published).

66. Cd: G. Toussaint and G. Champier, Phys. Stat. Sol. **54**, 165 (1972).

67. Tb: J. C. Glyden Houmann and R. M. Nicklow, Phys. Rev. **B1**, 3943 (1970).

68. Ho: J. A. Leake, V. J. Minkiewicz, and G. Shirane, Solid State Comm. **7**, 535 (1969).

69. Ho: R. M. Nicklow, N. Wakabayashi, and P. R. Vijayaraghavan, Phys. Rev. **B3**, 1229 (1971).

70. R. E. DeWames, T. Wolfram, and G. W. Lehman, Phys. Rev. **138**, A717 (1965).

71. A. P. Roy, B. A. Dasannacharya, C. L. Thaper, and P. K. Iyengar, Phys. Rev. Letters **30**, 906 (1973).

72. B. N. Brockhouse, E. D. Hallman, and S. C. Ng, in *Magnetic and Inelastic Scattering of Neutrons*, edited by T. J. Rowland and P. A. Beck (Gordon and Breach, New York, 1968), p. 161.

73. W. Cochran, Proc. Roy. Soc. (London) **A276**, 308 (1963).

74. N. F. Mott and H. Jones, *The Theory of the Properties of Metals and Alloys*, (Dover, New York, 1958), p. 13.

75. K. Fuchs, Proc. Roy. Soc. (London) **A153**, 622 (1936).

76. See Seitz, Chapter 10 and references cited therein.

77. J. de Launay, Solid State Phys. **2**, 219 (1956).

78. A. B. Bhatia, Phys. Rev. **97**, 363 (1955).

79. P. K. Sharma and S. K. Joshi, J. Chem. Phys. **39**, 2633 (1963).

80. T. Toya, J. Res. Inst. Catalysis, Hokkaido University **6**, 161, 183 (1958).

81. J. S. Langer and S. H. Vosko, J. Phys. Chem. Solids **12**, 196 (1960); see also J. Friedel, Nuovo Cim. Suppl. **7**, 287 (1958).

82. W. Harrison, *Pseudopotentials in the Theory of Metals* (Benjamin, New York, 1966), p. 44.

83. S. H. Koenig, Phys. Rev. **135**, A1693 (1964).

84. A. J. E. Foreman and W. M. Lomer, Proc. Phys. Soc. (London) **B70**, 1143 (1957).

85. W. Kohn, Phys. Rev. Letters **2**, 393 (1959); E. J. Woll Jr. and W. Kohn, Phys. Rev. **126**, 1693 (1962).

86. B. N. Brockhouse, K. R. Rao, and A. D. B. Woods, Phys. Rev. Letters **7**, 93 (1961).

87. A. Pashkin and R. J. Weiss, Phys. Rev. Letters **9**, 199 (1962).

88. C. Kittel, *Introduction to Solid State Physics* (Wiley, New York, 1956), second edition, Appendix B.

89. D. Pines, *Elementary Excitations in Solids* (Benjamin, New York, 1963), pp. 137–138.

90. W. Hanke and H. Bilz, in Grenoble proceedings, p. 3.

91. β-brass: G. Gilat and G. Dolling, Phys. Rev. **138**, A1053 (1965); see also Bombay proceedings, Vol. I, p. 343.

92. Fe₃Al: C. Van Dijk and J. Bergsma, in Copenhagen proceedings, Vol. I, p. 233.

93. α-Brass: E. D. Hallman and B. N. Brockhouse, Can. J. Phys. **47**, 1117 (1969).

94. Pb–Tl–Bi alloys: S. G. Ng and B. N. Brockhouse, in Copenhagen proceedings, Vol. I, p. 253. See also A. P. Roy, Ph. D. thesis, McMaster University (1970).

95. Mo–Nb alloys: A. D. B. Woods, in *Symposium on Inelastic Scattering of Neutrons by Condensed Systems*, BNL–940 (CAS) (Brookhaven National Laboratory, Upton, New York, 1966), BNL 940.

96. Ni–Pd alloy: W. A. Kamitakahara and B. N. Brockhouse, in Grenoble proceedings, p. 73.

97. Au in Cu: E. C. Svensson, B. N. Brockhouse, and J. M. Rowe, Solid State Comm. **3**, 245 (1965); E. C. Svensson and B. N. Brockhouse, Phys. Rev. Letters **18**, 858 (1967).

98. Al in Cu: R. M. Nicklow, P. R. Vijayaraghavan, H. G. Smith, G. Dolling, and M. K. Wilkinson, in Copenhagen proceedings, Vol. I, p. 47.

99. LiF: G. Dolling, H. G. Smith, R. M. Nicklow, P. R. Vijayaraghavan, and M. K. Wilkinson, Phys. Rev. **168**, 970 (1968).

100. NaF: J. D. Pirie and T. Smith, J. Phys. C **1**, 648 (1968); W. J. L. Buyers, Phys. Rev. **153**, 923 (1967).

101. NaCl: G. Raunio, L. Almqvist, and R. Stedman, Phys. Rev. **178**, 1496 (1969); R. E. Schmunk and D. R. Winder, J. Phys. Chem. Solids **31**, 131 (1970).

102. NaBr: J. S. Reid, T. Smith, and W. J. L. Buyers, Phys. Rev. **B1**, 1833 (1970).

103. NaI: A. D. B. Woods, B. N. Brockhouse, R. A. Cowley, and W. Cochran, Phys. Rev. **131**, 1025 (1963); A. D. B. Woods, W. Cochran, and B. N. Brockhouse, Phys. Rev. **119**, 980 (1960).

104. KCl: J. R. D. Copley, R. W. Macpherson, and T. Timusk, Phys. Rev. **182**, 965 (1969); G. Raunio and L. Almqvist, Phys. Stat. Sol. **33**, 209 (1969).

105. KBr: A. D. B. Woods, B. N. Brockhouse, R. A. Cowley, and W. Cochran, Phys. Rev. **131**, 1025 (1963).

106. KI: G. Dolling, R. A. Cowley, C. Schittenhelm, and I. M. Thorson, Phys. Rev. **147**, 577 (1966).

107. RbF: G. Raunio and S. Rolandson, J. Phys. C **3**, 1013 (1970).

108. RbCl: Ibid.

109. RbBr: S. Rolandson and G. Raunio, J. Phys. C **4**, 958 (1971).

110. RbI: G. Raunio and S. Rolandson, Phys. Stat. Sol. **40**, 749 (1970).

111. J. R. Tessman, A. H. Kahn, and W. Schockley, Phys. Rev. **92**, 890 (1953).

112. See, for example, D. H. Martin, *Magnetism in Solids* (Iliffe, London; MIT Press, Cambridge, Mass., 1967), p. 169.

113. L. Pauling, Proc. Roy. Soc. (London) **A114**, 181 (1927).

114. BH, p. 25.

115. R. A. Cowley, W. Cochran, B. N. Brockhouse, and A. D. B. Woods, Phys. Rev. **131**, 1030 (1963).

116. A. M. Karo and J. R. Hardy, Phys. Rev. **129**, 2024 (1963).

117. BH, p. 21.

118. Kittel, p. 105.

119. U. Schröder, Solid State Comm. **4**, 347 (1966).

120. J. S. Melvin, J. D. Pirie, and T. Smith, Phys. Rev. **175**, 1082 (1968).

121. S. K. Sinha, Critical Reviews in Solid State Sciences **3**, 273 (1973).

122. H. Hahn, in Chalk River proceedings, Vol. I, p. 37.

123. J. F. Nye, *Physical Properties of Crystals* (Clarendon Press, Oxford, 1957), p. 186.

124. MgO: M. J. L. Sangster, G. E. Peckham, and D. H. Saunderson, J. Phys. C **3**, 1026 (1970).

125. CaO: D. H. Saunderson and G. E. Peckham, in Nusimovici, p. 171.

126. CaO: P. R. Vijayaraghavan, Marsongkohadi, and P. K. Iyengar, in Grenoble proceedings, p. 95.

127. AgCl: P. R. Vijayaraghavan, R. M. Nicklow, H. G. Smith, and M. K. Wilkinson, Phys. Rev. **B1**, 4819 (1970).

128. S. K. Sinha, private communication.

129. K. Fischer, H. Bilz, R. Haberkorn, and W. Weber, Phys. Stat. Sol. **54**, 285 (1972).

130. PbTe: R. A. Cowley and G. Dolling, Phys. Rev. Letters **14**, 549 (1965).

131. PbTe: W. Cochran, R. A. Cowley, G. Dolling, and M. M. Elcombe, Proc. Roy. Soc. (London) **A293**, 433 (1966).

132. SnTe: G. S. Pawley, W. Cochran, R. A. Cowley, and G. Dolling, Phys. Rev. Letters **17**, 753 (1966).

133. SnTe: E. R. Cowley, J. K. Darby, and G. S. Pawley, J. Phys. C **2**, 1916 (1969).

134. TaC, HfC: H. G. Smith and W. Gläser, Phys. Rev. Letters **25**, 1611 (1970).

135. ZrC, UC, NbC: H. G. Smith, in *Superconductivity in d- and f-Band Metals*, see reference 34 above.

136. TaC: H. G. Smith, Phys. Rev. Letters **29**, 353 (1972) W. Weber, Phys. Rev. **B8**, 5082 (1973).

137. W. Weber, H. Bilz, and U. Schröder, Phys. Rev. Letters **28**, 600 (1972); W. Weber, Phys. Rev. **B8**, 5082 (1973).

138. J. Lindhard, Kgl. Danske Videnskab. Selskab, Mat-fys. Medd. **28**, 8 (1954).

139. J. Shah, T. C. Damen, J. F. Scott, and R. C. C. Leite, Phys. Rev. **B3**, 4238 (1971); J. F. Scott, T. C. Damen, R. C. C. Leite, and J. Shah, Phys. Rev. **B1**, 4330 (1970).

140. B. N. Ganguly and R. F. Wood, Phys. Rev. Letters **28**, 681 (1972).

141. M. Born, Ann. Phys. **44**, 605 (1914).

142. C. V. Raman, Proc. Indian Acad. Sci. **18A**, 237 (1943); **26**, 339 (1947).

143. H. M. J. Smith, Phil. Trans. Roy. Soc. London **A241**, 105 (1948).

144. Y. C. Hsieh, J. Chem. Phys. **22**, 306 (1954).

145. Ge: B. N. Brockhouse and P. K. Iyengar, Phys. Rev. **111**, 747 (1958).

146. Ge: A. Ghose, H. Palevsky, D. J. Hughes, I. Pelah, and C. M. Eisenhauer, Phys. Rev. **113**, 49 (1959).

147. Ge: B. N. Brockhouse and B. A. Dasannacharya, Solid State Comm. **1**, 205 (1963).

148. Ge and Si: G. Nilsson and G. Nelin, Phys. Rev. **B6**, 3777 (1972).

149. Si: B. N. Brockhouse, Phys. Rev. Letters **2**, 256 (1959).

150. Si: G. Dolling, in Chalk River proceedings, Vol. II, p. 37.

151. Diamond: J. L. Warren, J. L. Yarnell, G. Dolling, and R. A. Cowley, Phys. Rev. **158**, 805 (1967); J. L. Warren, R. G. Wenzel, and J. L. Yarnell, in Bombay proceedings, Vol. I, p. 361.

152. Diamond: G. Peckham, Solid State Comm. **5**, 311 (1967).

153. α-Sn and InSb: D. L. Price, J. M. Rowe, and R. M. Nicklow, Phys. Rev. **B3**, 1268 (1971).

154. T. I. Kucher, Fiz. Tverd. Tela **4**, 2385 (1962) [Sov. Phys.—Solid State **4**, 1747 (1963)].

155. M. Mostoller, J. Phys. Chem. Solids **31**, 307 (1970).

156. Kittel, p. 142.

157. F. Herman, J. Phys. Chem. Solids **8**, 405 (1959).

158. K. G. Aggarwal, Proc. Phys. Soc. (London) **91**, 381 (1967).

159. H. L. McMurry, A. W. Solbrig Jr., J. K. Boyter, and C. Noble, J. Phys. Chem. Solids **28**, 2359 (1967).

160. A. W. Solbrig Jr., J. Phys. Chem. Solids **32**, 1761 (1971).

161. B. D. Singh and B. Dayal, Phys. Stat. Sol. **38**, 141 (1970).

162. M. Lax, Phys. Rev. Letters **1**, 133 (1958).

163. W. Cochran, Phys. Rev. Letters **2**, 495 (1959).

164. W. Cochran, Proc. Roy. Soc. (London) **A253**, 260 (1959).

165. G. Dolling and R. A. Cowley, Proc. Phys. Soc. (London) **88**, 463 (1966).

166. K. B. Tolpygo, Fiz. Tverd. Tela **3**, 943 (1961) [Sov. Phys.—Solid State **3**, 685 (1961)].

167. T. I. Kucher, Fiz. Tverd. Tela **4**, 992 (1962) [Sov. Phys.—Solid State **4**, 729 (1962)].

168. Z. A. Demidenko, T. I. Kucher, and K. B. Tolpygo, Fiz. Tverd. Tela **4**, 104 (1962) [Sov. Phys.—Solid State **4**, 73 (1962)].

169. T. I. Kucher and V. V. Nechiporuk, Fiz. Tverd. Tela **8**, 317 (1966) [Sov. Phys.—Solid State **8**, 261 (1966)].

170. J. L. Warren, in *Inelastic Scattering of Neutrons by Condensed Systems*, see reference 95 above, p. 88.

171. R. M. Martin, Chem. Phys. Letters **2**, 268 (1968).

172. R. M. Martin, Phys. Rev. **186**, 871 (1969).

173. GaAs: G. Dolling and J. L. T. Waugh, in Wallis, p. 19; Phys. Rev. **132**, 2410 (1963).

174. GaP: J. L. Yarnell, J. L. Warren, R. G. Wenzel, and P. J. Dean, in Copenhagen proceedings, Vol. I, p. 301.

175. ZnS and ZnTe: L. A. Feldkamp, D. K. Steinman, N. Vagelatos, J. S. King, and G. Venkataraman, J. Phys. Chem. Solids **32**, 1573 (1971); L. A. Feldkamp, G. Venkataraman, and J. S. King, Solid State Comm. **7**, 1571 (1969); N. Vagelatos, D. Wehe, and J. S. King, J. Chem. Phys. **60**, 3613 (1974).

176. ZnS: J. Bergsma, Phys. Letters **32A**, 324 (1970); Reactor Centrum Nederland Report RCN-121 (1970).

177. ZnSe: B. Hennion, F. Moussa, G. Pepy, and K. Kunc, Phys. Letters **36A**, 376 (1971).

178. CuCl: C. Carabatos, B. Hennion, K. Kunc, F. Moussa, and C. Schwab, Phys. Rev. Letters **26**, 770 (1971).

179. CuI: B. Hennion, F. Moussa, B. Prevot, C. Carabatos, and C. Schwab, Phys. Rev. Letters **28**, 964 (1972).

180. M. A. Nusimovici, M. Balkanski, and J. L. Birman, Phys. Rev. **B1**, 595 (1970).

181. G. Arlt and P. Quadfleig, Phys. Stat. Sol. **25**, 323 (1968).

182. L. A. Feldkamp, J. Phys. Chem. Solids **33**, 711 (1972).

183. W. Cochran and R. A. Cowley, J. Phys. Chem. Solids **23**, 447 (1962).

184. R. A. Cowley, Proc. Roy. Soc. (London) **A268**, 121 (1962).

185. G. Lucovsky, R. M. Martin, and E. Burstein, Phys. Rev. **B4**, 1367 (1971).

186. J. C. Phillips, Rev. Mod. Phys. **42**, 317 (1970).

187. S. S. Mitra and R. Marshall, J. Chem. Phys. **41**, 3158 (1964).

188. J. Tateno, J. Phys. Chem. Solids **31**, 1641 (1970).

189. P. Lawaetz, Phys. Rev. Letters **26**, 697 (1971).

190. P. W. Anderson, Izv. Akad. Nauk. SSSR, 290 (1960).

191. W. Cochran, Phys. Rev. Letters **3**, 412 (1959).

192. BH, p. 105.

193. A. S. Barker Jr., in *Far Infrared properties of Solids*, edited by S. S. Mitra and S. Nudelman (Plenum, New York, 1970).

194. R. A. Cowley, Phys. Rev. Letters **9**, 159 (1962); Phys. Rev. **134**, A981 (1964).

195. W. Cochran, Advan. Phys. **9**, 387 (1960); **10**, 401 (1961).

196. K. Otnes, T. Riste, G. Shirane, and J. Feder, Solid State Comm. **9**, 1103 (1971).

197. B. Dorner, J. D. Axe, and G. Shirane, Phys. Rev. **B6**, 1950 (1972).

# 7 Some Recent Developments

In this volume we have been concerned mainly with the Born–von Kármán approach to lattice dynamics and in particular with various phenomenological models. Our attention has been so directed largely because such models have considerable practical utility and should remain useful for years to come. Nevertheless we must take cognizance of recent attempts to go beyond phenomenology. Particularly noteworthy are the efforts to develop

1. microscopic theories of lattice dynamics, and
2. a formalism applicable to quantum crystals, like helium, where the force-constant approach totally fails.

Since these topics represent the new horizons of lattice dynamics, it is appropriate to conclude with a brief glimpse of them so as to prepare the reader for future developments.

## 7.1 Microscopic theories of lattice dynamics

By a microscopic theory, we mean one that discusses dynamics starting from a consideration of the electronic wave functions and their readjustments to nuclear motions. In this sense, the foundations for a microscopic approach were laid long ago by Born and Oppenheimer,[1] though only recently has their program been given a concrete shape.

Microscopic theories have been formulated at various levels of sophistication. The least complicated views the crystal as a collection of ion cores plus valence electrons. The motions of the valence electrons in the perfect crystal are assumed to be known and the dynamics of the ion cores are investigated by supposing that (1) the ions move as a unit (either rigidly or with some distortion), and (2) that the valence electrons adjust adiabatically to the core motions (Sec. 2.1). This is the approach closest to the realm of practical calculations. At the next higher level of sophistication, the crystal is viewed as a collection of nuclei plus electrons. Once again the electron problem for the perfect crystal is assumed to have been solved, and the nuclear dynamics are investigated using the adiabatic approximation, the electronic readjustments to the nuclear motion being described formally in terms of *exact* many-electron wave functions pertaining to the perfect crystal. A still more fundamental approach discards the adiabatic assumption and treats the electron and nuclear motions simultaneously in a mutually consistent fashion.[2,3] This enables one to handle the lattice dynamics of superconductors, excitonic insulators, etc. All such microscopic theories involve sophisticated mathematical techniques outside our scope. Accordingly we restrict ourselves to highlighting their important features. Furthermore, attention will be confined to theories retaining the adiabatic approximation and corresponding to 0 K; the effects of retardation[4] will be ignored.

### 7.1.1 Microscopic longitudinal dielectric function

The basic problem in microscopic theories of lattice dynamics is to determine first how the energy of the electronic system changes with such changes in the nuclear configuration as occur during vibrations and then to evaluate the influence of this electronic response on the nuclear motions themselves. That consideration of electronic response is important may be appreciated from the following example. Figure 7.1 shows the dispersion curves for Al computed as if the crystal were an assembly of $Al^{+++}$ ions organized in a fcc lattice and immersed in a *nonresponsive* sea of conduction electrons. Physically, this implies that one ignores the change in the con-

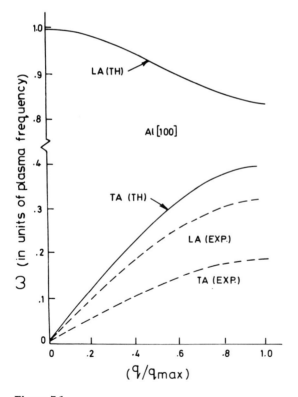

**Figure 7.1**
Dispersion curves for Al in the [100] direction computed without considering the effects of electronic polarization. Also shown for comparison are the experimental curves.

duction electron density caused by ion-core motions (that is, the electronic polarization) and thereby the contribution to the crystal potential energy resulting from the density fluctuation. In the language of the previous chapter this amounts to calculating the dynamics using for the interatomic potential $V(r)$ the bare Coulomb potential $(Z^2 e^2/r)$ and not the screened form in (6.5). Consequently the computed LA branch tends to a finite frequency in the limit $\mathbf{q} \to 0$, in contrast to experiment. In fact, the $\mathbf{q} = 0$ frequency calculated is that of the plasma oscillations of the ions. In reality such oscillations are not observed because of the screening arising from the polarization of the conduction electrons. Such polarization occurs in all types of crystals (see Chapter 6) and is conveniently described through a nonlocal *microscopic longitudinal dielectric function* $\varepsilon(\mathbf{r}, \mathbf{r}')$, to which we now direct attention.

Consider a system $S$ consisting of a large number of electrons and suppose that it is subjected to a static external scalar potential $V_{\text{ext}}(\mathbf{r})$, due to an appropriate set of external charges. We now explore the potential inside $S$ with a suitable probe. The latter experiences a total potential $V_t(\mathbf{r})$ given by

$$V_t(\mathbf{r}) = V_{\text{ext}}(\mathbf{r}) + V_{\text{ind}}(\mathbf{r}), \tag{7.1}$$

where $V_{\text{ind}}(\mathbf{r})$ is the potential induced in $S$ by the external potential. If the latter is small (as is often the case), then to first order $V_{\text{ind}}$ can be expressed as a linear function of $V_{\text{ext}}$, so that formally

$$V_t(\mathbf{r}) = V_{\text{ext}}(\mathbf{r}) + \int f(\mathbf{r}, \mathbf{r}') V_{\text{ext}}(\mathbf{r}') \, d\mathbf{r}', \tag{7.2}$$

where $f(\mathbf{r}, \mathbf{r}')$ is a nonlocal function expressing the potential induced at $\mathbf{r}$ by unit external potential at $\mathbf{r}'$. Usually (7.2) is written in the form[5]

$$V_t(\mathbf{r}) = \int \varepsilon^{-1}(\mathbf{r}, \mathbf{r}') V_{\text{ext}}(\mathbf{r}') \, d\mathbf{r}' \tag{7.3}$$

which defines the response function $\varepsilon^{-1}$ in terms of the applied and measured potentials. The dielectric function $\varepsilon$ referred to earlier is the inverse of the response function appearing here:

$$\int \varepsilon(\mathbf{r}, \mathbf{r}') \varepsilon^{-1}(\mathbf{r}', \mathbf{r}'') \, d\mathbf{r}' = \int \varepsilon^{-1}(\mathbf{r}, \mathbf{r}') \varepsilon(\mathbf{r}' \, \mathbf{r}'') \, d\mathbf{r}' = \delta(\mathbf{r} - \mathbf{r}''). \tag{7.4}$$

We must now be explicit about the nature of the external charges and the probe, that is, whether they are electrons (e) or test charges (t) *distinguishable* from electrons (for example, positrons or ions). This leads in effect to four different response functions[6-8] $\varepsilon_{AB}^{-1}$ ($A \to$ source $=$ t or e; $B \to$ probe $=$ t or e). Of these we shall be interested mainly in $\varepsilon_{tt}^{-1}$, since it enters in lattice dynamics. (From now on the subscript tt will be understood, though occasionally the subscript will be explicitly indicated for emphasis).

A formal expression for $\varepsilon^{-1}$ in terms of many-electron wave functions can be written as follows: Suppose that the system is at 0 K and that $\psi_0(\mathbf{r}_1, \ldots, \mathbf{r}_N)$ denotes the ground state wave function, $\mathbf{r}_1, \ldots, \mathbf{r}_N$ being the electron coordinates. Standard perturbation theory gives for the first-order change $\delta\psi_0$ the result†

$$\delta\psi_0 = \sum_{k(\neq 0)} (E_0 - E_k)^{-1} \left\{ \int \psi_k^* \left[ -e \sum_j \delta(\mathbf{r}'' - \mathbf{r}_j) V_{\text{ext}}(\mathbf{r}'') \right] \right.$$
$$\left. \times \; \psi_0 \, d\mathbf{r}'' \, d\mathbf{r}_1 \cdots d\mathbf{r}_N \right\} \psi_k, \tag{7.5}$$

where $E_0$ and $E_k$ are respectively the energies corresponding to the states $\psi_0$ and $\psi_k$. The corresponding first-order change in the electron density at $\mathbf{r}'$ is

$$\Delta\rho(\mathbf{r}') = \sum_i \int d\mathbf{r}_i \, \delta(\mathbf{r}' - \mathbf{r}_i)\{\delta\psi_0^*\psi_0 + \psi_0^*\delta\psi_0\}d\mathbf{r}_1 \cdots d\mathbf{r}_{i-1}d\mathbf{r}_{i+1} \cdots d\mathbf{r}_N \tag{7.6}$$

and induces at $\mathbf{r}$ the potential

$$V_{\text{ind}}(\mathbf{r}) = -\int \frac{e\Delta\rho(\mathbf{r}') \, d\mathbf{r}'}{|\mathbf{r} - \mathbf{r}'|}. \tag{7.7}$$

To write this in the same form as the second term on the right-hand side of (7.2), we make use of (7.5) which gives

$$V_{\text{ind}}(\mathbf{r}) = \sum_{k(\neq 0)} \int d\mathbf{r}' \frac{e^2}{|\mathbf{r} - \mathbf{r}'|} (E_0 - E_k)^{-1}\{\langle\psi_k|\rho(\mathbf{r}', \mathbf{r}_1, \ldots, \mathbf{r}_N)|\psi_0\rangle$$
$$\times \; \langle\psi_k|\rho(\mathbf{r}'', \mathbf{r}_1, \ldots, \mathbf{r}_N)|\psi_0\rangle^* + \text{c.c.}\}V_{\text{ext}}(\mathbf{r}'') \, d\mathbf{r}'', \tag{7.8}$$

where

$$\rho(\mathbf{r}, \mathbf{r}_1, \ldots, \mathbf{r}_N) = \sum_i \delta(\mathbf{r} - \mathbf{r}_i), \tag{7.9}$$

and $V_{\text{ext}}$ is recognized to be real.

We now introduce a quantity $\chi(\mathbf{r}', \mathbf{r}'')$ which is closely related to $\varepsilon^{-1}$ and the quantity $\chi_{sc}(\mathbf{r}', \mathbf{r}'')$ defined in Sec. 4.6.10. Formally, $\chi(\mathbf{r}', \mathbf{r}'')$ is

---

†Remember our convention that the charge on an electron is $-e$.

defined as the functional derivative†

$$\chi(\mathbf{r}', \mathbf{r}'') = -\frac{1}{e}\frac{\delta\Delta\rho(\mathbf{r}')}{\delta V_{\text{ext}}}\bigg|_{\mathbf{r}''}.$$   (7.10a)

In integral form this reads

$$\Delta\rho(\mathbf{r}') = -e\int\chi(\mathbf{r}', \mathbf{r}'')V_{\text{ext}}(\mathbf{r}'')\,d\mathbf{r}''.$$   (7.10b)

From (7.5), (7.6), and (7.10) we obtain

$$\chi(\mathbf{r}', \mathbf{r}'') = \sum_{k(\neq 0)}(E_0 - E_k)^{-1}\{\langle\psi_k|\rho(\mathbf{r}', \mathbf{r}_1, \ldots, \mathbf{r}_N)|\psi_0\rangle$$
$$\times\langle\psi_k|\rho(\mathbf{r}'', \mathbf{r}_1, \ldots, \mathbf{r}_N)|\psi_0\rangle^* + \text{c.c.}\}.$$   (7.11)

This enables us to write (7.8) as

$$V_{\text{ind}}(\mathbf{r}) = \int\frac{e^2}{|\mathbf{r} - \mathbf{r}'|}\chi(\mathbf{r}', \mathbf{r}'')V_{\text{ext}}(\mathbf{r}'')\,d\mathbf{r}'\,d\mathbf{r}'',$$   (7.12)

which, when combined with Eqs. (7.1)–(7.3), gives

$$\varepsilon^{-1}(\mathbf{r}, \mathbf{r}'') = \delta(\mathbf{r} - \mathbf{r}'') + \int v(\mathbf{r}, \mathbf{r}')\chi(\mathbf{r}', \mathbf{r}'')\,d\mathbf{r}',$$   (7.13)

where

$$v(\mathbf{r}, \mathbf{r}') = \frac{e^2}{|\mathbf{r} - \mathbf{r}'|}.$$   (7.14)

We next introduce the Fourier transform of $\varepsilon^{-1}(\mathbf{r}, \mathbf{r}')$ via the defini-

---

†If a quantity depends on the entire course of some function $\phi(x)$ of a parameter $x$, then it is called a functional of $\phi$ and is denoted by $F[\phi(x)]$. A typical example is

$$F[\phi] = \int_c [\text{some function of } \phi(x)]\,dx,$$

where $c$ denotes the course (domain) of $x$. If $F[\phi]$ is continuous in some sense for variations in $\phi$, the first-order variation in $F$ may be shown to be given by

$$\lim_{\varepsilon\to 0}\frac{F[\phi + \varepsilon f] - F[\phi]}{\varepsilon} = \int_c \frac{\delta F[\phi]}{\delta\phi}\bigg|_{x''}f(x'')\,dx'',$$

where $f(x'')$ is a smooth but arbitrary, integrable *test* function. The coefficient of $f(x'')$ under the integral sign is called the functional derivative of $F$ with respect to $\phi$ at the point $x''$.

In our problem, $\Delta\rho$ is not only a function of the *variable* $\mathbf{r}'$ but is also a *functional* of the function $V_{\text{ext}}$. Strictly speaking therefore we should write $\Delta\rho$ as $\Delta\rho(\mathbf{r}', V_{\text{ext}})$, although we shall not do so. In this sense, the quantity defined in (7.10) may be looked upon as the "partial" functional derivative with respect to $V_{\text{ext}}$ at $\mathbf{r}''$.

tions

$$\varepsilon^{-1}(\mathbf{r}, \mathbf{r}') = \frac{1}{\Omega} \sum_{\mathbf{Q}} \sum_{\mathbf{Q}'} \varepsilon^{-1}(\mathbf{Q}, \mathbf{Q}') e^{i\mathbf{Q}\cdot\mathbf{r}} e^{-i\mathbf{Q}'\cdot\mathbf{r}'}, \tag{7.15}$$

$$\varepsilon^{-1}(\mathbf{Q}, \mathbf{Q}') = \frac{1}{\Omega} \int d\mathbf{r}\, d\mathbf{r}'\, \varepsilon^{-1}(\mathbf{r}, \mathbf{r}') e^{-i\mathbf{Q}\cdot\mathbf{r}} e^{i\mathbf{Q}'\cdot\mathbf{r}'}. \tag{7.16}$$

The convention followed is discussed in Appendix 1. In the case of crystals, we can choose $\Omega = Nv$. Further, on account of the translational symmetry, we can write $\mathbf{Q} = \mathbf{q} + \mathbf{G}$ and $\mathbf{Q}' = \mathbf{q} + \mathbf{G}'$. The Fourier components $\varepsilon(\mathbf{Q}, \mathbf{Q}')$ of $\varepsilon(\mathbf{r}, \mathbf{r}')$ may be defined similarly. In view of (7.4) we also have

$$\sum_{\mathbf{Q}'} \varepsilon(\mathbf{Q}, \mathbf{Q}') \varepsilon^{-1}(\mathbf{Q}', \mathbf{Q}'') = \sum_{\mathbf{Q}'} \varepsilon^{-1}(\mathbf{Q}, \mathbf{Q}') \varepsilon(\mathbf{Q}', \mathbf{Q}'') = \delta_{\mathbf{Q}\mathbf{Q}''}. \tag{7.17}$$

Both $\varepsilon(\mathbf{Q}, \mathbf{Q}')$ and its inverse $\varepsilon^{-1}(\mathbf{Q}, \mathbf{Q}')$ can be viewed as (infinite) matrices with the values of $\mathbf{Q}$ and $\mathbf{Q}'$ defining the rows and columns.

It is appropriate to note here an important physical distinction between $\varepsilon^{-1}$ and $\varepsilon$. For this purpose we first write the latter as

$$\varepsilon(\mathbf{r}, \mathbf{r}'') = \delta(\mathbf{r} - \mathbf{r}'') - g(\mathbf{r}, \mathbf{r}''). \tag{7.18}$$

By combining this with Eqs. (7.1), (7.2), and (7.4) it is easily seen that

$$V_{\text{ind}}(\mathbf{r}) = \int g(\mathbf{r}, \mathbf{r}'')\, V_{\text{t}}(\mathbf{r}'')\, d\mathbf{r}'', \tag{7.19}$$

which is to be compared with the result

$$V_{\text{ind}}(\mathbf{r}) = \int f(\mathbf{r}, \mathbf{r}'') V_{\text{ext}}(\mathbf{r}'')\, d\mathbf{r}'' \tag{7.20}$$

obtainable from (7.1) and (7.2). This shows that whereas $f(\mathbf{r}, \mathbf{r}'')$ and hence $\varepsilon^{-1}(\mathbf{r}, \mathbf{r}'')$ can be defined in terms of the response to an *external* scalar potential, $g(\mathbf{r}, \mathbf{r}'')$ and hence $\varepsilon(\mathbf{r}, \mathbf{r}'')$ must be defined in terms of the response to the *total* scalar potential acting in the medium. By analogy with (7.13) we may write

$$g(\mathbf{r}, \mathbf{r}'') = \int v(\mathbf{r}, \mathbf{r}')\, \chi_{\text{sc}}(\mathbf{r}', \mathbf{r}'')\, d\mathbf{r}', \tag{7.21}$$

where $\chi_{\text{sc}}$ is the functional derivative

$$\chi_{\text{sc}}(\mathbf{r}', \mathbf{r}'') = -\frac{1}{e} \frac{\delta \Delta\rho(\mathbf{r}')}{\delta V_{\text{t}}}\bigg|_{\mathbf{r}''}. \tag{7.22a}$$

As in (7.10b) we may also write (see (4.382))

$$\Delta\rho(\mathbf{r}') = -e \int \chi_{\text{sc}}(\mathbf{r}', \mathbf{r}'') V_{\text{t}}(\mathbf{r}'')\, d\mathbf{r}''. \tag{7.22b}$$

The quantity $\chi_{sc}$ denotes the screened dielectric susceptibility and is to be distinguished from $\chi$ defined in (7.10). In fact, by combining (7.4), (7.13), (7.18), and (7.21) and judiciously manipulating the dummy indices, it can be shown that

$$\chi(\mathbf{r}, \mathbf{r}') = \int v^{-1}(\mathbf{r}, \mathbf{r}'')\varepsilon^{-1}(\mathbf{r}'', \mathbf{r}''')v(\mathbf{r}''', \mathbf{r}'''')\chi_{sc}(\mathbf{r}'''', \mathbf{r}') \, d\mathbf{r}'' \, d\mathbf{r}''' \, d\mathbf{r}'''',$$

(7.23)

where $v^{-1}$ is such that[5]

$$\int v^{-1}(\mathbf{r}, \mathbf{r}'')v(\mathbf{r}'', \mathbf{r}') \, d\mathbf{r}'' = \int v(\mathbf{r}, \mathbf{r}'')v^{-1}(\mathbf{r}'', \mathbf{r}') \, d\mathbf{r}'' = \delta(\mathbf{r} - \mathbf{r}'). \quad (7.24)$$

As defined above, the quantity $\chi_{sc}$ is identical with that introduced in Sec. 4.6.10.

It emerges from above (as pointed out by Pines and Nozieres[9] in a related context) that whereas $\varepsilon^{-1}$ can be directly studied by measuring experimentally the response to an external scalar source, $\varepsilon$ is merely a *construct* inasmuch as it measures the response to a *screened* potential internal to the system. From a calculational standpoint, however, it is more convenient to deal with $\varepsilon$ rather than its inverse since actual calculations are always done using a self-consistent approach which leads more naturally to $\varepsilon$. Having calculated $\varepsilon$, the desired inverse may be obtained via (7.4) or (7.17).

That $\varepsilon$ rather than its inverse emerges more naturally from a self-consistent calculation can be appreciated by recalling certain features of the Hartree approximation.* We note first that the electron density $\rho^{(0)}(\mathbf{r})$ before the application of $V_{ext}$ can be written

$$\rho^{(0)}(\mathbf{r}) = \sum_i |\psi_i^{(0)}(\mathbf{r})|^2,$$

(7.25)

where the wave function $\psi_i^{(0)}$ satisfies the equation

$$\left[ -\frac{\hbar^2}{2m}\nabla^2 - eV - eV_{sc}^{(0)}(\mathbf{r}) \right]\psi_i^{(0)}(\mathbf{r}) = E_i^{(0)}\psi_i^{(0)}(\mathbf{r}),$$

(7.26)

where $E_i^{(0)}$ denotes the electron energy appropriate to the single particle state $\psi_i^{(0)}$, and $V$ is the potential contribution that arises due to ions, etc., permanently embedded in the system. Here $V_{sc}^{(0)}$ is the self-consistent

---

*The usual method of dealing with a many-body system is via the single-particle approach. The basic techniques were developed by Hartree, Slater, and Fock in connection with atomic theory and are well described in the literature. See, for example, the books by Slater,[10] Seitz,[11] and Pines.[12]

potential and is related to $\rho^{(0)}(\mathbf{r})$ via Poisson's equation*

$$\nabla^2 V_{sc}^{(0)}(\mathbf{r}) = 4\pi e \rho^{(0)}(\mathbf{r}). \tag{7.27}$$

When the external potential is applied, the wave function is modified to

$$\psi_i(\mathbf{r}) \approx \psi_i^{(0)}(\mathbf{r}) + \delta\psi_i(\mathbf{r}), \tag{7.28}$$

the $\psi_i(\mathbf{r})$'s satisfying the modified Hartree equation

$$\left[ -\frac{\hbar^2}{2m}\nabla^2 - eV - eV_{sc}(\mathbf{r}) - eV_{ext} \right]\psi_i(\mathbf{r}) = E_i\psi_i(\mathbf{r}), \tag{7.29}$$

the self-consistent potential now being given by

$$V_{sc}(\mathbf{r}) = V_{sc}^{(0)}(\mathbf{r}) + V_{ind}(\mathbf{r}). \tag{7.30}$$

As before, $V_{ind}$ denotes the potential induced by $V_{ext}$ and satisfies the equation

$$\nabla^2 V_{ind}(\mathbf{r}) = 4\pi e \Delta\rho(\mathbf{r}), \tag{7.31}$$

where $\Delta\rho$ is the (first-order) change in the electron density and is given by

$$\Delta\rho(\mathbf{r}) \approx \sum_i |\psi_i(\mathbf{r})|^2 - \sum_i |\psi_i^{(0)}(\mathbf{r})|^2. \tag{7.32}$$

Using first-order perturbation theory to evaluate $\Delta\rho(\mathbf{r})$, one can solve (7.31) to establish a relation of the form (7.19) (see Eqs. (7.32) and (7.29)). This then leads to expressions for $g(\mathbf{r}, \mathbf{r}')$ and $\varepsilon(\mathbf{r}, \mathbf{r}')$.

Explicit calculation using the single-particle approach and ignoring exchange and correlation effects gives[13]

$$\varepsilon(\mathbf{r}, \mathbf{r}') = \delta(\mathbf{r} - \mathbf{r}') - e^2 \int \frac{d\mathbf{r}''}{|\mathbf{r} - \mathbf{r}''|} \left[ \sum_{\substack{1 \ 2 \\ (1 \ne 2)}} \langle 1(\mathbf{r}''')|\delta(\mathbf{r}'' - \mathbf{r}''')|2(\mathbf{r}''')\rangle \right.$$
$$\left. \times \langle 2(\mathbf{r}''')|\delta(\mathbf{r}' - \mathbf{r}''')|1(\mathbf{r}''')\rangle \frac{f(E_1) - f(E_2)}{E_1 - E_2} \right]. \tag{7.33}$$

$|1\rangle$ and $|2\rangle$ are the *single-particle* wave functions normalized over a volume $\Omega$ and $E_1$ and $E_2$ the energies of the corresponding states. The function $f(E)$ is the familiar Fermi-Dirac distribution function appropriate to 0 K, that is, just the occupation number for the state of energy $E$. The

---

*Strictly speaking, the self-consistent potential in (7.26) must exclude that part of the electron density which is associated with $\psi_i^{(0)}(\mathbf{r})$, that is, we must have

$$\nabla^2 V_{sc}^{(0)}(\mathbf{r}) = 4\pi e \sum_{j(\ne i)} |\psi_j^{(0)}(\mathbf{r})|^2.$$

However, this correction may be ignored when the number of electrons is large.

factor $f(E_1) - f(E_2)$ ensures that the matrix elements appearing in (7.33) connect a filled state to an unfilled state. Comparison with (7.18) and (7.21) shows

$$\chi_{\text{sc}}(\mathbf{r}, \mathbf{r}') = \sum_{\substack{1 \ 2 \\ (1 \neq 2)}} \langle 1(\mathbf{r}'') | \delta(\mathbf{r} - \mathbf{r}'') | 2(\mathbf{r}'') \rangle$$

$$\times \langle 2(\mathbf{r}'') | \delta(\mathbf{r}' - \mathbf{r}'') | 1(\mathbf{r}'') \rangle \frac{f(E_1) - f(E_2)}{E_1 - E_2}. \tag{7.34}$$

The $\mathbf{Q}$-space version of (7.33) is[13]

$$\varepsilon(\mathbf{Q}, \mathbf{Q}') = \delta_{\mathbf{Q}\mathbf{Q}'} - \frac{4\pi e^2}{\Omega Q^2} \sum_{\substack{1 \ 2 \\ (1 \neq 2)}} \langle 1(\mathbf{r}) | e^{-i\mathbf{Q}\cdot\mathbf{r}} | 2(\mathbf{r}) \rangle$$

$$\times \langle 2(\mathbf{r}) | e^{i\mathbf{Q}'\cdot\mathbf{r}} | 1(\mathbf{r}) \rangle \frac{f(E_1) - f(E_2)}{E_1 - E_2}. \tag{7.35}$$

Often this is written as

$$\varepsilon(\mathbf{Q}, \mathbf{Q}') = \delta_{\mathbf{Q}\mathbf{Q}'} - v(\mathbf{Q})\chi_{\text{sc}}(\mathbf{Q}, \mathbf{Q}') \tag{7.36}$$

where

$$v(Q) = \frac{4\pi e^2}{Q^2} \tag{7.37}$$

and

$$\chi_{\text{sc}}(\mathbf{Q}, \mathbf{Q}') = \frac{1}{\Omega} \sum_{\substack{1 \ 2 \\ (1 \neq 2)}} \langle 1(\mathbf{r}) | e^{-i\mathbf{Q}\cdot\mathbf{r}} | 2(\mathbf{r}) \rangle$$

$$\times \langle 2(\mathbf{r}) | e^{i\mathbf{Q}'\cdot\mathbf{r}} | 1(\mathbf{r}) \rangle \frac{f(E_1) - f(E_2)}{E_1 - E_2} \tag{7.38}$$

denotes the Fourier transform of $\chi_{\text{sc}}(\mathbf{r}, \mathbf{r}')$.

As it stands, (7.36) is sufficiently general as to be applicable to molecules, amorphous solids, and crystals. We shall, however, be concerned only with the latter.

Many metals, for example, the alkalis, can be visualized as a collection of ions immersed in a sea of electrons. For purposes of calculating the screening effects of the conduction electrons, one could to a first approximation replace the periodic distribution of the ions by a *uniform* background of positive ions, in which case the problem reduces to that for an electron gas. Several expressions have been derived for the dielectric function of the electron gas, based on different treatments of the electron-electron interactions. In every case, $\varepsilon(\mathbf{r}, \mathbf{r}')$ reduces to a local function

$\varepsilon(|\mathbf{r} - \mathbf{r}'|)$. Correspondingly, $\varepsilon(\mathbf{Q}, \mathbf{Q}')$ is diagonal and a function only of $|\mathbf{Q}|$; it is therefore usually represented simply as $\varepsilon(Q)$. It is worth noting that in this case

$$\varepsilon^{-1}(\mathbf{Q}, \mathbf{Q}') = \delta_{\mathbf{Q}\mathbf{Q}'}\varepsilon^{-1}(\mathbf{Q}, \mathbf{Q}) = \delta_{\mathbf{Q}\mathbf{Q}'}\frac{1}{\varepsilon(Q)}, \tag{7.39}$$

which is very convenient from a practical standpoint.

The most frequently employed expression for $\varepsilon(Q)$ is based on the *self-consistent Hartree* or *random-phase approximation* (RPA) and is given by[12,14]

$$\varepsilon(Q) = 1 - v(Q)\chi_{sc}(Q) \tag{7.40}$$

with

$$\chi_{sc}(Q) = -\frac{3\rho^{(0)}}{2E_F}\left(\frac{1}{2} + \frac{4k_F^2 - Q^2}{8k_FQ}\ln\left|\frac{Q + 2k_F}{Q - 2k_F}\right|\right), \tag{7.41}$$

where $k_F$ and $E_F$ denote the Fermi wave vector and energy and $\rho^{(0)}$ is the average electron density. In the RPA, exchange and correlation do not enter. Furthermore it turns out that $\chi_{sc}(Q)$ is identical with the "unscreened" susceptibility $\chi^0(Q)$ for a *noninteracting* electron gas.[15] For this reason (7.40) is frequently written

$$\varepsilon(Q) = 1 - v(Q)\chi^0(Q) \tag{7.42}$$

with $\chi^0$ also given by (7.41).

Exchange and correlation primarily decrease electron-electron interaction at short distances or, equivalently, reduce $v(Q)$ for large $Q$. Upon including their effects one has[16]

$$\varepsilon_{tt}(Q) = 1 - \frac{v(Q)\chi^0(Q)}{1 + v(Q)G(Q)\chi^0(Q)}, \tag{7.43}$$

where $G(Q)$ is the factor incorporating the effects of exchange and correlation. (In the RPA, $G(Q)$ vanishes). It is also useful to note that[17]

$$\varepsilon_{te}(Q) = 1 - v(Q)[1 - G(Q)]\chi^0(Q) = 1 - v^*(Q)\chi^0(Q), \tag{7.44}$$

whence

$$\varepsilon_{tt}^{-1}(Q) = \frac{1 + v(Q)G(Q)\chi^0(Q)}{\varepsilon_{te}(Q)}, \tag{7.45}$$

a relation which enables us to calculate $\varepsilon_{tt}^{-1}(Q)$ (the quantity required in lattice dynamics) from $\varepsilon_{te}(Q)$ (the quantity frequently calculated for model systems).* Table 7.1 summarizes some interrelations involving various microscopic susceptibilities and dielectric functions encountered in lattice dynamics.

Several expressions for $G(Q)$ have been proposed. One frequently used is that suggested by Hubbard[17] and adopted by Sham[18] in which

$$G(Q) = \frac{1}{2} \frac{Q^2}{Q^2 + k_F^2 + k_S^2},$$ (7.46)

where

$$k_S = \left(\frac{4me^2 k_F}{\pi \hbar^2}\right)^{1/2}$$ (7.47)

is the Thomas-Fermi wave vector.

Recently, Singwi and coworkers[19] have proposed a series of improved versions in which $G(\mathbf{Q})$ is determined self-consistently. The latest of these has

$$G(\mathbf{Q}) = \left(1 + a\rho^{(0)} \frac{\partial}{\partial \rho^{(0)}}\right)\left(-\frac{1}{(2\pi)^3 \rho^{(0)}}\right) \int \frac{\mathbf{Q} \cdot \mathbf{Q}'}{Q'^2} [S(\mathbf{Q} - \mathbf{Q}') - 1] \, d\mathbf{Q}',$$ (7.48)

where $S(\mathbf{Q})$ itself depends on the frequency- and wave-vector-dependent dielectric function $\varepsilon(\mathbf{Q}, \omega)$, and $a$ is a parameter taken as $\frac{2}{3}$; $S(\mathbf{Q})$ is also related to the electron-electron pair correlation function $g(r)$ by

$$g(r) = 1 + \frac{3}{2r} \int_0^\infty \frac{Q}{k_F^3} \sin(Qr)[S(Q) - 1] \, dQ.$$ (7.49)

Other versions of $G(Q)$ are also available.[16,20–24] Dispersion curves based on some of these have been calculated and discussed by Price et al.[25]

Figure 7.2 shows $\varepsilon(Q)$ in the RPA. A noteworthy feature is the infinite slope at $Q = 2k_F$ which results from the logarithmic singularity in the derivative of $\varepsilon(Q)$ (see Eq. (7.41)). As noted in Chapter 6, this singularity is responsible for the Kohn anomaly.

We have already alluded to the fact that the matrix $\varepsilon^{-1}(\mathbf{Q}, \mathbf{Q}')$ (required in lattice dynamics) must be obtained by inverting $\varepsilon(\mathbf{Q}, \mathbf{Q}')$. This inversion can be skirted in a few special cases, such as the alkali metals, via

---

*The quantity $v^*(Q) = v(Q)[1 - G(Q)]$ may be regarded as the effective electron-electron interaction with exchange and correlation corrections. In fact, it is the Fourier transform of the potential $V(\mathbf{r}, \mathbf{r}')$ (see (4.381)), evaluated in the local approximation.

**Table 7.1**
Interrelations between the different types of susceptibilities and dielectric functions.

| Quantity | Electron gas (RPA) | Electron gas (exchange and correlation) | Crystal (exchange and correlation) |
|---|---|---|---|
| $\chi(Q); \chi(\mathbf{Q}, \mathbf{Q}')$ | $\dfrac{\chi_{sc}(Q)}{1 - v(Q)\chi_{sc}(Q)}$ | $\dfrac{\chi_{sc}(Q)}{1 - v^*(Q)\chi_{sc}(Q)}$ | $\sum_{\mathbf{Q}''} \chi_{sc}(\mathbf{Q}, \mathbf{Q}'')[1_\infty - \mathbf{v}^*\boldsymbol{\chi}_{sc}]^{-1}(\mathbf{Q}'', \mathbf{Q}')$ |
| $\chi_{sc}(Q); \chi_{sc}(\mathbf{Q}, \mathbf{Q}')$ | $\dfrac{\chi(Q)}{1 + v(Q)\chi(Q)}$ | $\dfrac{\chi(Q)}{1 + v^*(Q)\chi(Q)}$ | $\sum_{\mathbf{Q}''} \chi(\mathbf{Q}, \mathbf{Q}'')[1_\infty + \mathbf{v}^*\boldsymbol{\chi}]^{-1}(\mathbf{Q}'', \mathbf{Q}')$ |
| $1/\varepsilon_{tt}(Q); \varepsilon_{tt}^{-1}(\mathbf{Q}, \mathbf{Q}')$ | $1 + v(Q)\chi(Q)$ | $1 + v(Q)\chi(Q)$ | $\delta_{\mathbf{Q}\mathbf{Q}'} + v(Q)\chi(\mathbf{Q}, \mathbf{Q}')$ |
| $\varepsilon_{tt}(Q); \varepsilon_{tt}(\mathbf{Q}, \mathbf{Q}')$ | $1 - v(Q)\chi_{sc}(Q)$ | $1 - \dfrac{v(Q)\chi_{sc}(Q)}{1 + [v(Q) - v^*(Q)]\chi_{sc}(Q)}$ | $\delta_{\mathbf{Q}\mathbf{Q}'} - v(Q)\sum_{\mathbf{Q}''}[1_\infty + \boldsymbol{\chi}_{sc}(\mathbf{v} - \mathbf{v}^*)]^{-1}(\mathbf{Q}, \mathbf{Q}'')\chi_{sc}(\mathbf{Q}'', \mathbf{Q}')$ |
| $1/\varepsilon_{te}(Q); \varepsilon_{te}^{-1}(\mathbf{Q}, \mathbf{Q}')$ | $1 + v(Q)\chi(Q)$ | $1 + v^*(Q)\chi(Q)$ | $\delta_{\mathbf{Q}\mathbf{Q}'} + v^*(Q)\chi(\mathbf{Q}, \mathbf{Q}')$ |
| $\varepsilon_{te}(Q); \varepsilon_{te}(\mathbf{Q}, \mathbf{Q}')$ | $1 - v(Q)\chi_{sc}(Q)$ | $1 - v^*(Q)\chi_{sc}(Q)$ | $\delta_{\mathbf{Q}\mathbf{Q}'} - v^*(Q)\chi_{sc}(\mathbf{Q}, \mathbf{Q}')$ |

$$\varepsilon_{tt}^{-1}(Q) - 1 = (1 - G(Q))^{-1}\{\varepsilon_{te}^{-1}(Q) - 1\}$$

$$\varepsilon_{tt}^{-1}(\mathbf{Q}, \mathbf{Q}') - \delta_{\mathbf{Q}\mathbf{Q}'} = (1 - G(Q))^{-1}\{\varepsilon_{te}^{-1}(\mathbf{Q}, \mathbf{Q}') - \delta_{\mathbf{Q}\mathbf{Q}'}\}$$

Notes:
1. As in the case of an electron gas, exchange and correlation effects for electrons in a crystal are usually accommodated within the single-particle approximation by the replacement $v(Q) \to v^*(Q)$ in the expression for $\varepsilon_{te}$.
2. Within the self-consistent approximation, $\chi_{sc}(Q)$ and $\chi_{sc}(\mathbf{Q}, \mathbf{Q}')$ are always given by (7.38) and thus do not depend on exchange and correlation effects.
3. Notation: $\mathbf{v}$ is a matrix with elements $v(Q)\delta_{\mathbf{Q}\mathbf{Q}'}$, $\boldsymbol{\chi}$ has elements $\chi(\mathbf{Q}, \mathbf{Q}')$, and $[\ldots]^{-1}(\mathbf{Q}, \mathbf{Q}')$ denotes the $\mathbf{Q}\mathbf{Q}'$th element of the inverse of the matrix in brackets.

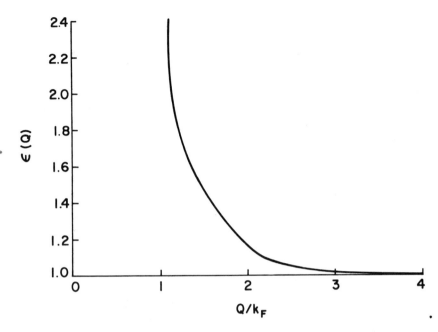

**Figure 7.2**
Dielectric function $\varepsilon(Q)$ computed with parameters appropriate to aluminium. The effects of the logarithmic singularity are barely visible. (After Harrison, ref. 37.)

the use of diagonal screening (see (7.39)) but in general cannot be avoided. Much effort has therefore been devoted to devising a form for $c$ which permits easy inversion.

The basic idea is as follows: Let us write $\varepsilon(\mathbf{Q}, \mathbf{Q'})$ in the form*

$$\varepsilon(\mathbf{q} + \mathbf{G}, \mathbf{q} + \mathbf{G'}) = \delta_{\mathbf{G},\mathbf{G'}} - v(\mathbf{q} + \mathbf{G})\chi_{sc}(\mathbf{q} + \mathbf{G}, \mathbf{q} + \mathbf{G'}). \qquad (7.50)$$

Suppose now that the infinite matrix $\mathbf{v}\chi_{sc}$ can be factored as

$$\mathbf{v}\chi_{sc} = \mathbf{ANB}, \qquad (7.51)$$

where the sizes of the matrices on the right-hand side are, respectively, $(\infty \times p)$, $(p \times p)$, $(p \times \infty)$, $p$ being a conveniently small number. It is then straightforward to show that

$$\varepsilon^{-1} = \mathbb{1}_{\infty} + \mathbf{A}[\mathbb{1}_p - \mathbf{NBA}]^{-1}\mathbf{NB}. \qquad (7.52)$$

---

*Expressions (7.50) and (7.53) imply the RPA. Exchange and correlation effects can be accomodated by the replacement $v(Q) \rightarrow v^*(Q)$, and the resulting dielectric function will be of the type $\varepsilon_{te}$ (see Table 7.1). The inverse function $\varepsilon_{tt}^{-1}$ required in lattice dynamics (see Sec. 7.1.3) can then be obtained as outlined earlier.

Noting that **NBA** is a $(p \times p)$ matrix, we see that the problem of inversion reduces to that of inverting a $(p \times p)$ matrix.

We may inquire as to when a factorization scheme as in (7.51) is possible. Sinha,[26] who proposed the idea first, showed it is workable for an ideal insulator with just two bands, a flat valence band and a flat conduction band. The electron energy $E(\mathbf{k})$ is thus independent of the wave vector $\mathbf{k}$ in both bands. Later Sinha, Gupta, and Price[27] suggested that such a factorization could have wider applicability and that the dielectric function could in general be approximated in the form

$$\varepsilon(\mathbf{Q}, \mathbf{Q}') = \varepsilon_0(\mathbf{Q})\delta_{\mathbf{Q}\mathbf{Q}'} + v(\mathbf{Q}) \sum_{\substack{\alpha\beta \\ ss'}} Q_\alpha Q'_\beta f_s^*(\mathbf{Q}) e^{i\mathbf{Q}\cdot\mathbf{r}_s}$$

$$\times f_{s'}(\mathbf{Q}') e^{-i\mathbf{Q}'\cdot\mathbf{r}_{s'}} a_{\alpha\beta}\!\begin{pmatrix}\mathbf{q}\\ss'\end{pmatrix}, \quad (\mathbf{Q} = \mathbf{q} + \mathbf{G},\ \mathbf{Q}' = \mathbf{q} + \mathbf{G}').$$
$$(7.53)$$

Here $\alpha$, $\beta$ are Cartesian indices and $\mathbf{r}_s$, $\mathbf{r}_{s'}$ run over suitably chosen sites in the primitive cell and can include not only the nuclear positions but others as well. These sites and the functions $\varepsilon_0(\mathbf{Q})$, $f_s(\mathbf{Q})$, and $a_{\alpha\beta}(\begin{smallmatrix}\mathbf{q}\\ss'\end{smallmatrix})$ are chosen to provide the best possible representation for the solid in question. In particular, it is possible by suitable choices to recover from (7.53) the form appropriate to an electron gas on the one hand and an ideal insulator on the other.

It should be noted that, as in (7.51), the second term on the right-hand side of (7.53) has the form of a product of three matrices of dimensionalities $(\infty \times 3p)$, $(3p \times 3p)$, $(3p \times \infty)$, where $p$ is the number of values the index $s$ can take.

Physical meaning may be attributed to the quantities $\mathbf{a}$ and $f_s(\mathbf{Q})$. The former denotes essentially the (electronic) polarizability of the primitive cell and in a loose sense describes the contribution from the "localized" charges. For example, in the case of transition metals this involves the $d$-electron polarization, while in alkali halides $\mathbf{a}$ effectively describes the ionic distortion. Correspondingly, for an electron gas $\mathbf{a}$ vanishes. The function $f_s(\mathbf{Q})$ is related to induced multipoles arising from the distortion and residing at the site $s$.

More recently Hanke and Bilz[28] have argued that the factorization in (7.51) can be obtained by expressing the electronic wave functions in the Wannier representation. The problem needs to be investigated further, especially with regard to constructing suitable Wannier functions appropriate to different types of solids and consistent with their respective band structures. From a practical standpoint, however, one need not necessarily try to build up a dielectric function from basic electronic wave functions. Instead one could adopt the form proposed in (7.53), in which case the problem reduces to that of fixing the various disposable quantities.

### 7.1.2 Some comments on the dielectric function

1. The term "longitudinal dielectric function" (LDF) used earlier was first employed by Lindhard[14] in his discussion of the electron gas, in the context of a net *scalar* potential $V_t$ resulting from a perturbation caused by a *charge-density* fluctuation. Lindhard's definition is based essentially on Eqs. (7.3) and (7.4), which yield for an electron gas the relation

$$\int \varepsilon(|\mathbf{r} - \mathbf{r}'|) V_t(\mathbf{r}') \, d\mathbf{r}' = V_{\text{ext}}(\mathbf{r}). \tag{7.54}$$

When Fourier transformed this gives

$$\varepsilon(Q) V_t(Q) = V_{\text{ext}}(Q), \tag{7.55}$$

which may be taken as the defining equation for $\varepsilon(Q)$. However, this is completely equivalent to defining $\varepsilon(Q)$ via the equation[29]

$$\varepsilon(Q) \mathbf{E}(Q) = \mathbf{D}(Q), \tag{7.56}$$

where $\mathbf{D}$ and $\mathbf{E}$ are *longitudinal* fields given by

$$\mathbf{D}(Q) = -i\mathbf{Q} V_{\text{ext}}(Q), \qquad \mathbf{E}(Q) = -i\mathbf{Q} V_t(Q).$$

Equation (7.56) is, of course, a natural generalization of Maxwell's constitutive equation $D = \varepsilon E$, which explains the rationale for terming $\varepsilon(Q)$ as well as $\varepsilon(\mathbf{r}, \mathbf{r}')$ and its Fourier transform as LDF.

2. We next wish to comment on the so-called *local-field corrections* (LFC). The local-field concept was originally introduced by Lorentz in his study of the polarization of dielectrics. He noted that the field acting on an ion in a dielectric placed in a homogeneous external electric field is not given simply by the macroscopic field but has an additional contribution arising from the (point) dipoles induced on other ions (see Sec. A2.3). In the theory of dielectrics, this additional contribution to the field is often referred to as the LFC and is taken into consideration while calculating the polarization and the dielectric constant.

The same jargon is employed in a similar spirit in the derivation of the microscopic dielectric function and relates to the "graininess" in the induced charge density. Consider first the case of electrons in a *crystal*. The charge distribution varies on a microscopic scale, consistent with the symmetry of the crystal. If now we apply a slowly varying potential of the

form $V_{ext}(\mathbf{q})e^{i\mathbf{q}\cdot\mathbf{r}}$, $(\mathbf{q} \to 0)$, then because of the periodicity of the charge
distribution the induced charge not only has the Fourier component
$\Delta\rho(\mathbf{q})$ but also components $\Delta\rho(\mathbf{q} + \mathbf{G})$.[30,31] In fact,

$$\Delta\rho(\mathbf{q} + \mathbf{G}) = - e\chi(\mathbf{q} + \mathbf{G}, \mathbf{q})V_{ext}(\mathbf{q})$$

(see (7.10)). While $\Delta\rho(\mathbf{q})$ may be identified as the macroscopic part of the
induced density, the components $\Delta\rho(\mathbf{q} + \mathbf{G})$, being of shorter wavelength,
vary more rapidly, indicative of greater localization. In this sense they are
analogous to the induced dipoles of the Lorentz problem. One may therefore, in a certain sense, distinguish between the effective field and the
average field and infer appropriate LFC. Observe that the appearance
of Fourier components $\Delta\rho(\mathbf{q} + \mathbf{G})$ due to a driving potential of wave
vector $\mathbf{q}$ implies that the dielectric function is a matrix with nonvanishing
off-diagonal elements. The appearance of the off-diagonal elements in $\varepsilon$
is thus a manifestation of LFC. It is important to note that this type of
LFC is a consequence of the periodic nature of the charge distribution and
would arise even in a RPA calculation for a crystal.

Another kind of LFC often considered has a different origin but may,
broadly speaking, be viewed as similar in spirit to the Lorentz case. To
simplify matters let us consider an electron gas. Here, the short-range correlations due to exchange and other many-body effects (as typified in the
correlation function $g(r)$) result in a contribution to the electric field
beyond that calculated in the Hartree approximation or RPA.[19] The
consequent modification to the dielectric function is also sometimes termed
as LFC. The $G(Q)$ in (7.43) and (7.44) is a manifestation of LFC in the
present sense. This type of LFC does not appear in the RPA.

3. We have as yet not discussed the precise relationship between the
microscopic dielectric function $\varepsilon(\mathbf{q} + \mathbf{G}, \mathbf{q} + \mathbf{G}')$ and the macroscopic
tensors $\boldsymbol{\varepsilon}(0)$ and $\boldsymbol{\varepsilon}(\infty)$ familiar to us from Chapter 4. That some connection must exist was apparent even from Lindhard's early work. The
details, however, have become clear only recently.

Now the microscopic dielectric function introduced in this chapter
describes basically the electronic polarization. We therefore expect some
connection with $\boldsymbol{\varepsilon}(\infty)$. Indeed the relationship is[5]

$$\lim_{\mathbf{q} \to 0} \frac{1}{\varepsilon_{tt}^{-1}(\mathbf{q}, \mathbf{q})} = \sum_{\alpha\beta} \hat{q}_\alpha \varepsilon_{\alpha\beta}(\infty)\hat{q}_\beta = \varepsilon_\infty(\hat{\mathbf{q}}). \tag{7.57}$$

Observe that it is $\varepsilon_{tt}$ which is related to the macroscopic dielectric constant. This is to be expected, for the definition of $\boldsymbol{\varepsilon}(\infty)$ as given in classical
electricity implies that both the source and the probe are distinguishable
charges.

Analogous to (7.57) we also have[5]

$$\sum_{\alpha\beta} \hat{q}_\alpha \varepsilon_{\alpha\beta}(0)\hat{q}_\beta = \varepsilon_0(\hat{\mathbf{q}})$$

$$= \lim_{q\to 0}\left[\frac{1}{\varepsilon_{tt}^{-1}(\mathbf{q},\mathbf{q})} + \frac{4\pi}{v}\sum_{\alpha\beta\gamma\delta}\sum_{kk'}\hat{q}_\alpha Z_{\alpha\beta}(k|0)\,[\overline{\mathbf{M}}(0)]_{\beta\gamma}^{-1}\,(kk')\right.$$

$$\left. \times\, Z_{\delta\gamma}(k'|0)\hat{q}_\delta\right],\tag{7.58}$$

where the second term on the right-hand side refers to the contribution from nuclear displacements, the notation being the same as in Chapter 4 except that the various quantities now have a microscopic basis, as will become clear later. Equations (7.57) and (7.58) are used to derive the LST relation from microscopic theory.

### 7.1.3 Formal aspects of microscopic theory, including the role of the dielectric function

Having reviewed at some length the nature and importance of the dielectric function in describing electron response, we are in a position to appreciate its role in microscopic lattice dynamical theory. Let us view the crystal as a periodic array of nuclei embedded in a system of electrons. In the adiabatic approximation, nuclear motions can be regarded as controlled by a net internuclear interaction consisting of (1) direct Coulomb interaction between the nuclei and (2) interaction brought about by the intervention of the electrons. The latter depends on the manner in which the electrons respond to nuclear motions.

The contribution from (1) to the force constant is given by $\phi_{\alpha\beta}^C\!\left(\begin{smallmatrix} l & l' \\ k & k' \end{smallmatrix}\right)$ defined in (4.24–4.25) with $Z_k e$ and $Z_{k'}e$ now denoting the *nuclear* charges. The contribution from (2) can be obtained by calculating the electronic energy $E_0(R)$ defined in Eq. (2.3). For an arbitrary nuclear configuration $R$, the energy is in general different from the energy $E_0(X)$ appropriate to the equilibrium configuration. If the nuclear displacements are small, $E_0(R)$ may be expressed as a power series in the displacements. The leading term of this series is $E_0(X)$; the first-order term can be shown to vanish, and for the harmonic theory it is sufficient to consider just the second-order term $E_0^{(2)}$. The latter can be written as[32] (see also (4.379))

$$E_0^{(2)} = \int \rho^{(0)}(\mathbf{r})V_{ne}^{(2)}(\mathbf{r},R)\,d\mathbf{r} + \tfrac{1}{2}\int \Delta\rho(\mathbf{r})V_{ne}^{(1)}(\mathbf{r},R)\,d\mathbf{r},\tag{7.59}$$

where $\rho^{(0)}$ represents the electron density in the undistorted crystal and $\Delta\rho$ is the first-order change brought about by nuclear motion; $V_{ne}$ denotes the electron-nucleus potential energy, the superscript indicating the order

of the nuclear displacements. The term involving $\rho^{(0)}$ guarantees translational invariance, while the term containing $\Delta\rho$ describes the effects of
electron density changes consequent to nuclear motion. As already seen,
density changes can be discussed via a dielectric formulation, and indeed
it is at this point that microscopic theory puts all the crystal types on a
common footing. (However, it must be noted that effects associated with
density changes can be treated in other ways, as we shall see later.)

Let us express $E_0^{(2)}$ formally as

$$E_0^{(2)} = \tfrac{1}{2} \sum_{lk\alpha} \sum_{l'k'\beta} \phi_{\alpha\beta}^{\text{el}}\binom{l\ l'}{k\ k'} u_\alpha\binom{l}{k} u_\beta\binom{l'}{k'}. \tag{7.60}$$

This shows that the electronic contribution $\phi_{\alpha\beta}^{\text{el}}$ becomes defined when
$E_0^{(2)}$ is calculated. Explicit evaluation using the apparatus of many-body
theory gives[5,33]

$$\phi_{\alpha\beta}^{\text{el}}\binom{l\ l'}{k\ k'} = \left[ \frac{\partial^2}{\partial r_\alpha\binom{l}{k}\partial r_\beta\binom{l'}{k'}} \int d\mathbf{r}\, d\mathbf{r}'\, \frac{Z_k e^2}{|\mathbf{r} - \mathbf{r}\binom{l}{k}|} \chi(\mathbf{r},\mathbf{r}') \frac{Z_{k'}e^2}{|\mathbf{r}' - \mathbf{r}\binom{l'}{k'}|} \right]_0,$$
$$\binom{l}{k} \neq \binom{l'}{k'}; \tag{7.61}$$

$$\phi_{\alpha\beta}^{\text{el}}\binom{l\ l}{k\ k} = \left[ \frac{\partial^{2\,'}}{\partial r_\alpha\binom{l}{k}\partial r_\beta\binom{l}{k}} \left\{ \int d\mathbf{r}\, \rho^{(0)}(\mathbf{r}) \frac{Z_k e^2}{|\mathbf{r} - \mathbf{r}\binom{l}{k}|} \right.\right.$$
$$\left.\left. + \int d\mathbf{r}\, d\mathbf{r}'\, \frac{Z_k e^2}{|\mathbf{r} - \mathbf{r}\binom{l}{k}|} \chi(\mathbf{r},\mathbf{r}') \frac{Z_k e^2}{|\mathbf{r}' - \mathbf{r}\binom{l}{k}|} \right\} \right]_0. \tag{7.62}$$

These constants satisfy the sum rule (2.35). By combining $\phi_{\alpha\beta}^{C}$ with $\phi_{\alpha\beta}^{\text{el}}$ it
can be shown that[5]

$$\phi_{\alpha\beta}\binom{l\ l'}{k\ k'} = \left[ \frac{\partial^2}{\partial r_\alpha\binom{l}{k}\partial r_\beta\binom{l'}{k'}} \int d\mathbf{r}\, \varepsilon_{\text{tt}}^{-1}\left(\mathbf{r}\binom{l}{k},\mathbf{r}\right) \frac{Z_k Z_{k'}e^2}{|\mathbf{r} - \mathbf{r}\binom{l'}{k'}|} \right]_0,$$
$$\binom{l}{k} \neq \binom{l'}{k'}. \tag{7.63}$$

Comparing the integrand with the familiar result $(Z_1 Z_2 e^2/r\varepsilon)$ for screened
interactions in electrostatics, we see that screening enters the present
problem very similarly.

The physical reason for the appearance above of $\varepsilon_{\text{tt}}^{-1}$ is not difficult to
see. Now the interaction between the nuclei $\binom{l}{k}$ and $\binom{l'}{k'}$ through the
electrons arises because the movement of $\binom{l'}{k'}$ produces a potential that is
the analogue of $V_{\text{ext}}$ considered earlier. The effect of this external potential
is then sampled by $\binom{l}{k}$, which acts like a test charge. The nature of $\varepsilon_{\text{tt}}^{-1}$ is
sufficiently comprehensive to include in $\phi_{\alpha\beta}$ many-body and volume-
dependent contributions.

Based on (7.61–7.63) the dynamical matrix can be manipulated to the form[5,33]

$$M_{\alpha\beta}\begin{pmatrix} \mathbf{q} \\ kk' \end{pmatrix} = \hat{M}_{\alpha\beta}\begin{pmatrix} \mathbf{q} \\ kk' \end{pmatrix} - \delta_{kk'} \sum_{k''} \hat{M}_{\alpha\beta}\begin{pmatrix} 0 \\ kk'' \end{pmatrix}, \tag{7.64}$$

where

$$\hat{M}_{\alpha\beta}\begin{pmatrix} \mathbf{q} \\ kk' \end{pmatrix} = \frac{v}{4\pi e^2} \sum_{\mathbf{G}} \sum_{\mathbf{G}'} \frac{(\mathbf{q} + \mathbf{G})_\alpha (\mathbf{q} + \mathbf{G}')_\beta}{|\mathbf{q} + \mathbf{G}'|^2} \left( \frac{4\pi Z_k e^2}{|\mathbf{q} + \mathbf{G}|^2 v} \right)$$

$$\times |\mathbf{q} + \mathbf{G}|^2 e^{i\mathbf{G}\cdot\mathbf{x}(k)} \varepsilon^{-1}(\mathbf{q} + \mathbf{G}, \mathbf{q} + \mathbf{G}') \left( \frac{4\pi Z_{k'} e^2}{|\mathbf{q} + \mathbf{G}'|^2 v} \right)$$

$$\times |\mathbf{q} + \mathbf{G}'|^2 e^{-i\mathbf{G}'\cdot\mathbf{x}(k')}, \tag{7.65}$$

and

$$\hat{M}_{\alpha\beta}\begin{pmatrix} 0 \\ kk'' \end{pmatrix} = \sum_{\mathbf{G}(\neq 0)} \sum_{\mathbf{G}'(\neq 0)} \frac{G_\alpha G_\beta'}{|\mathbf{G}'|^2} \frac{4\pi e^2}{v} Z_k Z_{k''} \varepsilon^{-1}(\mathbf{G}, \mathbf{G}') e^{i\mathbf{G}\cdot\mathbf{x}(k)} e^{-i\mathbf{G}'\cdot\mathbf{x}(k'')}.* \tag{7.66}$$

Sometimes it is preferable to write $\hat{M}_{\alpha\beta}$ explicitly as the sum of direct nuclear and electronic parts:

$$\hat{M}_{\alpha\beta} = \hat{M}_{\alpha\beta}^C + \hat{M}_{\alpha\beta}^{el} \tag{7.67}$$

with

$$\hat{M}_{\alpha\beta}^C\begin{pmatrix} \mathbf{q} \\ kk' \end{pmatrix} = \frac{4\pi e^2}{v} Z_k Z_{k'} \sum_{\mathbf{G}} \frac{(\mathbf{q} + \mathbf{G})_\alpha (\mathbf{q} + \mathbf{G})_\beta}{|\mathbf{q} + \mathbf{G}|^2} e^{i\mathbf{G}\cdot[\mathbf{x}(k) - \mathbf{x}(k')]}, \tag{7.68}$$

*The exclusions in (7.66) result from the overall charge neutrality of the system, which enables one to set $v(\mathbf{Q} \equiv 0) = 0$, leading in turn to the result

$$\varepsilon^{-1}(0, \mathbf{G}) = \varepsilon^{-1}(\mathbf{G}, 0) = 0 \text{ for } G \neq 0.$$

Consider a two-component charge distribution with densities $\rho_+(\mathbf{r})$ and $\rho_-(\mathbf{r})$ and having overall neutrality. The interaction energy is

$$\tfrac{1}{2} \int [\rho_+(\mathbf{r}) - \rho_-(\mathbf{r})] v(\mathbf{r}, \mathbf{r}') [\rho_+(\mathbf{r}') - \rho_-(\mathbf{r}')] \, d\mathbf{r}' \, d\mathbf{r},$$

its $\mathbf{Q} = 0$ Fourier component being

$$\tfrac{1}{2} [\rho_+(\mathbf{Q} = 0) - \rho_-(\mathbf{Q} = 0)] v(\mathbf{Q} = 0) [\rho_+(\mathbf{Q} = 0) - \rho_-(\mathbf{Q} = 0)].$$

Since

$$\rho_+(\mathbf{Q} = 0) - \rho_-(\mathbf{Q} = 0) = \int [\rho_+(\mathbf{r}) - \rho_-(\mathbf{r})] \, d\mathbf{r},$$

it follows immediately that the $\mathbf{Q} \equiv 0$ Fourier component of the interaction energy vanishes identically. We may recover this feature formally by following the convention usually adopted[5,33] for electrically neutral systems:

$$v(Q) = 0, \qquad Q = 0$$
$$= \frac{4\pi e^2}{Q^2}, \qquad Q \neq 0.$$

and

$$
\begin{aligned}
\hat{M}^{\text{el}}_{\alpha\beta}\!\left(\begin{matrix}\mathbf{q}\\kk'\end{matrix}\right) &= \frac{v}{4\pi e^2}\sum_{\mathbf{G}}\sum_{\mathbf{G}'}\frac{(\mathbf{q}+\mathbf{G})_\alpha(\mathbf{q}+\mathbf{G}')_\beta}{|\mathbf{q}+\mathbf{G}'|^2}\left(\frac{4\pi Z_k e^2}{|\mathbf{q}+\mathbf{G}|^2 v}\right)\\
&\quad\times |\mathbf{q}+\mathbf{G}|^2 e^{i\mathbf{G}\cdot\mathbf{x}(k)}[\varepsilon^{-1}(\mathbf{q}+\mathbf{G},\mathbf{q}+\mathbf{G}')-\delta_{\mathbf{GG}'}]\\
&\quad\times\left(\frac{4\pi Z_{k'}e^2}{|\mathbf{q}+\mathbf{G}'|^2 v}\right)|\mathbf{q}+\mathbf{G}'|^2 e^{-i\mathbf{G}'\cdot\mathbf{x}(k')}.
\end{aligned}
\tag{7.69}
$$

Similar expressions may be written for $\hat{M}^{\text{C}}_{\alpha\beta}(\begin{smallmatrix}0\\kk''\end{smallmatrix})$ and $\hat{M}^{\text{el}}_{\alpha\beta}(\begin{smallmatrix}0\\kk''\end{smallmatrix})$. It may be observed that the result (7.68) for the Coulombic lattice sum is in agreement with that given in Eq. (A2.21), provided we choose the parameter $\eta$ occurring there such that only the reciprocal-space series $\mathscr{B}_{\alpha\beta}$ survives.

As defined above, $M_{\alpha\beta}$ in general has a part that is regular at $\mathbf{q} = 0$ and a part that is not. For some purposes it is convenient to display these explicitly.[5,33] Let us write (7.65) as

$$
\begin{aligned}
\hat{M}_{\alpha\beta}\!\left(\begin{matrix}\mathbf{q}\\kk'\end{matrix}\right) &= \frac{4\pi e^2}{v}Z_k Z_{k'}\Bigg\{\frac{q_\alpha q_\beta}{q^2}\,\varepsilon^{-1}(\mathbf{q},\mathbf{q})\\
&\quad +\sum_{\mathbf{G}(\neq 0)}\frac{(\mathbf{q}+\mathbf{G})_\alpha q_\beta}{|\mathbf{q}|^2}\,\varepsilon^{-1}(\mathbf{q}+\mathbf{G},\mathbf{q})e^{i\mathbf{G}\cdot\mathbf{x}(k)}\\
&\quad +\sum_{\mathbf{G}'(\neq 0)}\frac{q_\alpha(\mathbf{q}+\mathbf{G}')_\beta}{|\mathbf{q}+\mathbf{G}'|^2}\,\varepsilon^{-1}(\mathbf{q},\mathbf{q}+\mathbf{G}')e^{-i\mathbf{G}'\cdot\mathbf{x}(k')}\\
&\quad +\sum_{\mathbf{G}(\neq 0)}\sum_{\mathbf{G}'(\neq 0)}\frac{(\mathbf{q}+\mathbf{G})_\alpha(\mathbf{q}+\mathbf{G}')_\beta}{|\mathbf{q}+\mathbf{G}'|^2}[\varepsilon^{-1}(\mathbf{q}+\mathbf{G},\mathbf{q}+\mathbf{G}')\\
&\quad -{}^{(r)}\varepsilon^{-1}(\mathbf{q}+\mathbf{G},\mathbf{q}+\mathbf{G}')]e^{i\mathbf{G}\cdot\mathbf{x}(k)}e^{-i\mathbf{G}'\cdot\mathbf{x}(k')}\Bigg\}\\
&\quad +\frac{4\pi e^2}{v}Z_k Z_{k'}\sum_{\mathbf{G}(\neq 0)}\sum_{\mathbf{G}'(\neq 0)}\frac{(\mathbf{q}+\mathbf{G})_\alpha(\mathbf{q}+\mathbf{G}')_\beta}{|\mathbf{q}+\mathbf{G}'|^2}\\
&\quad\times{}^{(r)}\varepsilon^{-1}(\mathbf{q}+\mathbf{G},\mathbf{q}+\mathbf{G}')e^{i\mathbf{G}\cdot\mathbf{x}(k)}e^{-i\mathbf{G}'\cdot\mathbf{x}(k')}.
\end{aligned}
\tag{7.70}
$$

Here ${}^{(r)}\varepsilon(\mathbf{q}+\mathbf{G},\mathbf{q}+\mathbf{G}')$ is the submatrix of $\varepsilon$ with both $\mathbf{G}$ and $\mathbf{G}' \neq 0$, and ${}^{(r)}\varepsilon^{-1}(\mathbf{q}+\mathbf{G},\mathbf{q}+\mathbf{G}')$ denotes the inverse of this submatrix. It is to be noted that ($\mathbf{G}, \mathbf{G}'$ both understood to be nonvanishing)

$$
\varepsilon^{-1}(\mathbf{q}+\mathbf{G},\mathbf{q}+\mathbf{G}') \neq {}^{(r)}\varepsilon^{-1}(\mathbf{q}+\mathbf{G},\mathbf{q}+\mathbf{G}');
$$

in fact,[5]

$$
\begin{aligned}
\varepsilon^{-1}(\mathbf{q}+\mathbf{G},\mathbf{q}+\mathbf{G}') &- {}^{(r)}\varepsilon^{-1}(\mathbf{q}+\mathbf{G},\mathbf{q}+\mathbf{G}')\\
&= \varepsilon^{-1}(\mathbf{q}+\mathbf{G},\mathbf{q})\frac{1}{\varepsilon^{-1}(\mathbf{q},\mathbf{q})}\varepsilon^{-1}(\mathbf{q},\mathbf{q}+\mathbf{G}').
\end{aligned}
\tag{7.71}
$$

Using (7.71) in (7.70) and rearranging, we may eventually write

$$
M_{\alpha\beta}\begin{pmatrix} \mathbf{q} \\ kk' \end{pmatrix} = \frac{4\pi e^2}{v} Z_k Z_{k'} \left[ S_{\alpha\beta}\begin{pmatrix} \mathbf{q} \\ kk' \end{pmatrix} - \delta_{kk'} \sum_{k''} S_{\alpha\beta}\begin{pmatrix} 0 \\ kk'' \end{pmatrix} \right]
$$
$$
+ \frac{4\pi}{v} \varepsilon^{-1}(\mathbf{q}, \mathbf{q}) \left[ \sum_{\gamma} \hat{q}_{\gamma} Z_{\gamma\alpha}^{*}(k|\mathbf{q}) \right] \left[ \sum_{\delta} \hat{q}_{\delta} Z_{\delta\beta}(k'|\mathbf{q}) \right], \qquad (7.72)
$$

a form that is convenient for discussing possible nonregular behavior of the dynamical matrix of insulators at $\mathbf{q} = 0$. The quantity $\mathbf{Z}(k|\mathbf{q})$ above is defined via the relation

$$
\sum_{\delta} \hat{q}_{\delta} Z_{\delta\beta}(k'|\mathbf{q}) = \sum_{\mathbf{G}} Z_{k'} e \frac{|\mathbf{q}|(\mathbf{q} + \mathbf{G})_{\beta}}{|\mathbf{q} + \mathbf{G}|^2} \frac{\varepsilon^{-1}(\mathbf{q}, \mathbf{q} + \mathbf{G})}{\varepsilon^{-1}(\mathbf{q}, \mathbf{q})} e^{-i\mathbf{G}\cdot\mathbf{x}(k')}, \qquad (7.73)
$$

while

$$
S_{\alpha\beta}\begin{pmatrix} \mathbf{q} \\ kk' \end{pmatrix} = \sum_{\mathbf{G}(\neq 0)} \sum_{\mathbf{G}'(\neq 0)} \frac{(\mathbf{q} + \mathbf{G})_{\alpha}(\mathbf{q} + \mathbf{G}')_{\beta}}{|\mathbf{q} + \mathbf{G}'|^2} {}^{(r)}\varepsilon^{-1}(\mathbf{q} + \mathbf{G}, \mathbf{q} + \mathbf{G}')
$$
$$
\times e^{i[\mathbf{G}\cdot\mathbf{x}(k) - \mathbf{G}'\cdot\mathbf{x}(k')]}. \qquad (7.74)
$$

Equation (7.72) applies to all types of solids. In every case, electronic response has a role to play via the inverse dielectric function $\varepsilon^{-1}$, the structure of which depends on the band structure and the electronic wave functions of the solid in question. Indeed, by an examination of the basic properties corresponding to various types of solids, the following conclusions may be drawn.

**Insulators.** In this case,[5,33]

$$
\lim_{\mathbf{q}\to 0} \frac{1}{\varepsilon^{-1}(\mathbf{q}, \mathbf{q})} = \hat{\mathbf{q}}^{tr} \mathbf{B} \hat{\mathbf{q}}, \qquad (7.75a)
$$

$$
\lim_{\mathbf{q}\to 0} \frac{\varepsilon^{-1}(\mathbf{q} + \mathbf{G}, \mathbf{q})}{\varepsilon^{-1}(\mathbf{q}, \mathbf{q})} = -\frac{4\pi e^2}{v} |\mathbf{q}| \hat{\mathbf{q}}^{tr} \mathbf{A}^{*}(\mathbf{G}) |\mathbf{G}|^{-2}, \qquad (7.75b)
$$

$$
\lim_{\mathbf{q}\to 0} \frac{\varepsilon^{-1}(\mathbf{q}, \mathbf{q} + \mathbf{G}')}{\varepsilon^{-1}(\mathbf{q}, \mathbf{q})} = -\frac{4\pi e^2}{v} \frac{1}{|\mathbf{q}|} \hat{\mathbf{q}}^{tr} \mathbf{A}(\mathbf{G}'), \qquad (7.75c)
$$

where the vector $\mathbf{A}$ and the tensor $\mathbf{B}$ are both independent of $\mathbf{q}$. Using these results it can be shown[5] that for insulators the last term on the right-hand side of (7.72) does in general not vanish in the $\mathbf{q} \to 0$ limit, leading to nonregular behavior. In fact this term produces all the effects discussed phenomenologically in Sec. 4.3 in terms of the macroscopic electric field.

Microscopic theory also provides an expression for the apparent charge tensor $\mathbf{Z}(k|\mathbf{q})$ via (7.73). In particular $Z_{\alpha\beta}(k|0)$ can be written as the sum of the nuclear charge and the charge contributed by the electrons dragged along with the nucleus.[5,33,34] From (7.73) we have

$$\lim_{\mathbf{q}\to 0} \sum_{\delta} \hat{q}_{\delta} Z_{\delta\beta}(k|\mathbf{q}) \;=\; \sum_{\delta} \hat{q}_{\delta} Z_{\delta\beta}(k|0)$$

$$= \sum_{\delta} \hat{q}_{\delta}\delta_{\delta\beta} Z_{k} e \;-\; \sum_{\delta} Z_{k} e \sum_{\mathbf{G}(\neq 0)} \frac{4\pi e^{2}}{v} \frac{G_{\beta}}{|\mathbf{G}|^{2}} \hat{q}_{\delta} A_{\delta}(\mathbf{G}) e^{-i\mathbf{G}\cdot\mathbf{x}(k)}$$

(using (7.75)), which gives

$$Z_{\alpha\beta}(k|0) \;=\; Z_{k} e\delta_{\alpha\beta} \;-\; Z_{k} e \sum_{\mathbf{G}(\neq 0)} \frac{4\pi e^{2}}{v} \frac{A_{\alpha}(\mathbf{G}) G_{\beta}}{G^{2}} e^{-i\mathbf{G}\cdot\mathbf{x}(k)}. \qquad (7.76)$$

This electronic dressing of the nuclear charge is determined by the response function for the electrons and is contributed by all the electrons. In phenomenological theory, on the other hand, we suppose that an entity termed the ion moves with a bare charge $Z_{k}e$ which subsequently acquires a moderate dressing due to the distortions of the valence electron cloud. The phenomenological description is contained in (7.76) and could, in principle, be recovered by writing the second term as the sum of core and valence electron contributions. Upon combining the former with the nuclear charge, the apparent charge could be expressed as the sum of the nominal ionic charge plus a valence deformation contribution. (Several approximations are involved, however.)

Microscopic theory is further able to *prove* that

$$\sum_{k} Z_{\alpha\beta}(k|0) = 0, \qquad (7.77)$$

a result insisted upon in the phenomenological theory (see (4.82)) in order to guarantee that an infinitesimal translation produces no macroscopic polarization. The proof of this "apparent charge neutrality" has been demonstrated[5,33] using the overall electrical neutrality of the primitive cell (in terms of the nuclei and the electrons contained therein) and the properties noted in (7.75), thus providing greater substance and justification to the phenomenological constraint.

The charge neutrality condition (7.77) can be reexpressed using (7.73). Let $\mathbf{q} \to 0$ along a particular direction $\hat{\mathbf{q}}$. Then from (7.73) and (7.77) we have

$$\sum_{\alpha} \hat{q}_{\alpha} \sum_{k} Z_{\alpha\beta}(k|0) = 0,$$

implying

$$\lim_{\mathbf{q}\to 0} \sum_k Z_k \left[ \sum_{\mathbf{G}} \frac{|\mathbf{q}|(\mathbf{q}+\mathbf{G})_\beta}{|\mathbf{q}+\mathbf{G}|^2} \varepsilon^{-1}(\mathbf{q}, \mathbf{q}+\mathbf{G}) e^{-i\mathbf{G}\cdot\mathbf{x}(k)} \right] = 0. \qquad (7.78)$$

This result, known as the *acoustic sum rule*,[5,13,33,34] applies to all insulators. It is a nontrivial requirement on the inverse dielectric function of any insulator, because in general every term in the sum over **G** in (7.78) is finite. The sum rule can be satisfied only if there is an appropriate cancellation between the diagonal and off-diagonal terms of $\varepsilon^{-1}$. It follows that, in insulators, $\varepsilon$ and its inverse are *never* diagonal matrices. The practical implications are that in numerical calculations one must contend with the inversion problem (this is where the ansatz (7.51) helps) and that (7.78) places a constraint on the off-diagonal terms. If this constraint is violated by any computed or model $\varepsilon$, then the $\mathbf{q} = 0$ acoustic modes acquire nonvanishing frequencies. In other words, violation of (7.78) implies *incomplete screening associated with the long-wavelength acoustic modes*, an unphysical situation.*

Microscopic theory is also able to deliver the generalized LST relation by exploiting the properties of the dielectric function, in particular Eqs. (7.57) and (7.58).[5]

**Metals and alloys.** In contrast to insulators, here[5]

$$\lim_{\mathbf{q}\to 0} \varepsilon^{-1}(\mathbf{q}, \mathbf{q}) \sim q^2 \qquad (7.79a)$$

$$\lim_{\mathbf{q}\to 0} \varepsilon^{-1}(\mathbf{q}+\mathbf{G}, \mathbf{q}) \sim q^2. \qquad (7.79b)$$

All the other elements of $\varepsilon^{-1}$ tend to a finite limit as $\mathbf{q} \to 0$. As a result the dynamical matrix does not have any nonregular terms, which is physically satisfying (see (7.70)). Also there is no constraint on the off-diagonal terms of $\varepsilon^{-1}$ as in (7.78). Indeed, if circumstances warrant one may even ignore the off-diagonal terms.

### 7.1.4 Theories adapted for practical calculations
The formal theory just reviewed is quite impressive in that it is able to account for the differences observed in the vibration spectra of different

---

*Recall Figure 7.1, which displays a similar example of lack of screening leading to a finite frequency for the $\mathbf{q} = 0$ LA mode. In that case, however, the correct limiting behavior could be obtained with an $\varepsilon$ appropriate to an electron gas, for which off-diagonal elements are not required. It may be noted that incomplete screening can occur for some long-wavelength *optic* modes in insulators.

types of solids. However, it is hardly suited for making practical calculations. Apart from the conceptual discomfort of having to treat the inner
shell and valence electrons on an equal footing, the theory involves obtaining $E_0^{(2)}$ (see (7.59)) by calculating large quantities which cancel anyway. Physically, this cancellation arises because the bulk of $\Delta\rho$ is contributed not by a genuine polarization effect but by the rigid displacement
with the nuclei of the tightly bound part of the electronic charge. To see
this more explicitly, let us formally divide $\rho^{(0)}$ into two parts, one rigidly
carried by the nuclei and the other capable of deformation during nuclear
motions:

$$\rho^{(0)} = \rho^{(0)}(\text{nondef}) + \rho^{(0)}(\text{def}) \tag{7.80}$$

One expects that the rigid part will have contributions from all the inner
shell electrons and perhaps a part of the valence/conduction electrons. The
deformation part, on the other hand, will be associated entirely with
electrons of the latter category. Corresponding to (7.80) one also has

$$\Delta\rho = \Delta\rho(\text{nondef}) + \Delta\rho(\text{def}). \tag{7.81}$$

We next write $\rho^{(0)}(\text{nondef})$ and $V_{\text{ne}}$ as sums of contributions associated
with the individual unit cells, assuming, for simplicity, a Bravais lattice.

$$\rho^{(0)}(\text{nondef}) = \sum_l \sigma^{(0)}(\mathbf{r} - \mathbf{r}(l)), \tag{7.82a}$$

$$V_{\text{ne}}(\mathbf{r}, R) = \sum_l U(\mathbf{r} - \mathbf{r}(l)). \tag{7.82b}$$

This gives

$$
\begin{aligned}
\Delta\rho(\text{nondef}) &= \sum_{l\alpha} \left.\frac{\partial\sigma^{(0)}(\mathbf{r} - \mathbf{r}(l))}{\partial r_\alpha(l)}\right|_0 u_\alpha(l) \\
&= -\sum_{l\alpha} \frac{\partial\sigma^{(0)}(\mathbf{r} - \mathbf{x}(l))}{\partial r_\alpha} u_\alpha(l),
\end{aligned}
\tag{7.83a}
$$

$$
\begin{aligned}
V_{\text{ne}}^{(1)}(\mathbf{r}, R) &= \sum_{l\alpha} \left.\frac{\partial U(\mathbf{r} - \mathbf{r}(l))}{\partial r_\alpha(l)}\right|_0 u_\alpha(l) \\
&= -\sum_{l\alpha} \frac{\partial U(\mathbf{r} - \mathbf{x}(l))}{\partial r_\alpha} u_\alpha(l),
\end{aligned}
\tag{7.83b}
$$

$$V_{\text{ne}}^{(2)}(\mathbf{r}, R) = \tfrac{1}{2}\sum_{l\alpha\beta} \frac{\partial^2 U(\mathbf{r} - \mathbf{x}(l))}{\partial r_\alpha \partial r_\beta} u_\alpha(l) u_\beta(l). \tag{7.83c}$$

The energy $E_0^{(2)}$ may now be written as

$$\tfrac{1}{2} \sum_{l\alpha} \sum_{l'\beta} u_\alpha(l) u_\beta(l') \underbrace{\int d\mathbf{r} \left[ \frac{\partial}{\partial r_\alpha} \sigma^{(0)}(\mathbf{r} - \mathbf{x}(l)) \frac{\partial}{\partial r_\beta} U(\mathbf{r} - \mathbf{x}(l')) \right]}_{A}$$

$$+ \tfrac{1}{2} \sum_{l\alpha} \sum_{l'\beta} u_\alpha(l) u_\beta(l) \underbrace{\int d\mathbf{r}\, \sigma^{(0)}(\mathbf{r} - \mathbf{x}(l')) \frac{\partial^2 U(\mathbf{r} - \mathbf{x}(l))}{\partial r_\alpha \partial r_\beta}}_{B}$$

+ terms involving the deformable part of
the charge density. (7.84)

Consider the integral

$$\int d\mathbf{r} \left[ \frac{\partial}{\partial r_\alpha} \sigma^{(0)}(\mathbf{r} - \mathbf{x}(l)) \frac{\partial}{\partial r_\beta} U(\mathbf{r} - \mathbf{x}(l')) \right].$$

Bearing in mind that $\sigma^{(0)}$ is bounded, we obtain on integrating by parts

$$- \int d\mathbf{r} \left[ \sigma^{(0)}(\mathbf{r} - \mathbf{x}(l)) \frac{\partial^2 U(\mathbf{r} - \mathbf{x}(l'))}{\partial r_\alpha \partial r_\beta} \right].$$

Therefore the part

$$\tfrac{1}{2} \sum_{l\alpha\beta} u_\alpha(l) u_\beta(l) \int d\mathbf{r} \left[ \frac{\partial}{\partial r_\alpha} \sigma^{(0)}(\mathbf{r} - \mathbf{x}(l)) \frac{\partial}{\partial r_\beta} U(\mathbf{r} - \mathbf{x}(l)) \right]$$

of $A$ exactly cancels the part

$$\tfrac{1}{2} \sum_{l\alpha\beta} u_\alpha(l) u_\beta(l) \int d\mathbf{r}\, \sigma^{(0)}(\mathbf{r} - \mathbf{x}(l)) \frac{\partial^2 U(\mathbf{r} - \mathbf{x}(l))}{\partial r_\alpha \partial r_\beta}$$

of $B$. When this cancellation is explicitly taken into account, the harmonic part of the energy can be manipulated[35,36] (making suitable approximations) into a sum of (see (4.380)) $\Phi_{ii}^{(2)}$ and

$$E_0^{(2)} = \int \rho^{(0)}(\mathbf{r}; \text{def})\, V_{ie}^{(2)}(\mathbf{r}, R)\, d\mathbf{r} + \tfrac{1}{2} \int \Delta\rho(\mathbf{r}; \text{def}) V_{ie}^{(1)}(\mathbf{r}, R)\, d\mathbf{r}, \quad (7.85)$$

where $V_{ie}$ is the potential presented to the "deformable part" of the electron charge cloud by a composite unit we term "ion." As in (7.82b) we may write

$$V_{ie}(\mathbf{r}, R) = \sum_l U^P(\mathbf{r} - \mathbf{r}(l)), \quad (7.86)$$

where $U^P$ is now the "effective ion-electron potential." The latter has the virtue of being weak compared to $V_{en}$, greatly facilitating the calculation of $\Delta\rho$ by self-consistent perturbation theory. In the case of metals and semiconductors, the effective ion-electron potential can be taken to be the same as the bare, local pseudopotential used extensively in band-structure calculations.[37] For ionic crystals, however, such an identification is not possible since pseudopotentials do not exist (in the same sense as they do for simple metals and semiconductors), although from a lattice-dynamical point of view one can meaningfully talk of a weak ion-electron potential.[36]

The problem of calculating (7.85) now reduces to a specification of $U^P$ and the response of the electronic medium to ionic motion. Naturally the manner in which these problems are handled depends on the type of solid, as discussed below.

### 7.1.5 Microscopic theories for metals

Starting with the pioneering work of Toya,[38] by far the largest number of first-principles calculations have been reported for metals, particularly the NFE metals. In such metals, one may to a good approximation replace the conduction electrons by an electron gas and use the explicit results for $\varepsilon(Q)$ discussed earlier. The only remaining problem is that of describing the ion-electron interaction. (The ion cores are usually treated as electronically rigid and nonoverlapping.) Here many avenues are available,[2,18,35,37,39,40] the most popular of which is the local-pseudopotential method.[18,37]

In the local-pseudopotential approach, the dynamical matrix splits into two parts. The first of these, $\hat{M}^C_{\alpha\beta}$, represents the contribution from the Coulomb interaction between the ions and is given as before by (7.68) with $Z_k e$ now representing the ionic charge. The electronic contribution is given by[19]*

$$\hat{M}^{el}_{\alpha\beta}\binom{\mathbf{q}}{kk'} = \frac{v}{4\pi e^2}\sum_{\mathbf{G}}\sum_{\mathbf{G}'}\frac{(\mathbf{q}+\mathbf{G})_\alpha(\mathbf{q}+\mathbf{G}')_\beta}{|\mathbf{q}+\mathbf{G}'|^2}\,U^P_k(|\mathbf{q}+\mathbf{G}|)$$
$$\times\,|\mathbf{q}+\mathbf{G}|^2 e^{i\mathbf{G}\cdot\mathbf{x}(k)}U^P_{k'}(|\mathbf{q}+\mathbf{G}'|)|\mathbf{q}+\mathbf{G}'|^2 e^{-i\mathbf{G}'\cdot\mathbf{x}(k')}$$
$$\times\,[\varepsilon_{tt}^{-1}(\mathbf{q}+\mathbf{G},\mathbf{q}+\mathbf{G}')-\delta_{\mathbf{G}\mathbf{G}'}], \tag{7.87}$$

where[37]

$$U^P_k(Q) = \langle\mathbf{k}+\mathbf{Q}|U^P_k(r)|\mathbf{k}\rangle = \frac{1}{v}\int e^{-i(\mathbf{k}+\mathbf{Q})\cdot\mathbf{r}}U^P_k(r)e^{i\mathbf{k}\cdot\mathbf{r}}\,d\mathbf{r}$$
$$= \frac{1}{v}\int e^{-i\mathbf{Q}\cdot\mathbf{r}}U^P_k(r)\,d\mathbf{r}.$$

*Equation (7.87) is in fact applicable to all situations for which Eq. (7.85) is valid.

Comparison with the earlier expression (7.69) of the formal theory shows that the structures of the two expressions are similar, with $U_k^P(Q)$ replacing $(-4\pi Z_k e^2/vQ^2)$ and $\varepsilon_{tt}$ now referring to the response due to the conduction electrons alone. In the diagonal approximation for $\varepsilon_{tt}$, (7.87) reduces to

$$
\hat{M}_{\alpha\beta}^{el}\binom{\mathbf{q}}{kk'} = \frac{v}{4\pi e^2} \sum_{\mathbf{G}} (\mathbf{q} + \mathbf{G})_\alpha (\mathbf{q} + \mathbf{G})_\beta |\mathbf{q} + \mathbf{G}|^2
$$
$$
\times\, U_k^P(|\mathbf{q} + \mathbf{G}|) U_{k'}^P(|\mathbf{q} + \mathbf{G}|)
$$
$$
\times\, e^{i\mathbf{G}\cdot[\mathbf{x}(k) - \mathbf{x}(k')]} \left( \frac{1}{\varepsilon_{tt}(|\mathbf{q} + \mathbf{G}|)} - 1 \right), \tag{7.88}
$$

a formula extensively used for the calculation of phonon spectra of alkali and hcp metals.

The pseudopotential $U^P(Q)$ may be obtained in several ways. A direct method is to compute it using proper wave functions for the core electrons.[18,41] A second method is to follow the approach pioneered by Heine, Abarenkov, and Animalu,[6,42] who assume a simple form for the potential $U^P(r)$ appropriate to an isolated ion in free space. The model parameters are adjusted to reproduce the observed spectroscopic levels, and the potential is then carried over for solid-state calculations with appropriate modifications where necessary. The philosophy is to produce a definite $U^P(Q)$ function for each *type* of ion which can be used in a variety of calculations. These $U^P(Q)$'s in effect determine the ion-electron scattering amplitude, using which the scattering effects for any *assembly* of ions may be discussed by taking the *geometry* of the assembly into account. Extensive tables[37]* of $U^P(Q)$ are available and have been used for the calculation of phonon dispersion curves, among other things.

Other models for the pseudopotential are available. Prominent among these are the following.

Bardeen's model:[39]

$$
U^P(Q) = \frac{1}{v}\left[ -\frac{4\pi Z e^2}{Q^2} + \alpha \right] g(Q r_s). \tag{7.89}
$$

Here $Z$ is the valence, $r_s$ the radius of the Wigner-Seitz sphere, and $\alpha$ an adjustable parameter. The function $g(x)$ is given by

$$
g(x) = 3(\sin x - x \cos x)/x^3. \tag{7.90}
$$

---

*The tables, mainly due to Animalu, are reproduced by Harrison[37] but contain the *screened* pseudopotential, $U^P(Q)/\varepsilon_{te}(Q)$.

Ashcroft's model: [43]

$$U^P(Q) = -\frac{4\pi Z e^2}{vQ^2} \cos Q r_c, \tag{7.91}$$

$r_c$ being an adjustable parameter.

Harrison[37]/Schneider-Stoll model: [44]

$$U^P(Q) = -\frac{4\pi Z e^2}{vQ^2} + \sum_{i=1,2,\cdots} \frac{\beta_i (Q r_i)^{i-1}}{[1 + (Q r_i)^2]^{i+1}}, \tag{7.92}$$

$\beta_i$ and $r_i$ being parameters.

These models are based on physical considerations and the parameters are determined by fitting to experimental data such as Fermi surface results. Once the pseudopotential has been determined thus, it can be used in (7.87–7.88) for computing dispersion curves. In some cases the parameters are determined from the measured phonon dispersion curves themselves, in which case the analysis assumes the same character as force-constant fitting, except that the fitting is done in terms of parameters which probably have better microscopic significance. Indeed, one may invert the procedure and use the phonon-spectrum-determined $U^P(Q)$ for computing such properties as the Fermi surface and the electrical resistivity.[41,45] Such a program is appealing since it establishes an intimate link between the vibrational and the electronic properties of metals.

In retrospect, the pseudopotential method may be looked upon as equivalent to using an interatomic potential $V(r)$ of the following form: [37]

$$\begin{aligned} V(r) &= V^C(r) - V^{el}(r) \\ &= \frac{Z^2 e^2}{r} - \frac{1}{(2\pi)^3} \int \left(\frac{Q^2 v^2}{4\pi e^2}\right)[1 - \varepsilon_{tt}^{-1}(Q)](U^P(Q))^2 e^{i\mathbf{Q}\cdot\mathbf{r}}\, d\mathbf{Q} \quad (7.93a) \\ &= \frac{1}{(2\pi)^3} \int e^{i\mathbf{Q}\cdot\mathbf{r}}\, d\mathbf{Q} \left\{\frac{4\pi Z^2 e^2}{Q^2} - \frac{v^2 Q^2}{4\pi e^2}[1 - \varepsilon_{tt}^{-1}(Q)](U^P(Q))^2\right\}. \end{aligned}$$
$$\tag{7.93b}$$

The first term in (7.93a) gives rise to the contribution $\hat{M}_{\alpha\beta}^C$, the second to $\hat{M}_{\alpha\beta}^{el}$ as in (7.87–7.88). Using one of the various parameterized forms for $U^P(Q)$ given in (7.89–7.92), one essentially has a parameterized form for the Fourier transform $V(Q)$ of the two-body potential:

$$V(Q) = \frac{4\pi Z^2 e^2}{Q^2} - \frac{v^2 Q^2}{4\pi e^2}[1 - \varepsilon_{tt}^{-1}(Q)](U^P(Q))^2 \tag{7.94}$$

(see Sec. 1.2).

As previously pointed out, two-body potentials of this type do not completely exhaust the crystal potential $\Phi$, since there are also volume-dependent forces. Use of the local pseudopotential approach (and, by implication, of (7.93)) is thus equivalent to employing an axially symmetric force-constant model. (See also Sec. 6.2.2.)

In contrast to those for the NFE metals, the calculation of the phonon spectra of the noble and transition metals is difficult, because here the distinction between the core and conduction electrons is not clear-cut for the $d$ electrons. Some allowance must be made for the $d$-electron contribution to the polarization while calculating the dielectric function, and herein lies the problem; not surprisingly very few calculations have been reported. One of the first is due to Prakash and Joshi[46] who employed the diagonal screening approximation and obtained the dispersion curves for paramagnetic nickel. Hanke and Bilz[28] have attempted to go further by using the form (7.50–7.51) for the dielectric function, assuming that the $3d$ functions are localized while the $4s$ functions are not. The calculations compare favorably with experiment.

A significant achievement of the microscopic theory is the successful explanation of the Kohn effect discussed in Chapter 6. This effect could not have been foreseen on the basis of early parameterized force-constant models.

### 7.1.6 Microscopic theories for insulators

There are several versions of the microscopic theory for insulators, differing mainly in the description of electronic polarization. Not all of these employ the dielectric formulation, and historically, in fact, theories of this type were proposed much ahead of those for metals. Irrespective of the treatment used, one generally encounters, as compared to metals, an order-of-magnitude increase in complexity arising mainly from the highly localized nature of the induced moments. In the language of the dielectric formulation, this springs from the need to deal with the off-diagonal elements of $\varepsilon^{-1}$, consistent with the acoustic sum rule (7.78).

**Theories not using the dielectric formulation.** The first attempt to include the effects of electronic polarization in lattice dynamics starting from basic electronic wave functions was made by Tolpygo[47] soon after Szigeti[48] and Odelevski[49] pointed out the limitations of the rigid-ion model. Tolpygo started by considering an isolated pair of closely situated ions and wrote an approximate wave funcion $\Psi$ as an antisymmetric product of $\psi$ and $\phi$, the wave functions of the individual ions. The latter are different from the ground state functions $\psi_0$ and $\phi_0$ appropriate to the isolated ions, the difference arising from electronic polarization. Mathematically this is described by expanding the perturbed functions $\psi$ and $\phi$ in terms of complete sets of respective atomic orbitals. The expectation value of the Hamiltonian for the ion pair is then minimized with respect to the expansion coefficients for specified internuclear separation, electric fields, and dipole moments on the ions. This result is then used to write down the potential energy for the crystal in terms of the nuclear displacements and induced dipole moments.

A somewhat similar procedure was followed in the case of diamond-type crystals.[50,51] The crystal wave function $\Psi$ is first expressed as a Slater determinant of symmetrized functions $\psi_{pq}$ built up from $\sigma$ bonding orbitals on each bond $pq$. In the undistorted crystal, the bonding orbitals arise from the usual $sp^3$ hybridization characteristic of tetrahedral bonds.[52] In a crystal distorted because of vibrations, the bonding orbitals are a mixture of $sp^3$ hybrids of the perfect crystal plus the excited atomic orbitals. Thus, as in alkali halides, admixture from excited states is used to describe electronic polarization. The coefficients of mixing are then determined so that the electronic energy of the distorted configuration is a minimum, subject to the condition that the dipole moments of the atoms have prescribed values. Eventually this leads to a potential function similar in form to that obtained for alkali halides (see also (4.321)).

As already noted (Sec. 4.6.5), Cochran[53] pointed out some deficiencies in the earlier version of this theory.[50] Also, Kaplan[54] has criticized Tolpygo's theory as it involves the need to evaluate infinite sums of integrals over atomic functions. He has suggested that this convergence difficulty could be overcome by using the Wannier representation.

Numerical calculations based on the Tolpygo-Mashkevich theory are never made by explicitly evaluating all microscopic quantities. Rather, the final equations are parameterized, at which point the theory becomes essentially phenomenological (see Sec. 4.6). Nevertheless, credit for visualizing the appearance of vibration-induced dipoles in insulators via wavefunction distortion goes to the Soviet school. The relationship of their work, in particular that of Mashkevich,[55] to the shell model was later established by Cowley.[56]

Cowley assumed that the (electronic) contribution to the potential $\Phi$ is a function not only of nuclear displacements but also of the induced dipole moments (and possibly the higher moments). These moments are determined from the first-order change in the electronic wave function, and their equations of motion are deduced assuming that their conjugate force is zero. This is equivalent to supposing that the induced dipoles have no inertia, a way of incorporating the adiabatic assumption.

The resulting equations of motion can then be formally identified with the usual shell equations and, in conjunction with the corresponding equations for the ion cores, provide the microscopic counterparts of the phenomenological equations of motion.

A somewhat different approach was taken by Lundqvist[57] but specifically restricted to NaCl-type crystals. Following Löwdin,[58] Lundqvist used the Heitler-London scheme together with several approximations to show that the crystal potential $\Phi$ can be expressed in the form given in (4.351). The basic assumption underlying the derivation of this result is that the electronic wave functions are *rigidly attached to the nuclei and move with the latter without deformation*. An important feature of the potential function so deduced is the three-body term (see the third term on the right-hand side of (4.351)) whose origin was explained by Lundqvist as follows: Let $\rho(\mathbf{r})$ represent the electron density in the crystal. This is given by

$$\rho(\mathbf{r}) = \sum_{lk} \rho_{lk}(\mathbf{r}) + \tfrac{1}{2} \sum_{lk} \sum_{\substack{l'k' \\ (lk \neq l'k')}} \Delta\rho(\mathbf{r}; lk, l'k')$$

$$= \sum_{lk} \rho_{lk}(\mathbf{r}) + \Delta\rho_{\text{overlap}}(\mathbf{r}). \tag{7.95}$$

Here $\rho_{lk}(\mathbf{r})$ is the electron density at $\mathbf{r}$ in a free ion located at $\mathbf{r}\binom{l}{k}$. It is to be noted that the total electron density is not merely the sum of those contributed by the individual ions but has an additional contribution $\Delta\rho_{\text{overlap}}$. The latter arises because of the antisymmetry requirement on the wave function and causes a charge depletion in the region of the overlap. As noted earlier (Sec. 4.6) this overlap charge is a function of internuclear separation and gives rise to the three-body term in (4.351) through its interactions with the ions. Using this approach, Lundqvist first demonstrated that in alkali halides (1) the Cauchy relations are violated, and (2) the Szigeti charge $e_S^*$ is different from $e$. Electronic polarization is completely disregarded since a rigid displacement of electronic wave functions during nuclear motions is assumed. He then sought to rectify this lacuna by incorporating the polarization effects via an effective field as in the Lyddane-Herzfeld model.[59] The resulting model is thus a hybrid of quantum and classical parts and hence, as acknowledged by Lundqvist himself, does not strictly fall in the category of a microscopic theory.

Lundqvist's work has been carried further by Gliss et al.[60] These authors represent the total electronic wave function by a product of single-particle wave functions associated with particular ions in real space (Wannier representation). The single-particle wave function in the distorted configuration is taken to consist of the rigidly displaced equilibrium function plus a deformation part described in terms of admixture from the excited states (as in Tolpygo's theory). Thus electronic polarization is built into the model from the beginning. The admixture coefficients are determined by a variational procedure leading finally to

$$\Phi^{(2)} = \Phi^{(2)}(\text{nondef}) + \Phi^{(2)}(\text{def}), \tag{7.96}$$

where the two terms on the right-hand side denote the contributions arising from the rigid displacement of wave functions and their distortion. In particular, these authors show that the Löwdin-Lundqvist result may be recovered, taking the atomic orbitals to be linear combinations of certain nonorthogonal functions centered around the displaced ions. This causes $\Phi^{(2)}(\text{def})$ to vanish, leaving only the first term which, like (4.351), has a pair part and a three-body part.

The dynamical equations derived from (7.96) should correspond roughly to those of the shell model. The exact correspondence is, however, difficult to establish as the polarization is not explicitly described in terms of a dipole density. No detailed numerical results using this formalism have been reported except for a preliminary note.[61] Its only other use is by Zeyher,[62] who ignored deformation and essentially rederived some of Lundqvist's results in an improved approximation. A hybrid model combining a quantum-mechanical treatment of three-body effects and a shell-model description of polarization effects has also been constructed by Zeyher.

**Theories using the dielectric formulation.** The first calculation using the dielectric formulation is due to Martin[63,64] who reported results for Si. A notable feature of this work is that it circumvented the problem of dealing explicitly with the off-diagonal elements by an interesting ansatz concerning the nature of the screening.

Martin's work may be understood most simply by considering Eq. (7.93) with specific reference to Si. Suppose we use in that equation a dielectric function appropriate to a semiconductor rather than for an electron gas, assuming for the moment that the semiconductor also can be characterized by a scalar function $\varepsilon(Q)$. A typical example of such a function, computed by Srinivasan[65] assuming that Si can be treated as an isotropic semiconductor, is shown in Figure 7.3; also shown is the RPA result for an electron gas. While for large values of $Q$ the two functions are alike, significant differences exist for small $Q$ values. In particular, as $Q \to 0$ Srinivasan's dielectric function tends to a finite value ($\sim 12$) which is in fact the value of the macroscopic dielectric constant $\varepsilon(0)$ for Si. When this dielectric function is used in (7.93) in place of the RPA result, one

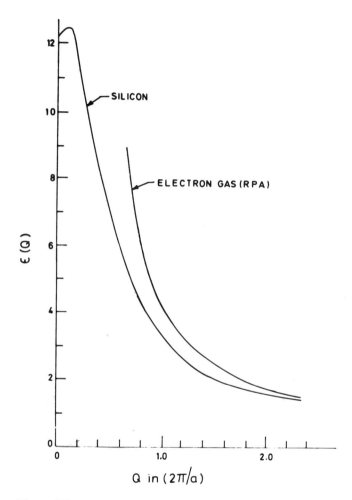

**Figure 7.3**
$\varepsilon(Q)$, the diagonal part of the dielectric matrix, for Si in the isotropic semiconductor approximation. Also shown for comparison is the corresponding function for an electron gas. (After Srinivasan, ref. 65.)

obtains a net two-body potential that does not rapidly die out to zero. Instead the potential behaves asymptotically as $Z^2 e^2/r\varepsilon(0)$, which clearly has a long tail. This, however, is contrary to our intuitive expectations which suggest that since every Si atom in the crystal is effectively neutral, one must have a completely screened interaction for large $r$ (as in a metal), especially as there are no macroscopic field effects associated with the $q = 0$ optic modes. The resolution of this paradox lies in the fact that $\varepsilon(Q)$ used in (7.93) does not characterize the full screening. What is missed can be described rigorously only by using the full dielectric matrix.* In more descriptive (though somewhat loose) language, only part of the electrons participate in the screening described by $\varepsilon(Q)$.

We may ask whether we can continue using $\varepsilon(Q)$ but change the description of $V(r)$ as given in (7.93) so that something like a neutral pseudoatom emerges. It turns out we can, provided we modify the picture as follows. We observe first that the crystal is built up of Si cores each of charge $4e$ (the valency of Si is taken as 4) arranged on a diamond-type lattice and linked by tetrahedrally coordinated covalent bonds. Since each bond is shared by two atoms, there are effectively two bonds per atom, each with a net charge of $-2e$. We now consider a cluster consisting of a silicon core and two point charges each of value $-2e$ located at the mid-points of the two bonds assigned to the core concerned. The negative charges residing on the bonds are referred to as *bond charges*. This cluster is now imagined to be introduced into the medium accomplishing the screening, and the net potential presented by it is sampled with a set of test charges which constitute an identical cluster. One may then imagine the effective Si-Si interaction to consist of the four interactions illustrated by Figure 7.4. Contribution $V_a$ is evaluated as in (7.93) and represents the counterpart of the metallic result. The additional contributions $V_b$, $V_c$, and $V_d$ compensate for incomplete screening in $V_a$. Martin estimated these by assuming that, as far as these interactions are concerned, the screened core appears as a point charge $[4e/\varepsilon(0)]$, while the screened bond charge appears to have a value $[-2e/\varepsilon(0)]$. In other words,

$$V_b = V_d = -\frac{8e^2}{r\varepsilon(0)} \quad \text{and} \quad V_c = \frac{4e^2}{r\varepsilon(0)}. \tag{7.97}$$

The effective Si-Si interaction $V = (V_a + V_b + V_c + V_d)$ is no longer a scalar function of the distance between the silicon cores (compare with (7.93)). It is in fact noncentral, as necessary for the diamond structure to be stable.[66]

---

*Use of a scalar function $\varepsilon(Q)$ implies $\varepsilon^{-1}(\mathbf{Q}, \mathbf{Q}') = \delta_{QQ'}\varepsilon^{-1}(\mathbf{Q}, \mathbf{Q})$ (see (7.39)).

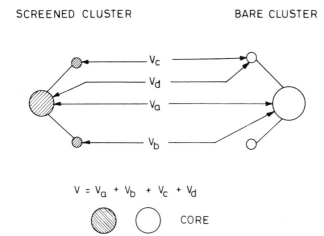

**Figure 7.4**
Contributions to the effective Si-Si interaction in Martin's model. The screened cluster represents the "external source" charges with effects of screening superposed. The bare cluster corresponds to the "probe." The interaction between the clusters gives the net Si–Si interaction.

The dynamical matrix may be calculated on the basis of this effective potential. The first of the resulting three terms arises from the Coulomb interaction between the Si cores and is given by (7.68) with $Z_k e$ denoting the core charge. The second delivers the electronic contribution associated with core-core interactions and is given by (7.88) with $\varepsilon(Q)$ appropriate to a semiconductor. These two terms together represent the contribution from $V_a$. The last term in the dynamical matrix represents the contributions from $V_b$, $V_c$, and $V_d$. In computing this term, the bond charges are assumed always to remain at the centers of their respective bonds (see (6.47)). Comparing with (7.87), we see that the role of the potentials $V_b$, $V_c$, and $V_d$ is to replace the effects of the off-diagonal elements of the dielectric matrix.

Martin's model has certain shortcomings which have been pointed out by Sinha.[67] Nevertheless it has the virtue of explaining at a microscopic level the success of the phenomenological bond-charge model (see Sec. 6.4.3).

Martin's model has also been applied to $MgO$[68] where, too, covalent effects are important.

Sinha and coworkers[27,36,67,69] have gone a step further by employing the full dielectric matrix in their treatment of the dynamics of insulating crystals. In the simplest version, the crystal is regarded as consisting of "ions" plus valence electrons, the latter in a sense distributed throughout the crystal. The ions are assumed to move without distortion and to interact with the valence electrons via a weak local potential. It is further supposed that electronic polarization is characterized entirely by valence electron readjustments to ion motions. The electronic contribution to the harmonic energy is then given by (7.85), with the Fourier component $\Delta\rho(\mathbf{q} + \mathbf{G})$ of $\Delta\rho(\mathbf{r})$ fulfilling the self-consistency equation (considering for simplicity a Bravais lattice)[26,70]

$$
\begin{aligned}
\Delta\rho(\mathbf{q} + \mathbf{G}) = \sum_{\mathbf{G}'} [&\Delta\rho(\mathbf{q} + \mathbf{G}')v^*(\mathbf{q} + \mathbf{G}') \\
&+ iN\{(\mathbf{q} + \mathbf{G}') \cdot \mathbf{U}(\mathbf{q})\}U^{\mathrm{P}}(\mathbf{q} + \mathbf{G}')]\chi_{\mathrm{sc}}(\mathbf{q} + \mathbf{G}, \mathbf{q} + \mathbf{G}'),
\end{aligned}
\tag{7.98}
$$

where $\mathbf{U}(\mathbf{q})$ is the displacement vector for wave vector $\mathbf{q}$. (This expression is the microscopic counterpart of (4.381) now expressed in $\mathbf{q}$ space).

Recognizing the overall success of the shell model in describing the dynamics of insulating crystals and the fact that the elimination of the induced dipole moments from the final equations involves only the inversion of a finite matrix (see the term $[\boldsymbol{\zeta} + \mathbf{Y}_{\mathrm{d}}\mathbf{C}\mathbf{Y}_{\mathrm{d}}]^{-1}$ in (4.292)), Sinha[26] argued that if microscopic theory is to deliver the above feature of the shell model, then $\chi_{\mathrm{sc}}$ must have the "factorized form"

$$
\begin{aligned}
\chi_{\mathrm{sc}}(\mathbf{q} + \mathbf{G}, \mathbf{q} + \mathbf{G}') = -\sum_{\alpha\beta} &(\mathbf{q} + \mathbf{G})_\alpha f(\mathbf{q} + \mathbf{G}) \\
&\times [a_{\alpha\beta}(\mathbf{q})](\mathbf{q} + \mathbf{G}')_\beta f^*(\mathbf{q} + \mathbf{G}')
\end{aligned}
\tag{7.99}
$$

which, as noted earlier, permits an easy inversion of $\varepsilon(\mathbf{Q}, \mathbf{Q}')$. Later Sinha, Gupta, and Price[27] suggested that the form (7.53), which incorporates the factorization ansatz (7.99), could be "adapted to calculate the electron response for almost any solid."

It is interesting to compare (7.99) with

$$
\chi_{\mathrm{sc}}(\mathbf{q} + \mathbf{G}, \mathbf{q} + \mathbf{G}') = -\frac{1}{e^2 v}\sum_{\alpha\beta} (\mathbf{q} + \mathbf{G})_\alpha [\alpha_{\alpha\beta}](\mathbf{q} + \mathbf{G}')_\beta,
\tag{7.100}
$$

which is the Fourier transform of the model function $\chi_{\mathrm{sc}}(\mathbf{r}, \mathbf{r}')$ in (4.387) (specialized to a Bravais lattice). We see that even from a purely phenomenological viewpoint the shell model does imply a dielectric susceptibility of the factorized form.

The significance of the factorization ansatz becomes clearer from the following arguments. Recall that the polarization density $\mathbf{P}(\mathbf{r})$ is related to the deformation charge density via

$$\mathbf{V} \cdot \mathbf{P}(\mathbf{r}) = 4\pi e \Delta \rho(\mathbf{r}).$$

Given a $\Delta\rho$, say from (7.98), a unique solution for $\mathbf{P}(\mathbf{r})$ is in general not possible. However, the specific form (7.99) leads to a solution[26]

$$\mathbf{P}(\mathbf{q} + \mathbf{G}) = \text{const} \times \mathbf{W}(\mathbf{q}) f(\mathbf{q} + \mathbf{G}) \qquad (\mathbf{W}(\mathbf{q}) \to \text{dipole amplitude}),$$

which is a generalization of the familiar shell model result

$$\mathbf{P}(\mathbf{Q}) = \text{const} \times \mathbf{W}(\mathbf{Q}).$$

This explains from a microscopic point of view how the shell model is able not only to describe lattice dynamics (which depends only on $\Delta\rho$), but also the dielectric properties (which depend on $\mathbf{P}$).

Numerical calculations such as those of Sinha and coworkers are made using (7.87) and the corresponding Coulomb matrix, in conjunction with the form (7.53) for $\varepsilon$. The quantities to be specified are the potentials $U_k^P(Q)$, the form factors $f_s(Q)$ and the polarization matrix $\mathbf{a}\binom{\mathbf{q}}{ss'}$. The latter may be written as the Fourier series

$$a_{\alpha\beta}\binom{\mathbf{q}}{ss'} = \frac{1}{e^2 v} \sum_{l'} \alpha_{\alpha\beta}\binom{0 \ l'}{s \ s'} \exp\left[i\mathbf{q}\cdot\mathbf{x}(l')\right], \qquad (7.101)$$

where $\alpha_{\alpha\beta}\binom{0 \ l'}{s \ s'}$ denotes the polarization parameters in real space. It is interesting that, in contrast to the conventional shell model, the above version permits cross polarizabilities $\alpha_{\alpha\beta}\binom{0 \ l'}{s \ s'}$, $\binom{0}{s} \neq \binom{l'}{s'}$.[67] Physically, this implies that the polarization at a given site $\binom{0}{s}$ could involve the excited states associated with a different site $\binom{l'}{s'}$ (see also Figure 7.5). Of course, such a possibility arises only if the electron orbitals are highly delocalized as in semiconductors.

By suitably parameterizing the various quantities mentioned above, dispersion curves have been calculated for Si[27], Ge[69], $\alpha$-Sn[71], KCl, and KBr.[36] In the case of Si, Sinha et al.[27] have also shown how the bond-charge model of Martin can be recovered by appropriate choices for the quantities entering in $\varepsilon$. It may be noted that at the level of calculations the theory becomes essentially phenomenological, although at a more basic level than in usual Born–von Kármán-type calculations.

## 7.2 Dynamics of quantum crystals

### 7.2.1 Nonapplicability of the Born–von Kármán approach

Among the rare-gas solids, helium occupies a very special place by virtue of its small atomic mass and the weakness of the interatomic interaction.

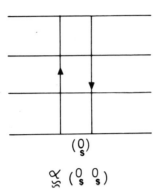

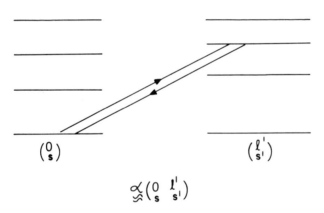

**Figure 7.5**
Schematic illustration of the virtual electronic transitions involved in the "self" polariz-
ability $\boldsymbol{a}\!\left(\begin{smallmatrix} 0 & 0 \\ s & s \end{smallmatrix}\right)$ and the "cross" polarizability $\boldsymbol{a}\!\left(\begin{smallmatrix} 0 & l' \\ s & s' \end{smallmatrix}\right)$.

As a result, its dynamics have several peculiar features among which are—
1. large-amplitude motions including those associated with zero-point energy and
2. correlated motion of nearest neighbors arising from hard-core part the of the LJ potential.
These are quantum effects and must be handled by special techniques.[72] Some of these techniques have wider applicability, and therefore the helium problem is of more than local interest.

The large-amplitude motions not only make the conventional Taylor series (2.19) unsuitable, but also do not permit us to ignore the zero-point effects. Furthermore, on account of their large excursions, neighboring atoms in the lattice would seem to have a good chance of "bumping" into each other. In practice they manage to avoid each other by suitably correlating their motions. This effect, known as *short-range correlations*, is related to the hard-core part of the two-body potential and is not allowed for in the conventional treatment of dynamics. In short, on account of the above features, the Born–von Kármán approach is unsuited for helium, and the dynamics must be handled quantum mechanically right from the start.* However, it turns out that the final result can in fact be reinterpreted in the Born–von Kármán language.

In this section, we present a brief summary of developments concerning quantum crystals and indicate their relevance to other situations.

### 7.2.2 Nature of quantum effects in helium

It is useful to understand first the origin and the role of quantum effects in helium. We begin with Figure 7.6 which displays the LJ potential for the entire rare-gas family (see also Sec. 1.1). It is at once evident that in helium the attraction is rather weak. Indeed, this weak binding allows the kinetic-energy effects to dominate, bringing in their wake the short-range correlations. To appreciate this better, let us consider solid argon at $p = 0$ and $T = 0$ K. The structure is fcc, with every atom having 12 nearest neighbors. The nearest-neighbor distance is almost the same as the distance $r_0$ of the potential minimum (see Table 1.1). Thus every atom sees a well depth

$$U_0 \approx Z V(r_0), \tag{7.102}$$

where $Z$ is the number of first neighbors and $V(r_0) = -\varepsilon$ is the depth parameter of the LJ potential. In the vicinity of $r_0$, the spatial variation of

---

*An additional factor favoring a quantum treatment is that it enables one to take account of statistics, that is, of the fact that He[4] is a Bose particle while He[3] is a Fermi particle. It emerges, however, that statistics can be ignored as far as lattice dynamics is concerned, although not necessarily for other solid state properties.

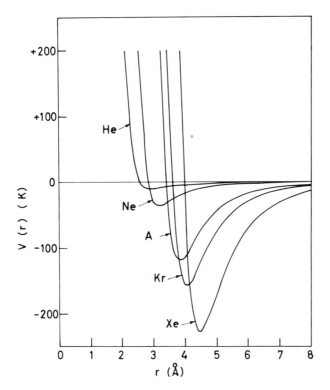

**Figure 7.6**
LJ potentials for the rare-gas family.

the total potential is approximately parabolic, being given by*

$$U(r) \sim \tfrac{1}{2}ZV''(r_0)u^2, \tag{7.103}$$

where

$$\mathbf{u} = \mathbf{r} - \mathbf{r}_0.$$

Treating the atom as an Einstein oscillator in this well, we can associate
with it a *dynamic energy*[73]†

---

*This approximation is crude but adequate for our present purposes. Strictly speaking one
must expand in powers of **u** the quantity

$$U(\mathbf{r}) = \sum_i V(\mathbf{r}_i - \mathbf{r}_0 - \mathbf{u}),$$

where the summation is over all the neighbors.
†The total energy of a three dimensional isotropic oscillator in the ground state is KE +
PE = $\tfrac{3}{2}\hbar\omega$, (KE = PE). Since the potential energy arises from the interaction with its
neighbors, we assign only half the PE to the atom concerned, the other half being distributed among the neighbors.

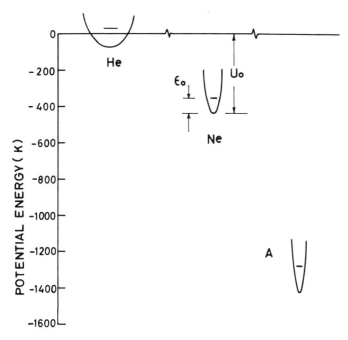

**Figure 7.7**
Potential well experienced by a given atom in fcc argon, fcc neon, and fcc helium (hypothetical), with a nearest neighbor spacing of $r_0$. The horizontal bars indicate the dynamic energy estimated from Eq. (7.104). (After Guyer, ref. 73.)

$$\varepsilon_0 = \mathrm{KE} + \tfrac{1}{2}\mathrm{PE}$$

$$= \tfrac{3}{4}(\tfrac{3}{2}\hbar\omega) = \tfrac{9}{8}\hbar\left(\frac{k}{m_A}\right)^{1/2}, \tag{7.104}$$

where $m_A$ is the mass of the argon atom and the spring constant $k$ equals $Z\,V''(r_0)$.

Figure 7.7 illustrates the potential well (7.103) and the energy corresponding to (7.104). Also shown are similar quantities for fcc neon and fcc helium on the assumption that such crystals exist with nearest-neighbor spacings of $r_0$. (This assumption is reasonably safe for neon but invalid for helium. Nevertheless we shall discuss the case for illustrative purposes.) It is clear from the figure that in both argon and neon the atom is well bound and localized around $r_0$, since $\varepsilon_0$ is less than the well depth $U_0$. In helium, on the other hand, the positive (kinetic) energy, which depends essentially on the extent to which the particle is confined around $r_0$, exceeds the negative potential energy gained by sitting there, and the crystal is unable to form as proposed.* In other words, the atom is too light and under too weak an interaction to reside at $r_0$.

To be bound, the atom's kinetic energy must be reduced, which would involve a broadening of its wave function. But this is not possible with the atom sitting at $r_0$, since it is already bumping into the hard core of its neighbors. The lattice therefore expands to achieve a nearest-neighbor spacing greater than $r_0$ (see Table 1.1). In spite of this, the kinetic effects are so strong that, even at 0 K, for $p < 25$ bars the system is a liquid. For $p > 25$ bars, external pressure finally overcomes the kinetic energy and the crystalline state is achieved with an enlarged lattice spacing. The phase diagram of He$^4$ is shown in Figure 7.8.

This discussion illustrates the importance of quantum effects in the

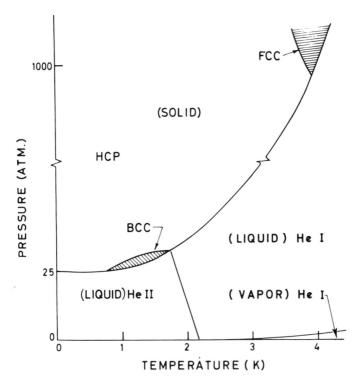

**Figure 7.8**
Phase diagram for He$^4$. (After Guyer, ref. 73.)

*This is reminiscent of the result that a particle of mass $m$ moving in a three-dimensional square well of radius $a$ has no bound state unless the well depth $U > (\pi^2 \hbar^2 / 8ma^2)$.

formation of crystalline helium.† Indeed, these effects are important also in the study of dynamics. Figure 7.9 shows a row of three atoms with a mean spacing $\Delta$ forming a part of the crystal lattice. Also shown is the potential felt by the central atom $B$ due to its two neighbors. The net potential is indicated by the dashed line and is a maximum at the position of $B$. The same is true of every atom in the crystal, so that we have the peculiar situation that the *mean positions of all atoms are local potential maxima*! If we were to accept this picture at face value, then clearly we would be dealing with a highly unstable situation. In fact, if one calculates the force constants by differentiating the pair potential as per (2.225) and then

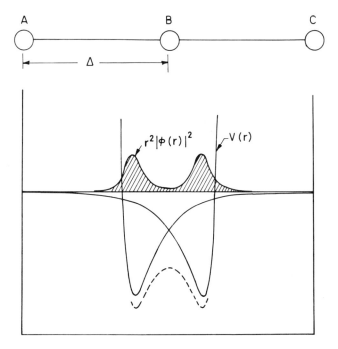

**Figure 7.9**
Sketch of the potential experienced by a helium atom due to its neighbors. Also shown is the (quantum) probability of finding the atom at various distances from its mean position. (After Guyer, ref. 73.)

† It is emphasized once again that it is the interplay between the small mass and the weakness of the attractive interaction which leads to large zero-point motions and consequently to quantum effects. By this standard, quantum effects are largest in solid helium 3. The molecular crystals $H_2$ and $D_2$ also have particles of small mass at each lattice site. However, the interaction is stronger than in helium, and the quantum effects may be treated as corrections.

computes the normal-mode frequencies as was done in Sec. 2.9 for argon, one finds all frequencies to be imaginary.[74] The flaw in this argument is that the conclusion of instability is reached via a classical picture. If the problem is treated quantum mechanically, then it is found that, in spite of the maxima associated with the mean atomic positions, phononlike excitations exist in solid helium, a result which has been confirmed experimentally (see Chapter 6). The instability disappears in the quantum picture in the following way: Let $\phi(r)$ describe the wave function of the atom $B$ in Figure 7.9. Then the probability of finding the atom at a distance $r$ from the mean position is given by

$$\rho(r) \, dr = |\phi(r)|^2 4\pi r^2 \, dr$$

and is also plotted in Figure 7.9. It is now clear that though the nominal position of $B$ is the maximum of the dotted curve, the atom actually spends most of its time in the potential minimum.

### 7.2.3 Single-particle approach

The formulation of the dynamics of quantum crystals has been the subject of much work.[72] Broadly speaking, two distinct treatments have emerged, namely the *single-particle* approach and the *collective* approach. We shall first review these theories without consideration of short-range correlations.

In the single-particle picture,[75–78] one starts by focusing attention on the Schrödinger equation for a single atom moving in the potential well provided by its neighbors ($m$ is here the mass of the atom):

$$\left[ -\frac{\hbar^2}{2m} \mathbf{V}^2 + v(u) \right] \phi_\mu^k(u) = \varepsilon_\mu^k \phi_\mu^k(u).  \tag{7.105}$$

Here $\phi_\mu^k(u)$ is the eigenfunction appropriate to the $\mu$th state having an energy $\varepsilon_\mu^k$ (as usual $k$ labels the atom in the cell).* The potential $v(u)$ is the *average* potential contributed by all the neighbors and is given by

$$v(u) = \sum_{l'k'}' \frac{\displaystyle\int d\mathbf{u}' \, V(|\mathbf{x} + \mathbf{u} - \mathbf{u}'|) \sum_\nu |\phi_\nu^{k'}(\mathbf{u}')|^2 \exp\{-\varepsilon_\nu^{k'}/k_\mathrm{B}T\}}{\displaystyle\sum_\lambda \exp(-\varepsilon_\lambda^{k'}/k_\mathrm{B}T)}  \tag{7.106}$$

with $\mathbf{x} = \mathbf{x}\binom{0 \; l'}{k \; k'}$, $\mathbf{u} = \mathbf{u}\binom{0}{k}$ and $\mathbf{u}' = \mathbf{u}\binom{l'}{k'}$; $V(|\mathbf{R}|)$ denotes the interatomic potential, and the prime on the summation implies the exclusion of the term $\binom{l'}{k'} = \binom{0}{k}$.

---

*The crystal however is regarded as monatomic. The index $k$ is necessary in the case of hcp helium.

Equations (7.105) and (7.106) constitute the Hartree self-consistent equations for the motion of a single atom. Self-consistency is involved because the eigenfunctions are determined by the Hartree potential $v(u)$ determined by the eigenfunctions themselves. An important feature of (7.106) is that the averaging is done over the single-particle states. Further, the thermal weighting is done by the usual Boltzmann factor. In other words, the atoms are regarded as distinguishable, the effects of quantum statistics being ignored.

Having analyzed the individual motions of all the atoms in their respective potential wells and having determined the eigenvalues and the eigenfunctions, one next determines the collective modes by imagining the system to be subjected to a weak external force which causes atomic displacements. A study of the displacement response then leads to a knowledge of the collective vibration frequencies. Specifically, the poles in the complex displacement-response function $R_{kk'}(\mathbf{q}, z)$ ($z$ denotes complex frequency) correspond to the frequencies of the collective modes. Some complications do arise though. Whereas for a crystal we must have only $3n$ normal-mode frequencies for each $\mathbf{q}$, an analysis of $R_{kk'}(\mathbf{q}, z)$ actually yields an infinite number of poles for each $\mathbf{q}$. It turns out, however, that $3n$ of these are the conventional phonon frequencies, while those remaining correspond roughly to multiphonon modes. Special techniques have been developed for extracting the $3n$ frequencies of the phonon modes.[72] In general these frequencies are complex, indicative of damping, but, to the extent that damping may be ignored, these frequencies may be obtained from the eigenvalue equation[72]

$$m\omega^2(\mathbf{q})\,U_\alpha(k|\mathbf{q}) = \sum_{k'\beta} M_{\alpha\beta}\binom{\mathbf{q}}{kk'}U_\beta(k'|\mathbf{q}), \tag{7.107}$$

where

$$M_{\alpha\beta}\binom{\mathbf{q}}{kk'} = \sum_{l'} \phi^{\text{eff}}_{\alpha\beta}\binom{0\ l'}{k\ k'}\exp\left\{i\mathbf{q}\cdot\mathbf{x}\binom{l'\ 0}{k'\ k}\right\}$$

and

$$\phi^{\text{eff}}_{\alpha\beta}\binom{0\ l'}{k\ k'} = \frac{-\sum_{\mu\nu}\exp\left\{-\dfrac{\varepsilon^k_\mu + \varepsilon^{k'}_\nu}{k_{\mathrm{B}}T}\right\}\left\langle\dfrac{kk'}{\mu\nu}\bigg|\dfrac{\partial^2 V(r)}{\partial r_\alpha \partial r_\beta}\bigg|\dfrac{kk'}{\mu\nu}\right\rangle}{\sum_{\lambda\lambda'}\exp\left\{-\dfrac{\varepsilon^k_\lambda + \varepsilon^{k'}_{\lambda'}}{k_{\mathrm{B}}T}\right\}}, \quad \binom{0}{k}\neq\binom{l'}{k'}, \tag{7.108}$$

with

$$r = \left|\mathbf{x}\binom{l'\ 0}{k'\ k} + \mathbf{u}\binom{l'}{k'} - \mathbf{u}\binom{0}{k}\right|.$$

The wave function corresponding to the pair state $|_{\mu\nu}^{kk'}\rangle$ is $\phi_\mu^k(\mathbf{u}\binom{0}{k}))\,\phi_\nu^{k'}(\mathbf{u}\binom{l'}{k'}))$, where the $\phi$'s satisfy the Hartree equation (7.105). The self term is given by

$$\phi_{\alpha\beta}^{\text{eff}}\begin{pmatrix}0 & 0 \\ k & k\end{pmatrix} = -\sum_{l'k'}{}' \ \phi_{\alpha\beta}^{\text{eff}}\begin{pmatrix}0 & l' \\ k & k'\end{pmatrix}.$$

It is clear that the vibrational frequencies are given by the same equations as in the Born–von Kármán formalism, with the force constants reinterpreted as in (7.108). Whereas the conventional force constants are defined via second derivatives in the equilibrium configuration, that is,

$$\phi_{\alpha\beta}\begin{pmatrix}0 & l' \\ k & k'\end{pmatrix} = -\left.\frac{\partial^2 V(r)}{\partial r_\alpha \partial r_\beta}\right|_{r=|\mathbf{x}\binom{l'\ 0}{k'\ k})|},$$

the force constants for quantum crystals involve an average over the quantum motions of the two interacting atoms. Observe that such averaging is necessary even at 0 K because of the large zero-point motion of the helium atoms ($\sim 30\%$ of interatomic spacing).

It is interesting to note that the single-particle theory corresponds roughly to considering successively the Einstein-oscillator aspects of the individual atoms and their coupling to form vibrational waves, a procedure that is analogous to that adopted for the internal vibrations of complex crystals in Sec. 2.7.

### 7.2.4 Collective approach

In the collective picture[79-83] one assumes at the outset that collective vibrational waves (phonons) exist and then investigates how these collective motions govern themselves, in particular their frequencies. Interestingly, this approach was pioneered by Born[84] himself.

One begins by assuming that the *true* Hamiltonian of the crystal is of the form

$$\mathscr{H} = -\sum_{lk}\frac{\hbar^2}{2m}\nabla_{lk}^2 + \tfrac{1}{2}\sum_{lk}\sum_{\substack{l'k' \\ (lk \neq l'k')}} V\left(\left|\mathbf{x}\begin{pmatrix}l & l' \\ k & k'\end{pmatrix} + \mathbf{u}\begin{pmatrix}l \\ k\end{pmatrix} - \mathbf{u}\begin{pmatrix}l' \\ k'\end{pmatrix}\right|\right). \quad (7.109)$$

Next one supposes that to a good approximation one may adopt a harmonic Hamiltonian $H_\text{h}$, where

$$H_\text{h} = -\sum_{lk}\frac{\hbar^2}{2m}\nabla_{lk}^2 + \tfrac{1}{2}\sum_{lk\alpha}\sum_{l'k'\beta}\phi_{\alpha\beta}^{\text{eff}}\begin{pmatrix}l & l' \\ k & k'\end{pmatrix}u_\alpha\begin{pmatrix}l \\ k\end{pmatrix}u_\beta\begin{pmatrix}l' \\ k'\end{pmatrix} \quad (7.110)$$

and $\phi_{\alpha\beta}^{\text{eff}}$ plays the role of a force constant. Restricting to 0 K, the aim now is to determine force constants such that $H_\text{h}$ represents a satisfactory approximation to $\mathcal{H}$. This is done via a variational procedure in which one adopts the ground-state wave function of $H_h$ as a trial wave function and seeks to minimize the expectation value of $\mathcal{H}$ with respect to this function. This leads to the requirement[82, 85]

$$\phi_{\alpha\beta}^{\text{eff}}\begin{pmatrix} 0 & l' \\ k & k' \end{pmatrix} = \int \rho_2\left[\mathbf{u}\begin{pmatrix} 0 & l' \\ k & k' \end{pmatrix}\right] \frac{\partial^2 V(r)}{\partial r_\alpha \partial r_\beta} \, d\mathbf{u}\begin{pmatrix} 0 & l' \\ k & k' \end{pmatrix}, \tag{7.111}$$

where

$$r = \left| \mathbf{r}\begin{pmatrix} 0 \\ k \end{pmatrix} - \mathbf{r}\begin{pmatrix} l' \\ k' \end{pmatrix} \right|$$

and

$$\mathbf{u}\begin{pmatrix} 0 & l' \\ k & k' \end{pmatrix} = \mathbf{u}\begin{pmatrix} 0 \\ k \end{pmatrix} - \mathbf{u}\begin{pmatrix} l' \\ k' \end{pmatrix}.$$

Further

$$\rho_2\left[\mathbf{u}\begin{pmatrix} 0 & l' \\ k & k' \end{pmatrix}\right] = \frac{\exp\{-\frac{1}{2}\sum_{\alpha\beta} u_\alpha\begin{pmatrix} 0 & l' \\ k & k' \end{pmatrix}(\mathbf{G}^{-1})_{\alpha\beta}\begin{pmatrix} 0 & l' \\ k & k' \end{pmatrix}u_\beta\begin{pmatrix} 0 & l' \\ k & k' \end{pmatrix}\}}{\int d\mathbf{u}\begin{pmatrix} 0 & l' \\ k & k' \end{pmatrix}\exp\{-\frac{1}{2}\sum_{\alpha\beta} u_\alpha\begin{pmatrix} 0 & l' \\ k & k' \end{pmatrix}(\mathbf{G}^{-1})_{\alpha\beta}\begin{pmatrix} 0 & l' \\ k & k' \end{pmatrix}u_\beta\begin{pmatrix} 0 & l' \\ k & k' \end{pmatrix}\}} \tag{7.112}$$

with

$$G_{\alpha\beta}\begin{pmatrix} 0 & l' \\ k & k' \end{pmatrix} = \frac{\hbar}{2Nm} \sum_{qj} \frac{1}{\omega_j(\mathbf{q})} \left\{ e_\alpha\left(k \middle| \begin{matrix} \mathbf{q} \\ j \end{matrix}\right) - e_\alpha\left(k' \middle| \begin{matrix} \mathbf{q} \\ j \end{matrix}\right) \exp i\mathbf{q}\cdot\mathbf{x}\begin{pmatrix} l' & 0 \\ k' & k \end{pmatrix} \right\}$$
$$\times \left\{ e_\beta^*\left(k \middle| \begin{matrix} \mathbf{q} \\ j \end{matrix}\right) - e_\beta^*\left(k' \middle| \begin{matrix} \mathbf{q} \\ j \end{matrix}\right) \exp\left[ -i\mathbf{q}\cdot\mathbf{x}\begin{pmatrix} l' & 0 \\ k' & k \end{pmatrix} \right] \right\}. \tag{7.113}$$

The quantities $\omega_j(\mathbf{q})$ and $\mathbf{e}\binom{\mathbf{q}}{j}$ appearing above are obtained from the eigenvalue equation

$$m\omega_j^2(\mathbf{q})e_\alpha\left(k \middle| \begin{matrix} \mathbf{q} \\ j \end{matrix}\right) = \sum_{\beta k'} M_{\alpha\beta}\begin{pmatrix} \mathbf{q} \\ kk' \end{pmatrix} e_\beta\left(k' \middle| \begin{matrix} \mathbf{q} \\ j \end{matrix}\right) \tag{7.114}$$

with

$$M_{\alpha\beta}\begin{pmatrix} \mathbf{q} \\ kk' \end{pmatrix} = \sum_{l'} \phi_{\alpha\beta}^{\text{eff}}\begin{pmatrix} 0 & l' \\ k & k' \end{pmatrix} \exp\left\{ i\mathbf{q}\cdot\mathbf{x}\begin{pmatrix} l' & 0 \\ k' & k \end{pmatrix} \right\}. \tag{7.115}$$

As is clear from the above equations, the effective constants themselves depend on the eigenvalues and eigenvectors so that one has coupled equations. A trial set of force constants is fed into (7.115) and the results of (7.114) used to construct the $\mathbf{G}$ tensor defined in (7.113) and thence a new set of force constants via (7.111). The process is iterated until the force constants converge.

The physical meaning of (7.111) becomes clearer by considering an alternate form of $\rho_2(\mathbf{u})$.[86] Let $(\mathbf{u}_1, \mathbf{u}_2, \ldots, \mathbf{u}_s)$ denote the set of displacements associated with the atoms in the crystal and $\psi_0(\mathbf{u}_1, \mathbf{u}_2, \ldots, \mathbf{u}_s)$ the ground-state wave function of (7.110) in coordinate space. In terms of $\psi_0$, $\rho_2(\mathbf{u}_i - \mathbf{u}_j)$ is given by

$$\int \psi_0^*(\mathbf{u}_1, \ldots, \mathbf{u}_s)\psi_0(\mathbf{u}_1, \ldots, \mathbf{u}_s)\, d\mathbf{u} \prod_{\substack{\lambda=1 \\ (\lambda \neq i \text{ and } j)}}^{s} d\mathbf{u}_\lambda, \quad (\mathbf{u} = \mathbf{u}_i + \mathbf{u}_j).$$

With this result it can be shown that the effective force constants correspond to the ground state average of the appropriate second derivative of the pair potential.

Essentially similar results[72] are obtained for finite temperatures, with $\mathbf{G}$ slightly modified to include temperature factors. Further, $\rho_2$ now denotes the diagonal part of the two-particle density matrix for a system of phonons characterized by (7.114) and corresponding to a temperature $T$. Thus, as before, the effective force constants involve an average, this time over the thermal equilibrium distribution of the relative displacements $\mathbf{u}\binom{0 \ l'}{k \ k'}$.

Further insight[87] into the concept of effective force constants is obtained by considering the usual equations of motion without, however, the harmonic assumption. We have (for a monatomic system)

$$\frac{d^2}{dt^2} u_\alpha\binom{l}{k}; t = -\frac{1}{m} \sum_{l'k'\beta} \left[ \frac{\partial^2 \Phi}{\partial u_\alpha\binom{l}{k}\partial u_\beta\binom{l'}{k'}}\bigg|_0 + \frac{1}{2}\sum_{l'k''\gamma} \frac{\partial^3 \Phi}{\partial u_\alpha\binom{l}{k}\partial u_\beta\binom{l'}{k'}\partial u_\gamma\binom{l''}{k''}}\bigg|_0 \right.$$

$$\times u_\gamma\binom{l''}{k''}; t + \frac{1}{3!}\sum_{\substack{l'k''\gamma \\ l''k'''\delta}} \frac{\partial^4 \Phi}{\partial u_\alpha\binom{l}{k}\partial u_\beta\binom{l'}{k'}\partial u_\gamma\binom{l''}{k''}\partial u_\delta\binom{l'''}{k'''}}\bigg|_0$$

$$\left. \times u_\gamma\binom{l''}{k''}; t\, u_\delta\binom{l'''}{k'''}; t + \cdots \right] u_\beta\binom{l'}{k'}; t, \tag{7.116}$$

where the $\mathbf{u}$'s are Heisenberg operators. We now linearize (7.116) using the following approximation commonly used in many-body theory for linearizing a product of commuting operators:

$$ABC\ldots \approx A\langle BC\ldots\rangle + B\langle AC\ldots\rangle + C\langle AB\ldots\rangle + \cdots,$$

$\langle \ldots \rangle$ denoting an average over states as yet unspecified. Using this approximation (7.116) may be written as

$$\frac{d^2}{dt^2} u_\alpha\binom{l}{k}; t = -\frac{1}{m}\sum_{l'k'\beta} \phi_{\alpha\beta}^{\text{eff}}\binom{l \ l'}{k \ k'} u_\beta\binom{l'}{k'}; t,$$

where

$$\phi_{\alpha\beta}^{\mathrm{eff}}\begin{pmatrix} l & l' \\ k & k' \end{pmatrix} = \left.\frac{\partial^2\Phi}{\partial u_\alpha\binom{l}{k}\partial u_\beta\binom{l'}{k'}}\right|_0 + \sum_{l''k''\gamma}\left.\frac{\partial^3\Phi}{\partial u_\alpha\binom{l}{k}\partial u_\beta\binom{l'}{k'}\partial u_\gamma\binom{l''}{k''}}\right|_0 \left\langle u_\gamma\binom{l''}{k''};t\right\rangle$$

$$+ \frac{1}{2}\sum_{\substack{l''k''\gamma\\l'''k'''\delta}}\left.\frac{\partial^4\Phi}{\partial u_\alpha\binom{l}{k}\partial u_\beta\binom{l'}{k'}\partial u_\gamma\binom{l''}{k''}\partial u_\delta\binom{l'''}{k'''}}\right|_0 \left\langle u_\gamma\binom{l''}{k''};t\right)u_\delta\binom{l'''}{k'''};t\right\rangle$$

$$+ \cdots,$$

and denotes a Taylor expansion of

$$\left\langle\frac{\partial^2\Phi}{\partial u_\alpha\binom{l}{k}\partial u_\beta\binom{l'}{k'}}\right\rangle$$

(see also (2.158)). Thus both the single-particle and collective theories are in a sense quasi-harmonic and seek to include a part of the anharmonic effect to all orders. (In the diagrammatic language of many-body theory, this corresponds to summing a select subset of diagrams.)

The physical difference between the two theories lies in the type of averaging done to evaluate the effective force constants. In the single-particle picture, the averaging is done assuming Boltzmann factors of the form $\exp\{-(\text{single particle excitation energy})/k_B T\}$. On the other hand, in the collective picture the averaging is done assuming factors of the form $\exp\{-(\text{phonon excitation energy})/k_B T\}$ (see Sec. 2.6). In actual fact, the energy states associated with the motion of an atom in a crystal must be described in terms of both phonon-occupation and single-particle states. Whereas the low-lying states (associated mainly with small-amplitude motions) can be described in terms of population of phonon states, it is not convenient to do so for the highly excited states which involve considerable excursion of the atoms from their mean positions. Such excitations are reminiscent of vacancy-interstitial pairs and are best described via single-particle states. Thus one really requires a suitable combination of both pictures, a synthesis that has not yet been achieved. At low temperatures, however, the collective picture is probably adequate.

### 7.2.5 Short-range correlations
The theories reviewed thus far do not take account of short-range correlations. They may therefore be applied to all anharmonic crystals where the atoms interact via pair potentials and where the averaging over particle motions does not require an integration over the hard-core part of the potential. This covers practically all crystals except solid helium, and in fact the self-consistent theory was formulated by Ranninger[79] and Choquard[80] for hot solids, independently of the helium problem.

In helium, however, the oscillation amplitudes are so large that one

would expect at first sight that adjacent atoms could come close enough
to involve the hard-core part of the potential in the averaging process.
Fortunately, this does not happen as the atoms suitably correlate their
motions as to be "out of each other's way," and the mathematical complications of dealing with a nonintegrable potential do not arise. However,
this requires the use of a correlated wave function of the form

$$\Psi(\mathbf{r}_1, \ldots, \mathbf{r}_N) = \Psi_0(\mathbf{r}_1, \ldots, \mathbf{r}_N) \prod_{i<j\leqslant N} f_{ij}(\mathbf{r}_{ij}), \qquad (7.117)$$

where $\Psi_0$ is the wave function in the absence of correlations and $f_{ij}(\mathbf{r}_{ij})$
takes account of correlations; $f(r)$ is chosen such that $f \to 0$ as $r \to 0$ and
$f \to 1$ as $r \to \infty$. By considering the example of a two-particle system, it is
easily seen that with $f$ built into the wave function the two particles have
vanishingly small probability of being found close to each other.

   Short-range correlations are important in the consideration of many
properties of solid helium, including its ground-state energy. The usually
adopted treatment of these correlations relies heavily on the work of
Nosanow[88] and Hetherington et al.[89] who assume a trial wave function
as in (7.117) and make a variational calculation of the ground-state energy
using cluster-expansion techniques. In these calculations, the correlation
function $f$ is chosen to be of the form

$$f = \exp\{-KV(r)\}, \quad (V(r) \to \text{LJ potential}), \qquad (7.118)$$

and $K$ is left free as a variational parameter. The net effect of introducing
correlations is to alter the pair potential into an effective potential given by

$$V_{\text{eff}}(r) = f^2(r)\left[ V(r) - \frac{\hbar^2}{2m}\mathbf{V}^2 \ln f(r) \right]\frac{1}{\langle f^2(r)\rangle_0}, \qquad (7.119)$$

where $\langle f^2(r)\rangle_0$ is a normalization factor.

   Many authors incorporate this result in lattice dynamical calculations
for solid helium by simply making the heuristic replacement

$$V(r) \to V_{\text{eff}}(r). \qquad (7.120)$$

For instance, computations have been made in this way by de Wette et
al.[78] in the single-particle picture and by Horner[81] and Benin[90] in the
collective picture. It has, however, been pointed out by Koehler[91] and
by Gillis et al.[83] that the frequencies calculated for example from (7.111–
7.115) in conjunction with the replacement (7.120) do not represent the
phonon frequencies. The latter are required to be the poles of the displacement autocorrelation function which leads to a fresh eigenvalue equation[72,83]

$$\Omega_j^{-1}(\mathbf{q})\mathbf{E}\!\begin{pmatrix}\mathbf{q}\\j\end{pmatrix} = \mathbf{R}(\mathbf{q})\,\mathbf{E}\!\begin{pmatrix}\mathbf{q}\\j\end{pmatrix}, \qquad\qquad (7.121)$$

where the $3n$-dimensional matrix $\mathbf{R}$ involves $\omega_j(\mathbf{q})$ and $\mathbf{e}\binom{\mathbf{q}}{j}$ derived from (7.114), with $V_{\text{eff}}$ replacing $V(r)$. It is observed[82] that the use of (7.121) rather than (7.114) to define phonon frequencies produces a significant difference near the zone boundary.

The major achievement of the quantum theories just reviewed is that they are able to predict real frequencies for solid helium without any adjustable parameters. Stimulated by these theoretical developments, a number of experimental measurements have been made, and a review of these was given earlier (Chapter 6). In general, the theories predict frequencies reasonably close to those observed. However, the volume dependence of the frequencies is not predicted satisfactorily, as has become evident with the availability of data for He$^4$ at several molar volumes. For instance, at a molar volume of 21.1 cm$^3$ (hcp He) the theoretical curves lie consistently *above* the experimental curves,[92] while at a molar volume of 16.00 cm$^3$ (hcp He) the theoretical curves lie *below* the experimental curves.[93] This has led to refinements, particularly with regard to residual anharmonic corrections. Results for bcc He$^4$ obtained with such refinements are shown in Figure 7.10.[94] Plotted here are the phonon line profiles for various $\mathbf{q}$ values along [100]. In contrast to the harmonic theory (Eqs. (7.107) and (7.114)), the lines are not $\delta$-functions. Their complex shapes raise the problem of identifying the phonon frequency. In Figure 7.11 two sets of curves are shown, corresponding to the main maximum and to the mean of the spectral shape. For comparison, the experimental frequencies[95] are also displayed. Experiments have also revealed broad and anomalous line shapes, but so far multiple peaks as in Figure 7.10 have not been resolved.

### 7.2.6 Application to other systems
As already mentioned, the concepts and techniques developed for quantum crystals have been applied in other areas. For instance, the notion of an average force constant, necessary in helium because of the large zero-point excursions, is useful in disordered systems like glass. Though the atoms in the latter do not make large movements, the averaging is necessary to handle the distribution of interparticle distances. The problem has been considered by Hubbard and Beeby[96] who follow a displacement-response approach and show that the collective frequencies of a monatomic disordered system are given by the roots of the secular equation

$$\|m\omega^2(\mathbf{q})\mathbb{1}_3 - \sigma\mathbf{\Psi}(\mathbf{q})\| = 0. \qquad\qquad (7.122)$$

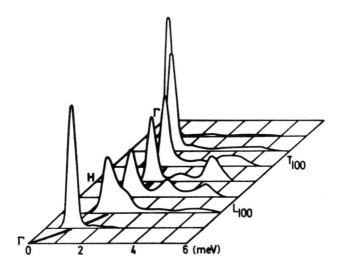

**Figure 7.10**
Line shapes for longitudinal phonons along [100] in bcc He⁴ at a molar volume of 20.18 cm³ computed by including residual anharmonic effects. (After Horner, ref. 94.)

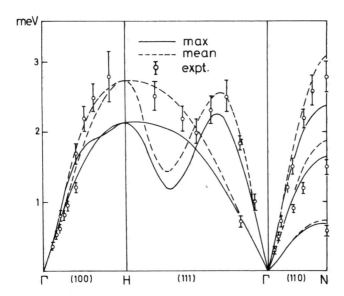

**Figure 7.11**
Dispersion curves for bcc He⁴ at 20.18 cm³ molar volume. The solid lines are based on maxima in the phonon line shapes while the dashed lines are based on their centers of gravity. The experimental points are due to Osgood et al. (ref. 95). (After Horner, ref. 94.)

Here

$$\Psi(\mathbf{q}) = \int \nabla\nabla V(R)[1 - e^{-i\mathbf{q}\cdot\mathbf{R}}]g(\mathbf{R})\, d\mathbf{R}, \tag{7.123}$$

and the pair distribution function $\sigma g(\mathbf{R})$ measures the probability of finding two (distinct) atoms a distance $\mathbf{R}$ apart. As usual, $V(R)$ is the two-body potential. The factor $\sigma$ is included so that $g(\mathbf{R}) \to 1$ for large $\mathbf{R}$. For a crystal (monatomic and with one atom per cell),

$$g(\mathbf{r}) = \frac{1}{\sigma}\sum_{l(\neq 0)} \delta(\mathbf{r} - \mathbf{x}(l)), \quad (\text{origin at } \mathbf{x}(l=0)), \tag{7.124}$$

and, using the inversion symmetry of the crystal, (7.122) reduces to

$$\|m\omega^2(\mathbf{q})\mathbb{1}_3 + \sum_{l'} \boldsymbol{\phi}(0, l')\,[1 - e^{i\mathbf{q}\cdot\mathbf{x}(l')}]\| = 0,$$

$$\phi_{\alpha\beta}(0, l') = -\frac{\partial^2 V(r)}{\partial r_\alpha \partial r_\beta}\bigg|_{r=|\mathbf{x}(l')|}, \tag{7.125}$$

which delivers the usual phonon dispersion relations. From Eq. (7.123) it is clear that the force constants governing the collective modes of a disordered system are obtained as the average (over the pair distribution function) of the derivatives of the pair potential.

The treatment of the soft mode in crystals like $KH_2PO_4$ (KDP) also follows closely the techniques discussed earlier. It is widely believed[97-101] that the ferroelectric transition in KDP is closely associated with the collective mode built up from the tunneling of the various protons in the double-well potential provided by their respective H bonds. From the standpoint of lattice dynamics, the problem is to describe collective vibrations arising from the coupling of tunneling motions which, by nature, are of large amplitude. This prompted Miller and Kwok[102] to adopt the techniques of the single-particle theory described earlier. These authors first studied the motion of a single proton in its bond via a suitable Hartree Hamiltonian and then investigated the collective modes via the displacement-response function. Quite conceivably, other applications along these lines will be discovered in the future.

## References

1. M. Born and J. R. Oppenheimer, Ann. Physik **84**, 457 (1927).

2. S. K. Joshi and A. K. Rajagopal, Solid State Phys. **22**, 159 (1968).

3. A. K. Rajagopal and M. H. Cohen, Collective Phenomena **1**, 9 (1972).

4. R. M. Pick, Advan. Phys. **19**, 269 (1970).

5. R. M. Pick, M. H. Cohen, and R. M. Martin, Phys. Rev. **B1**, 910 (1970).

6. V. Heine and I. Abarenkov, Phil. Mag. **9**, 451 (1964).

7. V. Heine, P. Nozieres, and J. W. Wilkins, Phil. Mag. **13**, 741 (1966).

8. V. Heine and D. Weaire, Solid State Phys. **24**, 249 (1970).

9. D. Pines and P. Nozieres, *Theory of Quantum Liquids* (Benjamin, New York, 1966), Vol. I, p. 206.

10. J. C. Slater, *Quantum Theory of Molecules and Solids* (McGraw-Hill, New York, 1963).

11. See Seitz, Chapter VI.

12. D. Pines, *Elementary Excitations in Solids* (Benjamin, New York, 1963).

13. M. H. Cohen, R. M. Martin, and R. M. Pick, in Copenhagen proceedings, Vol. I, p. 119.

14. J. Lindhard, Kgl. Danske Videnskab. Selskab, Mat.-Fys. Medd **28**, 8 (1954).

15. Reference 9 above, p. 281.

16. L. Kleinman, Phys. Rev. **160**, 585 (1967).

17. J. Hubbard, Proc. Roy. Soc. (London) **A240**, 539 (1957); **A243**, 336 (1958).

18. L. J. Sham, Proc. Roy. Soc. (London) **A283**, 33 (1965).

19. K. S. Singwi, M. P. Tosi, R. H. Land, and A. Sjolander, Phys. Rev. **176**, 589 (1968); K. S. Singwi, A. Sjolander, M. P. Tosi, and R. H. Land, Phys. Rev. **B1**, 1044 (1970); P. Vashishta and K. S. Singwi, Phys. Rev. **B6**, 875 (1972).

20. D. J. W. Geldart and S. H. Vosko, J. Phys. Soc. Japan **20**, 20 (1965).

21. D. J. W. Geldart, R. Taylor, and Y. P. Varshni, Can. J. Phys. **48**, 183 (1970).

22. L. Kleinman, Phys. Rev. **172**, 383 (1968).

23. P. R. Antoniewicz and L. Kleinman, Phys. Rev. **B2**, 2808 (1970).

24. F. Toigo and T. O. Woodruff, Phys. Rev. **B2**, 3959 (1970).

25. D. L. Price, K. S. Singwi, and M. P. Tosi, Phys. Rev. **B2**, 2983 (1970).

26. S. K. Sinha, Phys. Rev. **177**, 1256 (1969).

27. S. K. Sinha, R. P. Gupta, and D. L. Price, Phys. Rev. Letters **26**, 1324 (1971).

28. W. Hanke and H. Bilz, in Grenoble proceedings, p. 3.

29. Reference 9 above, p. 155.

30. S. L. Adler, Phys. Rev. **126**, 413 (1962).

31. N. Wiser, Phys. Rev. **129**, 62 (1963).

32. P. D. DeCicco and F. A. Johnson, Proc. Roy. Soc. (London) **A310**, 111 (1969); see also F. A. Johnson, ibid., **A310**, 79, 89, 101 (1969).

33. L. J. Sham, Phys. Rev. **188**, 1431 (1969).

34. P. N. Keating, Phys. Rev. **175**, 1171 (1968).

35. S. K. Sinha, Phys. Rev. **169**, 477 (1968).

36. N. Wakabayashi and S. K. Sinha, Phys. Rev. **B10**, 745 (1974).

37. W. A. Harrison, *Pseudopotentials in the Theory of Metals* (Benjamin, New York, 1966).

38. T. Toya, J. Res. Inst. Catalysis, Hokkaido University, **6**, 161, 183 (1958).

39. L. J. Sham and J. M. Ziman, Solid State Phys. **15**, 221 (1963).

40. S. H. Vosko, R. Taylor, and G. H. Keech, Can. J. Phys. **43**, 1187 (1965).

41. V. C. Sahni and G. Venkataraman, Phys. Rev. **185**, 1002 (1969).

42. A. O. E. Animalu and V. Heine, Phil. Mag. **12**, 1249 (1965).

43. N. W. Ashcroft, Phys. Letters **4**, 202 (1963).

44. T. Schneider and E. Stoll, Phys. Kondens Materie **5**, 331 (1966).

45. T. Schneider and E. Stoll, in Copenhagen proceedings, Vol. I, p. 101.

46. S. Prakash and S. K. Joshi, Phys. Rev. **B4**, 1770 (1971).

47. K. B. Tolpygo, Zh. Eksp. Teor. Fiz. **20**, 497 (1950).

48. B. Szigeti, Proc. Roy. Soc. (London) **A204**, 51 (1950).

49. V. I. Odelevski, Izv. Akad. Nauk. Sev. Fiz. SSSR **14**, 232 (1950).

50. V. S. Mashkevich and K. B. Tolpygo, Zh. Eksp. Teor. Fiz. **32**, 520 (1957) [Sov. Phys.—JETP **5**, 435 (1957)].

51. K. B. Tolpygo, Fiz. Tverd. Tela **3**, 943 (1961) [Sov. Phys.—Solid State **3**, 685 (1961)].

52. L. S. Pauling, *The Nature of the Chemical Bond* (Cornell, Ithaca, N. Y., 1945).

53. W. Cochran, Proc. Roy. Soc. (London) **A253**, 260 (1959).

54. H. Kaplan, in Wallis, p. 615.

55. V. S. Mashkevich, Fiz. Tverd. Tela **2**, 2629 (1960) [Sov. Phys.—Solid State **2**, 2345 (1961)].

56. R. A. Cowley, Proc. Roy. Soc. (London) **A268**, 109 (1962).

57. S. O. Lundqvist, Ark. Fysik **6**, 25 (1952).

58. P. O. Löwdin, Advan. Phys. **5**, 1 (1956).

59. R. H. Lyddane and K. F. Herzfeld, Phys. Rev. **54**, 846 (1938).

60. B. Gliss, R. Zeyher, and H. Bilz, Phys. Stat. Sol. **44**, 747 (1971).

61. B. Gliss and H. Bilz, Phys. Rev. Letters **21**, 884 (1968).

62. R. Zeyher, Phys. Stat. Sol. **48**, 711 (1971).

63. R. M. Martin, Phys. Rev. Letters **21**, 536 (1968).

64. R. M. Martin, Phys. Rev. **186**, 871 (1969).

65. G. Srinivasan, Phys. Rev. **178**, 1244 (1969).

66. M. Born, Ann. Physik **44**, 605 (1914).

67. S. K. Sinha, Critical Reviews in Solid State Sciences **3**, 273 (1973).

68. N. S. Gillis, Phys. Rev. **B3**, 1482 (1971).

69. S. K. Sinha, R. P. Gupta, and D. L. Price, Phys. Rev. **B9**, 2564, 2573 (1974).

70. W. Cochran, Critical Reviews in Solid State Sciences **2**, 1 (1971).

71. R. P. Gupta, S. K. Sinha, J. P. Walter, and M. L. Cohen, Solid State Commun. **14**, 1313 (1974).

72. N. R. Werthamer, Am. J. Phys. **37**, 763 (1969). This is a tutorial review which gives a comprehensive discussion of the lattice dynamics of quantum crystals.

73. R. A. Guyer, Solid State Phys. **23**, 413 (1969).

74. F. W. de Wette and B. R. A. Nijboer, Phys. Letters **18**, 19 (1965).

75. W. Brenig, Z. Physik **171**, 60 (1963).

76. D. R. Fredkin and N. R. Werthamer, Phys. Rev. **138**, A1527 (1965).

77. N. S. Gillis and N. R. Werthamer, Phys. Rev. **167**, 607 (1968).

78. F. W. de Wette, L. H. Nosanow, and N. R. Werthamer, Phys. Rev. **162**, 824 (1967).

79. J. Ranninger, Phys. Rev. **140**, A2031 (1965).

80. P. Choquard, *The Anharmonic Crystal* (Benjamin, New York, 1967).

81. H. Horner, Z. Physik **205**, 72 (1967).

82. T. R. Koehler, Phys. Rev. Letters **17**, 89 (1966); Phys. Rev. **165**, 942 (1968).

83. N. S. Gillis, T. R. Koehler, and N. R. Werthamer, Phys. Rev. **165**, 951 (1968); **175**, 1110 (1968).

84. M. Born, Festschrift der Akademie der Wissenschaften, Gottingen (1951).

85. T. R. Koehler, Phys. Rev. **144**, 789 (1966). Although the spirit of equations (7.111), etc., are based on this and ref. 82, the actual forms are closer to those quoted in reference 72 above, apart from small notational changes.

86. T. R. Koehler, Phys. Rev. **139**, A1097 (1965); see Eq. (12) in this paper.

87. S. K. Sinha, in *Nuclear Physics and Solid State Physics (India)* (Dept. of Atomic Energy, Bombay, 1973), Vol. **15A**, p. 155.

88. L. H. Nosanow, Phys. Rev. **146**, 120 (1966).

89. J. H. Hetherington, W. J. Mullin and L. H. Nosanow, Phys. Rev. **154**, 175 (1967).

90. D. B. Benin, Phys. Rev. Letters **20**, 1352 (1968).

91. T. R. Koehler, Phys. Rev. Letters **18**, 654 (1967).

92. V. J. Minkiewicz, T. A. Kitchens, F. P. Lipschultz, R. Nathans, and G. Shirane, Phys. Rev. **174**, 267 (1968).

93. R. A. Reese, S. K. Sinha, T. O. Brun, and C. R. Tilford, Phys. Rev. **A3**, 1688 (1971).

94. H. Horner, in Grenoble proceedings, p. 119.

95. E. B. Osgood, V. J. Minkiewicz, T. A. Kitchens, and G. Shirane, Phys. Rev. **A5**, 1537 (1972).

96. J. Hubbard and J. L. Beeby, J. Phys. C **2**, 556 (1969).

97. P. D. de Gennes, Solid State Comm. **1**, 132 (1963).

98. R. Brout, K. A. Muller, and H. Thomas, Solid State Comm. **4**, 507 (1966).

99. M. Tokunaga and T. Matsubara, Progr. Theoret. Phys. (Kyoto) **35**, 581 (1966); **36**, 857 (1966).

100. J. Villain and S. Stamenković, Phys. Stat. Sol. **15**, 585 (1966).

101. B. D. Silverman, Phys. Rev. Letters **20**, 443 (1968).

102. P. B. Miller and P. C. Kwok, Phys. Rev. **175**, 1062 (1968).

## Appendix 1
## Some Results Pertaining to Cyclic Crystals

We collect in this appendix a few useful results pertaining to cyclic crystals.

Consider a crystal subjected to periodic boundary conditions as discussed in Sec. 2.2.9. This enables us to introduce a wave vector $\mathbf{Q}$ defined by

$$\mathbf{Q} = \sum_{i=1}^{3} \zeta_i \mathbf{b}_i, \qquad (\zeta_i = n_i/L), \tag{A1.1}$$

where $n_i$ is an integer $(+, 0, -)$ and the $\mathbf{b}_i$'s denote the basis vectors of the reciprocal lattice. The vectors $\mathbf{G}$ of the latter are special cases of $\mathbf{Q}$ defined above. We shall use the symbol $\mathbf{q}$ if $\mathbf{Q}$ is contained in the central Brillouin Zone (abbreviated BZ).

One frequently used theorem of considerable importance is

$$\sum_{\substack{l \\ \text{over cyclic} \\ \text{crystal}}} \exp\{i\mathbf{Q} \cdot \mathbf{x}(l)\} = N\Delta(\mathbf{Q}), \tag{A1.2}$$

where

$$\Delta(\mathbf{Q}) = 1 \text{ if } \mathbf{Q} = \mathbf{G}$$
$$= 0 \text{ if } \mathbf{Q} \neq \mathbf{G}. \tag{A1.3}$$

*Proof.* Using (A1.1) and Eqs. (2.7) and (2.9), we can write

$$\sum_{\substack{l \\ \text{over cyclic} \\ \text{crystal}}} \exp\{i\mathbf{Q} \cdot \mathbf{x}(l)\} = \prod_{i=1}^{3} \sum_{l_i=0}^{L-1} \exp\{i2\pi\zeta_i l_i\}$$

$$= \prod_{i=1}^{3} \frac{1 - \exp(2\pi i \zeta_i L)}{1 - \exp(2\pi i \zeta_i)}. \tag{A1.4}$$

If $\mathbf{Q} \neq \mathbf{G}$, then the $\zeta_i$'s are not integers but the $\zeta_i L$'s are, whence from the right-hand side we see that the required sum vanishes. On the other hand if $\mathbf{Q} = \mathbf{G}$, then recalling $\mathbf{G} \cdot \mathbf{x}(l) = 2\pi \times$ integer, we note that every term in the sum contributes unity, giving in all $N$.

A useful corollary of (A1.2) is

$$\sum_{\substack{l \\ \text{over cyclic} \\ \text{crystal}}} \exp\{i\mathbf{q} \cdot \mathbf{x}(l)\} \exp\{-i\mathbf{q}' \cdot \mathbf{x}(l)\} = N\delta_{\mathbf{q},\mathbf{q}'}. \tag{A1.5}$$

The inverse of the above result is

$$\sum_{\substack{\mathbf{q} \\ \text{over BZ}}} \exp\{i\mathbf{q} \cdot \mathbf{x}(l)\} \exp\{-i\mathbf{q} \cdot \mathbf{x}(l')\} = N\delta_{l,l'}, \tag{A1.6}$$

which is easily proved as above, using the rule for the sum of a geometric progression.

Closely related to (A1.2) is the result

$$| \sum_{\substack{l \\ \text{over cyclic} \\ \text{crystal}}} \exp\{i\mathbf{Q}\cdot\mathbf{x}(l)\}|^2 = N^2\Delta(\mathbf{Q}). \tag{A1.7}$$

The above results are *exact*. Sometimes Eqs. (A1.2) and (A1.7) appear in the literature in slightly different forms. To deduce the alternate expressions, we make the approximation that owing to the high density in the mesh of points defined by (A1.1), we may replace a summation over $\mathbf{Q}$ over any specified region of reciprocal space by a corresponding integral over the same region, that is,

$$\sum_{\mathbf{Q}} \rightarrow \frac{Nv}{(2\pi)^3} \int d\mathbf{Q}, \tag{A1.8}$$

where the factor $[Nv/(2\pi)^3]$ specifies the density of mesh points (A1.1). It is well to remember that, though an excellent approximation, (A1.8) is in fact *not* exact.

Define now the *periodic delta function* $\sum_{\mathbf{G}} \delta(\mathbf{Q} - \mathbf{G})$ by

$$\int_{\substack{\text{over any} \\ \text{particular BZ}}} d\mathbf{Q} \sum_{\mathbf{G}} \delta(\mathbf{Q} - \mathbf{G}) = 1. \tag{A1.9}$$

In other words, $\sum_{\mathbf{G}} \delta(\mathbf{Q} - \mathbf{G})$ delivers unity whenever integrated around any reciprocal lattice point.

To rewrite (A1.2) in the alternate form, let us now sum the left-hand side over $\mathbf{Q}$, restricting it to a particular zone in reciprocal space. We then get

$$\sum_{\substack{\mathbf{Q} \\ \text{over specified} \\ \text{zone}}} \left\{ \sum_{\substack{l \\ \text{over cyclic} \\ \text{crystal}}} \exp[i\mathbf{Q}\cdot\mathbf{x}(l)] \right\}$$

$$\approx \frac{Nv}{(2\pi)^3} \int_{\substack{\text{over specified} \\ \text{zone}}} d\mathbf{Q} \left\{ \sum_{\substack{l \\ \text{over cyclic} \\ \text{crystal}}} \exp[i\mathbf{Q}\cdot\mathbf{x}(l)] \right\}.$$

On the other hand, using (A1.2) we have

$$\sum_{\substack{\mathbf{Q} \\ \text{over specified} \\ \text{zone}}} \left\{ \sum_{\substack{l \\ \text{over cyclic} \\ \text{crystal}}} \exp[i\mathbf{Q}\cdot\mathbf{x}(l)] \right\} = \sum_{\substack{\mathbf{Q} \\ \text{over specified} \\ \text{zone}}} N\Delta(\mathbf{Q})$$

$$= N = N \int_{\substack{\text{over specified} \\ \text{zone}}} d\mathbf{Q} \sum_{\mathbf{G}} \delta(\mathbf{Q} - \mathbf{G}).$$

Comparing the last two equations and remembering that the reciprocal lattice has inversion symmetry, we obtain

$$\sum_{\substack{l \\ \text{over cyclic} \\ \text{crystal}}} \exp[i\mathbf{Q}\cdot\mathbf{x}(l)] \approx \frac{(2\pi)^3}{v} \sum_{\mathbf{G}} \delta(\mathbf{Q} - \mathbf{G})$$

$$\approx \frac{(2\pi)^3}{v} \sum_{\mathbf{G'}} \delta(\mathbf{Q} + \mathbf{G'}). \tag{A1.10}$$

Similarly,

$$\left| \sum_{\substack{l \\ \text{over cyclic} \\ \text{crystal}}} \exp[i\mathbf{Q}\cdot\mathbf{x}(l)] \right|^2 \approx N\frac{(2\pi)^3}{v} \sum_{\mathbf{G}} \delta(\mathbf{Q} - \mathbf{G}). \tag{A1.11}$$

We next wish to discuss some applications of Fourier integrals and Fourier series to cyclic crystals.

Any well-behaved function $f(\mathbf{r})$ can formally be expressed as a Fourier integral:

$$f(\mathbf{r}) = \frac{1}{(2\pi)^3} \int f(\mathbf{Q})e^{i\mathbf{Q}\cdot\mathbf{r}} \, d\mathbf{Q}, \tag{A1.12}$$

where the Fourier transform $f(\mathbf{Q})$ is given by

$$f(\mathbf{Q}) = \int f(\mathbf{r})e^{-i\mathbf{Q}\cdot\mathbf{r}} \, d\mathbf{r}. \tag{A1.13}$$

In (A1.12) and (A1.13) the integrations are over the whole of $\mathbf{Q}$ space and the whole of $\mathbf{r}$ space, respectively.

For convenience, the *integral* in (A1.12) is sometimes expressed as a *summation*. One thus writes*

$$f(\mathbf{r}) = \frac{1}{(2\pi)^3} \times \text{const} \times \sum_{\mathbf{Q}_i} f(\mathbf{Q}_i)e^{i\mathbf{Q}_i\cdot\mathbf{r}}, \tag{A1.14}$$

where $\mathbf{Q}_i$ defines a suitable mesh covering the whole of $\mathbf{Q}$ space. Though not necessary, it is convenient to define the mesh $\mathbf{Q}_i$ by considering a certain volume $\Omega$ in $\mathbf{r}$ space and applying periodic boundary conditions to it. The choice of $\Omega$ is arbitrary, but in the context of a function $f(\mathbf{r})$ which occurs in crystal physics it is advantageous to choose the macrocell as the normalization volume, giving $\Omega = Nv$. The mesh $\mathbf{Q}_i$ then becomes identical to that defined in (A1.1), and the constant in (A1.14) equals

---

*The reader familiar with the theory of integrals will recognize that this is simply an expression of the Riemann integral as a sum in terms of a suitable "partition" of the region of integration.

$(2\pi)^3/Nv$. Thus we may write

$$f(\mathbf{r}) = \frac{1}{Nv} \sum_{\mathbf{Q}} f(\mathbf{Q}) e^{i\mathbf{Q}\cdot\mathbf{r}}, \tag{A1.15}$$

with $f(\mathbf{Q})$ given by (A1.13). For example, using this result the function $(1/r)$ is often expressed as

$$\frac{1}{r} = \frac{1}{Nv} \sum_{\mathbf{Q}} \frac{4\pi}{Q^2} e^{i\mathbf{Q}\cdot\mathbf{r}}, \tag{A1.16}$$

since $(4\pi/Q^2)$ is the Fourier transform of $(1/r)$.

Suppose now $f(\mathbf{r})$ has macrocell periodicity, that is,

$$f(\mathbf{r}) = f(\mathbf{r} + n_1 L\mathbf{a}_1 + n_2 L\mathbf{a}_2 + n_3 L\mathbf{a}_3), \tag{A1.17}$$

where $n_1$, $n_2$, $n_3$ are integers. In this case, by combining (A1.12), (A1.13), and (A1.17), we get

$$\begin{aligned}
f(\mathbf{r}) = \frac{1}{(2\pi)^3} \int d\mathbf{Q}\, e^{i\mathbf{Q}\cdot\mathbf{r}} \Bigg[ & \int_{\mathbf{r}'\in\Omega} d\mathbf{r}'\, e^{-i\mathbf{Q}\cdot\mathbf{r}'} f(\mathbf{r}') \\
& \times \sum_{n_1}\sum_{n_2}\sum_{n_3}^{\infty}{}_{-\infty} \exp\{-i\mathbf{Q}\cdot[n_1 L\mathbf{a}_1 + n_2 L\mathbf{a}_2 + n_3 L\mathbf{a}_3]\} \Bigg],
\end{aligned} \tag{A1.18}$$

where $\Omega = Nv$ is the volume of the macrocell.

To simplify this formula further it is useful to note some correspondences between the properties of the lattice generated by the macrocell and those of the familiar Bravais lattice. Analogous to the basis vectors $\mathbf{a}_i$ of the latter, we have for the former the basis vectors $\mathbf{A}_i = L\mathbf{a}_i$. This enables us to define the basis vectors $\mathbf{B}_i$ of the corresponding reciprocal lattice (see (2.9)). The vectors $\mathbf{Q}_m$ of the reciprocal lattice are then given by

$$\mathbf{Q}_m = m_1 \mathbf{B}_1 + m_2 \mathbf{B}_2 + m_3 \mathbf{B}_3, \tag{A1.19}$$

where $m_i$ $(i = 1, 2, 3)$ are integers. The vector $\mathbf{Q}$ introduced in (A1.12–A1.13) can be expressed in terms of $\mathbf{B}_i$ via the relation

$$\mathbf{Q} = \sum_i \zeta_i \mathbf{B}_i,$$

where $\zeta_i$ is a continuous variable. With these definitions, we can now write in analogy with (A1.10)

$$\sum_{n_1}\sum_{n_2}\sum_{n_3}^{\infty}{}_{-\infty} \exp\{-i\mathbf{Q}\cdot L(n_1\mathbf{a}_1 + n_2\mathbf{a}_2 + n_3\mathbf{a}_3)\}$$

$$= \frac{(2\pi)^3}{\Omega} \sum_{\mathbf{Q}_m} \delta(\mathbf{Q} - \mathbf{Q}_m). \tag{A1.20}$$

Using (A1.20) in (A1.18) we get

$$f(\mathbf{r}) = \frac{1}{\Omega} \sum_{\mathbf{Q}_m} e^{i\mathbf{Q}_m \cdot \mathbf{r}} f(\mathbf{Q}_m),\tag{A1.21}$$

where

$$f(\mathbf{Q}_m) = \int_{\mathbf{r}' \in \Omega} f(\mathbf{r}') e^{-i\mathbf{Q}_m \cdot \mathbf{r}'}\, d\mathbf{r}'.\tag{A1.22}$$

Consider next the special case where $f(\mathbf{r})$ is lattice periodic:

$$f(\mathbf{r}) = f(\mathbf{r} + \mathbf{x}(l)).\tag{A1.23}$$

Then by using arguments similar to those leading to (A1.21) we get

$$f(\mathbf{r}) = \frac{1}{v} \sum_{\mathbf{G}} f(\mathbf{G}) e^{i\mathbf{G} \cdot \mathbf{r}}\tag{A1.24}$$

with

$$f(\mathbf{G}) = \int_{\mathbf{r}' \in v} f(\mathbf{r}') e^{-i\mathbf{G} \cdot \mathbf{r}'}\, d\mathbf{r}'.\tag{A1.25}$$

Note that Eqs. (A1.21) and (A1.24) represent Fourier *series* for $f(\mathbf{r})$. Such representations are possible because of the intrinsic periodicity of the function, as reflected in the definition of the Fourier coefficients in (A1.22) and (A1.25), where the integrations are over $\Omega$ and $v$ respectively. By contrast, (A1.15) does not represent a Fourier series even though $f(\mathbf{r})$ is expressed as a sum, as may be seen by examining the definition of $f(\mathbf{Q})$ in that case (see Eq. (A1.13)). We also remark that Eqs. (2.60) and (2.170) represent the counterparts of (A1.21) and (A1.22), respectively, for the special case of a function which is defined only at a discrete set of points in every primitive cell and which has an overall macrocell periodicity. In this sense, Eq. (2.60) may be regarded as a Fourier series representation for the lattice displacement.

Finally, let us consider a few results pertinent to a function $g(\mathbf{r}, \mathbf{r}')$ of two variables. As in (A1.12) and (A1.13), we can write formally

$$g(\mathbf{r}, \mathbf{r}') = \frac{\Omega}{(2\pi)^6} \int g(\mathbf{Q}, \mathbf{Q}') e^{i\mathbf{Q} \cdot \mathbf{r}} e^{-i\mathbf{Q}' \cdot \mathbf{r}'}\, d\mathbf{Q}\, d\mathbf{Q}',\tag{A1.26}$$

$$g(\mathbf{Q}, \mathbf{Q}') = \frac{1}{\Omega} \int g(\mathbf{r}, \mathbf{r}') e^{-i\mathbf{Q} \cdot \mathbf{r}} e^{i\mathbf{Q}' \cdot \mathbf{r}'}\, d\mathbf{r}\, d\mathbf{r}',\tag{A1.27}$$

where $\Omega$ is at present arbitrary. As earlier, the integrations are over the wholes of the appropriate spaces. Observe also the opposite signs in the two exponents. This sign convention and the introduction of $\Omega$ are not necessary but are incorporated here in order to conform to the practice followed in some papers cited in Chapter 7.

Suppose now that $g(\mathbf{r}, \mathbf{r}')$ has the following type of (macrocell) periodicity:

$$g(\mathbf{r}, \mathbf{r}') = g\left(\mathbf{r} + L\sum_{i=1}^{3} n_i\mathbf{a}_i, \mathbf{r}' + L\sum_{i=1}^{3} n_i\mathbf{a}_i\right). \tag{A1.28}$$

In this case, $g(\mathbf{Q}, \mathbf{Q}')$ can be manipulated to an alternate form which displays interesting features.

Define $\mathbf{x}_n = L\sum_i n_i\mathbf{a}_i$ where the $n_i$'s are integers. Any vector $\mathbf{r}$ can then be always written as $\mathbf{r} = \mathbf{R} + \mathbf{x}_n$, where $\mathbf{R}$ is restricted to the macrocell at the origin. Equation (A1.27) may thus be rewritten (taking $\Omega$ as the macrocell volume) as

$$\begin{aligned}
g(\mathbf{Q}, \mathbf{Q}') &= \frac{1}{\Omega}\sum_{\mathbf{x}_n}\int_{\mathbf{R}\in\Omega} d\mathbf{R}\int d\mathbf{r}'\, g(\mathbf{R} + \mathbf{x}_n, \mathbf{r}')e^{-i\mathbf{Q}\cdot(\mathbf{R}+\mathbf{x}_n)}e^{i\mathbf{Q}'\cdot\mathbf{r}'} \\
&= \frac{1}{\Omega}\sum_{\mathbf{x}_n}\int_{\mathbf{R}\in\Omega} d\mathbf{R}\int d\mathbf{r}'\, g(\mathbf{R} + \mathbf{x}_n, \mathbf{r}' + \mathbf{x}_n - \mathbf{x}_n)e^{-i\mathbf{Q}\cdot(\mathbf{R}+\mathbf{x}_n)}e^{i\mathbf{Q}'\cdot\mathbf{r}'} \\
&= \frac{1}{\Omega}\sum_{\mathbf{x}_n}\int_{\mathbf{R}\in\Omega} d\mathbf{R}\int d\mathbf{r}'\, g(\mathbf{R}, \mathbf{r}' - \mathbf{x}_n)e^{-i\mathbf{Q}\cdot(\mathbf{R}+\mathbf{x}_n)}e^{i\mathbf{Q}'\cdot\mathbf{r}'}.
\end{aligned}$$

Upon suitable change of variables, this last result may be recast as

$$\begin{aligned}
g(\mathbf{Q}, \mathbf{Q}') = \frac{1}{\Omega}\sum_{n_1}\sum_{n_2}\sum_{n_3}^{\infty}\int_{\mathbf{r}\in\Omega} d\mathbf{r}\int d\mathbf{r}'\, \exp\{-i\mathbf{Q}\cdot[\mathbf{r}+L(n_1\mathbf{a}_1+n_2\mathbf{a}_2+n_3\mathbf{a}_3)]\} \\
\times \exp\{i\mathbf{Q}'\cdot[\mathbf{r}' + L(n_1\mathbf{a}_1 + n_2\mathbf{a}_2 + n_3\mathbf{a}_3)]\}g(\mathbf{r}, \mathbf{r}'), \quad (A1.29)
\end{aligned}$$

where the $\mathbf{r}'$ integration, unlike that over $\mathbf{r}$, is over the whole of coordinate space.

Since as in (A1.20) we may write

$$\sum_{n_1}\sum_{n_2}\sum_{n_3}^{\infty}\exp\{i(\mathbf{Q}' - \mathbf{Q})\cdot L(n_1\mathbf{a}_1 + n_2\mathbf{a}_2 + n_3\mathbf{a}_3)\}$$

$$= \frac{(2\pi)^3}{\Omega}\sum_{\mathbf{Q}_m}\delta(\mathbf{Q}' - \mathbf{Q} - \mathbf{Q}_m), \tag{A1.30}$$

(A1.29) becomes

$$g(\mathbf{Q}, \mathbf{Q}') = \frac{(2\pi)^3}{\Omega^2}\sum_{\mathbf{Q}_m}\delta(\mathbf{Q}' - \mathbf{Q} - \mathbf{Q}_m)\int_{\mathbf{r}\in\Omega} d\mathbf{r}\int d\mathbf{r}'\, g(\mathbf{r}, \mathbf{r}')e^{-i\mathbf{Q}\cdot\mathbf{r}}e^{i\mathbf{Q}'\cdot\mathbf{r}'}. \tag{A1.31}$$

The interesting feature here is that the $\delta$-function allows $\mathbf{Q}'$ to differ from $\mathbf{Q}$ only to the extent of a vector $\mathbf{Q}_m$ belonging to the mesh (A1.19), if $g(\mathbf{Q}, \mathbf{Q}')$ is to be nonvanishing.

Analogously, if $g(\mathbf{r}, \mathbf{r}')$ has lattice periodicity, that is,

$$g(\mathbf{r}, \mathbf{r}') = g(\mathbf{r} + \mathbf{x}(l), \mathbf{r}' + \mathbf{x}(l)),$$

then by similar arguments it can be shown that $g(\mathbf{Q}, \mathbf{Q}')$ vanishes unless

$$\mathbf{Q}' = \mathbf{Q} + \mathbf{G}.$$

Remembering this, we may as in (A1.15) express $g(\mathbf{r}, \mathbf{r}')$ as a summation:

$$
\begin{aligned}
g(\mathbf{r}, \mathbf{r}') &= \frac{1}{Nv} \sum_{\mathbf{Q}} \sum_{\mathbf{Q}'} g(\mathbf{Q}, \mathbf{Q}') e^{i\mathbf{Q}\cdot\mathbf{r}} e^{-i\mathbf{Q}'\cdot\mathbf{r}'} \\
&= \frac{1}{Nv} \sum_{\substack{\mathbf{q} \\ \text{over BZ}}} \sum_{\mathbf{G}} \sum_{\mathbf{G}'} g(\mathbf{q} + \mathbf{G}, \mathbf{q} + \mathbf{G}') e^{i(\mathbf{q}+\mathbf{G})\cdot\mathbf{r}} e^{-i(\mathbf{q}+\mathbf{G}')\cdot\mathbf{r}'} \qquad \text{(A1.32)}
\end{aligned}
$$

with $g(\mathbf{Q}, \mathbf{Q}')$ as in (A1.27).

## Appendix 2
## Concerning Coulombic Lattice Sums and Dipolar Fields

This appendix is devoted to the evaluation of certain quantities arising from the Coulomb interactions between ions in a lattice.

### A2.1 Fourier sums of Coulombic force constants

We begin by considering the problem of evaluating Fourier sums of force constants associated with Coulomb interactions between *point ions*. The sums of interest are of the form

$$B_{\alpha\beta}^{C}\begin{pmatrix} \mathbf{q} \\ kk' \end{pmatrix} = \sum_{l'} \phi_{\alpha\beta}^{C}\begin{pmatrix} 0 & l' \\ k & k' \end{pmatrix} \exp\left[i\mathbf{q}\cdot\mathbf{x}(l')\right], \tag{A2.1}$$

where the Coulombic force constant $\phi_{\alpha\beta}^{C}$ is explicitly given by

$$\phi_{\alpha\beta}^{C}\begin{pmatrix} 0 & l' \\ k & k' \end{pmatrix} = -Z_k Z_{k'} e^2 \lim_{\mathbf{r}\to 0} \left\{ \frac{\partial^2}{\partial r_\alpha \partial r_\beta} \frac{1}{|\mathbf{r} - [\mathbf{x}\binom{l'}{k'}) - \mathbf{x}\binom{0}{k}]|} \right\} \tag{A2.2a}$$

$$= Z_k Z_{k'} e^2 \left\{ \frac{\delta_{\alpha\beta}}{r^3} - \frac{3 r_\alpha r_\beta}{r^5} \right\}_{\mathbf{r}=|\mathbf{x}\binom{l' \; 0}{k' \; k})|}, \quad \binom{l'}{k'} \neq \binom{0}{k}. \tag{A2.2b}$$

Here $Z_k e$ is the charge of the ion in the $k$th sublattice. The value of $e$ is taken as $+4.8 \times 10^{-10}$ esu. As usual, the self term is given by (see Eq. (2.35))

$$\phi_{\alpha\beta}^{C}\begin{pmatrix} 0 & 0 \\ k & k \end{pmatrix} = -\sum_{l'k'}' \phi_{\alpha\beta}^{C}\begin{pmatrix} 0 & l' \\ k & k' \end{pmatrix}. \tag{A2.3}$$

The direct evaluation of (A2.1) using (A2.2) and (A2.3) is difficult because of the slow convergence of the $l'$ series, an annoying feature characteristic of problems involving long-range Coulombic forces. Fortunately, this difficulty can be overcome by employing a method due to Kellerman[1] based on techniques previously devised by Ewald[2] in connection with problems involving potentials and fields in periodic lattices.

The essence of the Ewald technique is the replacement of the quantity $(1/|\mathbf{r} - \mathbf{x}\binom{l}{k})|)$, characteristic of Coulomb interactions, by the equivalent quantity

$$\frac{2}{\sqrt{\pi}} \int_0^\infty \exp\left\{ -\left|\mathbf{r} - \mathbf{x}\binom{l}{k}\right|^2 \rho^2 \right\} d\rho. \tag{A2.4}$$

The range of integration is next split into two domains $\int_0^\eta$ and $\int_\eta^\infty$, where $\eta$ is arbitrary, and the associated integrands are subsequently handled differently. The sum

$$\sum_{l'} \frac{1}{|\mathbf{r} - \mathbf{x}\binom{l' \; 0}{k' \; k})|} \tag{A2.5}$$

then becomes[3]

$$
\frac{2\pi}{v} \sum_{\mathbf{G}} \int_0^{\eta} \frac{1}{\rho^3} \exp\{i\mathbf{G}\cdot[\mathbf{r} + \mathbf{x}(k) - \mathbf{x}(k')]\} \exp(-G^2/4\rho^2)\,d\rho
$$

$$
+ \frac{2}{\sqrt{\pi}} \sum_{l'} \int_{\eta}^{\infty} \exp\left\{ -\left| \mathbf{r} - \mathbf{x}\left(\begin{smallmatrix} l' & 0 \\ k' & k \end{smallmatrix}\right) \right|^2 \rho^2 \right\}\,d\rho
$$

$$
= \frac{4\pi}{v} \sum_{\mathbf{G}} \frac{1}{G^2} \exp\{i\mathbf{G}\cdot[\mathbf{r} + \mathbf{x}(k) - \mathbf{x}(k')]\} \exp(-G^2/4\eta^2)
$$

$$
+ \sum_{l'} [\operatorname{erfc}(\eta x)/x]_{x = |\mathbf{r} - \mathbf{x}\left(\begin{smallmatrix} l' & 0 \\ k' & k \end{smallmatrix}\right)|}, \tag{A2.6}
$$

where

$$
\operatorname{erfc}(x) = 1 - \operatorname{erf}(x) = 1 - \frac{2}{\sqrt{\pi}} \int_0^x \exp(-\lambda^2)\,d\lambda
$$

$$
= \frac{2}{\sqrt{\pi}} \int_x^{\infty} \exp(-\lambda^2)\,d\lambda. \tag{A2.7}
$$

From a physical point of view, the Ewald trick is tantamount to replacing an array of point ions by two arrays, the nature of which is schematically illustrated in Figure A2.1. In the first array, every point is assigned a Gaussian charge distribution having the same sign and total charge as the ion corresponding to that point. In the second array, one associates with every lattice site a Gaussian distribution opposite in sign and in magnitude to that in the first array, superposed on which is a point charge equal to the ionic charge.[4] Note that whereas the first array consists of charged entities, the second consists of nearly neutral entities. The reciprocal-space part of (A2.6) is helpful in describing interactions associated with the first array, while the real-space part is suited to describing interactions associated with the second array.

Analogous to (A2.6) we have

$$
\sum_{l'} \frac{\exp[i\mathbf{q}\cdot\mathbf{x}(l')]}{|\mathbf{r} - \mathbf{x}\left(\begin{smallmatrix} l' & 0 \\ k' & k \end{smallmatrix}\right)|} = \frac{4\pi}{v} \sum_{\mathbf{G}} \frac{1}{|\mathbf{G} + \mathbf{q}|^2} \exp\{i(\mathbf{G} + \mathbf{q})\cdot[\mathbf{r} + \mathbf{x}(k) - \mathbf{x}(k')]\}
$$

$$
\times \exp\{-(\mathbf{G} + \mathbf{q})^2/4\eta^2\}
$$

$$
+ \sum_{l'} \exp[i\mathbf{q}\cdot\mathbf{x}(l')]\left(\frac{\operatorname{erfc}(\eta x)}{x}\right)_{x = |\mathbf{r} - \mathbf{x}\left(\begin{smallmatrix} l' & 0 \\ k' & k \end{smallmatrix}\right)|}. \tag{A2.8}
$$

Using this result in Eq. (A2.1) together with the form (A2.2a) for $\phi_{\alpha\beta}^C$ and then performing the differentiation,[5] we get

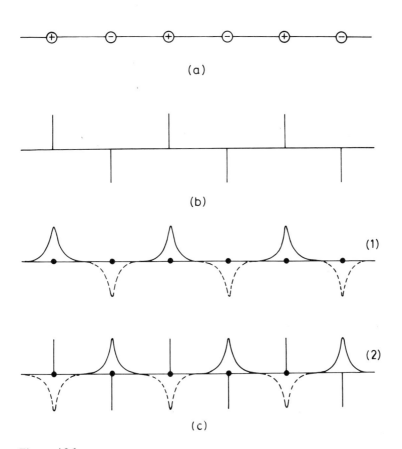

**Figure A2.1**
(a) shows a schematic representation of a diatomic ionic crystal and (b) the corresponding
charge distribution. In (c) are shown two charge distributions which together simulate the
delta-function array of b. (1) corresponds to a Gaussian array of the same sign as the array
in b. (2) corresponds to a delta-function array as in b plus a Gaussian array opposite in
sign to that in 1. The Ewald trick amounts to replacing the charge distribution b by the
distributions 1 and 2.

$$B_{\alpha\beta}^{C}\begin{pmatrix} \mathbf{q} \\ kk' \end{pmatrix} = Z_k Z_{k'} e^2 \{ \mathscr{B}_{\alpha\beta}(\mathbf{q}, \mathbf{x}(k) - \mathbf{x}(k'))$$

$$- \mathscr{C}_{\alpha\beta}(\mathbf{q}, \mathbf{x}(k) - \mathbf{x}(k')) \}, \, k \neq k', \qquad (A2.9)$$

where*

$$\mathscr{B}_{\alpha\beta}(\mathbf{q}, \mathbf{x}) = \frac{4\pi}{v} \sum_{\mathbf{G}} [(\mathbf{G} + \mathbf{q})_\alpha (\mathbf{G} + \mathbf{q})_\beta / |\mathbf{G} + \mathbf{q}|^2]$$

$$\times \exp[-|\mathbf{G} + \mathbf{q}|^2 / 4\eta^2] \exp[i(\mathbf{G} + \mathbf{q}) \cdot \mathbf{x}], \qquad (A2.10)$$

and

$$\mathscr{C}_{\alpha\beta}(\mathbf{q}, \mathbf{x}) = \sum_{l'} \exp[i\mathbf{q} \cdot \mathbf{x}(l')] \left\{ \frac{[\mathbf{x}(l') - \mathbf{x}]_\alpha [\mathbf{x}(l') - \mathbf{x}]_\beta}{|\mathbf{x}(l') - \mathbf{x}|^5} \right.$$

$$\times \left( 3 \text{erfc}(y) + \frac{\exp(-y^2)}{\sqrt{\pi}} (4y^3 + 6y) \right) - \frac{\delta_{\alpha\beta}}{|\mathbf{x}(l') - \mathbf{x}|^3}$$

$$\left. \times \left( \text{erfc}(y) + \frac{2y}{\sqrt{\pi}} \exp(-y^2) \right) \right\}, \qquad (A2.11a)$$

with

$$y = \eta |\mathbf{x}(l') - \mathbf{x}|. \qquad (A2.11b)$$

The series $\mathscr{B}_{\alpha\beta}$ and $\mathscr{C}_{\alpha\beta}$ describe, respectively, the reciprocal-space and the real-space series occurring in the Fourier sum of the Coulombic force constants. Rapid convergence may be obtained for both series by a proper choice of $\eta$. In practice, the choice

$$\eta \sim \frac{1}{\text{unit-cell dimension}}$$

is satisfactory.†

From (A2.10) and (A2.11) the following properties of $\mathscr{B}_{\alpha\beta}$ and $\mathscr{C}_{\alpha\beta}$ may be noted:

---

*We follow Venkataraman and Sahni[6] in introducing the $\mathscr{B}$ and $\mathscr{C}$ series. However, our $\mathscr{B}$ differs by a factor of 4.
†Instead of replacing the point charges by two arrays as described above, one could also view the ions as nonoverlapping *spherical* charge distributions. In this case, lattice sums of Coulombic-type quantities can be expressed purely in terms of reciprocal-space series. The convergence of this series can be controlled by appropriate choice of the radius of the sphere (S. K. Sinha, private comm.). See also W. Cochran in Wallis, p. 75. However, it must also be pointed out that from the standpoint of relating to the local fields acting in crystals, it is preferable to express the Coulombic contribution to the dynamical matrix in terms of both real- and reciprocal-space series.

$$\mathscr{B}_{\alpha\beta}(\mathbf{q}, \mathbf{x}) = \mathscr{B}_{\beta\alpha}(\mathbf{q}, \mathbf{x}), \qquad \mathscr{B}_{\alpha\beta}^*(\mathbf{q}, \mathbf{x}) = \mathscr{B}_{\alpha\beta}(\mathbf{q}, -\mathbf{x}),$$
$$\mathscr{C}_{\alpha\beta}(\mathbf{q}, \mathbf{x}) = \mathscr{C}_{\beta\alpha}(\mathbf{q}, \mathbf{x}), \qquad \mathscr{C}_{\alpha\beta}^*(\mathbf{q}, \mathbf{x}) = \mathscr{C}_{\alpha\beta}(\mathbf{q}, -\mathbf{x}). \tag{A2.12}$$

In the series for $\mathscr{B}_{\alpha\beta}(\mathbf{q}, \mathbf{x}(k) - \mathbf{x}(k'))$, the leading ($\mathbf{G} = 0$) term is of the form

$$\frac{4\pi}{v} \frac{q_\alpha q_\beta}{|\mathbf{q}|^2} \exp\{i\mathbf{q} \cdot [\mathbf{x}(k) - \mathbf{x}(k')]\} \exp(-q^2/4\eta^2). \tag{A2.13}$$

The factor $(q_\alpha q_\beta/|\mathbf{q}|^2)$ has no unique limit as $\mathbf{q} \to 0$, the limiting value depending entirely on the direction by which the point $\mathbf{q} = 0$ is approached. Hence when evaluating $B_{\alpha\beta}^C(\substack{\mathbf{q}\\kk'})$ for the $\mathbf{q} = 0$ case, one must first decide along which direction the point $\mathbf{q} = 0$ is to be approached and then evaluate the dynamical matrix by taking the limit along the direction of interest. This nonregular behavior leads to the peculiarities associated with the IR-active modes discussed in Chapter 4.

Let us now consider $B_{\alpha\beta}^C(\substack{\mathbf{q}\\kk})$. We have

$$B_{\alpha\beta}^C\begin{pmatrix}\mathbf{q}\\kk\end{pmatrix} = \sum_{l'(\neq 0)} \phi_{\alpha\beta}^C\begin{pmatrix}0 & l'\\k & k\end{pmatrix} \exp[i\mathbf{q}\cdot\mathbf{x}(l')] + \phi_{\alpha\beta}^C\begin{pmatrix}0 & 0\\k & k\end{pmatrix}$$

$$= \sum_{l'(\neq 0)} \phi_{\alpha\beta}^C\begin{pmatrix}0 & l'\\k & k\end{pmatrix} \exp[i\mathbf{q}\cdot\mathbf{x}(l')] - \sum_{l'(\neq 0)} \phi_{\alpha\beta}^C\begin{pmatrix}0 & l'\\k & k\end{pmatrix}$$

$$- \sum_{k'(\neq k)} \sum_{l'} \phi_{\alpha\beta}^C\begin{pmatrix}0 & l'\\k & k'\end{pmatrix},$$

using (A2.3). We add formally the infinity

$$(-Z_k e)^2 \lim_{r \to 0} \frac{\partial^2}{\partial r_\alpha \partial r_\beta}\left(\frac{1}{r}\right)$$

to the first term on the right-hand side of the above equation and subtract it from the second. This gives

$$B_{\alpha\beta}^C\begin{pmatrix}\mathbf{q}\\kk\end{pmatrix} = -(Z_k e)^2 \sum_{l'} \lim_{\mathbf{r} \to 0} \frac{\partial^2}{\partial r_\alpha \partial r_\beta}\left\{\frac{1}{|\mathbf{r} - \mathbf{x}(l')|}\right\} \exp[i\mathbf{q}\cdot\mathbf{x}(l')]$$

$$+ (Z_k e)^2 \sum_{l'} \lim_{\mathbf{r} \to 0} \frac{\partial^2}{\partial r_\alpha \partial r_\beta}\left\{\frac{1}{|\mathbf{r} - \mathbf{x}(l')|}\right\}$$

$$+ \sum_{k'(\neq k)} Z_k Z_{k'} e^2 \sum_{l'} \lim_{\mathbf{r} \to 0} \frac{\partial^2}{\partial r_\alpha \partial r_\beta}\left\{\frac{1}{|\mathbf{r} - \mathbf{x}(\substack{l' \ 0\\k' \ k})|}\right\}$$

$$= (Z_k e)^2\{\mathscr{B}_{\alpha\beta}(\mathbf{q}, 0) - \mathscr{C}_{\alpha\beta}(\mathbf{q}, 0)\}$$

$$- (Z_k e)^2\{\mathscr{B}_{\alpha\beta}(\mathbf{q} \equiv 0, 0) - \mathscr{C}_{\alpha\beta}(\mathbf{q} \equiv 0, 0)\}$$

$$- \sum_{k'(\neq k)} Z_k Z_{k'} e^2\{\mathscr{B}_{\alpha\beta}(\mathbf{q} \equiv 0, \mathbf{x}(k) - \mathbf{x}(k'))$$

$$- \mathscr{C}_{\alpha\beta}(\mathbf{q} \equiv 0, \mathbf{x}(k) - \mathbf{x}(k'))\}. \tag{A2.14}$$

This last result requires two comments. First, in arriving at it we have used the results

$$-\sum_{l'} \frac{\partial^2}{\partial r_\alpha \partial r_\beta} \frac{1}{|\mathbf{r} - \mathbf{x}(l')|} = \mathcal{B}_{\alpha\beta}(\mathbf{q} \equiv 0, \mathbf{r}) - \mathcal{C}_{\alpha\beta}(\mathbf{q} \equiv 0, \mathbf{r}) \qquad \text{(A2.15a)}$$

and

$$-\sum_{l'} \lim_{\mathbf{r} \to 0} \frac{\partial^2}{\partial r_\alpha \partial r_\beta} \frac{1}{|\mathbf{r} - \mathbf{x}(l')|} = \mathcal{B}_{\alpha\beta}(\mathbf{q} \equiv 0, 0) - \mathcal{C}_{\alpha\beta}(\mathbf{q} \equiv 0, 0). \qquad \text{(A2.15b)}$$

In the above equations, $\mathbf{q} \equiv 0$ means that $\mathbf{q}$ is to be set identically equal to zero in the relevant expression and not allowed to approach zero by a limiting process.

The next comment concerns the occurrence of ill-defined terms in (A2.14), specifically the leading terms in $\mathcal{B}_{\alpha\beta}(\mathbf{q} \equiv 0, \mathbf{x})$ and $\mathcal{C}_{\alpha\beta}(\mathbf{q}, 0)$. The former is of the form

$$\frac{4\pi}{v} \left[ \frac{G_\alpha G_\beta}{G^2} \right]_{\mathbf{G} \equiv 0} \qquad \text{(A2.16a)}$$

and is clearly ill defined. The same is true of

$$\left\{ \frac{x_\alpha(l) x_\beta(l)}{|\mathbf{x}(l)|^5} \left[ 3 + \frac{4\eta^3 |\mathbf{x}(l)|^3 + 6\eta |\mathbf{x}(l)|}{\sqrt{\pi}} \right] \right.$$
$$\left. - \frac{\delta_{\alpha\beta}}{|\mathbf{x}(l)|^3} \left[ 1 + \frac{2\eta |\mathbf{x}(l)|}{\sqrt{\pi}} \right] \right\}_{l=0}, \qquad \text{(A2.16b)}$$

which occurs in the leading terms of $\mathcal{C}_{\alpha\beta}(\mathbf{q}, 0)$. The total ill-defined contribution of the type (A2.16a) to $B_{\alpha\beta}^C\binom{\mathbf{q}}{kk}$ is

$$-Z_k e \left( \sum_{k'} Z_{k'} e \right) \frac{4\pi}{v} \left[ \frac{G_\alpha G_\beta}{G^2} \right]_{\mathbf{G} \equiv 0}.$$

This clearly vanishes since $\sum_{k'} Z_{k'} e = 0$ by virtue of the overall charge neutrality of the primitive cell. Thus there is a net cancellation of the various ill-defined terms which arise in the reciprocal-space series. Similarly, terms of the type (A2.16b) arise from $(Z_k e)^2 \mathcal{C}_{\alpha\beta}(\mathbf{q}, 0)$ and $-(Z_k e)^2 \mathcal{C}_{\alpha\beta}(\mathbf{q} \equiv 0, 0)$ and cancel each other.

We indicate explicitly that the various ill-defined terms may safely be omitted during practical calculations by rewriting (A2.14) in the form

$$B_{\alpha\beta}^C \binom{\mathbf{q}}{kk} = (Z_k e)^2 \{ \mathcal{B}_{\alpha\beta}(\mathbf{q}, 0) - \mathcal{C}_{\alpha\beta}'(\mathbf{q}, 0) \}$$
$$- (Z_k e)^2 \{ \mathcal{B}_{\alpha\beta}'(\mathbf{q} \equiv 0, 0) - \mathcal{C}_{\alpha\beta}'(\mathbf{q} \equiv 0, 0) \}$$
$$- \sum_{k'(\neq k)} Z_k Z_{k'} e^2 \{ \mathcal{B}_{\alpha\beta}'(\mathbf{q} \equiv 0, \mathbf{x}(k) - \mathbf{x}(k'))$$
$$- \mathcal{C}_{\alpha\beta}(\mathbf{q} \equiv 0, \mathbf{x}(k) - \mathbf{x}(k')) \}, \qquad \text{(A2.17)}$$

the primes on $\mathscr{B}_{\alpha\beta}$ and $\mathscr{C}_{\alpha\beta}$ denoting the omission of the leading terms in the corresponding series.

Expressions for $D_{\alpha\beta}^C\binom{\mathbf{q}}{kk'}$ and $M_{\alpha\beta}^C\binom{\mathbf{q}}{kk'}$ can be written down via the relations

$$D_{\alpha\beta}^C\binom{\mathbf{q}}{kk'} = (m_k m_{k'})^{-1/2} B_{\alpha\beta}^C\binom{\mathbf{q}}{kk'},$$

$$M_{\alpha\beta}^C\binom{\mathbf{q}}{kk'} = \exp\{-i\mathbf{q}\cdot[\mathbf{x}(k) - \mathbf{x}(k')]\} B_{\alpha\beta}^C\binom{\mathbf{q}}{kk'}. \qquad (A2.18)$$

Of these, the matrix $\mathbf{M}^C(\mathbf{q})$ is particularly suitable for a discussion of the long-wavelength properties. It may be written in the form

$$M_{\alpha\beta}^C\binom{\mathbf{q}}{kk'} = \bar{M}_{\alpha\beta}^C\binom{\mathbf{q}}{kk'} + Z_k Z_{k'} \frac{4\pi e^2}{v} \frac{q_\alpha q_\beta}{|\mathbf{q}|^2}. \qquad (A2.19)$$

Inspection of (A2.19) using Eqs. (A2.9), (A2.17), and (A2.18) shows that the leading term in the reciprocal-space series for $\bar{\mathbf{M}}^C$ is of the form

$$-\frac{4\pi}{v} \frac{q_\alpha q_\beta}{|\mathbf{q}|^2} [1 - \exp(-q^2/4\eta^2)],$$

which vanishes at $\mathbf{q} = 0$. Consequently $\bar{\mathbf{M}}^C$ is regular in $\mathbf{q}$ at $\mathbf{q} = 0$, all the nonregularity having been absorbed in the second term on the right-hand side of (A2.19). It is evident from this that if $\bar{\mathbf{M}}^C$ is used in place of $\mathbf{M}^C$ in a calculation, then effects associated with the nonregular contribution disappear. (An example of this appears in Figure 4.2.)

In the limit $\mathbf{q} \to 0$, the second term on the right-hand side of (A2.19) may be related to a slowly varying macroscopic polarization. This contribution to the Coulombic part $\mathbf{M}^C$ of the dynamical matrix is sometimes referred to as arising from the "long-range part" of the Coulomb interactions. The rest of the electric contribution is absorbed in $\bar{\mathbf{M}}^C$ and is considered to result from the "short-range part." The rationale for such descriptions will become clearer shortly.

To facilitate comparison with the literature, particularly Born and Huang, we introduce the following definitions:

$$Q_{\alpha\beta}\binom{\mathbf{q}}{kk'} = -\exp\{-i\mathbf{q}\cdot[\mathbf{x}(k) - \mathbf{x}(k')]\}$$
$$\times [\mathscr{B}_{\alpha\beta}(\mathbf{q}, \mathbf{x}(k) - \mathbf{x}(k')) - \mathscr{C}_{\alpha\beta}(\mathbf{q}, \mathbf{x}(k) - \mathbf{x}(k'))]$$
$$+ \frac{4\pi}{v} \frac{q_\alpha q_\beta}{|\mathbf{q}|^2}, \quad (k \neq k'), \qquad (A2.20a)$$

$$Q_{\alpha\beta}\binom{\mathbf{q}}{kk} = -\{\mathscr{B}_{\alpha\beta}(\mathbf{q}, 0) - \mathscr{C}_{\alpha\beta}'(\mathbf{q}, 0)\}$$
$$+ \frac{4\pi}{v} \frac{q_\alpha q_\beta}{|\mathbf{q}|^2} - \frac{4\eta^3}{3\sqrt{\pi}} \delta_{\alpha\beta}, \qquad (A2.20b)$$

$$Q_{\alpha\beta}\begin{pmatrix} 0 \\ kk' \end{pmatrix} = -\{\mathscr{B}'_{\alpha\beta}(\mathbf{q} \equiv 0, \mathbf{x}(k) - \mathbf{x}(k'))$$

$$- \mathscr{C}_{\alpha\beta}(\mathbf{q} \equiv 0, \mathbf{x}(k) - \mathbf{x}(k'))\}, \quad (k \neq k'), \quad \text{(A2.20c)}$$

$$Q_{\alpha\beta}\begin{pmatrix} 0 \\ kk \end{pmatrix} = -\{\mathscr{B}'_{\alpha\beta}(\mathbf{q} \equiv 0, 0) - \mathscr{C}_{\alpha\beta}(\mathbf{q} \equiv 0, 0)\} - \frac{4\eta^3}{3\sqrt{\pi}} \delta_{\alpha\beta}. \quad \text{(A2.20d)}$$

In terms of these quantities we may write

$$M_{\alpha\beta}^C\begin{pmatrix} \mathbf{q} \\ kk' \end{pmatrix} = Z_k Z_{k'} \frac{4\pi e^2}{v} \frac{q_\alpha q_\beta}{|\mathbf{q}|^2}$$

$$- Z_k Z_{k'} e^2 Q_{\alpha\beta}\begin{pmatrix} \mathbf{q} \\ kk' \end{pmatrix} + \delta_{kk'} \sum_{k''} (Z_k Z_{k''} e^2) Q_{\alpha\beta}\begin{pmatrix} 0 \\ kk'' \end{pmatrix}, \quad \text{(A2.21)}$$

a form popular with some authors. This result becomes further simplified for crystals in which *every* ion is in an environment having *at least* tetrahedral symmetry. Such structures are called *diagonally cubic* and include the NaCl, CsCl, and ZnS (sphalerite) structures. For such crystals[7]

$$Q_{\alpha\beta}\begin{pmatrix} 0 \\ kk' \end{pmatrix} = \frac{4\pi}{3v} \delta_{\alpha\beta}, \quad \text{(A2.22)}$$

which leads to the vanishing of the third term on the right-hand side of (A2.21), owing to the electrical neutrality of the primitive cell. For diagonally cubic crystals, $\mathbf{M}^C(\mathbf{q})$ is somewhat more conveniently calculated from (A2.21) and the definitions (A2.20) rather than by use of (A2.9), (A2.17), and (A2.18).

We shall also find it convenient to introduce the following definitions which are used in Chapter 4.

$$\hat{C}_{\alpha\beta}\begin{pmatrix} \mathbf{q} \\ kk' \end{pmatrix} = -\sum_{V} \frac{\partial^2}{\partial r_\alpha \partial r_\beta}\left(\frac{1}{r}\right)\bigg|_{r=|\mathbf{x}\begin{pmatrix} l'\,0 \\ k'\,k \end{pmatrix}|} \exp\left\{i\mathbf{q}\cdot\mathbf{x}\begin{pmatrix} l'\,0 \\ k'\,k \end{pmatrix}\right\}, \quad (k \neq k');$$

$$\text{(A2.23a)}$$

$$\hat{C}_{\alpha\beta}\begin{pmatrix} \mathbf{q} \\ kk \end{pmatrix} = -\sum_{l'(\neq 0)} \frac{\partial^2}{\partial r_\alpha \partial r_\beta}\left(\frac{1}{r}\right)\bigg|_{r=|\mathbf{x}\begin{pmatrix} l'\,0 \\ k\,k \end{pmatrix}|} \exp\left\{i\mathbf{q}\cdot\mathbf{x}\begin{pmatrix} l'\,0 \\ k\,k \end{pmatrix}\right\}. \quad \text{(A2.23b)}$$

We further introduce $\hat{C}_{\alpha\beta}\begin{pmatrix} 0 \\ kk' \end{pmatrix}$ $(k \neq k')$ and $\hat{C}_{\alpha\beta}\begin{pmatrix} 0 \\ kk \end{pmatrix}$ which are the $\mathbf{q} \equiv 0$ values of the corresponding quantities in (A2.23). We next define

$$C_{\alpha\beta}\begin{pmatrix} \mathbf{q} \\ kk' \end{pmatrix} = \hat{C}_{\alpha\beta}\begin{pmatrix} \mathbf{q} \\ kk' \end{pmatrix} - \delta_{kk'} \sum_{k''} \frac{Z_{k''}}{Z_k} \hat{C}_{\alpha\beta}\begin{pmatrix} 0 \\ kk'' \end{pmatrix}. \quad \text{(A2.24)}$$

It is then straightforward to show that

$$M_{\alpha\beta}^C\begin{pmatrix} \mathbf{q} \\ kk' \end{pmatrix} = Z_k e C_{\alpha\beta}\begin{pmatrix} \mathbf{q} \\ kk' \end{pmatrix} Z_{k'} e, \quad \text{(A2.25a)}$$

which is sometimes written in matrix form as

$$\mathbf{M}^C(\mathbf{q}) = \mathbf{Z}_d \mathbf{C} \mathbf{Z}_d, \tag{A2.25b}$$

where $\mathbf{Z}_d$ is a $(3n \times 3n)$ real matrix with elements

$$(\mathbf{Z}_d)_{\alpha\beta}(kk') = Z_k e \delta_{\alpha\beta} \delta_{kk'}, \tag{A2.26}$$

and $\mathbf{C}$ is a $(3n \times 3n)$ matrix with elements $C_{\alpha\beta}\binom{\mathbf{q}}{kk'}$. Note that both $\hat{\mathbf{C}}$ and $\mathbf{C}$ are Hermitian.

We call attention to the factor $(Z_{k''}/Z_k)$ in the second term on the right-hand side of (A2.24), which is necessary in order to guarantee the consequences of the translational sum rule for the Coulombic force constants. Corresponding to (A2.22) we have

$$\hat{C}_{\alpha\beta}\binom{0}{kk'} = -\frac{4\pi}{3v}\delta_{\alpha\beta} \tag{A2.27}$$

in diagonally cubic crystals, whence $C_{\alpha\beta}\binom{\mathbf{q}}{kk'} = \hat{C}_{\alpha\beta}\binom{\mathbf{q}}{kk'}$ for every $(kk')$.

We emphasize that the results (A2.21) and (A2.25) are appropriate to a lattice of *point* ions, that is, ions that cannot undergo electronic distortion. In such a case, the dynamical matrix $M_{\alpha\beta}\binom{\mathbf{q}}{kk'}$ can be written as

$$M_{\alpha\beta}\binom{\mathbf{q}}{kk'} = M_{\alpha\beta}^C\binom{\mathbf{q}}{kk'} + M_{\alpha\beta}^{SR}\binom{\mathbf{q}}{kk'}$$

$$= \overline{M}_{\alpha\beta}^C\binom{\mathbf{q}}{kk'} + \frac{4\pi e^2}{v} Z_k Z_{k'} \frac{q_\alpha q_\beta}{|\mathbf{q}|^2} + M_{\alpha\beta}^{SR}\binom{\mathbf{q}}{kk'},$$

where $M_{\alpha\beta}^{SR}$ is the contribution from the short-range forces. Since $M_{\alpha\beta}^{SR}$ is always regular at $\mathbf{q} = 0$, it can be combined with $\overline{M}_{\alpha\beta}^C$ and written as $\overline{M}_{\alpha\beta}$, so that

$$M_{\alpha\beta}\binom{\mathbf{q}}{kk'} = \overline{M}_{\alpha\beta}\binom{\mathbf{q}}{kk'} + \frac{4\pi e^2}{v} Z_k Z_{k'} \frac{q_\alpha q_\beta}{|\mathbf{q}|^2}.$$

If now we allow for the polarization of the ions, then once again $M_{\alpha\beta}$ can be written as the sum of a regular part $\overline{M}_{\alpha\beta}$ and a nonregular part. However, the latter will involve *charge tensors* $\mathbf{Z}(k|\mathbf{q})$ rather than scalar charges $Z_k e$. The manipulation of the dynamical matrix into this form requires a detailed examination of the effects associated with electronic distortion and is considered in Chapter 4 with reference to specific models.

## A2.2 Microscopic and macroscopic fields due to a dipolar array

Closely related to the problem just discussed is that of the Coulombic field due to a periodic array of point (electric) dipoles.* A familiar context

---

*A discussion of the potential due to periodic arrays of higher-order multipoles has been given by W. E. Rudge, Phys. Rev. **181**, 1020 (1969).

in which such a problem arises is that considered originally by Lorentz, namely the polarization of an ion in an alkali halide due to the field produced by polarization dipoles located at other lattice sites.

From electrostatics we know that the field at $\mathbf{r}$ due to a dipole of moment $\mathbf{p}$ placed at the origin is given by

$$\mathcal{E}(\mathbf{r}) = \frac{1}{r^5}[3(\mathbf{r}\cdot\mathbf{p})\mathbf{r} - r^2\mathbf{p}] \tag{A2.28a}$$

which can also be expressed as

$$\mathcal{E}_\alpha(\mathbf{r}) = \sum_\beta p_\beta \frac{\partial^2}{\partial r_\alpha \partial r_\beta}\left(\frac{1}{r}\right). \tag{A2.28b}$$

Consider now a periodic dipolar array in which the moment associated with the site $\binom{l'}{k'}$ is given by

$$\mathbf{p}\binom{l'}{k'} = \mathbf{p}(k')\exp\left[i\mathbf{q}\cdot\mathbf{x}\binom{l'}{k'}\right]. \tag{A2.29}$$

In view of (A2.28), the field $\mathcal{E}(\mathbf{r})$ at $\mathbf{r}$ due to this dipole distribution is given by

$$\mathcal{E}_\alpha(\mathbf{r}) = \sum_{k'\beta} p_\beta(k')\cdot\sum_{l'}\frac{\partial^2}{\partial r_\alpha \partial r_\beta}\frac{1}{|\mathbf{r} - \mathbf{x}\binom{l'}{k'}|}\exp\left[i\mathbf{q}\cdot\mathbf{x}\binom{l'}{k'}\right]. \tag{A2.30}$$

Straightforward application of the Ewald method[8] then gives (see also Eqs. (A2.8) and (A2.9))

$$
\begin{aligned}
\mathcal{E}_\alpha(\mathbf{r}) = &\sum_{k'}\sum_\beta p_\beta(k')\exp[i\mathbf{q}\cdot\mathbf{x}(k')]\frac{\partial^2}{\partial r_\alpha \partial r_\beta} \\
&\times \left(\frac{4\pi}{v}\sum_\mathbf{G}\frac{1}{|\mathbf{G}+\mathbf{q}|^2}\exp[-(\mathbf{G}+\mathbf{q})^2/4\eta^2]\right. \\
&\times \exp\{i(\mathbf{q}+\mathbf{G})\cdot[\mathbf{r}-\mathbf{x}(k')]\} + \sum_{l'}\left[\frac{\mathrm{erfc}(\eta x)}{x}\right]_{x=|\mathbf{r}-\mathbf{x}(l')-\mathbf{x}(k')|} \\
&\left.\times \exp[i\mathbf{q}\cdot\mathbf{x}(l')]\right) \\
= &\sum_{k'}\sum_\beta p_\beta(k')\exp[i\mathbf{q}\cdot\mathbf{x}(k')] \\
&\times \{-\mathcal{B}_{\alpha\beta}(\mathbf{q},\mathbf{r}-\mathbf{x}(k')) + \mathcal{C}_{\alpha\beta}(\mathbf{q},\mathbf{r}-\mathbf{x}(k'))\}.
\end{aligned}
\tag{A2.31}
$$

For convenience let us rewrite the above result as

$$
\begin{aligned}
\mathcal{E}_\alpha(\mathbf{r}) = &E_\alpha(\hat{\mathbf{q}})\exp(i\mathbf{q}\cdot\mathbf{r}) + \sum_{k'}\sum_\beta p_\beta(k')\left\{-\exp[i\mathbf{q}\cdot\mathbf{x}(k')]\right. \\
&\times \mathcal{B}_{\alpha\beta}(\mathbf{q},\mathbf{r}-\mathbf{x}(k')) + \exp[i\mathbf{q}\cdot\mathbf{x}(k')]\mathcal{C}_{\alpha\beta}(\mathbf{q},\mathbf{r}-\mathbf{x}(k')) \\
&\left.+ \frac{4\pi}{v}\hat{q}_\alpha\hat{q}_\beta\exp(i\mathbf{q}\cdot\mathbf{r})\right\},
\end{aligned}
\tag{A2.32}
$$

where

$$E_\alpha(\hat{\mathbf{q}}) = -\frac{4\pi}{v}\hat{q}_\alpha\left(\hat{\mathbf{q}}\cdot\sum_k \mathbf{p}(k)\right), \tag{A2.33}$$

$\hat{\mathbf{q}}$ being a unit vector in the direction of $\mathbf{q}$. The first term on the right-hand side of (A2.32) is not a regular function of $\mathbf{q}$ while the second term is.

From a physical point of view, $\mathcal{E}(\mathbf{r})$ represents the microscopic field at $\mathbf{r}$ due to the dipolar array and can fluctuate rapidly from point to point. This field has a contribution $\mathbf{E}(\mathbf{r}) = \mathbf{E}(\hat{\mathbf{q}})\exp(i\mathbf{q}\cdot\mathbf{r})$ which, in the $\mathbf{q} \to 0$ limit, represents a macroscopic field. In other words, unlike $\mathcal{E}(\mathbf{r})$, $\mathbf{E}(\mathbf{r})$ shows a smooth and slowly varying behavior as a function of $\mathbf{r}$, being uniform over regions of the size of a unit cell and varying only over a much larger scale.[9] The contribution to the microscopic field given by the second term of (A2.32) is often termed the *inner field* and, as pointed out by Born and Huang[10], gives that part of the electric interactions which has effectively a limited range. The macroscopic field, on the other hand, represents the contribution associated with the long-range part of the electrical interactions. Later we shall see that the inner field is closely related to $\overline{\mathbf{M}}^C$, while $\mathbf{E}(\hat{\mathbf{q}})$ is related to the term $(4\pi e^2/v)Z_k Z_k\cdot\hat{q}_\alpha\hat{q}_\beta$ in (A2.19). This explains the basis for referring to these terms as, respectively, the "short-range" and the "long-range" parts of the Coulomb interactions.

To demonstrate that $\mathbf{E}(\mathbf{r})$ refers to a macroscopic field, we proceed as follows: Suppose that $\mathbf{q}$ is very small compared to the inverse dimensions of the unit cell (the long-wavelength limit), so that the value of the dipole moment in any given sublattice changes very little from one cell to another. In other words, we can imagine the lattice as a polarized continuum with a macroscopic polarization $\mathbf{P}(\mathbf{r})$ given by*

$$\mathbf{P}(\mathbf{r}) = \mathbf{P}_0 \exp(i\mathbf{q}\cdot\mathbf{r}) = \frac{1}{v}\left(\sum_k \mathbf{p}(k)\right)\exp(i\mathbf{q}\cdot\mathbf{r}). \tag{A2.34}$$

The corresponding macroscopic field $\mathbf{E}(\mathbf{r})$ may be found from Poisson's equation

$$\mathbf{V}\cdot\mathbf{D} = \mathbf{V}\cdot[\mathbf{E}(\mathbf{r}) + 4\pi\mathbf{P}(\mathbf{r})] = 0, \tag{A2.35}$$

where it is to be noted that $\mathbf{E}(\mathbf{r})$ is an irrotational field.† We now wish to use (A2.34) in (A2.35) and then solve for $\mathbf{E}(\mathbf{r})$.

---

*The quantity $\mathbf{P}(\mathbf{r})$ is the polarization density and is usually treated as having only a dipolar character (as we have done). Strictly speaking, however, the polarization density includes contributions from the higher multipoles also. See L. Rosenfeld, *Theory of Electrons* (Dover, New York, 1965), Chap. I, Sec. 4; J. D. Jackson, *Classical Electrodynamics* (Wiley, New York, 1962) p. 101.

†A vector field whose curl is zero is referred to as irrotational.

We first split $\mathbf{P}(\mathbf{r})$ into $\mathbf{P}_{\|}(\mathbf{r})$ and $\mathbf{P}_{\perp}(\mathbf{r})$, respectively parallel and perpendicular to $\hat{\mathbf{q}}$. It is easy to verify that $\mathbf{P}_{\|}(\mathbf{r})$ is irrotational and that $\mathbf{P}_{\perp}(\mathbf{r})$ is solenoidal.* Therefore, from (A2.35) we have

$$\mathbf{V}\cdot[\mathbf{E}(\mathbf{r}) + 4\pi\mathbf{P}_{\|}(\mathbf{r})] + \mathbf{V}\cdot 4\pi\mathbf{P}_{\perp}(\mathbf{r}) = \mathbf{V}\cdot[\mathbf{E}(\mathbf{r}) + 4\pi\mathbf{P}_{\|}(\mathbf{r})] = 0. \quad \text{(A2.36)}$$

On the other hand, since both $\mathbf{E}(\mathbf{r})$ and $\mathbf{P}_{\|}(\mathbf{r})$ are irrotational,

$$\mathbf{V} \times [\mathbf{E}(\mathbf{r}) + 4\pi\mathbf{P}_{\|}(\mathbf{r})] = 0. \quad \text{(A2.37)}$$

It is a well-known theorem in vector analysis that if both the curl and the divergence of a vector vanish everywhere, then the vector field itself must vanish, whence

$$
\begin{aligned}
\mathbf{E}(\mathbf{r}) &= -4\pi\mathbf{P}_{\|}(\mathbf{r}) \\
&= -4\pi\hat{\mathbf{q}}(\hat{\mathbf{q}}\cdot\mathbf{P}(\mathbf{r})) \\
&= -\frac{4\pi}{v}\,\hat{\mathbf{q}}\left(\hat{\mathbf{q}}\cdot\sum_{k}\mathbf{p}(k)\right)\exp(i\mathbf{q}\cdot\mathbf{r}) \\
&= -4\pi\hat{\mathbf{q}}(\hat{\mathbf{q}}\cdot\mathbf{P}_{0})\exp(i\mathbf{q}\cdot\mathbf{r}) \\
&= \mathbf{E}(\hat{\mathbf{q}})\exp(i\mathbf{q}\cdot\mathbf{r}),
\end{aligned}
$$

from which (A2.33) follows. This provides the basis for our assertion that the first term on the right-hand side of (A2.32) describes (in the long-wavelength limit) the macroscopic field contributed by the dipolar array.

### A2.3 Relationship between electric fields of dipolar arrays and Coulombic sums

Result (A2.32) expresses the microscopic field at an arbitrary space point $\mathbf{r}$ due to a dipolar array. This expression may now be used to calculate the field acting on any particular dipole in the array itself, say the dipole at $\binom{0}{k}$, due to all the *other* dipoles. This field, called the *effective* or the *local* field,† is given by

$$\mathscr{E}\binom{0}{k} = \lim_{\mathbf{r}\to\mathbf{x}(k)}\left[\mathscr{E}(\mathbf{r}) - \text{field at }\mathbf{r}\text{ due to dipole at }\binom{0}{k}\right]. \quad \text{(A2.38)}$$

The evaluation of the right-hand side has been discussed by Born and Huang[11] and the final result is

$$\mathscr{E}_{\alpha}\binom{0}{k} = \exp[i\mathbf{q}\cdot\mathbf{x}(k)]\left[E_{\alpha}(\hat{\mathbf{q}}) + \sum_{k'\beta}p_{\beta}(k')Q_{\alpha\beta}\binom{\mathbf{q}}{kk'}\right]. \quad \text{(A2.39)}$$

---

*A vector field whose divergence is zero is termed solenoidal.
†Born and Huang refer to this as the exciting field after the German terminology *erregendes Feld* introduced by Ewald.

In many phenomenological models one visualizes the charge fluctuations (produced by the vibrations) in terms of appropriate dipoles located at the atomic sites. For example, these could be purely displacement dipoles or a combination of displacement and electronic distortion dipoles. In such dipolar models, the dynamical equations can often be manipulated in terms of local fields, and in this context (A2.39) proves useful. Whereas we shall be concerned mainly with local fields due to *point dipoles*, the concept has been enlarged recently to treat fields acting on electrons due to all the other charges in the medium. The latter viewpoint is useful in setting up a microscopic theory of lattice dynamics. It should also be noted that charge fluctuations arising from electronic polarization can be regarded as point dipoles only in solids where the electronic wave function is highly localized, for example, alkali halides. Though frequently used, the treatment of charge fluctuations via point dipoles is not appropriate for III-V and II-VI compounds.

As a simple illustration of the use of (A2.39) in lattice dynamics, let us consider the rigid-ion model discussed in Sec. 4.2. The equations of motion are

$$
\begin{aligned}
m_k \ddot{u}_\alpha\binom{l}{k} &= -\sum_{l'k'\beta} \left\{ \phi_{\alpha\beta}^{SR}\binom{l\ \ l'}{k\ \ k'} + \phi_{\alpha\beta}^{C}\binom{l\ \ l'}{k\ \ k'} \right\} u_\beta\binom{l'}{k'} \\
&= -\sum_{l'k'\beta} \phi_{\alpha\beta}^{SR}\binom{l\ \ l'}{k\ \ k'} u_\beta\binom{l'}{k'} - \sum_{\substack{l'k'\beta \\ (lk \neq l'k')}} \phi_{\alpha\beta}^{C}\binom{l\ \ l'}{k\ \ k'} u_\beta\binom{l'}{k'} \\
&\quad + \sum_{\substack{l'k'\beta \\ (lk \neq l'k')}} \phi_{\alpha\beta}^{C}\binom{l\ \ l'}{k\ \ k'} u_\beta\binom{l}{k}.
\end{aligned}
\tag{A2.40}
$$

Remembering (A2.2) and (A2.30), we may write Eq. (A2.40) as

$$
m_k \ddot{u}_\alpha\binom{l}{k} = -\sum_{l'k'\beta} \phi_{\alpha\beta}^{SR}\binom{l\ \ l'}{k\ \ k'} u_\beta\binom{l'}{k'} + Z_k e\left\{ \mathscr{E}_\alpha^{(1)}\binom{l}{k} + \mathscr{E}_\alpha^{(2)}\binom{l}{k} \right\}, \tag{A2.41}
$$

where $\mathscr{E}^{(1)}\binom{l}{k}$ and $\mathscr{E}^{(2)}\binom{l}{k}$ are, respectively, the local fields at $\binom{l}{k}$ due to the displacement dipoles

$$
\mathbf{p}^{(1)}\binom{l'}{k'} = Z_{k'} e \mathbf{u}\binom{l'}{k'}, \quad \mathbf{p}^{(2)}\binom{l'}{k'} = -Z_{k'} e \mathbf{u}\binom{l}{k}, \tag{A2.42}
$$

and are given explicitly as

$$
\mathscr{E}_\alpha^{(1)}\binom{l}{k} = \sum_{\substack{l'k'\beta \\ (lk \neq l'k')}} \frac{\partial^2}{\partial r_\alpha \partial r_\beta}\left(\frac{1}{r}\right)\bigg|_{r=|\mathbf{x}(\frac{l'\ l}{k'\ k})|} Z_{k'} e u_\beta\binom{l'}{k'} \tag{A2.43a}
$$

$$
\mathscr{E}_\alpha^{(2)}\binom{l}{k} = -\sum_{\substack{l'k'\beta \\ (lk \neq l'k')}} \frac{\partial^2}{\partial r_\alpha \partial r_\beta}\left(\frac{1}{r}\right)\bigg|_{r=|\mathbf{x}(\frac{l'\ l}{k'\ k})|} Z_{k'} e u_\beta\binom{l}{k}. \tag{A2.43b}
$$

In passing, it may be noted that (A2.41) could have been obtained by a consideration of the harmonic potential energy function

$$\Phi^{(2)} = \tfrac{1}{2} \sum_{l k \alpha} \sum_{l' k' \beta} \phi_{\alpha\beta}^{SR}\begin{pmatrix} l & l' \\ k & k' \end{pmatrix} u_\alpha\begin{pmatrix} l \\ k \end{pmatrix} u_\beta\begin{pmatrix} l' \\ k' \end{pmatrix} - \tfrac{1}{2} \sum_{l k \alpha} Z_k e u_\alpha\begin{pmatrix} l \\ k \end{pmatrix} \mathscr{E}_\alpha\begin{pmatrix} l \\ k \end{pmatrix}, \quad \text{(A2.44)}$$

where

$$\mathscr{E}\begin{pmatrix} l \\ k \end{pmatrix} = \mathscr{E}^{(1)}\begin{pmatrix} l \\ k \end{pmatrix} + \mathscr{E}^{(2)}\begin{pmatrix} l \\ k \end{pmatrix}. \quad \text{(A2.45)}$$

Let us now write the nuclear displacements in wavelike form as

$$\mathbf{u}\begin{pmatrix} l \\ k \end{pmatrix} = \mathbf{U}(k|\mathbf{q}) \exp\left\{ i \left[ \mathbf{q} \cdot \mathbf{x}\begin{pmatrix} l \\ k \end{pmatrix} - \omega(\mathbf{q})t \right] \right\}. \quad \text{(A2.46)}$$

The two dipolar arrays are then as illustrated in Figure A2.2. The corresponding local fields are obtainable from (A2.39):

$$\mathscr{E}_\alpha^{(1)}\begin{pmatrix} l \\ k \end{pmatrix} = \left[ -\frac{4\pi}{v} \hat{q}_\alpha \left\{ \sum_{k'\beta} \hat{q}_\beta Z_{k'} e U_\beta(k'|\mathbf{q}) \right\} \right.$$
$$\left. + \sum_{k'\beta} Z_{k'} e U_\beta(k'|\mathbf{q}) Q_{\alpha\beta}\begin{pmatrix} \mathbf{q} \\ kk' \end{pmatrix} \right] \exp\left\{ i \left[ \mathbf{q} \cdot \mathbf{x}\begin{pmatrix} l \\ k \end{pmatrix} - \omega t \right] \right\},$$
$$\text{(A2.47a)}$$

$$\mathscr{E}_\alpha^{(2)}\begin{pmatrix} l \\ k \end{pmatrix} = - \sum_{k'\beta} Z_{k'} e U_\beta(k|\mathbf{q}) Q_{\alpha\beta}\begin{pmatrix} 0 \\ kk' \end{pmatrix} \exp\left\{ i \left[ \mathbf{q} \cdot \mathbf{x}\begin{pmatrix} l \\ k \end{pmatrix} - \omega t \right] \right\}. \quad \text{(A2.47b)}$$

Two points are worthy of attention here. Firstly, the local field $\mathscr{E}^{(2)}$ has no contribution from the macroscopic field. This is because

$$\sum_{k'} Z_{k'} e U_\beta(k|\mathbf{q}) = U_\beta(k|\mathbf{q}) \sum_{k'} Z_{k'} e$$
$$= 0,$$

from charge neutrality. Hence the analogue of the macroscopic field term in (A2.47a) is absent in (A2.47b). Secondly the dipole distribution $\{\mathbf{p}^{(2)}\binom{l}{k'}\}$ is associated with *equal displacements* $\mathbf{U}(k|\mathbf{q}) \exp\{i[\mathbf{q} \cdot \mathbf{x}\binom{l}{k}) - \omega t]\}$ and *not* a wavelike set of displacements characterized by wave vector $\mathbf{q}$ as is the set $\{\mathbf{p}^{(1)}\binom{l}{k'}\}$. For this reason, it is $Q_{\alpha\beta}\binom{0}{kk'}$ which enters (A2.47b). Note that on account of (A2.22), $\mathscr{E}^{(2)}$ vanishes in diagonally cubic crystals.

Using (A2.46) and (A2.47), we can now write (A2.41) as

$$m_k \omega^2(\mathbf{q}) U_\alpha(k|\mathbf{q}) = \sum_{k'\beta} M_{\alpha\beta}^{SR}\begin{pmatrix} \mathbf{q} \\ kk' \end{pmatrix} U_\beta(k'|\mathbf{q}) - Z_k e \mathscr{E}_\alpha(k|\mathbf{q}), \quad \text{(A2.48)}$$

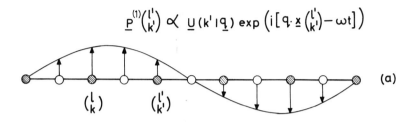

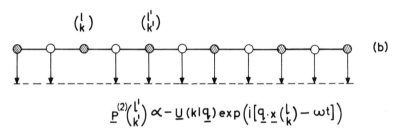

**Figure A2.2**
Schematic drawings (appropriate to a diatomic chain) illustrating the dipolar arrays giving rise to the local fields $\mathscr{E}^{(1)}$ and $\mathscr{E}^{(2)}$ discussed in the text. In (a) the displacements associated with the dipoles are undulatory while in (b) the associated displacements are equal. Thus, whereas in a the parameter $\mathbf{q}$ characterizes the wavelength of the modulation of the dipoles, in b it just characterizes the constant amplitude. Remember that, in computing the local field on $\binom{l}{k}$, the contribution from that ion is excluded.

where $\mathscr{E}_\alpha(k|\mathbf{q})$ is defined by

$$\mathscr{E}_\alpha\binom{l}{k} = \mathscr{E}_\alpha(k|\mathbf{q}) \exp\left\{i\left[\mathbf{q}\cdot\mathbf{x}\binom{l}{k} - \omega(\mathbf{q})t\right]\right\}$$

$$= \exp\left\{i\left[\mathbf{q}\cdot\mathbf{x}\binom{l}{k} - \omega(\mathbf{q})t\right]\right\}\left\{-\frac{4\pi}{v}\hat{q}_\alpha\sum_{k'\beta}\hat{q}_\beta Z_{k'}e U_\beta(k'|\mathbf{q})\right.$$

$$\left. + \sum_{k'\beta} Z_{k'}e U_\beta(k'|\mathbf{q}) Q_{\alpha\beta}\binom{\mathbf{q}}{kk'} - \sum_{k'\beta} Q_{\alpha\beta}\binom{0}{kk'} Z_{k'}e U_\beta(k|\mathbf{q})\right\}.$$

$$(A2.49)$$

In the notation of (A2.24),

$$\mathscr{E}_\alpha(k|\mathbf{q}) = -\sum_{k'\beta} C_{\alpha\beta}\binom{\mathbf{q}}{kk'} Z_{k'}e U_\beta(k'|\mathbf{q}). \qquad (A2.50)$$

This demonstrates how Coulomb interactions may, if necessary, be described in terms of local fields. In this context it is important to remember the role of $\mathscr{E}^{(2)}$, which is closely connected with the self terms in the Coulomb force constants. Further, in view of Eqs. (A2.46–A2.47) and (A2.21), we could also write the equations of motion (A2.41) as

$$m_k\omega^2(\mathbf{q})U_\alpha(k|\mathbf{q}) = \sum_{k'\beta} \left\{ M_{\alpha\beta}^{SR}\begin{pmatrix} \mathbf{q} \\ kk' \end{pmatrix} + M_{\alpha\beta}^{C}\begin{pmatrix} \mathbf{q} \\ kk' \end{pmatrix} \right\} U_\beta(k'|\mathbf{q}) - Z_k e E_\alpha(\hat{\mathbf{q}}),$$

where

$$E_\alpha(\hat{\mathbf{q}}) = -\frac{4\pi}{v} \hat{q}_\alpha \left[ \hat{\mathbf{q}} \cdot \sum_k Z_k e \mathbf{U}(k|\mathbf{q}) \right]$$

denotes the macroscopic field associated with the displacements in the vibrating lattice. Equation (A2.48) can thus be rewritten in the familiar form by expressing the local field as the sum of an inner field and a macroscopic field.

In the case of diagonally cubic crystals, (A2.39) acquires a particularly simple form in the $\mathbf{q} \to 0$ limit:

$$\mathscr{E} = \mathbf{E} + \frac{4\pi}{3}\mathbf{P}. \tag{A2.51}$$

Here $\mathscr{E}$ denotes the local field acting on a dipole and has the same amplitude for all the sublattices; $\mathbf{P}$ is the macroscopic polarization and in a dipole approximation can be defined as the dipole moment per unit volume. In ZnS, for example, $\mathbf{P}$ would be given by (in the notation of Sec. 4.1)

$$\mathbf{P} = \frac{1}{v}Ze[\mathbf{U}(1) - \mathbf{U}(2)],$$

if one were to consider only displacement dipoles. If one were to include the electronic polarizability effects associated with the local field, then

$$\mathbf{P} = \frac{1}{v}\{Ze[\mathbf{U}(1) - \mathbf{U}(2)] + (\alpha_+ + \alpha_-)\mathscr{E}\},$$

and so on. The macroscopic field $\mathbf{E}$ is related to $\mathbf{P}$ as usual by Eq. (A2.33). Result (A2.51) is often called the *Lorentz relation*. It can be derived directly in a simple manner by starting with a lattice that is diagonally cubic.[12]

The Lorentz relation is often quoted in elementary solid-state texts in a somewhat different context. The problem discussed therein is that of the electric field inside a *finite* dielectric specimen placed in a *uniform* electric field $\mathbf{E}_0$. As a result dipoles are created within, and the local field acting on any one of these dipoles is given by[13]

$\mathscr{E} = \mathbf{E}_0$

   $+ \mathbf{E}_1$ (depolarization field; depends on sample shape)

   $+ \mathbf{E}_2$ (field due to charge on inner surface of a spherical cavity centered on the concerned dipole; this contribution is always $\frac{4}{3}\pi \mathbf{P}$)

   $+ \mathbf{E}_3$ (field due to dipoles within the imaginary sphere; this contribution vanishes for sites having tetrahedral symmetry).    (A2.52)

The various contributions are illustrated in Figure A2.3. For diagonally cubic crystals, (A2.52) reduces to

$$\mathscr{E} = \mathbf{E}_0 + \mathbf{E}_1 + \frac{4\pi}{3}\mathbf{P}$$

$$= \mathbf{E} + \frac{4\pi}{3}\mathbf{P},$$    (A2.53)

where $\mathbf{E} = \mathbf{E}_0 + \mathbf{E}_1$ is the macroscopic field.

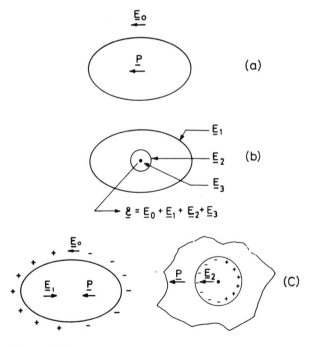

**Figure A2.3**
Contributions to the local field in the case of a dielectric specimen placed in a uniform external field $\mathbf{E}_0$. (a) depicts the direction of the induced polarization $\mathbf{P}$ with respect to $\mathbf{E}_0$. (b) shows the various contributions to the local field as discussed in Eq. (A2.52). (c) shows the directions of $\mathbf{E}_1$ and $\mathbf{E}_2$ relative to $\mathbf{P}$.

It is important to remember that, in spite of their similarity, (A2.51) and (A2.53) have been obtained under different assumptions. For example, the macroscopic field in (A2.53) involves surface charges resulting from the *uniform* polarization of the sample. On the other hand, $\mathbf{E}$ in (A2.51) is the result of the spatial variation of $\mathbf{P(r)}$ in an infinitely extended crystal. We draw attention to this point because in the literature there is sometimes a tendency to discuss fields in infinite crystals in the long-wavelength limit by borrowing the results for finite crystals placed in uniform fields. For example, the Lyddane-Sachs-Teller relation in cubic diatomic crystals (Eq. (4.5)) is sometimes considered along these lines. In our view, such a grafting procedure is not desirable. However, it is a different matter if one is explicitly considering long-wavelength modes in finite crystals. In this case surface charges do come into the picture and, further, the field may be regarded as truly uniform if $q \ll d^{-1}$, where $d$ refers to the smallest linear dimension of the crystal. A different treatment is required for dealing with the consequences of the macroscopic fields (the LST relation, for example) in such cases,[14] in order to allow room for specific surface effects.

## A2.4  Coulombic lattice sums in external-mode calculations

The method of Sec. A2.1 can be extended in a straightforward manner to calculate the elements $B_{\alpha\beta}^{C,ii'}(\genfrac{}{}{0pt}{}{\mathbf{q}}{\kappa\kappa'})$ of the matrix $\mathbf{B}^C(\mathbf{q})$ representing the Coulombic contribution to the dynamical matrix associated with external modes in complex crystals. The essential idea is to use the result (2.248) to express intermolecular force constants in terms of interatomic force constants. In this way, the problem of evaluating $B_{\alpha\beta}^{C,ii'}$ can be reduced to that of computing lattice sums involving $\phi_{\alpha\beta}^C\genfrac{(}{)}{0pt}{}{l\ \ l'}{\kappa\ \kappa'}{k\ k'}$ by methods considered in Sec. A2.1. The details appear in recent literature[6] and therefore will not be repeated here.

## References

1. E. W. Kellermann, Phil. Trans. Roy. Soc. London **A238**, 513 (1940). See also BH, Chapter V.

2. P. P. Ewald, Thesis, Univ. of Munich (1912); Ann. Physik **49**, 1, 117 (1916); **54**, 519, 557 (1917); **64**, 253 (1921); Nachr. Ges. Wiss. Gottingen, N. F. II, **55**, (1938).

3. BH, p. 250.

4. C. Kittel, *Introduction to Solid State Physics* (Wiley, New York, 1956), second edition, Appendix A.

5. A similar problem in a related context is discussed by E. K. Broch in Proc. Cambridge Phil. Soc. **33**, 485 (1938). The same techniques can be applied to the present case.

6. G. Venkataraman and V. C. Sahni, Rev. Mod. Phys. **42**, 409 (1970).

7. BH, p. 400.

8. BH, p. 252.

9. W. K. H. Panofsky and M. Phillips, *Classical Electricity and Magnetism* (Addison-Wesley, Reading, Mass., 1962), Chapter 2. For a comprehensive discussion concerning electric fields in dielectrics, see R. H. Cole, in *Progress in Dielectrics* (Heywood and Co., London, 1961), Vol. 3, p. 49.

10. BH, pp. 250, 252.

11. BH, p. 254.

12. BH, p. 102.

13. Kittel, Chapter 12. See also A. R. von Hippel, *Dielectrics and Waves* (Wiley, New York, 1954); R. P. Feynman, R. B. Leighton, and M. Sands, *The Feynman Lectures on Physics* (Addison-Wesley, Reading, Mass., 1964) Vol. II, Chapters 10 and 11; E. M. Purcell, *Electricity and Magnetism—Berkeley Physics Course* (McGraw-Hill, New York, 1965), Chapter 9.

14. A. A. Maradudin and G. H. Weiss, Phys. Rev. **123**, 1968 (1961).

# Glossary

Here we list the more important or often used symbols which occur in this volume. Chapter, section, or equation references are provided as a pointer to fuller definitions or to examples of usage. Many symbols are also used for other quantities, as the context should make clear.

| | |
|---|---|
| $\mathbf{a}_i$ | basis vector of the direct lattice, (2.7) |
| $a_{\mathbf{q}j}^{\dagger}$, $a_{\mathbf{q}j}$ | creation and annihilation operators, (2.175–2.176) |
| $A\binom{\mathbf{q}}{j}$ | amplitude associated with $\mathbf{e}\binom{\mathbf{q}}{j}$, (2.48) |
| $\mathbf{b}_i$ | basis vector of the reciprocal lattice, (2.8) |
| BZ | Brillouin zone |
| $B_{\alpha\beta}\binom{\mathbf{q}}{kk'}$ | element of alternate form of dynamical matrix, (2.77) |
| $B_{\alpha\beta}^{C}\binom{\mathbf{q}}{kk'}$ | Coulombic part of $B_{\alpha\beta}\binom{\mathbf{q}}{kk'}$, (A2.1) |
| $\mathscr{B}_{\alpha\beta}(\mathbf{q}, \mathbf{x})$ | reciprocal-space series in specification of $B_{\alpha\beta}^{C}\binom{\mathbf{q}}{kk'}$, (A2.10) |
| $c_j(\hat{\mathbf{q}})$ | velocity of acoustic waves of $j$th branch in direction $\hat{\mathbf{q}}$, (2.118) |
| $C_{\alpha\beta,\gamma\lambda}$ | elastic constant, (2.104) |
| $C_{ij}$ | elastic constant in Voigt notation, (2.110) |
| $C_{\alpha\beta}\binom{\mathbf{q}}{kk'}$ | $(Z_k Z_{k'} e^2)^{-1} M_{\alpha\beta}^{C}\binom{\mathbf{q}}{kk'}$, (4.29), (A2.24) |
| $\mathscr{C}_{\alpha\beta}(\mathbf{q}, \mathbf{x})$ | real-space series in specification of $B_{\alpha\beta}^{C}\binom{\mathbf{q}}{kk'}$, (A2.11a) |
| $D_{\alpha\beta}\binom{\mathbf{q}}{kk'}$ | element of the dynamical matrix, (2.43) |
| $\mathscr{D}(\mathbf{q})$ | block-diagonal dynamical matrix, (3.38) |
| $\mathbf{D}$ | electric displacement, Sec. 4.2 |
| $e$ | $4.8 \times 10^{-10}$ esu |
| $e^*$, $e_S^*$, $e_L^*$, $e_T^*$ | effective charges, Chapters 4 and 6 |
| $e_{\alpha,\beta\gamma}$ | piezoelectric constant, (4.92) |
| $e_{\alpha i}$ | piezoelectric constant in Voigt notation, Sec. 4.3.2 |
| $\mathbf{e}\binom{\mathbf{q}}{j}$ | normalized eigenvector, (2.48) |
| $\mathbf{e}(\mathbf{q})$ | matrix assembled from the $\mathbf{e}\binom{\mathbf{q}}{j}$, (2.50) |
| $E_F$ | the Fermi energy |
| $\mathscr{E}_m$ | translation group element, Sec. 2.2.2 |

| $\mathscr{E}\left(\begin{smallmatrix}l\\k\end{smallmatrix}\right)$ | local field at $\left(\begin{smallmatrix}l\\k\end{smallmatrix}\right)$, Sec. 4.6, (A2.38) |
|---|---|
| $\mathscr{E}(k\|\mathbf{q})$ | amplitude of the local field acting on ions of the $k$th sublattice, (4.30) |
| $\mathscr{E}(\mathbf{r})$ | microscopic field at $\mathbf{r}$, Sec. 4.1, Sec. A2.2 |
| $\mathbf{E}(\mathbf{r})$ | macroscopic field at $\mathbf{r}$, Sec. 4.1, Sec. A2.2 |
| $\mathbf{E}(\mathbf{q})$ | amplitude of the electric field vector, identifiable with the macroscopic field in the long-wavelength limit, (4.76), Sec. A2.2 |
| $e_\alpha(\mathbf{q})$ | $E_\alpha(\mathbf{q})e^{-i\omega t}$, (4.85) |
| $\mathbf{E}_{\mathrm{mac}}$ | $3n$-component macroscopic field vector used in the shell model, (4.298) |
| $g(\omega)$ | frequency distribution function, (2.90) |
| $g_j(\omega)$ | partial frequency distribution, corresponding to branch $j$, (2.94) |
| $\mathbf{G}$, $\mathbf{G}(h)$ | reciprocal lattice vector, (2.8) |
| $G$ | space group of the crystal, Sec. 2.2.1 |
| $G_0$ | point group of the crystal, Sec. 2.2.1 |
| $G(\mathbf{q})$ | group of the wave vector $\mathbf{q}$, Sec. 3.1 |
| $G_0(\mathbf{q})$ | point group of the wave vector $\mathbf{q}$, Sec. 3.1 |
| $H$ | Hamiltonian of the crystal, (2.1) |
| IMR | irreducible multiplier representation, Sec. 3.1 |
| $k$ | sublattice index, (2.6) |
| $k_{\mathrm{B}}$ | the Boltzmann constant |
| $k_{\mathrm{F}}$ | the Fermi wave vector |
| $\mathbf{K}(k)$ | core-shell interaction of an ion in the $k$th sublattice in the shell model, (4.237) |
| $l$ | cell index, (2.6) |
| $m$ | electron mass |
| $\bar{m}$ | reduced mass, Sec. 4.1.3 |
| $m_i$ | mass of $i$th entity, Sec. 2.2.2 |

| | |
|---|---|
| $\mathbf{m}$ | $3n \times 3n$ diagonal matrix of masses $m_i$, (2.78) |
| $M$ | mass of the primitive cell, (2.85) |
| $M_{\alpha\beta}\binom{\mathbf{q}}{kk'}$ | element of alternate form of the dynamical matrix, (2.81) |
| $\mathbf{M}^C(\mathbf{q})$, $\mathbf{M}^{SR}(\mathbf{q})$ | Coulombic and short-range parts of $\mathbf{M}(\mathbf{q})$, (4.27) |
| $\overline{\mathbf{M}}^C(\mathbf{q})$ | regular part of $\mathbf{M}^C(\mathbf{q})$, (4.32) |
| $\overline{\mathbf{M}}(\mathbf{q})$ | $\mathbf{M}^{SR}(\mathbf{q}) + \overline{\mathbf{M}}^C(\mathbf{q})$, (4.32), (4.76) |
| $n$ | number of atoms in the primitive cell, Sec. 2.2.1 |
| $N$ | number of primitive cells in the crystal, Sec. 2.2.9 |
| $n(\omega)$ | refractive index, Sec. 4.3.6 |
| $\bar{n}(\hat{\mathbf{q}}, \omega)$ | $n(\hat{\mathbf{q}}, \omega) + i\mathcal{K}(\hat{\mathbf{q}}, \omega)$, the complex refractive index, (5.22) |
| $\mathbf{p}\binom{l}{k}$ | electronic dipole moment of the $l k$th ion, Sec. 4.6 |
| $\mathbf{P}$ | 1. polarization density (3 dimensional), Sec. 4.1, Sec. A2.2 2. $3n$-component polarization vector used in the shell model, formed from 3-component vectors $\mathbf{P}(k)$, (4.297) |
| $\mathbf{P}(\mathbf{r})$ | macroscopic polarization at $\mathbf{r}$, Sec. 4.1 |
| $\mathbf{P}(\mathbf{q})$ | amplitude of the polarization density vector, (4.35), (4.77) |
| $\mathbf{q}$ | reduced wave vector, (2.41) |
| $\hat{\mathbf{q}}$ | $\mathbf{q}/|\mathbf{q}|$ |
| $\mathbf{Q}$ | a vector in reciprocal space |
| $Q\binom{\mathbf{q}}{j}$ | a normal coordinate, (2.60), (2.164) |
| $\hat{Q}_{\alpha\beta}$ | $\hat{q}_\alpha \hat{q}_\beta$, (4.36) |
| $\mathbf{r}(i)$ | position of $i$th entity, (1.1) |
| $\mathbf{r}\binom{l}{k}$ | $\mathbf{x}\binom{l}{k} + \mathbf{u}\binom{l}{k}$, (2.17) |
| $\mathbf{r}\binom{l'\ l}{k'\ k}$ | $\mathbf{r}\binom{l'}{k'} - \mathbf{r}\binom{l}{k}$ |
| $\mathscr{R}_m$ | $\{\mathbf{R}|\mathbf{x}(m) + \mathbf{v}(R)\}$, an element of $G(\mathbf{q})$, Sec. 3.1 |
| $\mathbf{R}$ | rotation matrix corresponding to $\mathscr{R}_m$, (3.5) |
| $\mathbf{s}$ | symmetric strain tensor, (2.103) |

| | |
|---|---|
| **S** | 1. rotation matrix corresponding to $\mathscr{S}_m$, (2.13) |
| | 2. stress tensor, Sec. 2.4.1 |
| $\mathscr{S}_m$ | $\{\mathbf{S}\|\mathbf{x}(m) + \mathbf{v}(S)\}$, space group element, (2.13) |
| $\mathscr{T}$ | translation group symbol, Sec. 2.2.2 |
| $\mathbf{T}(\mathbf{q}; \mathbf{R})$ | $3n$-dimensional multiplier representation matrix for element $\mathbf{R}$ of $G_0(\mathbf{q})$, (3.25) |
| $\mathbf{u}\binom{l}{k}$ | displacement of $\binom{l}{k}$, (2.17) |
| $u_\alpha(k\|\mathbf{q})$ | $U_\alpha(k\|\mathbf{q})e^{-i\omega t}$, (4.85) |
| $U_\alpha(k\|\mathbf{q})$, $U_\alpha(k\|_j^{\mathbf{q}})$ | component of wave-amplitude vector, (2.41) |
| $U(Q)$ | $(1/v) \times$ Fourier transform of ion-electron pseudopotential, (6.5) |
| $v$ | primitive cell volume, (2.11) |
| $V(r)$ | pair potential energy function, (1.3) |
| $V_{kk'}(r)$ | potential energy function for pair $kk'$, (2.225) |
| $V(Q)$ | Fourier transform of $V(r)$, Sec. 1.2, (7.94) |
| $v(Q)$ | formal Fourier transform of $e^2/r$, (7.37), Sec. 7.1.3, (A1.16) |
| $v^*(Q)$ | modification of $v(Q)$ to allow for exchange and correlation, (7.44) |
| $V_{kk'}(Q)$ | Fourier transform of $V_{kk'}(r)$, (2.230) |
| $V_{\text{ext}}(\mathbf{r})$ | external electrical potential, (4.381), (7.1) |
| $V_{\text{ind}}(\mathbf{r})$ | induced electrical potential, (7.1) |
| $V_t(\mathbf{r})$ | $V_{\text{ext}}(\mathbf{r}) + V_{\text{ind}}(\mathbf{r})$, (4.382), (7.1) |
| $V^{\text{el}}(r)$ | conduction electron modification to the potential energy, (6.5), (7.93) |
| $\mathbf{w}\binom{l}{k}$ | relative shell-core displacement in the shell model, (4.250) |
| $\mathbf{W}(k\|\mathbf{q})$ | wave amplitude for relative shell-core displacements, (4.276) |
| $\mathbf{x}(l)$ | equilibrium position of the $l$th primitive cell, (2.7) |
| $\mathbf{x}(k)$ | equilibrium position of the $k$th atom in the primitive cell relative to the cell origin, (2.6) |

$\mathbf{x}\binom{l}{k}$       equilibrium position of the $k$th atom in the primitive cell, $\mathbf{x}(l) + \mathbf{x}(k)$, (2.6)

$\mathbf{x}\binom{l\ l'}{k\ k'}$       $\mathbf{x}\binom{l}{k} - \mathbf{x}\binom{l'}{k'}$, (2.124)

$X_k e$       charge on the core of the $k$th type of ion in the shell model, Sec. 4.6

$Y_k e$       charge on the shell of the $k$th type of ion in the shell model, Sec. 4.6

$Z_k e$       charge on the $k$th type of ion, (4.24)

$\mathbf{Z}_d$       $3n \times 3n$ diagonal matrix of charges $Z_k e$, (A2.26), (4.285)

$\mathbf{Z}$       $3 \times 3n$ array of charges $Z_k e$, (4.34)

$Z_{\alpha\beta}(k|\mathbf{q})$       element of the $3 \times 3n$ apparent charge matrix $\mathbf{Z}(\mathbf{q})$, (4.77)

$\alpha$       1. a Cartesian coordinate index
2. crystal polarizability, Chapter 4

$\alpha_k$       electronic polarizability of an ion in the $k$th sublattice, Sec. 4.6

$\boldsymbol{\alpha}$       1. $3n \times 3n$ diagonal matrix involving the $\alpha_k$, Sec. 4.6
2. second rank polarizability tensor, Sec. 5.3.1

$\alpha(0), \alpha(\infty)$       static and high-frequency polarizabilities

$\beta$       1. a Cartesian coordinate index
2. $1/k_\mathrm{B} T$
3. compressibility, Sec. 4.6

$\boldsymbol{\gamma}^s(\mathbf{q}, \mathcal{R}_m)$       matrix in the $s$th allowable representation of $G(\mathbf{q})$, corresponding to element $\mathcal{R}_m$, Sec. 3.1.2

$\boldsymbol{\Gamma}(\mathbf{q}, \mathcal{R}_m)$       $3n$-dimensional representation matrix for element $\mathcal{R}_m$ of $G(\mathbf{q})$, (3.22)

$\Delta(\mathbf{Q})$       see (A1.3)

$\varepsilon$       scalar dielectric constant

$\varepsilon_{\alpha\beta\gamma}$       Levi-Civita symbol, (2.39)

$\varepsilon(0), \varepsilon(\infty)$       static and high-frequency dielectric constants, (4.5)

$\boldsymbol{\varepsilon}(\infty)$       high-frequency dielectric tensor, (4.80)

$\boldsymbol{\varepsilon}(\omega)$       frequency-dependent dielectric tensor, (4.155)

| | |
|---|---|
| $\varepsilon(\mathbf{q}, \omega)$ | dielectric tensor, dependent on wave vector and frequency, (6.37) |
| $\varepsilon^{-1}(\mathbf{r}, \mathbf{r}')$ | nonlocal response function relating total and external potentials, (7.3) |
| $\varepsilon^{-1}(\mathbf{Q}, \mathbf{Q}')$ | Fourier transform of $\varepsilon^{-1}(\mathbf{r}, \mathbf{r}')$, (7.16) |
| $\varepsilon(Q)$ | screening function in diagonal approximation, (7.39) |
| $\varepsilon(\mathbf{r}, \mathbf{r}')$ | microscopic longitudinal dielectric function; inverse of $\varepsilon^{-1}(\mathbf{r}, \mathbf{r}')$, (7.4) |
| $\varepsilon(\mathbf{Q}, \mathbf{Q}')$ | inverse of $\varepsilon^{-1}(\mathbf{Q}, \mathbf{Q}')$, (7.17) |
| $\zeta_\alpha(k|\mathbf{q})$ | displacement coordinate in Bloch form, (2.159) |
| $\Theta_D$ | Debye temperature |
| $\nu$ | frequency, $\omega/2\pi$ |
| $\rho$ | mass density, Sec. 2.4.3 |
| $\rho(\mathbf{r})$ | electron density at $\mathbf{r}$, Sec. 4.6.10, (7.95) |
| $\Sigma\binom{\mathbf{q}}{s a \lambda}$ | symmetry-adapted vector, (3.36) |
| $\tau^s(\mathbf{q}; \mathbf{R})$ | matrix in the $s$th IMR of $G_0(\mathbf{q})$, corresponding to element $\mathbf{R}$, (3.8) |
| $\Phi$ | crystal potential energy, (1.1) |
| $\phi_{\alpha\beta}\binom{l\ l'}{k\ k'}$ | element of the force-constant matrix, (2.20) |
| $\phi^C, \phi^{SR}$ | Coulombic and short-range parts of $\phi$, Sec. 4.2.1 |
| $\chi(\mathbf{q}; \mathbf{R})$ | trace of $\mathbf{T}(\mathbf{q}; \mathbf{R})$, (3.30) |
| $\chi^s(\mathbf{q}; \mathbf{R})$ | trace of $\tau^s(\mathbf{q}; \mathbf{R})$, (3.30) |
| $\chi$ | dielectric susceptibility |
| $\chi(0), \chi(\infty)$ | static and high-frequency susceptibility tensors, Sec. 4.3.1 |
| $\chi(\omega)$ | frequency-dependent susceptibility tensor, Sec. 4.3.5 |
| $\chi(\mathbf{r}, \mathbf{r}')$ | nonlocal dielectric susceptibility, (7.10) |
| $\chi(\mathbf{Q}, \mathbf{Q}')$ | Fourier transform of $\chi(\mathbf{r}, \mathbf{r}')$ |
| $\chi_{sc}(\mathbf{r}, \mathbf{r}')$ | nonlocal screened dielectric susceptibility, (4.381), (7.22) |
| $\chi_{sc}(\mathbf{Q}, \mathbf{Q}')$ | Fourier transform of $\chi_{sc}(\mathbf{r}, \mathbf{r}')$ |

| | |
|---|---|
| $\omega$ | angular frequency, $2\pi\nu$ |
| $\omega_j(\mathbf{q})$ | angular frequency of $j$th mode, (2.5) |
| $\omega_{\lambda,l}$ | a longitudinal frequency of an ionic crystal, Sec. 4.2 |
| $\omega_{\lambda}(0)$ | a dispersion frequency of an ionic crystal, Sec. 4.2 |
| $\mathbf{\Omega}(\mathbf{q})$ | diagonal matrix of squared eigenvalues, (2.53) |
| $\mathbb{1}_n$ | $n$-dimensional unit matrix |
| $[\alpha\beta,\ \gamma\lambda]$, $(\alpha\gamma,\ \beta\lambda)$ | "brackets," used in method of long waves, (2.142–2.143) |

# Bibliography

This bibliography is intended as a guide to further reading rather than as an exhaustive compilation of references.

## Chapter 1

Discussion of binding may be found in most introductory texts on solid state physics, for example, Kittel. Highly readable accounts pertinent to the present chapter are also available in Seitz and in

P. W. Anderson, *Concepts in Solids* (Benjamin, New York, 1963), Chapter 1.

In this context, Chapter 1 of BH is practically mandatory reading. Chemical background to the subject of binding is provided by

L. Pauling, *The Nature of the Chemical Bond* (Cornell University Press, Ithaca, N. Y., 1960), third edition,

while a very nice account of hydrogen bonds, especially in relation to their role in biological systems, has been presented by

P. O. Löwdin, in *Advances in Quantum Chemistry*, edited by P. O. Löwdin (Academic Press, New York, 1965), Vol. 2, p. 213.

A general survey of two-body potentials commonly used is available in

J. O. Hirschfelder, C. F. Curtiss, and R. B. Bird, *Molecular Theory of Gases and Liquids* (Wiley, New York, 1954).

See also

L. Jansen, in Löwdin, ed., *Advances in Quantum Chemistry*, p. 119

who gives a good discussion of three-body forces in rare-gas solids and alkali halides.

A useful source book is

H. Margenau and N. R. Kestner, *Theory of Intermolecular Forces* (Pergamon, New York, 1969).

## Chapter 2

### Section 2.1
The adiabatic approximation was originally suggested by

M. Born and J. R. Oppenheimer, Ann. Physik **84**, 457 (1927)

and is discussed in detail by BH in Chapter IV as well as in Appendix VII. A pedagogic treatment is given by

J. C. Slater, *Quantum Theory of Molecules and Solids* (McGraw-Hill, New York, 1963), Vol. I, Chapter 1 and App. 2,

while the relevance to metals is discussed by

G. V. Chester, Advan. Phys. **10**, 357 (1961) and

E. G. Brovman and Yu. Kagan, Zh. Eksp. Teor. Fiz. **52**, 557 (1967) [Sov. Phys.—JETP **25**, 365 (1967)].

### Section 2.2
Elementary versions of the Born–von Kármán theory, especially via one-dimensional chains, are available in most introductory solid state texts. More detailed versions are given in

Seitz,

R. E. Peierls, *Quantum Theory of Solids* (Clarendon Press, Oxford, 1965), Chapter 1,

J. M. Ziman, *Principles of the Theory of Solids* (Cambridge University Press, London, 1972), second edition, Chapter 2, and

A. Huag, *Theoretical Solid State Physics* (Pergamon, New York, 1972), Vol. I.

A number of review articles have also appeared on the subject. These include

M. Blackman, in *Handbuch der Physik*, edited by S. Flügge (Springer, Berlin, 1955), Vol. 7/1, p. 325,

G. Leibfried, ibid., p. 104,

J. de Launey, Solid State Phys. **2**, 219 (1956),

W. Cochran, Rept. Progr. Phys. **26**, 1 (1963),

W. Cochran and R. A. Cowley, in *Handbuch der Physik*, edited by S. Flügge (Springer, Berlin, 1967) Vol. 25/2a, p. 59,

G. Leibfried, in *Theory of Condensed Matter* (IAEA, Vienna, 1968), p. 175,

J. E. Parrott, in *Solid State Theory*, edited by P. T. Landsberg (Wiley, New York, 1969), p. 325, and

W. Cochran, Critical Reviews in Solid State Sciences **2**, 1 (1971).

Text books and monographs (other than BH) which discuss the topic are

J. M. Ziman, *Electrons and Phonons* (Clarendon Press, Oxford, 1960),

B. Donovan and J. F. Angress, *Lattice Vibrations* (Chapman and Hall, London, 1971),

A. A. Maradudin, E. W. Montroll, G. H. Weiss, and I. P. Ipatova, *Theory of Lattice Dynamics in the Harmonic Approximation* (Academic Press, New York, 1971), second edition,

A. K. Ghatak and L. S. Kothari, *An Introduction to Lattice Dynamics* (Addison-Wesley, Reading, Mass., 1972), and

W. Cochran, *The Dynamics of Atoms in Crystals* (Crane, Russak, New York, 1974).

**Section 2.3**
The concept of the frequency distribution is introduced in an illustrative fashion by

C. Kittel, in Stevenson, p. 1.

An elementary version is given by the same author in his introductory solid state text.

Analytical methods of calculating the frequency distribution are discussed by

Maradudin et al., *Theory of Lattice Dynamics*, Chapter 4,

while a comprehensive review of computed-based methods, including references to the literature on this subject, is given by

G. Gilat, J. Comput. Phys. **10**, 432 (1972).

The major references on critical points and singularities are the papers of van Hove and Phillips cited in Chapter 2. A review of this topic is available in

Maradudin et al., *Theory of Lattice Dynamics*, Chapter 4.

Introductory accounts of the origin and consequences of analytical critical points may be found in

G. H. Wannier, *Elements of Solid State Theory* (Cambridge University Press, London, 1959), Chapter 3,

G. Weinreich, *Solids* (Wiley, New York, 1965), Chapter 4, and

J. M. Ziman, *Principles of the Theory of Solids* (Cambridge University Press, London, 1972), Chapter 2.

A very useful report on the subject is

J. F. Veverka, *The Morse Theory and its Application to Solid State Physics*, Queen's papers in Pure and Applied Mathematics No. 3 (Kingston University, Ontario, 1966).

**Section 2.4**
The basic concepts of elasticity are well explained by

J. F. Nye, *Physical Properties of Crystals* (Clarendon Press, Oxford, 1957), Chapter VIII.

The best reference on the long-wavelength method is BH, Chapter V. An alternate method of relating elastic constants and force constants is that of *homogeneous deformation*. In this, one calculates the energy stored consequent to a static strain produced by a homogeneous deformation, using the macroscopic as well as the microscopic equations, and then compares them to obtain the desired interrelations. Born and Huang (Chapter III) use this method for crystals bound by central forces. For other cases, the method cannot be straightaway applied to infinite crystals because of the divergence of some terms. In fact, it was to overcome this difficulty that Born developed the long-wavelength method. However, in recent years it has been shown that the homogeneous deformation method can also be used for such cases, provided sufficient care is taken. The important references are

G. Leibfried and W. Ludwig, Z. Physik **169**, 80 (1960),

L. T. Hedin, Ark. Fysik **18**, 369 (1960),

M. Lax, in Wallis, p. 583, and

W. Ludwig, *Recent Developments in Lattice Theory* (Springer, Berlin, 1967).

An item of historical interest in the field of elasticity is the "Laval controversy." Laval argued that under dynamic conditions such as occur during sound propagation

$$C_{\alpha\gamma,\beta\lambda} \neq C_{\gamma\alpha,\beta\lambda}.$$

By an extension of the argument, Laval rejected also the Born-Huang relations (2.147). These objections have now been laid to rest, and a critical review is given by Lax in the paper cited above.

**Section 2.5**
The quasi-harmonic approximation is comprehensively reviewed by

G. Leibfried and W. Ludwig, Solid State Phys. **12**, 275 (1961).

A short summary is given by

G. Leibfried, in Wallis, p. 237.

To pursue further the subject of anharmonicity, one may start with the following references:

Peierls, *Quantum Theory of Solids*,

G. Leibfried and W. Ludwig, Solid State Phys. **12**, 275 (1961),

R. A. Cowley, Advan. Phys. **12**, 421 (1963),

Ludwig, *Recent Developments*, and

Duane C. Wallace, *Thermodynamics of Crystals* (Wiley, New York, 1972).

**Section 2.6**
The quantization of lattice vibrational oscillators is discussed in most of the references cited above under Sec. 2.2. The particle aspects of phonons are explored by

H. H. Jensen, in Bak, p. 1.

See also

J. A. Reissland, *The Physics of Phonons* (Wiley-Interscience, New York, 1973).

**Section 2.7**
External vibrations are comprehensively reviewed by

G. Venkataraman and V. C. Sahni, Rev. Mod. Phys. **42**, 409 (1970),

who reference earlier literature on this subject.

The problem of internal modes is discussed to some extent by

A. S. Davydov, *Theory of Molecular Excitons* (McGraw-Hill, New York, 1962).

See also

A. B. Zahlen, *Excitons, Magnons and Phonons in Molecular Crystals* (Cambridge University Press, London, 1968).

A highly mathematical but nevertheless pertinent article on the subject is

Yu. I. Polyakov and Yu. Ya. Kharitonov (Eng. Trans.) Optics and Spectroscopy **29**, 293 (1970).

**Section 2.8**
A useful computer code relating valence and Born–von Kármán force constants is discussed in the report

J. H. Schachtschneider, *Vibrational Analysis of Polyatomic Molecules VI*, Tech. Rep. 57–65, Shell Development Company.

**Section 2.9**
The determination of the structure of the force-constant matrices for arbitrary neighbors using symmetry considerations alone is illustrated for cubic structures by

G. L. Squires, Ark. Fysik **25**, 21 (1963),

and for hcp structures by

A. Czachor, in Bombay Proceedings, Vol. I, p. 181.

Several examples of force-constant calculations may be found in the IAEA proceedings, Wallis, and Nusimovici.

**Chapter 3**

A study of this chapter requires a knowledge of group theory. Basic texts on this subject have already been cited in the reference section of Chapter 3. Also useful in this context is

C. J. Bradley and A. P. Cracknell, *The Mathematical Theory of Symmetry in Solids* (Clarendon Press, Oxford, 1972).

Multiplier-representation theory, on which the chapter is based, has been discussed by

G. Ya. Lyubarskii, *The Application of Group Theory in Physics* (Pergamon, New York, 1960), and

M. Hamermesh, *Group Theory and its Applications to Physical Problems* (Addison-Wesley, Reading, Mass., 1962).

Tables of irreducible multiplier representations required in practical calculations can be had from

O. V. Kovalev, *Irreducible Representations of Space Groups* (Gordon and Breach, New York, 1964),

S. C. Miller and W. F. Love, *Tables of Irreducible Representations of Space Groups and Co-representations of Magnetic Space Groups* (Pruett, Boulder, Colorado, 1967), and

J. Zak, A. Casher, M. Gluck and Y. Gur, *Irreducible Representations of Space Groups* (Benjamin, New York, 1969).

If these tables are not available, then the required IMR's can be deduced following the procedure described by

V. C. Sahni and G. Venkataraman, Phys. Kondens Materie **11**, 119, 212 (1970).

Several detailed examples of the application of group theory to lattice dynamics have appeared recently. These include

NaCl, CsCl and hcp structures

J. L. Warren, Rev. Mod. Phys. **40**, 38 (1968)

Diamond structure

A. A. Maradudin and S. H. Vosko, Rev. Mod. Phys. **40**, 1 (1968)

Zinc blende structure

G. Venkataraman, L. A. Feldkamp, and J. S. King, Univ. of Michigan report 01358-1-T (1968),
H. Montgomery, Proc. Roy. Soc. (London) **A309**, 521 (1969),
J. L. Yarnell, J. L. Warren, R. G. Wenzel, and P. J. Dean, in Copenhagen proceedings, Vol. 1, p. 301.

Tin

S. H. Chen, Phys. Rev. **163**, 532 (1967).

$TiO_2$

R. S. Katiyar, J. Phys. C **3**, 1087 (1970).

$CaWO_4$

G. Venkataraman and V. C. Sahni, Rev. Mod. Phys. **42**, 409 (1970).

$NaNO_3$

K. R. Rao, S. F. Trevino and K. W. Logan, J. Chem. Phys. **53**, 4645 (1970).

$KN_3$

K. R. Rao and S. F. Trevino, J. Chem. Phys. **53**, 4661 (1970).

NaN$_3$

K. R. Rao and S. F. Trevino, J. Chem. Phys. **53**, 4624 (1970).

Gallium

W. B. Waeber, J. Phys. C **2**, 882 (1969).

CO$_2$

H. Montgomery and G. Dolling, J. Phys. Chem. Solids **33**, 1201 (1972).

CaCO$_3$

E. R. Cowley, Can. J. Phys. **47**, 1381 (1969).

Cu$_2$O

W. Ungier, Acta Physica Polonica **A43**, 747 (1973).

Orthorhombic FeS$_2$

W. Ungier, Acta Physica Polonica **A44**, 217 (1973).

The application of computer techniques to group-theoretic analysis of lattice vibrations has been considered by

T. G. Worlton and J. L. Warren, Computer Phys. Commun. **3**, 88 (1972); Computer Phys. Commun. **4**, 382 (1972) and

T. G. Worlton, Computer Phys. Commun. **4**, 249 (1972).

Results obtained by computer are compiled in

J. L. Warren and T. G. Worlton, *Symmetry Properties of the Lattice Dynamics of Twenty-Three Crystals*, Argonne National Laboratory Report ANL-8053 and LA-5465-MS (1973).

**Chapter 4**

Two recent volumes containing discussions of many topics of interest to us here are

*Optical Properties of Solids*, edited by F. Abeles (North Holland, Amsterdam, 1972) and

*Atomic Structure and Properties of Solids*, edited by E. Burstein (Academic Press, New York, 1972).

**Section 4.1**
A study of the effects of macroscopic electric fields in ionic crystals is best approached via cubic diatomic crystals as we have done in Sec. 4.1. Several books offer introductory treatments on this subject, for example, Kittel. See also

J. C. Slater, Quarterly Progress Report of Solid State and Molecular Theory Group, MIT, Oct. 15 (1960), p. 57, and

J. E. Parrott, in *Solid State Theory*, edited by P. T. Landsberg (Wiley, New York, 1969).

Chapter II of BH gives an excellent exposition on this subject. For probing further the effects of retardation consult also

K. Huang, Proc. Roy. Soc. (London) **A208**, 352 (1951) and

J. J. Hopfield, in *Proceedings of the International Conference on the Physics of Semiconductors* (J. Phys. Soc. Japan Suppl. **21**, 1966), p. 77.

**Section 4.2**
The best reference on the rigid-ion model is BH, §31. See also

K. Huang, Phil. Mag. **40**, 733 (1949).

Papers useful in connection with the evaluation of $M_{\alpha\beta}^{C}({}^{q}_{kk'})$ are

E. W. Kellerman, Phil. Trans. Roy. Soc. London **238**, 513 (1940); Proc. Roy. Soc. (London. **A178**, 17 (1941).

In this connection see also

W. Cochran and R. A. Cowley, in *Handbuch der Physik*, edited by S. Flügge (Springer, Berlin, 1967), Vol. 25/2a, p. 59.

The formulation of the Coulomb coefficients as a series entirely in reciprocal space is discussed by

W. Cochran, in Wallis, p. 75, and

W. Cochran, Critical Reviews in Solid State Sciences **2**, 1 (1971).

**Section 4.3**
A comprehensive formulation of the long-wavelength equations along the lines of our Eqs. (4.76–4.78) is given by

R. M. Pick, Advan. Phys. **19**, 269 (1970).

From his corresponding equations, Pick projects out individually the various features pertaining to the acoustic and optic modes, instead of dealing with them via separate sets of equations as we have done.

In §33 of BH there is a phenomenological discussion of the long-wavelength optic modes along the lines of Sec. 4.3.4. However, Born and Huang do not relate their parameters to lattice dynamical quantities generally, although this is done for specific models in their §34 and §35. A model-independent discussion along the lines of 4.3.4–4.3.6 was first given by

W. Cochran and R. A. Cowley, J. Phys. Chem. Solids **23**, 447 (1962).

For a review of polaritons and plasmaritons, see

E. Burstein, in *Dynamical Processes in Solid State Optics*, edited by R. Kubo and H. Kamimura (Benjamin, New York, 1967), p. 1 and p. 34, and

E. Burstein, in *Elementary Excitations in Solids*, edited by A. A. Maradudin and G. F. Nardelli (Plenum, New York, 1969), p. 367.

**Section 4.4**
The discussion in this section is largely based on the paper by

V. C. Sahni and G. Venkataraman, in Nusimovici, p. 121.

**Section 4.5**
Further discussion of the role of macroscopic fields in relation to external modes is available in

G. Venkataraman and V. C. Sahni, Rev. Mod. Phys. **42**, 409 (1970).

**Section 4.6**
An excellent review of early developments in regard to the shell model (with detailed references) has been given by

B. G. Dick, in Wallis, p. 159.

The early Soviet work in this area has been summarized by

K. B. Tolpygo, Usp. Fiz. Nauk. **74**, 269 (1961) [Sov. Phys.—Usp. **4**, 485 (1961)].

A critical survey of recent advances in such models has been presented by

W. Cochran, Critical Reviews in Solid State Sciences **2**, 1 (1971) and

S. K. Sinha, Critical Reviews in Solid State Sciences **3**, 273 (1973).

J. R. Hardy and A. M. Karo, Solid State Phys. (to be published), have given a similar review pertaining mainly to the alkali halides.

A very useful compilation of general references pertaining to insulators is available in the bibliography section of

J. C. Slater, *Quantum Theory of Molecules and Solids* (McGraw-Hill, New York, 1967), Vol. 3.

**Chapter 5**

**Section 5.1**
The basic theory of neutron scattering including the scattering of neutrons by phonons is discussed in detail by

W. Marshall and S. Lovesey, *Theory of Thermal Neutron Scattering* (Oxford University Press, London, 1971).

See also

I. I. Gurevich and L. V. Tarasov, *Low Energy Neutron Physics* (North Holland, Amsterdam, 1968).

Briefer accounts of the theory are available in the articles by

L. S. Kothari and K. S. Singwi, Solid State Phys. **8**, 109 (1959) and

W. Lomer and G. G. Low, in *Thermal Neutron Scattering*, edited by P. A. Egelstaff (Academic Press, New York, 1965), p. 2.

Short reports of the experimental study and results obtained have been given by

B. N. Brockhouse, in Bak, p. 221,

B. N. Brockhouse, in Stevenson, p. 110,

G. Dolling and A. D. B. Woods, in *Thermal Neutron Scattering*, edited by P. A. Egelstaff (Academic Press, New York, 1965), p. 193,

M. F. Collins, in *Elementary Excitations in Solids*, edited by A. A. Maradudin and G. V. Nardelli (Plenum, New York, 1969), p. 156, and

R. A. Cowley, in *Phonons and their Interactions*, edited by R. H. Enns and R. R. Haering (Gordon and Breach, New York, 1969), p. 43.

See also

H. Boutin and S. Yip, *Molecular Spectroscopy with Neutrons* (MIT Press, Cambridge, Mass., 1968).

## Sections 5.2–5.4

Several reviews address the study of lattice vibrations by optical techniques. Frequently they discuss both infrared absorption and Raman scattering, covering one-phonon as well as multiphonon processes.

A good starting point is the article by

T. Kurosawa, in *Dynamical Processes in Solid State Optics*, edited by R. Kubo and H. Kamimura (Benjamin, New York, 1967), p. 76,

which discusses the fundamentals of dielectric dispersion. Infrared spectra (due to single-phonon and multiphonon processes) are reviewed by

S. S. Mitra and P. J. Gieliesse, in *Progress in Infrared Spectroscopy*, edited by H. A. Szymanski (Plenum, New York, 1964), Vol. 2, p. 47

from a spectroscopist's viewpoint. A similar review is given by

G. H. Wilkinson, in *Molecular Dynamics and Structure of Solids*, edited by R. S. Carter and J. J. Rush (NBS, Washington, D. C., 1969), p. 77.

First-order reflection spectra of III–V compounds are discussed by

M. Haas, in *Semiconductors and Semimetals*, edited by R. K. Willardson and A. C. Beer (Academic Press, New York, 1967), Vol. 3, p. 3.

Useful theoretical background to the subject of infrared spectra in general is provided by

H. Bilz, in Stevenson, p. 208.

Burstein also covers the subject of optical spectra in several articles. See

E. Burstein, in Bak, p. 276,

E. Burstein, in Wallis, p. 315,

and the articles by him cited earlier under Sec. 4.3. These articles discuss both infrared and Raman spectra. Similar ground is also covered by

S. S. Mitra, in *Optical Properties of Solids*, edited by S. Nudelman and S. S. Mitra (Plenum, New York, 1969), p. 333.

A survey of a similar nature but with particular reference to ferroelectric crystals has been presented by

A. S. Barker Jr., in *Far-Infrared Properties of Solids*, edited by S. S. Mitra and S. Nudelman (Plenum, New York, 1970), p. 247.

An extensive review of second-order infrared spectra has been given by

F. A. Johnson, in *Progress in Semiconductors*, edited by A. F. Gibson and R. E. Burgess (Temple Press, London, 1965), Vol. 9, p. 181.

A similar discussion but restricted to III–V compounds is available in

W. Spitzer, in *Semiconductors and Semimetals*, edited by R. K. Willardson and A. C. Beer (Academic Press, New York, 1967), Vol. 3, p. 17.

The subject of Raman scattering has been comprehensively discussed by

R. Loudon, Advan. Phys. **13**, 423 (1964)

and more recently by

R. A. Cowley, in *The Raman Effect*, edited by A. Anderson (Dekker, New York, 1971), Vol. 1, p. 95; G. R. Wilkinson, ibid., Vol. 2 (1973), p. 811.

See also

R. Claus, Phys. Stat. Sol. **50**, 11 (1972).

Raman scattering with particular reference to two-phonon spectra has been reviewed by

J. R. Hardy, in Stevenson, p. 245.

See also

V. Gorelik and M. M. Sushchinskii, Usp. Fiz. Nauk. **98**, 237 (1969) [Sov. Phys.—Usp. **12**, 399 (1970)].

For a discussion of Brillouin scattering, see BH, §50 and

G. B. Benedek and K. Fritsch, Phys. Rev. **149**, 647 (1966),

R. S. Krishnan, in *The Raman Effect*, edited by A. Anderson (Dekker, New York, 1971), Vol. 1, p. 343, and

H. Z. Cummins and P. E. Schoen, in *Laser Handbook*, edited by F. T. Arecchi and E. D. Schulz-Dubois (North Holland, Amsterdam, 1972), Vol. II, p. 1029.

Useful source books on light scattering are the conference proceedings entitled

*Light Scattering Spectra of Solids*, edited by G. B. Wright (Springer, New York, 1969) and

*Light Scattering in Solids*, edited by M. Balkanski (Flammarion Sciences, Paris, 1971).

**Section 5.5**
The influence of lattice vibrations on x-ray scattering has been known for a long time. From the standpoint of extracting explicit lattice dynamical information from x-ray experiments, the first attempts were to determine the elastic constants from a study of the diffuse spots. This is discussed by

W. A. Wooster, *Diffuse X-ray Reflections from Crystals* (Clarendon Press, Oxford, 1962).

X-ray scattering with specific reference to the possibility of reconstructing phonon dispersion curves is considered by

W. Cochran, in Bak, p. 102,

W. Cochran, in Stevenson, p. 153,

T. Smith, in Stevenson, p. 161, and

J. L. Amaros and M. Amaros, *Molecular Crystals: Their Transforms and Diffuse Scattering* (Wiley, New York, 1968).

A useful article to study in this context is

J. C. Slater, Rev. Mod. Phys. **30**, 197 (1958).

For a discussion of the interactions involved in neutron scattering, infrared absorption, light scattering and x-ray scattering, see

A. Sjolander, in Bak, p. 76.

**Chapter 6**

Papers on lattice dynamical experiments are scattered in the literature but many are collected in the proceedings of the various IAEA Symposia, as well as in Wallis and Nusimovici. Also useful are the following reports which include a bibliography of papers on lattice dynamics:

*Bibliography for Thermal Neutron Scattering* JAERI-4043;—M.4666 (Japan Atomic Energy Research Institute, Tokai-Mura, 1971); JAERI—M.5395 (4th ed., 1973).

*Bibliography of papers relevant to the Scattering of Thermal Neutrons*, edited by A. Larose and J. Vanderwal (McMaster University, Hamilton, Ont., Canada, 1973).

Several reviews of the experimental results have appeared:

B. N. Brockhouse, in Bak p. 221, and in Stevenson p. 110, gives a survey of the work up to 1964/65.

The application of force-constant models to metals has been surveyed by

S. K. Joshi and A. K. Rajagopal, Solid State Phys. **22**, 159 (1968).

Reviews pertaining to alloys are

A. D. B. Woods, in *Symposium on Inelastic Scattering of Neutrons by Condensed Systems*. BNL-940 (CAS) (Brookhaven National Laboratory, Upton, New York, 1966), p. 8, and

B. N. Brockhouse, E. D. Hallman, and S. C. Ng, in *Magnetic and Inelastic Scattering of Neutrons by Metals*, edited by T. J. Rowland and P. A. Beck (Gordon and Breach, New York, 1968), p. 161.

Some useful references pertaining to theoretical aspects of the Kohn anomaly (in addition to those already cited in Chapter 5) are

P. L. Taylor, Phys. Rev. **131**, 1995 (1963),

A. M. Afanas'ev and Yu. Kagan, Zh. Eksp. Teor. Fiz. **43**, 1456 (1962) [Sov. Phys.—JETP **16**, 1030 (1963)],

M. I. Kaganov and A. I. Semenenko, Zh. Eksp. Teor. Fiz. **50**, 630 (1966) [Sov. Phys.—JETP **23**, 419 (1966)], and

A. Ya. Blank and E. A. Kaner, Zh. Eksp. Teor. Fiz. **50**, 1013 (1966) [Sov. Phys.—JETP **23**, 673 (1966)].

Early results pertaining to ionic and valence crystals are discussed by

B. G. Dick, in Wallis, p. 159,

especially in the context of the evolution of the shell model. A comprehensive analysis of the results for diamond structure crystals has been given by

G. Dolling and R. A. Cowley, Proc. Phys. Soc. (London) **88**, 463 (1966).

More recent discussions of the dynamics of insulators and semiconductors are available in

W. Cochran, Critical Reviews in Solid State Sciences **2**, 1 (1971),

S. K. Sinha, Critical Reviews in Solid State Sciences **3**, 273 (1973), and

J. R. Hardy and A. M. Karo, Solid State Phys. (to be published).

A critical discussion pertaining to effective charge has been presented by S. K. Sinha in the article cited.

The role of lattice dynamics in relation to ferroelectricity has been reviewed mainly by Cochran. The relevant articles are

W. Cochran, Advan. Phys. **9**, 387 (1960); **10**, 401 (1961); **18**, 157 (1969).

See also,

E. Fatuzzo and W. J. Merz, *Ferroelectricity* (North Holland, Amsterdam, 1967),

*Ferroelectricity*, edited by E. F. Weller (Elsevier, Amsterdam, 1967), and

*Structural Phase Transitions and Soft Modes*, edited by E. J. Samuelsen, E. Andersen, and J. Feder (Universitetsforlaget, Oslo, Norway, 1971).

Results for the dispersion curves of external vibrations have begun to appear only recently. The situation to 1970 has been summarized by

G. Venkataraman and V. C. Sahni, Rev. Mod. Phys. **42**, 409 (1970).

**Chapter 7**

For a proper understanding of this chapter, it is necessary to have at least a rudimentary knowledge of Hartree and Hartree-Fock theories. For this one may consult Seitz and

J. C. Slater, *Quantum Theory of Atomic Structure* (McGraw-Hill, New York, 1960), Vol. I, Chapter 9 and Vol. II, Chapter 17.

**Section 7.1**

To probe further into the literature on microscopic theories of lattice dynamics, a knowledge of many-body theory is necessary. Simple expositions are available in

D. Pines, *The Many-Body Problem* (Benjamin, New York, 1961),

D. Falkoff, in *The Many-Body Problem*, edited by C. Fronsdal (Benjamin, New York, 1962), p. 1,

T. D. Schultz, *Quantum Field Theory and the Many-Body Problem* (Gordon and Breach, New York, 1963),

S. Lundqvist, in *Theory of Condensed Matter* (IAEA, Vienna, 1968), p. 25, and

G. E. Brown, *Many-Body Problems* (North Holland, Amsterdam, 1972).

Some of the basic papers on the microscopic theories of lattice dynamics are

M. H. Cohen, R. M. Martin, and R. M. Pick, in Copenhagen Proceedings, Vol. 1, p. 119,

P. N. Keating, Phys. Rev. **175**, 1171 (1968); **187**, 1190 (1969),

L. J. Sham, Phys. Rev. **188**, 1431 (1969),

R. M. Pick, M. H. Cohen, and R. M. Martin, Phys. Rev. **B1**, 910 (1970),

S. K. Sinha, R. P. Gupta, and D. L. Price, Phys. Rev. **B9**, 2564 (1974),

D. L. Price, S. K. Sinha, and R. P. Gupta, Phys. Rev. **B9**, 2573 (1974), and

W. R. Hanke, Phys. Rev. **B8**, 4585, 4591 (1973).

At the heart of the microscopic theory lies the dielectric function, good discussions of which are given by

L. J. Sham and J. M. Ziman, Solid State Phys. **15**, 221 (1963), and

S. Lundqvist, in *Electrons in Crystalline Solids* (IAEA, Vienna, 1973), p. 281.

Two other useful references in this context are

S. L. Adler, Phys. Rev. **126**, 413 (1962) and

N. Wiser, Phys. Rev. **129**, 62 (1963).

The distinction between the various types of dielectric functions such as $\varepsilon_{tt}$ and $\varepsilon_{te}$ is clearly pointed out by

L. Kleinman, Phys. Rev. **160**, 585 (1967) and

V. Heine and D. Weaire, Solid State Phys. **24**, 249 (1970).

The dielectric function for the electron gas has received much attention. A useful introduction to the subject is available from

D. Pines, *Elementary Excitations in Solids* (Benjamin, New York, 1963).

See also

D. Pines and P. Nozieres, *The Theory of Quantum Liquids* (Benjamin, New York, 1966), Vol. I.

A review of the various versions proposed for $\varepsilon(Q)$ of the electron gas has been given by

M. P. Tosi, Nuovo Cimento **1**, 160 (1969).

The application of microscopic theories to the lattice dynamics of metals has been reviewed by

S. K. Joshi and A. K. Rajagopal, Solid State Phys. **22**, 159 (1968).

The pseudopotential version is covered by Harrison in two of his books:

W. A. Harrison, *Pseudopotentials in the Theory of Metals* (Benjamin, New York, 1966) and

W. A. Harrison, *Solid State Theory* (McGraw-Hill, New York, 1970).

For some recent developments concerning model potentials, see

R. W. Shaw Jr., Phys. Rev. **174**, 769 (1968).

Two theses which are helpful in becoming acquainted with the details of the algebra are

L. J. Sham, Ph. D. Thesis, Univ. of Cambridge (1963) and

J. U. Koppel, GA-8745, Gulf General Atomic Inc. (1968).

Topical reviews covering various stages in the evolution of microscopic theories are:

W. Cochran, in Wallis, p. 75,

W. A. Harrison, in Wallis, p. 85,

W. A. Harrison, in Stevenson, p. 73,

W. Cochran, in Bombay proceedings, Vol. I, p. 3,

Yu. Kagan and E. G. Brovman, in Copenhagen proceedings, Vol. I, p. 3,

A. Sjolander, in *Symposium on Inelastic Scattering of Neutrons by Condensed Systems* BNL-940 (CAS), (Brookhaven National Laboratory, Upton, New York, 1966), p. 29,

W. Hanke and H. Bilz, in Grenoble proceedings, p. 3,

W. Cochran, Critical Reviews in Solid State Sciences **2**, 1 (1971),

L. J. Sham, in *Phonons and their Interactions*, edited by R. H. Enns and R. R. Haering (Gordon and Breach, New York, 1969), p. 143, and

S. K. Sinha, Critical Reviews in Solid State Sciences **3**, 273 (1973).

The effects of retardation are discussed on a microscopic basis by

R. M. Pick, Advan. Phys. **19**, 269 (1970).

**Section 7.2**
Three comprehensive reviews have appeared on quantum crystals. The first by

R. A. Guyer, Solid State Phys. **23**, 413 (1969),

deals with the broad spectrum of problems pertaining to quantum crystals and gives a highly readable introduction to the underlying physics. The second, by

N. R. Werthamer, Am. J. Phys. **37**, 763 (1969),

deals specifically with phonons in quantum crystals and gives a tutorial review of the single-particle and collective theories. A knowledge of many-body theory is, however, essential for a full understanding of these articles. The third, by

S. K. Sinha, in *Nuclear Physics and Solid State Physics* (*India*) (Dept. of Atomic Energy, Bombay) **15A**, 155 (1973)

offers a critical comparison of the results of the various theories in relation to experimental data. These articles provide considerable background information and source material.

The adventurous reader may consult

N. S. Gillis, Phys. Rev. **B1**, 1872 (1970)

in which the dielectric formulation is combined with the techniques of the helium problem.

**Recent Work**

Here we list some publications which came to our attention while this volume was in press.

**Experimental results (simple crystals)**
Xe

B. J. Palmer, D. H. Saunderson, and D. W. Batchelder, J. Phys. C **6**, L 313 (1973).

Xe

N. A. Lurie, G. Shirane, and J. Skalyo Jr., Phys. Rev. **B9**, 5300 (1974).

Kr

J. Skalyo Jr., Y. Endoh, and G. Shirane, Phys. Rev. **B9**, 1797 (1974).

Cu

G. Nilsson and S. Rolandson, Phys. Rev. **B7**, 2393 (1973).

Au

J. W. Lynn, H. G. Smith, and R. M. Nicklow, Phys. Rev. **B8**, 3493 (1973).

Rb

J. R. D. Copley, C. A. Rotter, H. G. Smith, and W. A. Kamitakahara, Phys. Rev. Letters **33**, 365 (1974).

CsF

W. Bührer, J. Phys. C **6**, 2931 (1973).

CsI

B. I. Gorbachev, P. G. Ivanitskii, V. T. Krotenko, and M. V. Paschenik, Ukr. Fiz. Zh. **18**, 92 (1973).

CuBr

B. Prevot, C. Carabatos, C. Schwab, B. Hennion, and F. Moussa, Solid State Comm. **13,** 1725 (1973).

CdTe

J. M. Rowe, R. M. Nicklow, D. L. Price, and K. Zanio, Phys. Rev. **B10,** 671 (1974).

Se

W. C. Hamilton, B. Lassier, and M. I. Kay, J. Phys. Chem. Solids **35,** 1089 (1974).

GaSe

J. L. Brebner, S. Jandl, and B. M. Powell, Solid State Comm. **13,** 1555 (1973).

**Phenomenological model calculations**

K. B. Tolpygo and E. P. Troitskaya, Fiz. Tverd. Tela **14,** 2867 (1973) (Sov. Phys.—Solid State **14,** 2480 (1973)).

M. M. Shukla and J. P. Salzberg, J. Phys. Soc. Japan **35,** 996 (1973).

A. A. Ahmadieh and H. A. Rafizadeh, Phys. Rev. **B7,** 4527 (1973).

K. K. Mani and R. Ramani, Phys. Stat. Sol. **61B,** 659 (1974).

G. Bose, B. B. Tripathi, and H. C. Gupta, J. Phys. Chem. Solids **34,** 1867 (1973).

W. Weber, Phys. Rev. **B8,** 5082 (1973); Phys. Rev. Letters **33,** 371 (1974).

B. P. Pandey and B. Dayal, J. Phys. C **6,** 2943 (1973).

A. N. Basu and S. Sengupta, Phys. Rev. **B8,** 2982 (1973).

A. Rastogi, J. P. Hawranek, and R. P. Lowndes, Phys. Rev. **B9,** 1938 (1974).

W. Kress, Phys. Stat. Sol. **62B,** 403 (1974).

M. J. L. Sangster, J. Phys. Chem. Solids **35,** 195 (1974).

G. C. Cran and M. J. L. Sangster, J. Phys. C **7,** 1937 (1974).

R. Almairac and C. Benoit, J. Phys. C **7,** 2614 (1974).

M. P. Verma and S. K. Agarwal, Phys. Rev. **B8,** 4880 (1973).

$\phi$. Ra, J. Chem. Phys. **59,** 6009 (1973).

K. S. Upadhyaya and R. K. Singh, J. Phys. Chem. Solids **35,** 1175 (1974).

**Microscopic theory calculations**

P. L. Srivastava, N. R. Mitra, and N. Mishra, J. Phys. F **3,** 1388 (1973).

W. F. King III and P. H. Cutler, Phys. Rev. **B8,** 1303 (1973).

C. M. Bertoni, V. Bortolani, C. Calandra, and F. N. Nizzoli, J. Phys. F **4,** 19 (1974); Phys. Rev. Letters **31,** 1466 (1973).

C. M. Bertoni, V. Bortolani, C. Calandra, and B. Tosatti, Phys. Rev. **B9,** 1710 (1974).

P. V. S. Rao, J. Phys. Chem. Solids **35,** 669 (1974).

J. Hafner and P. Schmunk, Phys. Rev. **B9,** 4138 (1974).

N. Singh and S. Prakash, Phys. Rev. **B8,** 5532 (1973).

T. Soma, J. Phys. Soc. Japan **36,** 1301 (1974).

A. O. E. Animalu, Phys. Rev. **B8,** 3542, 3555 (1973).

D. C. Wallace, Aust. J. Phys. **26,** 217 (1973).

V. Mitskevich, Phys. Stat. Sol. **61B,** 675 (1974).

K. B. Tolpygo and E. P. Troitskaya, Ukr. Fiz. Zh. **19,** 428 (1974).

F. A. Johnson, Proc. Roy. Soc. (London) **A339,** 73 (1974); F. A. Johnson and K. Moore, ibid., **A339,** 85 (1974).

Y. Kurihara, Y. Kuroda, and N. Ishimura, Prog. Theor. Phys. (Japan) **51,** 959 (1974).

# Author Index

# Subject Index

Acoustic branch, 46
Acoustic sum rule, 423
Acoustic waves, 59–75
Adiabatic approximation, 20–23, 241, 401
Al, 341, 402, 413
Alkali halides, 359–375
Alkali metals, 339
Alloys, 11, 355–358
Anharmonicity, 76, 84
  electrical, 293, 301–302
  mechanical, 302
Annihilation operator, 83–84
Ar, 106–121, 441
Atomic coordinates, 1

$BaTiO_3$, 311
$Bi_{0.2}Tl_{0.8}$, 358
Bond angle, 14
Bond bending, 14, 99–104
Bond stretching, 99–104
Born-Huang relations, 70–73
Born identity, 380–381
Born-Mayer potential, 3, 12
Born model. *See* Model, rigid-ion
Born-Oppenheimer approximation. *See* Adiabatic approximation
Bragg scattering, 281
Branch, 41
  acoustic, 46
  optic, 46
Brillouin scattering, 306, 314–315
Brillouin zone, 39

Cauchy relations. *See* Elastic constants, Cauchy relations for
$CaWO_4$, 214–216, 221, 298, 310
Cell
  macro-, 38
  primitive, 23–24
  proximity, 272
  unit, 23
Charge
  apparent, 193, 385–386, 422, 473
  bond, 434
  Born (transverse), 387–388
  Callen (longitudinal), 387–388
  effective, 385–389
  exchange, 234
  ionic, 178, 386
  Szigeti, 230, 387–389, 431
Clausius-Mossotti formula, 248, 366
Compatibility relations, 140
Conduction electrons, 9, 348, 350

Coordinates
  atomic, 1
  bond-bending, 14, 99–104
  normal, 81–83
  nuclear, 1
  valence, 99–104
Corepresentations, 148
Covalent bond, 12–14
Creation operator, 83–84
Critical points, 51–59, 111–117
Crystal
  covalent, 2, 12–14
  cubic diatomic, 166–175, 186, 235–236
  cyclic. *see* Cyclic crystal
  diagonally cubic, 241, 472, 480
  hydrogen-bonded, 2, 14–18
  ionic, 2, 11–12
  metallic, 2, 9–11
  molecular, 2–6
  piezoelectric, 75
Cu, 356
CuZn ($\beta$-brass), 355
$Cu_3Zn$ ($\alpha$-brass), 356
Cyclic boundary conditions, 38–39
Cyclic crystal, 38, 126
  results for, 458–464

Davydov splitting, 95
Debye approximation, 50
Debye-Waller factor, 281
Degeneracy, 141
Diagonally cubic crystals, 241, 472, 480
Diamond, 312–313
Dielectric constant
  clamped; free, 196
  static; high-frequency, 170
Dielectric function, 193, 411–412, 415
  factorization of, 414, 436
  in Hartree approximation, 350, 408
  microscopic longitudinal, 402–415
Dielectric susceptibility, 171, 197
  nonlocal screened, 271–274
  nonlocal unscreened, 407
  tensor, high–frequency, 193
Dielectric tensor, 208–213
  clamped static, 196
  high-frequency, 193
Dipole moment operator, 291–293
Dispersion curves, 41
Dispersion frequency, 182
Dispersion relations, 41
Drude formula, 379
Dynamical matrix, 34–35,
  alternate forms of, 44–45

# Date Due

| | | | |
|---|---|---|---|
| | | | |
| | | | |
| | | | |
| | | | |
| | | | |
| | | | |
| | | | |
| | | | |
| | | | |
| | | | |
| | | | |
| | | | |
| | | | |
| | | | |
| | | | |
| | | | |
| | | | |
| | | | |
| | | | |
| | | | |
| | | | |

UML 735